PROCESS BIOTECHNOLOGY

THEORY AND PRACTICE

PROCESS BIOTECHNOLOGY

THEORY AND PRACTICE

S N Mukhopadhyay

teri

The Energy and Resources Institute

ISBN 978-81-7993-307-7

Suggested citation

Mukhopadhyay, S. N. *Process Biotechnology: Theory and Practice.* New Delhi: The Energy and Resources Institute (TERI), 2012.

Published by
The Energy and Resources Institute (TERI)
TERI Press
Darbari Seth Block
IHC Complex, Lodhi Road
New Delhi – 110 003
India

Tel. 2468 2100 or 4150 4900
Fax 2468 2144 or 2468 2145
India +91 • Delhi (0)11
E-mail teripress@teri.res.in
Website www.teriin.org

Printed in India

This book has been written to serve scholars and researchers in the discipline of biotechnology. It is dedicated to my parents, teachers, parents-in-law, and well-wishers.

Acknowledgements

My parents, teachers, and in-laws have supported me throughout my professional life. All of them have enabled me to develop and author books, which serve as valuable resources to scholars.

Tutorial practice in problem solving is key in the area of fundamental, developmental, and experimental process biotechnology for both scholars and faculty. My research and technical papers in many national and international journals, lectures in many national and international conferences, symposia, and workshops, and short-term training courses sponsored by the Department of Science and Technology and Department of Biotechnology, Government of India, and international organizations for researchers in the field of biotechnology aided in setting, formulating, and collating tutorial problems and multiple choice questions and in finding solutions to many model sets. I have also consulted various international journals, books, and reports and many publishers.

My learning at schools, universities of Calcutta and Jadavpur, Indian Institute of Science (IISc), Bangalore, and Indian Institute of Technology (IIT) Delhi was of immense help to build up my foundation in science, engineering, and technology and in handling living and non-living biosystems. My associations, visits, and exposures to industries such as Indian Drugs and Pharmaceuticals Ltd, Shaw Waales and Co., Standard Pharmaceuticals Ltd, Reckitt and Colman (barley factory), National Sugar Institute, Vam Organic Chemicals, National Dairy Research Institute, Mohan Meakin Breweries, National Chemical Laboratory, and Central Food Technological Research Institute gave insight to comprehend unit operations and unit processes involved in production, recovery, and purification of biomaterials.

I received immense help from various participants of the short-term training courses I conducted, including scholar participants, faculty from IITs and other universities and institutes. Their discussions also helped me in preparing this book.

Appreciation is extended to world-renowned laboratories, institutes, and universities in which I served or visited, including Laboratorios Nacionales de Fomento Industrial, Mexico; Instituto Politécnico Nacional, Mexico; Swiss Federal Institute of Technology (SFIT), Zurich; Bioengineering AG, Switzerland; DSM Culture Collection Centre, Berlin; University of Gottingen, Germany; SEM Culture Collection Centre, the Netherlands; University of Western Ontario, London; University of Waterloo, Canada; Osaka University, Osaka; Department of Applied Microbiology, Tokyo; University of Connecticut Health

Center (UCHC), Connecticut; Pfizer Central Research Division, Connecticut; Lehigh University, Pennsylvania; Centocor, Pennsylvania; B Braun Co., Pennsylvania; Roche Co., Pennsylvania; Hoffman LaRoche, Pennsylvania; The Bio-processing Resource Center, Pennsylvania; Merck Sharp and Dohme, New Jersey; Pennsylvania State University, Pennsylvania; Massachusetts Institute of Technology, Cambridge; Duke University, North Carolina; KBI Biopharma, North Carolina; Rockefeller University, New York; University of Compiegne, France; Tomas Bata University, Czech Republic; and Amgen, California.

In my professional career, the kindness of the following individuals are greatly appreciated: A Bose, A E Humphrey, A Fiechter, A K Sengupta, A N Bose, B D Sanwal, B K Bachawat, B N Srimani, BoMattiason, C S Jha, C V Seshadri, D K Das, E L Gaden Jr, G P Talwar, H Flores, J E Bailey, J M Lebault, J V Bhat, K L Chopra, K S Gopalkrishnan, K Toda, K Venkata Subramanian, M G K Menon, M K Bhan, M K Sarkar, M M Sharma, M MooYoung, M S Sodha, M S Swaminathan, M Sirsi, M Sharma, N Blakebrough, N D Banerjee, N M Swani, O P Jain, P K Bhattacharya, Peter Saha, R A Mashelkar, R K Mishra, R K Pachauri, R K Sachdeva, R N Dogra, R Rojas, S Aiba, S K Bramachari, S K Guha, S K Majumdar, S Nagai, S Ramachandran, Sisir K Chatterjee, Sunit Mukherjee, T Imanaka, T K Ghose, T K Roy, T Sen, V Ramaligaswamy, Viswa Rai, among others.

In preparing this book, I benefited a great deal through consultation of books and journals from the following libraries: Vidyasagar College (University of Calcutta), Central, Food Technology and Bio-Chemical Engineering and Chemical Engineering departments (Jadavpur University), Harcourt Butler Technological Institute (HBTI Kanpur), Biochemical Engineering Research Centre/Department of Biochemical Engineering and Biotechnology (DBEB) (IIT Delhi), SFIT Zurich, Universidad Nacional Autónoma de México, Osaka University, UCHC, California Institute of Technology (or Caltech), Duke University, Rockefeller University, University of Calgary, TBU Czech Republic, Indian Agricultural Research Institute, All India Institute of Medical Sciences (AIIMS), and Jawaharlal Nehru University.

My eternal gratitude to the medical staff who nursed me back to health. I am able to continue my active academic activities due to the timely diagnosis of Dr S Anila Khosla, Dr Lilly Kosa, and Dr S K Agarwal (CMO) of IIT Delhi, surgery and treatment by Prof. Dr Arvind Kumar and his team of doctors, chemotherapy by Prof. Dr P K Julka and his team of doctors (specially Dr Manoj), and help received from Dr T S Roy, AIIMS, among others.

Cooperation and support received from the faculty and staff of DBEB, IIT Delhi, is gratefully acknowledged.

Finally, I thank Prof. B N Jain, Vice Chancellor, Birla Institute of Technology and Science Pilani; Prof. Suresh Chandra; Prof. S K Dube, former Director, IIT Kharagpur;

and Mr S R Lakha, IAS (Retd), Vice Chancellor, Gautam Buddha University (Uttar Pradesh), whose active roles enabled me to be in my present position after I retired from DBEB, IIT Delhi.

I am particularly thankful to the entire active team of TERI Press for their involvement and keen interest in bringing out this book.

I am indebted to my wife Sakuntala, daughter Paramita, son Parag, and daughter-in-law Pooja for their love, support, appreciation, and frequent reminders to complete this book. Pooja and Parag took active interest to formulate a numerical that I had conceptualized and I am sincerely grateful for their help. I would also like to acknowledge the active help of my son-in-law, Arvind Nair, whenever it was required.

Preface

There has been rapid advancement in the field of process biotechnology during the past five decades. It has expanded in fundamental and experimental aspects and is growing further in terms of biochemical engineering and biotechnology. Theory and practice are still keywords in all research-based teaching and learning, and process biotechnology is no exception.

For teaching research-based subjects in academic institutions, universities, and colleges and handling problems in biotechnology industry, the author has written a few other books, including *Advanced Process Biotechnology* and *Experimental Process Biotechnology Protocols*. However, it is essential for beginners to have exposure to tutorials and design calculations. This book provides undergraduates, postgraduates, and pre-PhD students with a scope to understand theory-based numerical aspects of process biotechnology.

The field of process biotechnology has become more promising as many government, private, and multinational organizations have opened new scopes for biochemical engineering, biotechnology, and related areas. Thus, the need to provide scholars and teachers with relevant and comprehensive knowledge resource has become a necessity. In this regard, the author hopes that this book will be a comprehensive practice treatise for learners.

This book is divided into two sections, comprising four chapters in each section. Section I refreshes the theoretical background to a large extent within the first four chapters. Section II deals with bioprocess engineering tutorials with many model problems and solutions for candidates to practice.

Answers to the first 200 mulitiple choice questions given in *Process Biotechnology Fundamentals*, first and second edition, 2000 and 2004, respectively, are provided at the end of the book.

The author earnestly requests all readers to communicate any misprints, erroneous statements, and suggestions for improvement in the content. Moreover, he considers it a privilege to have the cooperation of all in keeping the text up to date.

Contents

Acknowledgements vii

Preface xi

Section I: Background on Life Formation

1. Prebiotic Science of Chemicals for Life 3

1.1 Introduction 3

1.2 Timescale of Terrestrial Evolution 4

1.3 Heat Death in Cosmos 5

1.4 Modern Biochemical Engineering and Biotechnology Viewpoints of Evolution of Life Forms 8

1.5 Ecology and Environment 13

1.6 Asymbiotic Non-rhizobial MPI in Eco-biotechnology 19

1.7 Overproducer Regenerated Plants 20

1.8 Human Body Design Evolution 23

1.9 Importance of Updated Human Body Design 26

1.10 Conclusion 27

References 28

Bibliography 29

2. Biotic Chemistry 31

2.1 Introduction 31

2.2 Hydrogen Bond 39

2.3 Water 40

2.4 Biochemistry and Immunochemistry: A Concise Treatise 55

2.5 Protein Kinesis Engineering 80

2.6 Tutorial Problem 90

References 93

Bibliography 94

3. Cell Biology and Microbiology 95

3.1 Definition of a Biocell 95

3.2 Biocell Types, Positioning, Characteristics, and Identification Tests 95
3.3 Biocellular Design and Activity 116
3.4 Archaebacteria: The Newer Biocells 116
3.5 Ultrastructural Features 118
3.6 Yeasts 128
3.7 Cell and Tissue System Biology 129
3.8 Cellular Materials and Metabolism in Biotechnology 136
3.9 Microbes and Microbiology 137
References 222
Bibliography 226

4. Chemical and Biochemical Engineering Principles **229**
4.1 Units and Dimensions 229
4.2 Dimensional Analysis 232
4.3 Fundamentals of Material Balance 241
4.4 Concept of Modelling of Biological Reaction Systems 249
4.5 Fluid Statics and Applications 279
4.6 Fluid Flow Phenomena: Viscosity and the Momentum Flux 284
4.7 Weirs 294
4.8 Laminar and Turbulent Flows 295
4.9 Basic Equations of Fluid Flow 302
4.10 Mass Transfer Principles 307
4.11 Mass Transfer Principles Advances 312
4.12 Heat Transfer by Conduction 318
4.13 Tutorials 321
4.14 Heat Transfer by Convection 322
4.15 Heat Transfer by Radiation 324
4.16 Evaporation 330
4.17 Refrigeration 358
4.18 Heat Exchangers 361
4.19 Adsorption 369
4.20 Extraction 378
4.21 Leaching and Bioleaching 385
4.22 Distillation 402
4.23 Drying of Biological Products 413

4.24 Freeze Drying of Biological Materials 420
4.25 Automatic Process Control Principles 424
4.26 Properties and Functions of Polymeric and Biopolymeric Engineering Materials 432
4.27 Basic of Computer and Its Use in Biosystems 453
4.28 Genetic Engineering Principles and Considerations in Handling Recombinant Cells in Large Scale 458
4.29 Modern Instrumental Techniques in Process Biotechnology 463
References 468
Bibliography 471

Section II: Bioprocess Engineering System: tutorials

5. Upstream Process Biotechnology: Solved Numerical Problems **475**
5.1 Pure Culture Microbial Biotechnology 475
5.2 Mixed Culture Microbial Biotechnology 514
5.3 Immobilized/Anchored Biosystems 519
References 527
Bibliography 527

6. Enzyme Engineering and Technology: Solved Numerical Problems **529**
6.1 Enzyme Kinetics 529
6.2 Enzyme Reactors 539
6.3 Enzyme Production Strategies 542
Bibliography 546

7. Product Recovery and Purification: Solved Numerical Problems **547**
7.1 Introduction 547
7.2 Filtration 548
7.3 Ultrafiltration 550
7.4 Reverse Osmosis 552
7.5 Foam Fractionation 553
7.6 Adsorption 557
7.7 Extraction 559
7.8 Novel Bioproduct Recovery Design Problem 560
7.9 Possible Contaminants 561
Bibliography 573

8. Unsolved Numerical Problems 575
8.1 Microbial Cell System 575
8.2 Reactor Kinetics and Stability 579
8.3 Plant Cell System 584
8.4 Animal Cell System 585
8.5 Miscellaneous Biosystems 586
Bibliography 591

Index 593
About the Author 601
Answer Keys 603

Section I

Background on Life Formation

1

Prebiotic Science of Chemicals for Life

1.1 INTRODUCTION

The myriad aspects of science, engineering, and technology in the universe are commanded by the Almighty Supra Nature (ASN). The earth came into existence in the universe through a gigantic explosion called the big bang. The science of ASN was responsible for cooling the super-hot, newly formed planet by heat death and creating an environment around the earth's surface, which is termed "nature". Natural science emerged through different stages by natural unit operations and processes. So, the earth's nature may be said to be the architect or engineer of the formation of all natural materials and designs through the evolution of science, leading to naturally acquired design engineering (NADE) materials. NADE materials, in the presence of existing natural conditions, undergo recombination processes. This leads to the formation of generations of recombinatorially acquired design engineering (RADE) materials on the earth. Besides non-living materials, NADE and RADE created life or living materials through evolution. Among the creations of NADE and RADE, the most spectacular is the creation of life. Of all life forms, one of the most significant is the *Homo sapiens*. Nature caused the emergence of anatomical technology advancement through NADE and RADE and other species transformed the primitive human design to modern sophisticated human design. The transformation involved positive transfectional fermentation process biotechnology (TFPB) in biological timescales. These advancements in features are readily visible in all biological components on earth. From the above discussion, one might say that NADE formed human beings.

The cause of creation/change in certain features of life forms on earth from known to unknown species is still a mystery. That is, evolution of all life forms has not yet been discovered. Human beings have spent unaccountable time and money in search of the cause and basis of this unknown phenomenon. Scientists and engineers alike have speculated on the specific origin of the creation of life. According to some, life forms appeared on earth during the cooling process of the planet's formation.

It is known that life forms consist of molecules that are naturally able to form large molecular compounds of life. Structurally and morphologically, these molecules have been shown by scientists and engineers to possess specific structures.

However, the natural environment has been changing since its creation. In order to cope with altered environments, molecular products also went on natural recombination for the survival of the fittest. Thus, newer environments compelled the RADE process to occur.

The atoms of natural elements must have existed on earth when life was being formed from molecules. Perhaps, over eons, these elements under existing environmental conditions combined by natural prebiotic chemistry and arranged or rearranged naturally designed forms of a living cell. Such living cells are assumed to exist in both anoxic and oxic environments. It is also conjectured that NADE could take place in the presence of water at sea. From the study of prebiotic chemistry, it was proposed that nature's important elements include carbon, hydrogen, and nitrogen. Oxygen was considered an important element only at a later stage. These elements possess life-giving properties and are present in all organisms. The presence of carbon makes all the difference between the presence or absence of life in a particular matter.

A modern living cell can be considered a compound made up of carbon, nitrogen, hydrogen, and oxygen, along with an energy or entropy generation device that forms the basis of life. In microbes, animals, and plants, prebiotic chemistry of chemical evolution derived life, imbibing energy/soul to it (Figures 1.1 and 1.2).

1.2 TIMESCALE OF TERRESTRIAL EVOLUTION

The formation of life on earth and its evolution through NADE and RADE presumably occurred through cosmos scale stages, requiring the creation of free energy. The cosmos scale comprises six major stages as per modern biochemical engineering viewpoints on the evolution of life (Figures 1.1 and 1.2). Through these stages, abiotic conditions gradually become congenial to biotic conditions.

The earth was surrounded by stars and planets after the Big Bang through heat death phases (Figure 1.2). This was the first stage after the Big Bang. Through billions of years, various processes on the earth's surface continued forming varied elements, including carbon—the central element of life. This could be termed the terrestrial evolution stage (Figure 1.1).

The second stage consisted of prebiotic natural chemistry stage for designing organic molecular compounds, which served as the stepping stones for chemicals of life and were discovered through current non-space and space probes. The third prebiotic chemical stage (Figure 1.3) led to the natural designing of the building blocks—the

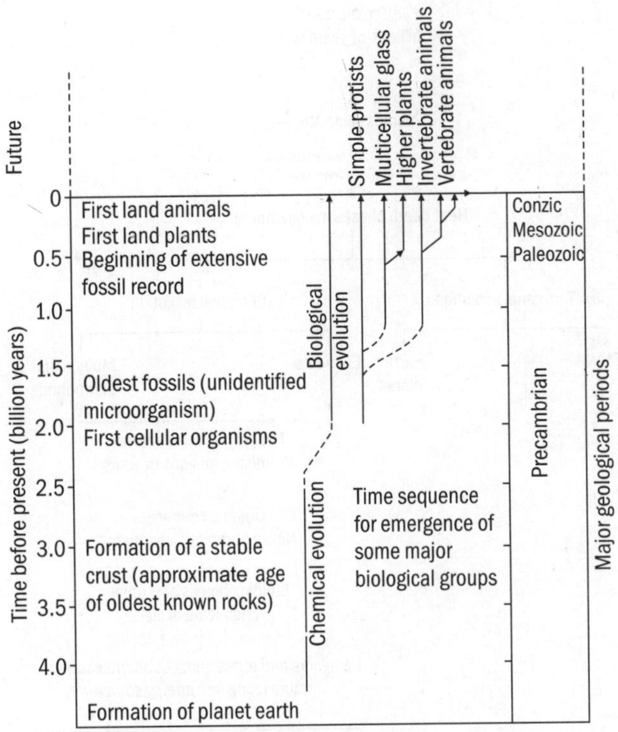

Figure 1.1 Timescale of terrestrial evolution

engineering rungs of the deoxyribonucleic acid (DNA) double helix or amino acids that initiate protein design. Thus, protein molecules are well-designed molecular devices by nature.

1.3 HEAT DEATH IN COSMOS

1.3.1 Need for Creating Free Energy

It is known that the early universe was hot and dense, with frequent collisions. Consequently, the free energy of the universe became zero when it reached thermal equilibrium. Our present universe, however, is not in a state of thermal equilibrium; it is locally alive as it has innumerable life forms. And life needs a steady source of energy.

In search of life forms in the universe, one has to look at places where free energy is being created or stored. Astronomists and physicists have found that the creation of free energy is impossible in a closed system of fixed volume. A unique characteristic of our universe is that its galaxies accelerate due to gravity (G). Physicists arrived at a mathematical expression to calculate the mass density of universe with a dust of galaxies at any time (Equation 1.1).

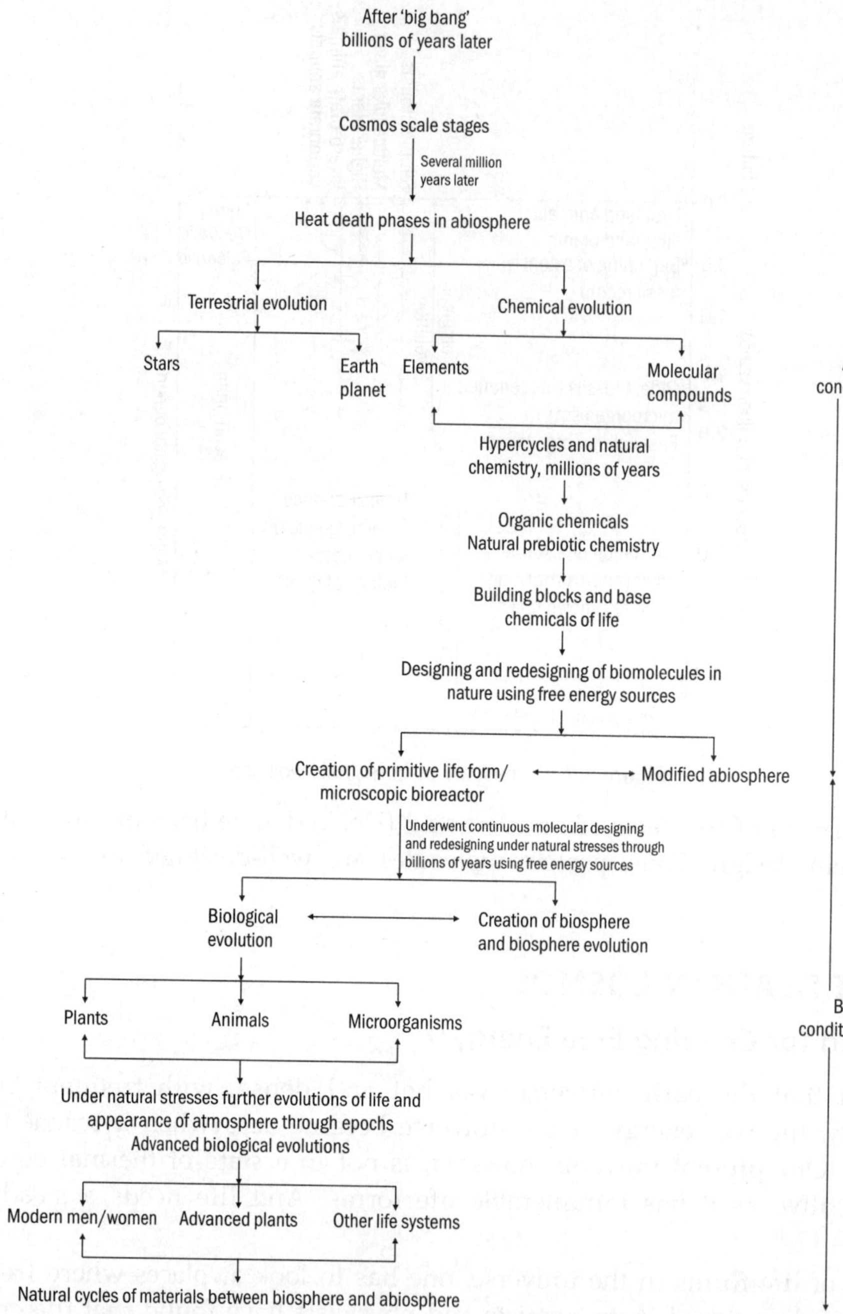

Figure 1.2 Modern biochemical engineering and biotechnological viewpoints on evolution of life forms
Source Mukhopadhyay (1996)

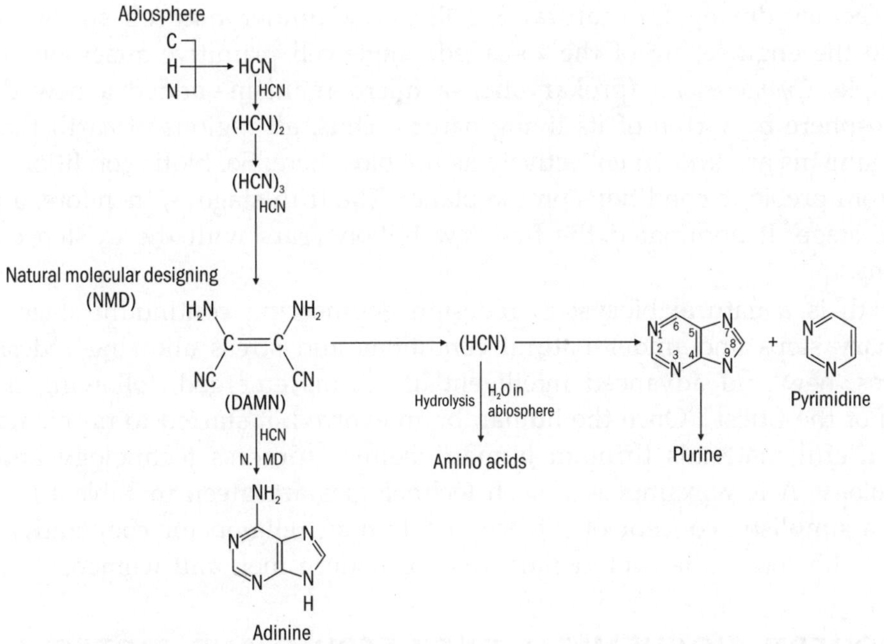

Figure 1.3 Prebiotic chemistry leading to the natural designing of base chemicals of life on earth

$$\rho(t) = \frac{1}{6}\pi G t^2 \tag{1.1}$$

where ρ is the density and *t* is the time.

In a similar way, they arrived at the conclusion that a radiation-dominated universe behaves alike.

$$\rho(t) = \frac{3}{4}\pi G t^2 \tag{1.2}$$

As a non-empty universe cannot be static, its density must have a singularity on the timescale. This singularity has been termed Big Bang. Scientists have known that the present overall temperature (*T*) of the universe is 2.73 K. Radiation light has been considered to be a sine wave drawn on a rubber sheet of length *R*. As the size *R* of the sheet increases, the wavelength (λ) stretches; correspondingly, *T* drops. This means that the universe was hot and dense in the past but has cooled down. The heat transfer from warmer regions to the cooler regions is irreversible, so the entropy increases. In later stages, at lower temperatures and densities, the coupling gradually faded away, producing expansion. The resulting temperature difference created free energy. It was also responsible for the survival of different cooling rates. Ultimately, the conditions for the creation of life forms and the required free energy materialized.

The third stage was crucial and can be regarded as a chemical evolution period that caused biomolecular system designing. The fourth stage is the stage of evolution of biomolecules during the natural RADE period under nature's stress conditions, leading to the engineering of the so-called, single-cell primitive anaerobic life forms. For example, *Cyanobacteria* (prokaryotic)—a microorganism—added a new dimension to the biosphere by virtue of its living nature. Thus, all regions of earth that contain living organisms are known collectively as the biosphere. So, biotic conditions began to appear from prebiotic conditions on the planet. The fifth stage is, therefore, a primitive biological stage. It dominated the first few billion years with the existence of living biosystems.

The sixth is a natural biosystem redesign engineering continuum stage. Through evolutionary steps and under natural conditions and stress allowing redesigning of biosystems, new and advanced intelligent life forms emerged, following the law of "survival of the fittest". Once the human brain evolved, it started to mimic nature and produce useful materials through human chemical/process technology and process biotechnology. A few examples of such technologies are given in Table 1.1. Figure 1.4 presents a simplistic concept of science and human religion for cooperative creation, and Figure 1.5 shows the evolutionary tree of modern men and women.

1.4 MODERN BIOCHEMICAL ENGINEERING AND BIOTECHNOLOGY VIEWPOINTS OF EVOLUTION OF LIFE FORMS

"It is better to be approximately correct than exactly wrong" is the first principle of biochemical engineering. Elucidation of evolution of life has always been based on this

Table 1.1 Major biotechnology/chemical technology of man-made bio-business areas for new millennium at international markets

Electronics
Gallium arsenude, photoresist, high purity gallium, printed circuit board, gallium phosphide, gallium arsenide, hybrid IC substrate, laminated board, electroceramics polyimide, film for PCB, photoresist LSI, optical disk, picrofishe photoresist, multilayer board,
Materials
Carbon fibre, rare metals, ion exchange membrane, carbon fibre, reverse osmosis, magnetic powder, carbon fibre, silicon carbide, hydrophilic polymer, alloys for hydrogen storage, silicon nitrides, metal fibre reinforced nylon, carbon fibre membrane, carbon fibre fine ceramics, membrane, functional polymer
Biotechnologies
Plant seed, TPA, human MAB, beta-endophene by *Escherichia coli,* stress proteins (example, HSP), interferon, human growth factor, bioreactor-assessing genomic technologies, biosensors, biofuels, planted seed, amino acids, MAB, vaccines, biodegradable polymers, interferon TNF, rapid bioforensics, bioinformatics interferons, L-lysine, T-cell growth factor

TPA – Tissue plasminogen activator; MAB – monoclonal antibody; HSP – heat shock protein; HGF – human growth factor; TNF – total nitrogen fixation; LSI – light sensitivity indicator; PCB – polychlorobenzene; IC – integrated circuit

Source SRI International, Menlo Park, California

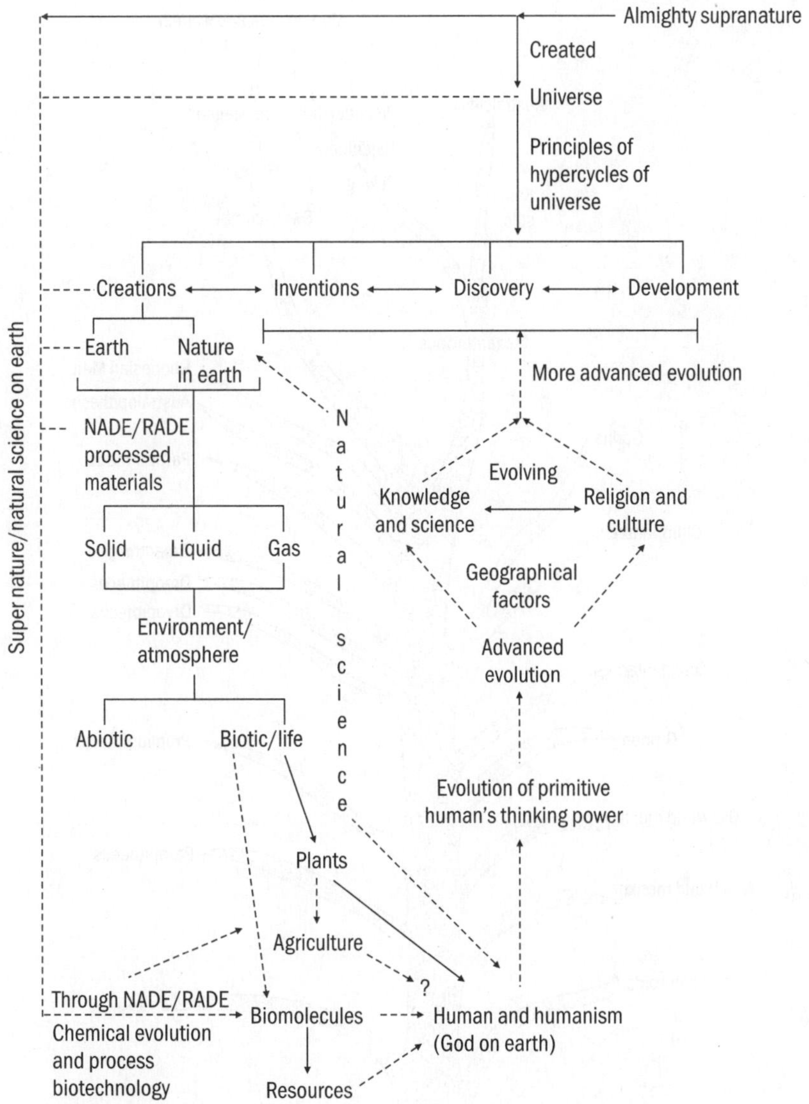

Figure 1.4 Simplistic concept of science and religion for cooperative creation in advancing biochemical engineering and biotechnology-caring environment and ecology

principle. The prebiotic universe was hot and dense, with random collisions, fusion, and fission of matter. It evolved sources of hydrogen, carbon, nitrogen, and oxygen. Scientists of the present biotic epoch believe that the early universe was in thermal equilibrium. They found evidence of this from the relic black body radiation. Its consequence was that the free energy of the universe was zero. Therefore, from biochemical engineering viewpoints, in the timescale of terrestrial evolution of engineered life and for its

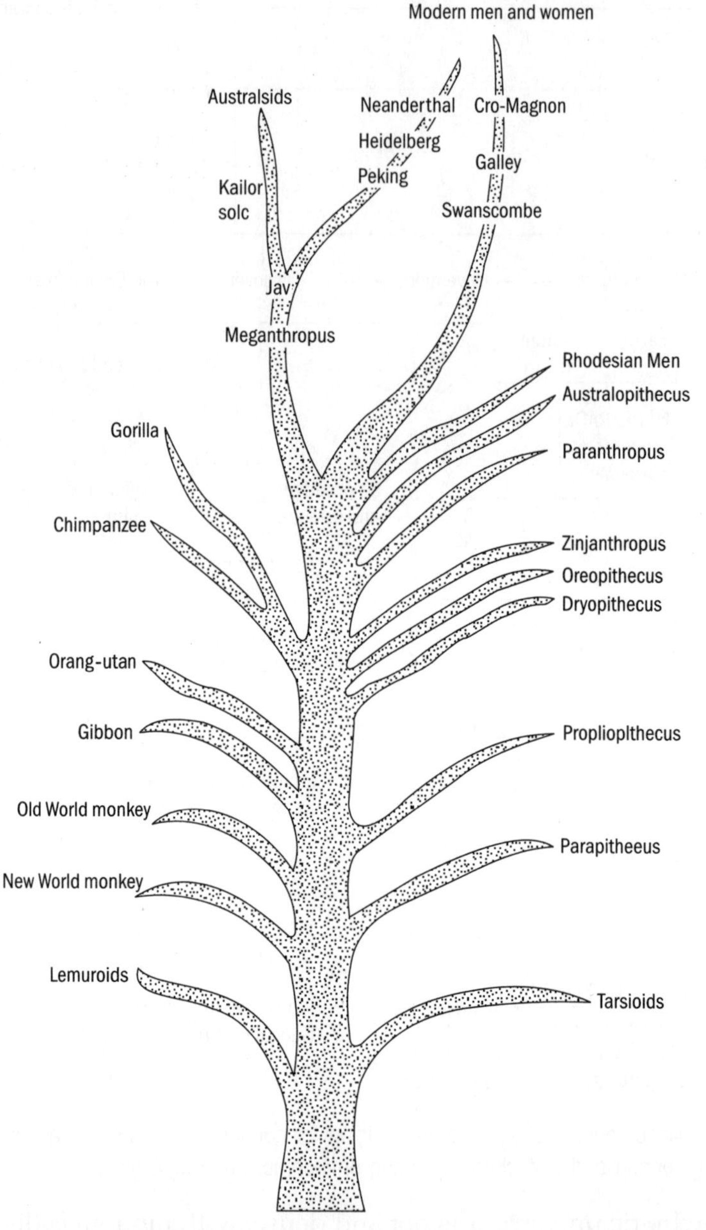

Figure 1.5 An evolutionary tree of the primates, beginning with the primitive tree insectivores

evolutionary natural biotechnological redesigning, it was necessary to create free energy/ entropy. This is because life needs a steady state source of free energy and shows a clear evolution towards a thermodynamic steady state.

As prebiotic universe was considered non-empty, it was obvious that a non-empty universe cannot be static, signifying that its density must have a singularity on timescale. Natural chemistry based on timescale singularity caused the formation of water on earth. In turn, natural molecular design engineering formed purines and pyrimidines from the prevailing carbon, hydrogen, and nitrogen elements in the windy environment. In addition, in hydrolytic condition of the universe, amino acids were formed in dynamically unstable universe, being the principal "uphill" source of free energy. This was utilized in the sequencing of amino acids, forming polymeric protein biomolecules.

Following this, natural protein redesign engineering, in coordination with molecular engineering, produced biomolecules in the primitive molecules system entity—the so-called single-cell anaerobic "life" or microorganism. It underwent adaptation and further redesigning through eras and biogenic weathering, resulting in historic changes in the universe and its life forms. This enabled to define a living being as an open, self-regulating, self-replicating, and developing heterogeneous system whose most important functional molecules are protein and nucleic acids. Thus, present life is the result of the phylogenetic evolution. It covers the path of ontogenetic development from zygote to old age and death, explaining why the process of evolution of life is irreversible. These modern biochemical engineering viewpoints on the evolution of life are shown in Figures 1.4 and 1.5.

In the evolution of life forms, the hypercycles of the universe (Figure 1.6) involved ecological cycles function as a component. Accordingly, life may be defined as a unit of properly, sequentially, and naturally ensembled chemicals of life imbibed with a soul having metabolic reactions; it is distributed in ecological domains. In the ecological distribution of living beings, the microbe-plant-animal interactions with nature are a common feature (Figure 1.7). One may preclude that by this interactive ecological engineering function from single-cell to multicellular organisms, formation has been possible. For example, multicellularity enabled a plant to become physically large and to have roots in the ground and leaves attached to the stems, where one set of cells possesses the capacity of uptake of water and nutrients, while another set efficiently captures radiant energy from the sun.

The ecological engineering principle (EEP) aims at deriving benefits from natural and man-made biosystems through the interaction design of sustainable ecosystems that integrates society with its environment. In the production of plant food for human beings, the growth of *Rhizobium* sp. in the roots of a food-producing plant acts as a biofertilizer and plays an important role in the microbe–plant interaction (MPI) system. The interaction between microbes and plants occurs in many ways. In nature, for ecological balancing, this interaction occurs in cycles. The hypercycles of the universe involve ecological cycles as a component. In ecology, the distribution of living beings

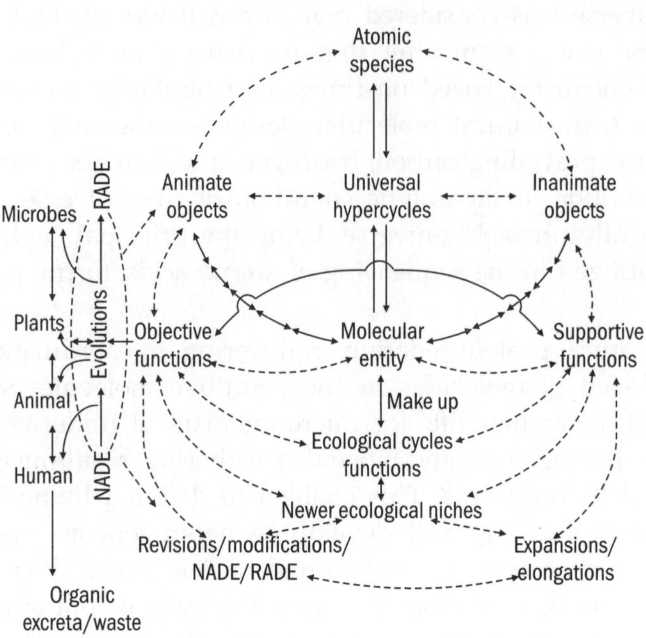

Figure 1.6 Involvement of ecological cycles in the domain of universal hypercycles in the creation of life
Source Mukhopadhyay and Mukhopadhyay (2000)

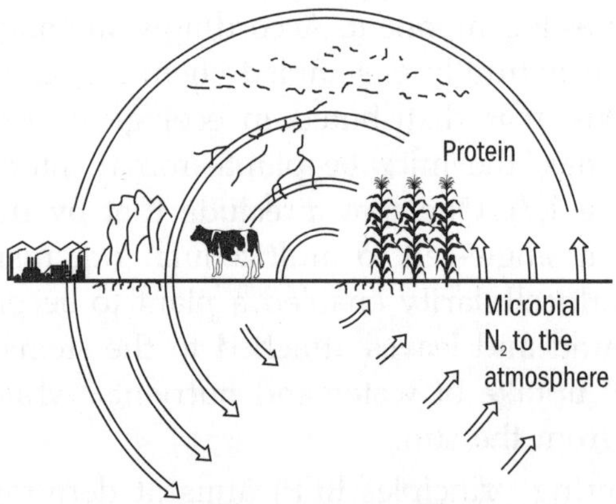

Figure 1.7 Microbe–plant–animal ecological interaction cycle

for MPIs with nature is a common feature. For MPI in the ecological engineering domain, many microbes act as donor systems. The oxygen cycle participation in MPI is inevitable, and the benefits and disadvantages in such interactions are many. Important beneficial contributions from the interaction of nitrogen-fixing microbes (as biofertilizer) to the amount of nitrogen concentrations in the growth of cereal plants have not been examined thoroughly from ecological engineering viewpoints. In deriving these benefits, engineering/biotechnological concerns regarding the interactive ecological engineering functions from single cell to multicellular organisms formation has been a predominant aspect. Many research scientists, therefore, aimed at describing the aspects of ecological engineering domains in MPI products of industrial and agricultural importance. Examples of several product formations in ecological engineering domain are well known.

Specialized cells in the stem of a plant form channels that transport water and nutrients between the roots and leaves. In addition, another set of specialized cells forms a layer of epidermis to prevent water loss and provide a protected internal environment. The plant does not compete directly with single-cell organisms for its ecological niche; it has found a radically different way to survive and propagate. However, in an ecological domain, single cells do associate to form colonies under aquatic environments. Like soil and air/wind, water streams and water resources of nature serve as a major environment surrounding all ecological habitats. Whether it is a habitat with a hot spring water resource or one at normal temperature, microorganisms/algal or phytoplankton flora are found in large numbers in such habitats (Bailey and Ollis 1986; Mukhopadhyay and Baisya 1997). Habitats like hot spring algal-bacterial mat ecosystems are useful environments for microbial ecologists to model a thermophilic carbon cycle or for industrial biotechnologists to isolate potentially useful microorganisms for bioconversion processes. For research in the field of ecological engineering to be successful, a general understanding of living beings in aquatic ecosystems and their distribution in nature is necessary. Ecological balance through distribution in nature demands that the structure of a living cell/organism is characterized by fitness and adaptation to a particular ecological system. The complexity of habitats is determined by the natural selection of genotypes. Fixing ecological scale-up criteria is, therefore, difficult for synthetic engineering biologists working in the area of ecological engineering without the maintenance of an eco-friendly environment.

1.5 ECOLOGY AND ENVIRONMENT

Biospheric ecology in relation to aquatic biosystems is important. In the balance of ecology and environment, an aquatic ecosystem plays a key role because the chemistry and properties of water are of great significance. Chemical structural properties lead to its cohesive/associative nature. This, in turn, is responsible for its unusual properties such as high surface tension, specific heat, and heat of vaporization (Table 1.2). In context to the progress of modern industrialization, all manufacturing industries like chemical, biotechnological, pharmaceutical, petrochemical, and fermentation always attempt not to upset or downgrade the ecology and environment. During the evolution of life on earth and evolutionary selection of oxygen because of its distribution in the gaseous state, water and soluble bio-media under normal conditions have gradually evolved in the oxygen cycle. A schematic of the oxygen cycle involving ecological domains of prokaryotic, eukaryotic, and archaebacterial biosphere (Figure 1.8) provides for realization of the effectiveness of ecological engineering (Mukhopadhyay and Das 1994).

Table 1.2 Pure water properties and their values

Property	Value	Unit
Colour/odour (ideal)	Colourless/odourless	—
Density (4°C)	1.00	g/cm^3
Density (at BP)	0.958	g/cm^3
Boiling point (BP)	100.0	°C
Freezing point (1 atm pressure)	0.0023	°C
Triple point (at 4.58 Hg)	+0.0098	°C
pH (potential Hydrogenii)	7.00	—
Ion product (25°C) constant	1.008×10^{-14}	mol^2/dm^{-6}
Viscosity (at 20°C)	0.001	Pas
Permittivity (18°C)	81.7	—
Specific conductance (18°C)	4×10^{-4}	kJ/kg/K
Thermal conductivity	0.537	W/m/K
Heat of fusion	1.434	kJ/mol
Heat capacity (liquid)	0.9986	cal/g/°C
Heat capacity (vapour)	0.484	cal/g/°C
Heat of formation (20°C)	68.35	kCal/mol
Heat of vaporization	9.719	kCal/mol
Latent heat of vaporization	576.6	cal/g
Critical temperature	374.1	°C
Surface tension (at BP)	58.8	dyn/cm
Electrolytic dissociation constant (18°C)	—	—
K_1 ($H_2O \rightarrow H^+ + OH^-$)	1.07×10^{-16}	—
K_2 ($OH \rightarrow H^+ + O^{-2}$)	$< 10^{-36}$	—
Dipole moment of vapour	1.85	cgs units × 10^{48}
Strength of a hydrogen bond	5.0	kCal/mol
Hydrogen bond length in water	0.26–0.31	nm
H–O–H valence angle	104.5	degree
Redox potential (E_h)	2.13	volt

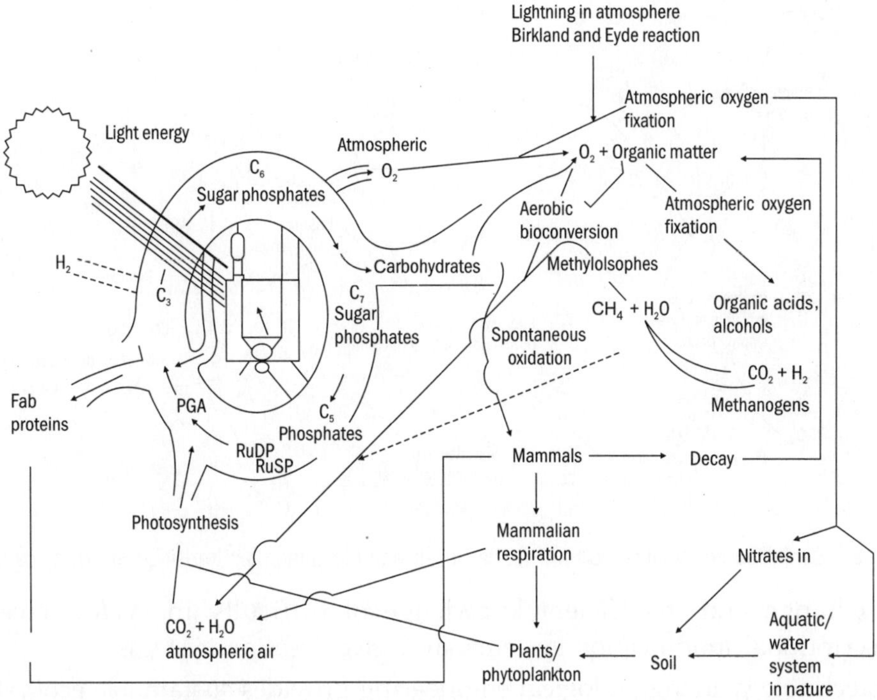

Figure 1.8 Schematic diagram of nature's oxygen cycle involving prokaryotic, eukaryotic, and archaebacterial biosphere in aquatic ecological domains under a dynamic coexistence

Source Mukhopadhyay and Mukhopadhyay (2000)

From this scheme, it is easy to visualize that there exists in the ecological engineering domains the cycles of matter in both aerobic and anaerobic environments; for example, in natural oxic atmosphere and in abyssal depths of the oceans. A simplified scheme of the interactive cycles of matter in aerobic–anaerobic ecosystems is shown in Figure 1.9 to indicate ecological linkages. From this scheme, it is evident that in the ecological species interactions, microorganisms are the primary agents of organic matter in freshwater and marine environments.

1.5.1 Benefits and Constraints

NADE has been manifested in harnessing ecological engineering to provide renewable resources of nature through solar energy participation. Many microbes, such as prokaryotes, act as a donor system for MPI in producing fruit/biomass, which in turn is consumed by mammals/animals. In ecological engineering terms, the benefits achieved by this interaction are also associated with certain scientific disadvantages. Benefits are concerned with in vivo biosafety and a wide range of product formation in some ecological domain of low oxygen requirement. The disadvantages include complex

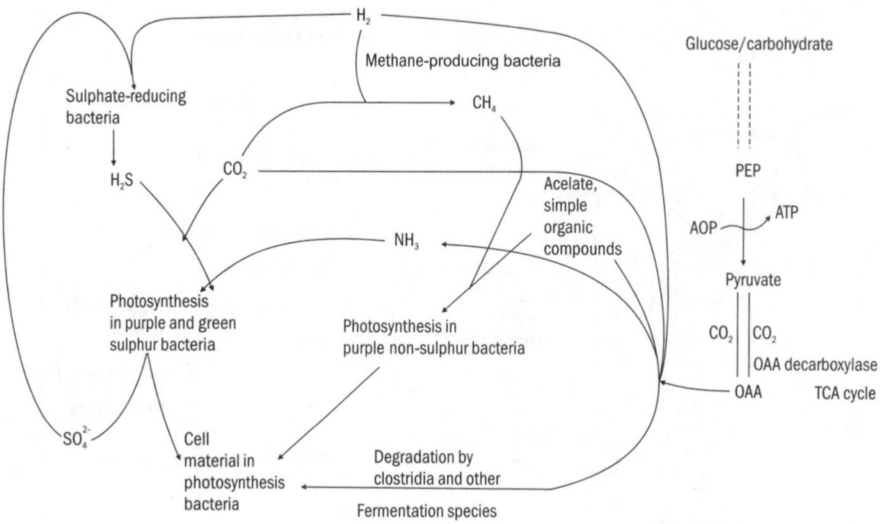

Figure 1.9 Simplified schematic cycles of matter in an aerobic–anaerobic transition of ecological linkages

genetic exchange and insufficient knowledge of difficulty in vector development, post-transcriptional modification systems, and glycolysation systems.

Eco-biotechnology using ecological engineering provides sustainable ecosystems that integrate society with its environment. The growth of *Rhizobium* sp. in the roots of food-producing plants acts as an eco-friendly biofertilizer and plays an important role in the MPI systems.

1.5.2 MPI Advances

As already mentioned, hypercycles of the universe involve ecological cycles functioning as a component. The MPI is an integral part of eco-biotechnology using ecological engineering principles. Its dimension has further enhanced as human beings have developed many genetically modified microorganisms (GMOs), which are usually disposed of in natural ecosystems.

In eco-biotechnology, biological nitrogen fixation (BNF), both symbiotic and asymbiotic, is significant. The possibilities of extending the use of BNF to non-legumes as an eco-friendly approach to overcome various environmental problems have also been explored. Moreover, interactions between microbes and plants through eco-biotechnological avenues could be used as a tool in ecological engineering to counteract problems relating to pollution. *Rhizobia* undergo symbiotic relationship with legumes in vivo. In in vitro addition, *Rhizobia* leave in soil a fair amount of nitrogen, which benefits subsequent crops. By intraspecies transfectional MPI, nitrogen-fixing ability of leguminous plants infected with appropriate species and a suitable strain of *Rhizobium* has been found to vary between 50 kg/ha and 150 kg/ha, depending on the characteristics of the plant crop as well as cultivation conditions prevailing in the region. The following sections provide progress of MPI towards the benefits of human beings.

1.5.3 Rhizobium-participated MPI in Eco-biotechnology

1.5.3.1 Molecular biology aspect

The BNF in ecological domain occurs in species of more than 100 genera distributed among several of the major phylogenetic divisions of prokaryotes (*Eubacteria* and *Archaea*). Sequence and mutational analyses of genes (*nif*) necessary for nitrogen fixation in many diazotrophs indicate that their products have common structures and functions, while the degree of linkage and arrangements of specific *nif* and associated genes vary considerably. In addition, *nif* genes and genes involved in plant invasion and the effectiveness of nitrogen fixation, such as *nod* and *fix* in *Rhizobium* species, are often linked in eco-biotechnological cooperation.

1.5.3.2 Role of environmental oxygen species

Symbiotic MPI occurs widely through most of the principal food-producing plants in the ecological engineering domain. *Rhizobium* provides legumes with substantial oxygen-containing species (nitrates). Mutualism in lichens is another example of ecological engineering for MPI in nature. Lichens consist of highly modified fungi that harbour blue-green and green algae along their hyphae (filaments). The association of these organisms forms a compact and highly efficient metabolic unit in the ecological niche and provides benefit in water. They also form most of the supporting structure and provide nutrients for both components via photosynthesis. Thus, the twofold benefits of keeping ecological balance and water purification by MPI in ecological engineering are obvious. For this balance, however, oxygen transport phenomena in MPI are important.

1.5.3.3 Oxygenation in plants

Leghemoglobin components, which are synthesized by the participation of *nod* D and *nif* A genes as per gene interaction scheme (Figure 1.10), facilitate in vitro–in vivo flow of oxygen to nitrogen-fixing rhizobial bacteroids. In this interactive phenomenon, red colouration is seen in the cross-section of root nodules. The in vivo oxygen partial pressure (pO_2) plays an important role. In addition, the effective symbiotic biomass production controls and coordinates the expression of host and prokaryotic genes. In the ecological engineering systemic biomass production, *nif* genes of bacteroids are activated only when the nodules have already grown sufficiently that the nitrogenase A (*ntr* A) enzyme system can meet proper condition for nitrogen fixation (compartmentalization and low pO_2). Thus, in ecological engineering, the regulation of specific genes is of crucial importance for an effective rhizobial growth through symbiotic MPI.

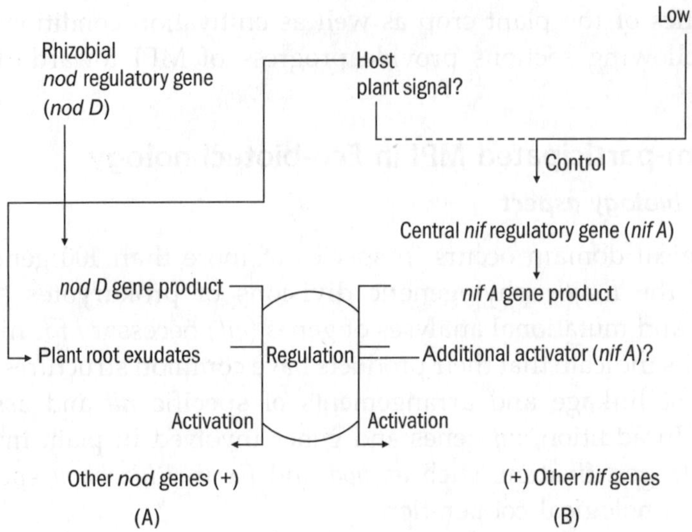

Figure 1.10 Schematic of role of oxygen in genes interaction in MPI meeting eco-biotechnological cooperation

1.5.3.4 Environmental pollution and alien species stresses on eco-biotechnology

Mismanagement of nitrogenous biofertilizers may cause stresses of nitrate contamination in flood-irrigated regions of ecological engineering handling agricultural produce. Besides agriculture, such environmental pollution stresses may exert impacts on other domains, which put a load on eco-biotechnology system boundaries (Table 1.3). Although the global nitrogen cycle in nature is dependent on microorganisms, it is also likely to be affected by pollution stresses on natural environment of ecological engineering. The maximum fertility of soils is derived from the capacity of resident microbial populations to interact with their environment and produce forms of nitrogen compounds available for plant growth and, subsequently, for its produce. Thus, besides rhizobial plant symbiosis, the presence of beneficial, asymbiotic, non-rhizobial microorganisms may play an important role in advancing ecological engineering benefits through MPI. The use of enrichment and pure culture techniques with the development of selective media and, in the case of symbiotic organisms, the use of plant hosts as a trap have allowed detection, isolation, and enumeration of the most abundant fraction of soil microorganisms involved in ecological engineering processes. It has revealed metabolic diversity and crossover present in soil flora. Physico-chemical, biochemical, and biological properties of soil microhabitats, in which microbial stresses exist, are difficult to ascertain and, therefore, difficult to reproduce in laboratory. Quantifying environmental damage and loss of biodiversity in ecology due to invasions of alien species worldwide are complicated by the fact that some 1.5 million of an estimated 10

Table 1.3 Environmental pollution stresses on ecological engineering boundaries

Ecological engineering boundary	Pollution sources impact	Evidence of stress
Agriculture	Uncontrolled use of surface and groundwater for irrigation	Seasonal water shortages
	Excessive use of agrochemicals	Increased soil salinity leading to reduced crop productivity Contamination of groundwater by pesticides and nitrates leading to health risks
Industry	Discharge of liquid water	Highly polluted contaminated rivers and coastal waters
	Uncontrolled dumping of solid wastes, including heavy metals	Possible pollution of river and groundwater by leachate seepage
	Disposal of waste oils and leakage from underground tanks	Waste oils/pollutants detection in rivers, wells, and sea, leading to long-term pollution and eco-biological imbalance
	Uncontrolled discharge of gaseous effluents	Highly polluted atmospheric pollution
Energy	Thermal power stations discharging cooling waters into coastal waters	Elevated temperature of water may lead to changes in marine ecology and possible damage to fisheries
Human dwellings	Unregulated sewage disposal and inadequate sewerage network	Microbial/viral contamination of surface and groundwater possibly poses health risk
	Excessive use of groundwater resources for domestic supply	Salt water influence in waste liquefiers

million species on earth are identified and described. The total number of introduced species in the USA, UK, Australia, India, South Africa, and Brazil is in the range of 2000–50 000. Competitors and predators threaten many native species, while others are endangered by hybridization/recombination with alien species and/or major ecosystem changes caused by these species through RADE in the ecological engineering system. Many such introduced species like rice, corn, wheat, forests, domestic chicken, and cattle have been beneficial and now provide more than 98% of the world's food supply, with a value of more than $5 trillion per year.

1.6 ASYMBIOTIC NON-RHIZOBIAL MPI IN ECO-BIOTECHNOLOGY

1.6.1 Prokaryotic Genes Interacting in Eukaryotic Plants

Vector-mediated transformations in plants are common in tissue culture. Co-integrating binary vectors plays an important role in MPI redesigning for ecological engineering transformation gene (of prokaryote) addition into natural cultivars in the ecological domain. Thus, the presence of transgenic plants in the ecological engineering domain is becoming immensely important. The disposal of transgenic plants on lands/receptacles is impacting ecology and posing long-term stress in balancing ecological domains. Plants redesigned for resistance to herbicides and insecticides are typical examples. Genes for herbicide resistance have been derived from *Pseudomonas* sp. and those for insect resistance have been extracted form *Bacillus thuringensis.*

1.6.2 MPI-based Eco-biotechnology for Industrial Production of Agrochemicals

Crown gall, a neoplastic disease of many dicotyledonous plants, is caused by a soil bacterium named *Agrobacterium tumefaciens.* A large plasmid in this bacterium was found to be responsible for its tumour-inducing capacity and was, therefore, called Ti plasmid. Bacteria-free crown gall cells have been cultivated in the absence of phytohormones like auxins and cytokinins. This hormone, independent of growth, defines tumorous growth in plants. Low molecular weight compounds called opines are produced in crown galls, which are actopine, nopaline, and agropine. Opines are utilized by *A. tumefaciens* selectively as sources of carbon, nitrogen, and energy. Thus, MPI can be seen as a species parasitic relationship that benefits bacteria (Figure 1.11).

1.7 OVERPRODUCER REGENERATED PLANTS

Technologically redesigning plant cells using Ti plasmid from *A. tumefaciens* could overproduce regenerated plants. In principle, a foreign gene is inserted into the Ti plasmid, which is used to infect plants, producing a crown gall. Plant cells in gall

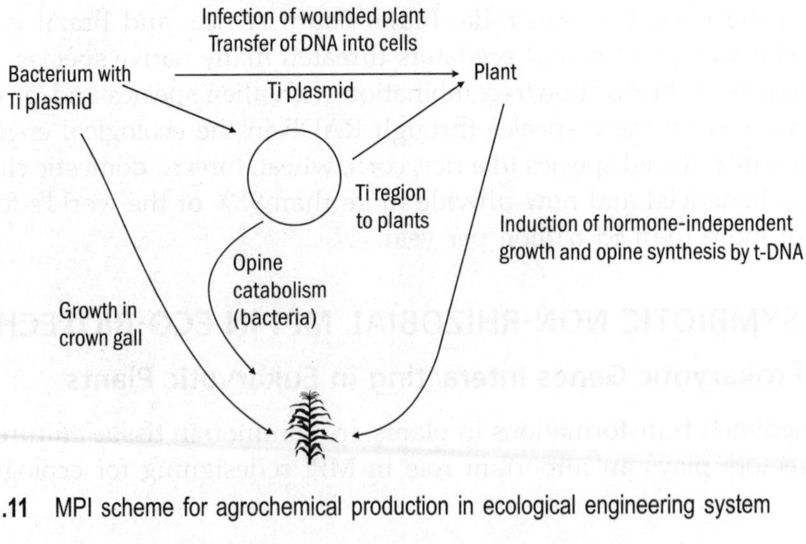

Figure 1.11 MPI scheme for agrochemical production in ecological engineering system

contain Ti plasmid and a piece of foreign DNA. Gall cells are then grown in a culture bioreactor to produce plantlets, which behave like single cells. Thus, the mass cultivation of plant cells can be carried out in a bioreactor and then transferred to soil where they grow into mature regenerated plants (Figure 1.12). The beneficial scope of this interaction relates to biological scale-up and bioreactor design for high cell density systems, considering rheological properties. Based on the concept of overproducer regenerated plants, it has been possible to cultivate regenerated plants in conventional or in modified MPI photobioreactors in high cell densities (Figure 1.13). Development of integrate/suspension and immobilized cell culture bioreactors has been reported.

1.7.1 Rhizobial–non-rhizobial Factors in MPI

In the cultivation of rice plant, paddy fields are usually waterlogged. If rhizobial biomass is used as a biofertilizer (Figure 1.14), it further improves the MPI activity by supplementing loss of applied chemical nitrogen fertilizer in paddy soil with biocompatible nitrogenous fertilizers.

Experiments show that the use of *Rhizobium* in rice growth under laboratory conditions in nitrogen-free media enhances the branching of roots and fresh weight of leaves. However, unlike legumes, it does not cause any curling of root hair or produce nodules where nitrogen fixation takes place. No change in protein pattern was observed in roots, while an overexpression of ~50 kilodaltons (kDa) protein was observed in leaves on the 10th day. When the seedlings were treated with a cell wall degrading enzyme mixture, the overexpression of the same protein was observed on the fifth day. The possible role of protein in the process of infection is yet to be known, but the possibility of *Rhizobium* use is suggested as a nitrogen supplement.

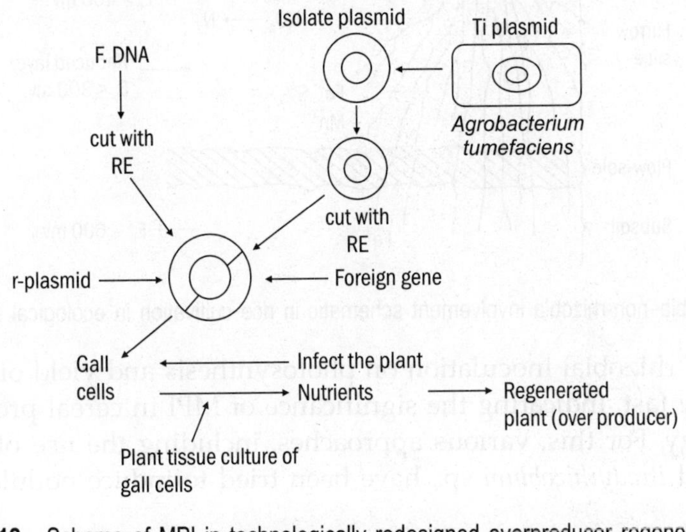

Figure 1.12 Scheme of MPI in technologically redesigned overproducer regenerated plant

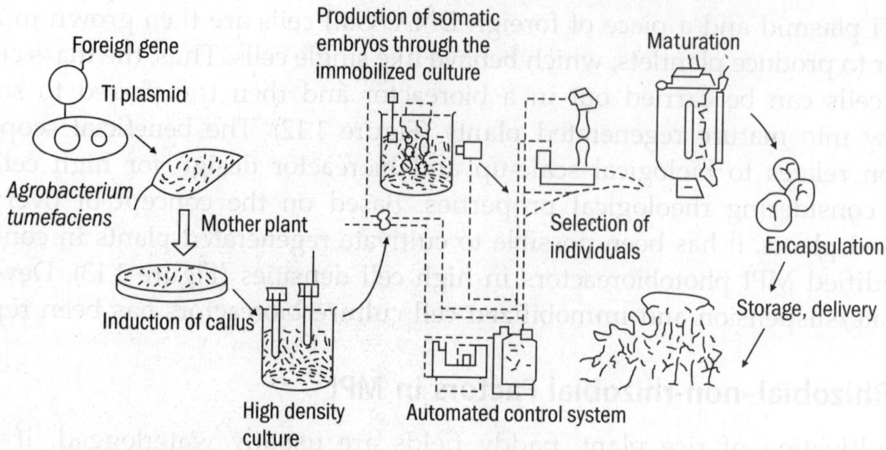

Figure 1.13 Conceptual scheme of artificial seed system using somatic embryos indicating microbes—plant system interaction in a bioreactor

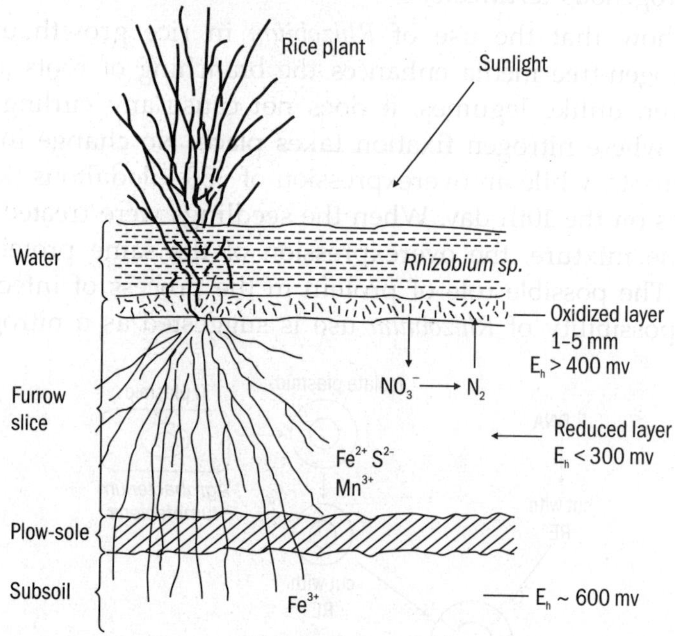

Figure 1.14 Rhizobia–non-rhizobia involvement schematic in rice cultivation in ecological engineering domains

The influence of rhizobial inoculation on photosynthesis and yield of low-land rice is progressing very fast, indicating the significance of MPI in cereal production process eco-biotechnology. For this, various approaches, including the use of various strains of *Rhizobium* and *Bradyrhizobium* sp., have been tried to induce nodule-like structures or pseudonodules in rice and other cereals. Such MPI is either spontaneous or uses various cell wall degrading enzymes that could remove the host specifically.

1.7.2 Human LBRE Evolution

The evolution of living body reactor engineering (LBRE) occurred through millennium after millennium. The following points are relevant to the evolution of human LBRE.

- Primitive human was ugly with primitive brain.
- Brain evolution started.
- Used functions of neurons of brain cells in innovative thinking.
- Living cell's natural neural network (NNNW) started functioning for development of humanism.

Thus, the beginning of natural process biotechnology (NPB) is imaginable from the day of existence of life.

1.8 HUMAN BODY DESIGN EVOLUTION

1.8.1 Belief and Biblical Records

The earth/world was created in the universe. This creation is believed, as described in Genesis 1:1 to 2:3, to have taken place in six days. On the first day, light was separated from darkness. On the second day, the sky and oceans were separated. The Almighty God separated the land from the oceans on the third day. The fourth day saw the creation of the sun, moon, and stars. On the fifth day, sea animals and birds were created. The creation of land animals and God's perfect image, man, was on the sixth day as depicted in Michelangelo's Creation of Adam. On Sabbath (seventh day), the Almighty rested. The most controversial debate over this creation belief/story relates to its time span, as scientists have found history of evolution of 14 billion years, which is defined as the natural science of descent.

1.8.2 Scientific Darwinian Theory

The Darwin's theory of evolution propounds the theory of common descent, the thinking that every living creature can trace an unbroken lineage back to the same primitive life forms. It was reasoned that all creatures were continually developing minor changes that made them either more or less fit to survive in their ecological niches. Those with good changes survived, while those with less propitious ones were weeded out by natural processes. Over long periods of time, these small changes accumulated until an entirely new species was established. Based on these proponents, Charles Darwin said that *Homo sapiens* evolved over millions of years through complexities in the universe/

nature. Presently, many scientists conjecture that humans share an ancestry with ugly apes (Figures 1.15–1.17).

1.8.3 Intelligent Descent Theory

The Intelligent Descent theory is one of the modern theories. According to this theory, the natural world is best explained by an intelligent cause, not an indirect process. This theory argues that in human evolutions, mutations happening over billions of years are impossible, citing the example of eye. Evolutionary bioscientists counter that even a light-sensitive spot in a primitive eye has a survival value. They have tried to show how a series of small modifications/improvements could eventually build the complete organ. Evolutionary biologists/bioscientists call it irreducible complicity. It means that all parts of an organ are necessary for them to function at all. However, the Intelligent Descent Theory proponents say that evolution is a theory, not a fact. Believers of this theory consider that there is a divine hand at work when it comes to evolution of all species.

1.8.4 Biocosm Theory

The Biocosm theory considers that the universe is life friendly and that life on earth was started by a very intelligent being. According to the believers of this theory, life and intelligence will evolve to a point where it is capable of finding out how to create, or give birth to, another universe, just as an earlier universe gave birth to our universe. Being progeny of an earlier universe, the cosmic equivalent of DNA stitched

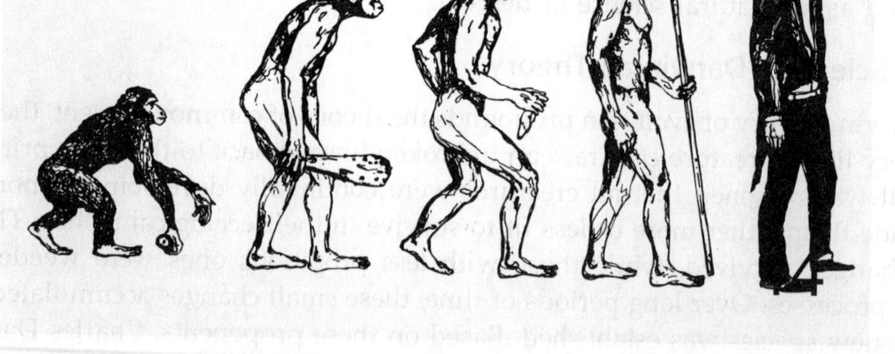

Figure 1.15 Evolutionary research engineering progress of crawler *Homo sapiens* sp. to standing human

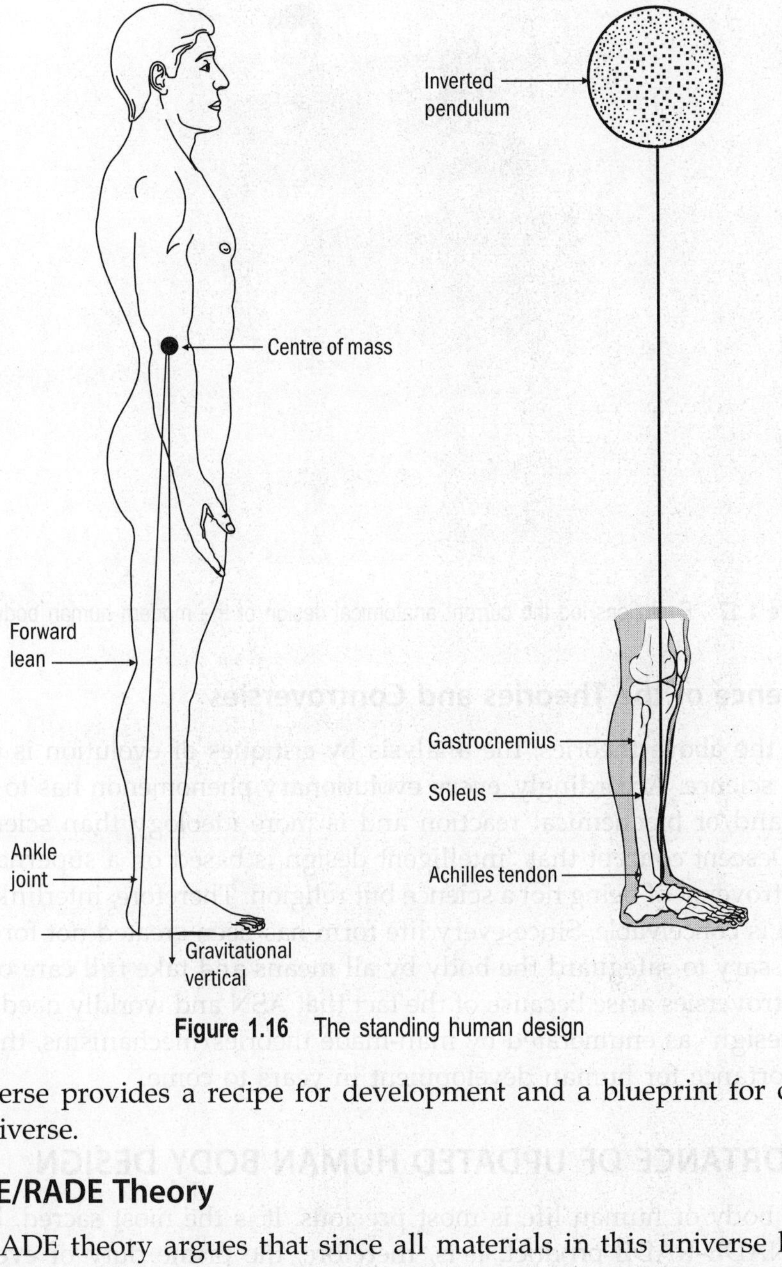

Figure 1.16 The standing human design

into our universe provides a recipe for development and a blueprint for constructing a younger universe.

1.8.5 NADE/RADE Theory

The NADE/RADE theory argues that since all materials in this universe have unique structural designs, they appeared naturally. This theory says that universal nature is the architect of natural design engineering. This is the concept of NADE. The NADE materials interacted with suitable recombination process, producing RADE materials. The NADE–RADE unit operations and processes went on through evolutionary stages to a more stable form using phylogenic and ontogenic characters and principles.

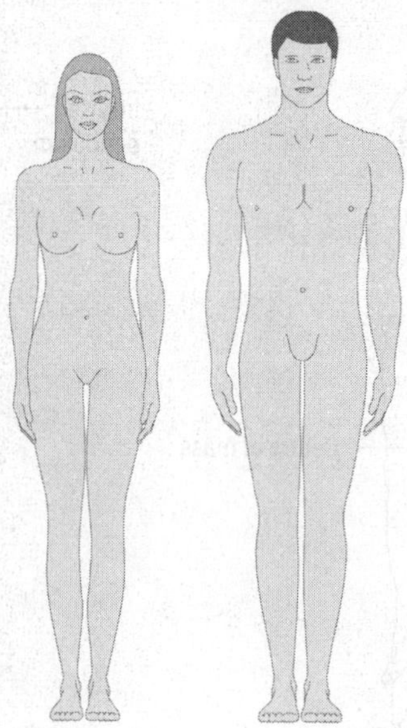

Figure 1.17 Evolutions led the current anatomical design of the modern human body reactor

1.8.6 Essence of the Theories and Controversies

From all of the above theories, the analysis by critiques of evolution is present in the language of science. Accordingly, every evolutionary phenomenon has to be reduced to a chemical and/or biochemical reaction and is more ideology than science. Thus, the Intelligent Descent concept that "intelligent design is based on a supernatural creator" ran into controversy of being not a science but religion. Therefore, interlinkage of science and religion is conceivable. Since every life form has been created not for indulgence, it is thus necessary to safeguard the body by all means and take full care of one's health. Though controversies arise because of the fact that ASN and worldly needs have created the "body design" as enumerated by man-made theories/mechanisms, this design is of utmost importance for human development in years to come.

1.9 IMPORTANCE OF UPDATED HUMAN BODY DESIGN

The human body or human life is most precious. It is the most sacred, beautiful, and invaluable NADE–RADE product. It is, therefore, the prime duty of every individual to take the greatest possible care of one's body as it is an indispensable medium for experiences. Nothing can be done, not even a simple task, without a healthy human body. That is why saints/sages and holy people believe that the human body is more precious than worldly wealth and other material acquisitions. Therefore, life derived through the assemblage of chemicals of life imbibed with soul producing a living body must be sanctified by making proper use of the body and engaging it in innovative actions. Once life ebbs out and the living body is gone, it cannot be recovered.

The LBRE is made in such a way that it becomes a clock marking time (circadian clock) and passage of life. It ticks away seconds, minutes, hours, days, weeks, months, and years. The living process of a living body that passes through the normal passage is termed apoptosis. No one can predict when the body clock will stop. When it stops by accident, the living path is termed necrosis. It is, therefore, most essential to make proper use of life and body while it is still available. The body is the vehicle for the journey/passage of life and requires proper maintenance. If it is not properly maintained, one must be prepared to face serious problems like diseases on the way. A biological human is a multicellular living being. About 300 trillion cells are present in the human body, and each is bound to another by tissues. The nucleus of a human cell is composed of 46 chromosomes functioning as about 100 000 units known as genes. These 100 000 genes compose the human genome. The human genome is responsible for the human body design and growth of a living body. A typical NADE- and RADE-based living body reactor (LBR) design product can be realized as given in Plates 1.1–1.12 (Bauer 1965).

Thus, current cornerstones of modern structural and evolutionary biology entail that the field as a whole has been transformed. If one could go back in a time machine and fetch the original *Homo sapiens*, one would find much evolutionary biology unintelligible—at least until one had time to study genetics, statistics, computer science, and engineering. This led evolutionary success to be measured, at present, in terms of the number of genes an individual contributes to the next generation. Human beings who die without reproducing may not directly contribute any role to the progress of phylogenicity. Phylogenicity and ontogenicity led to the evolution of the human brain, providing advanced intelligence and thinking power. These properties helped human beings to assess/estimate world markets for biotechnology products as given in Table 1.4.

1.10 CONCLUSION

From biochemical engineering and process biotechnology viewpoints, NADE and RADE played a major role in the evolution of life forms through fundamental and advanced process biotechnology of nature, including eco-biotechnology.

Increasing cost of inorganic fertilizers has encouraged many governments to support R&D on beneficial microorganisms for rice and other crops through MPI in the eco-biotechnology domain. Harnessing bacteria will not only fix nitrogen for cereals like rice but also promote its growth under minimal combined nitrogen input, fostering green revolution in agricultural countries like India.

Table 1.4 Estimated world markets for biotechnology products

Biotechnology product	1980s (million dollars)*	1990s (million dollars)*	2020s (million dollars)*
Medical products/healthcare			
Pharmaceuticals		3 500	20 000–30 000
Diagnostic	100	1 500	5 000
Veterinary products	100	1 500	5 000
Others (materials, sensors, and so on)	—	75	500
Chemicals			
Fine/speciality	—	500	2 000–4 000
Bulk (including methanol low molecular weight hydrocarbon)	—	—	1 000
Agricultural products			
Chemical/biological/biopolymer	20	500	2 000–4 000
Plants/seeds	25	1 500	1 000
Animal breeding (ET and so on)	20	500	
Food/feed/biofuels			
Additives/supplements	200	1 500	
Flavour/fragrances, bihydrogen, biomethane, and biodiesel	10	100	
Associated equipment and engineering system	1 500	4 000	10 000
Total	1 975	15 175	56000–69000

*At manufacturer's level, estimate 6%–8% per year inflation

Source SRI International, Menlo Park, California

REFERENCES

Bailey J E and Ollis D F. 1986. ***Biochemical Engineering Fundamentals,*** 2nd edn. New York: McGraw Hill Book Co. 918 pp.

Bauer W W (ed). 1965. ***Today's Health Guide: a manual of health information and guidance for the American family,*** pp. 80–81. American Medical Association

Mukhopadhyay S N. 1996. **Biochemical engineering: viewpoints on the evolution of life in prebiotic to biotic conditions**. Paper presented at the national symposium on *Evolution of Life,* 5–6 April 1996, Department of Biophysics, University of Delhi

Mukhopadhyay S N and Baisya R K. 1997. **Role of water composition in relation to food biotechnology**. In *Biochemical Engineering: marching toward the century of biotechnology,* edited by Z Y Shen, F Ouyang, J T Yu, and Z A Cao, pp. 1354–1356. Beijing: Tsinghua University Press

Mukhopadhyay S N and Das D K. 1994. ***Oxygen Responses, Reactivities, and Measurements in Biosystems.*** Boca Raton, Florida: CRC press. 224 pp.

Mukhopadhyay S N and Mukhopadhyay P. 2000. **Water: properties, pollution, and management**. In *Waste Recycling and Resource Management in Developing World: ecological engineering approach,* edited by B B Jana, R D Banerjee, B Guterstam, and J Heebs, pp. 551–557. Switzerland: India and International Ecological Engineering Society, University of Kalyani

BIBLIOGRAPHY

Alazard D and Duhoux E. 1990. **Development of Stem Nodules in a Tropical Forage Legume, *Aeschynomene afraspera*.** *Indian Journal of Experimental Botany* **41**: 1199–1206. [doi:10.1093/jxb/41.9.1199]

Al-Mallah M K, Davey M R, and Cocking E C. 1987. **Enzymatic treatment of clover root hairs removes a barrier to *Rhizobium*-host specificity**. *Biotechnology* **5**: 1319–1322

Al-Mallah M K, Davey M R, and Cocking E C. 1989. **Formation of nodular structures on rice seedlings by rhizobia**. *Indian Journal of Experimental Botany* **40**(4): 473–478

Armarger N. 1998. ***Ecology and Taxonomy: introductory remarks in biological nitrogen fixation for the 21st century***, pp. 587–558. The Netherlands: Eluver Aead Pub.

Bergersen F J and Goodchild D J. 1973. **Location and concentration of leghaemoglobin in soybean root nodules**. *Australian Journal of Biological Sciences* **26**: 741–756

Bhattacharjee R B, Singh A, and Mukhopadhyay S N. 2008. **Use of nitrogen-fixing bacteria as biofertilizer for non-legumes: prospects and challenges**. *Applied Microbiology and Biotechnology* **80**: 199–209

Bhojwani S S and Razdan M K. 1996. ***Plant Tissue Culture: theory and practice***. Amsterdam: Elsevier

Biswas J C, Chen Y, Peng S, Ladha J K. 2003. **Influence of Rhizobial inoculation on photosynthesis and yield of lowland rice**. In *Report of Third Working Group Meeting and Review of the Project on Assessing Opportunities of Nitrogen Fixation in Rice*

Biswas R and Mukhopadhyay S N. 2008. **Eco-biotechnology advances in bioproduct formation through microbe-plant interactions (MPI)**. *Journal of Scientific and Industrial Research* **67**: 821–826

Biswas R, Singh A, and Mukhopadhyay S N. 2003. **Potential use of biological nitrogen fixation in developing an eco-friendly fertilizer**, pp. 121–122. [Proceedings of the *CHEMCON 2003*, Bhubaneswar, Orissa, 19–23 December 2003]

Chatterjee R and Yuan L. **Directed evolutions of metabolic pathways**. *Trends in Biotechnology* **24**(1): 28–38

Critchlow S E and Jackson S P. 1998. **DNA end-joining: from yeast to man.** *Trends in Biochemical Sciences* **23**(274): 394–389

Culotta E and Pennisi E. 2005. **Evolution in action**. *Science* **310**: 1878–1879

Ghose T K. 1989. **Commercialization of biotechnology.** Delhi: Biochemical Engineering Research Centre, IIT Delhi. [Indian Survey Report, Document No. R069 1989]

Heidmann J (ed.). 1991. ***Lecture Notes in Physics 390***. Heidelberg, Berlin: Springer-Verlag

Honda I I, *et al*. 1997. **Development of bioreactors for production of plant embryos from immobilized carrot callas**. In *Biochemical Engineering: marching towards the century of biotechnology*, vol. 2, edited by Z Yshen *et al.*, pp. 683–686. Beijing: Tsinghua University Press

Jeans J H. 1930. **The dying Sun**. In *The Mysterious Universe.* Cambridge, UK: Cambridge University Press

Lee S, Reith A, Meletzus D, Sevitta M, Kennedy C. 2000. **Characterization of major cluster of nif, fix, and associated genes in a sugar cane endophyte *Acetobacter diazotrophicus*.** *Journal of Bacteriology* **182**(2000): 7088–7091.

Lodish H, Berk A, Kaiser C A, Krieger M, Scott M P, Brescher A, Ploegh H, Matsudaira P. 2008. ***Molecular Cell Biology***. New York: W H Freeman and Co.

Mitsch W J and Jørgensen S E. 1989. **Ecological engineering: an introduction to ecotechnology**. New York: John Wiley and Sons Inc.

Mukhopadhyay S N. 2001. **Achivements of NADE and RADE: molecules to human, Avidha (in Bengali)**. IIT Delhi: Basanta Utsav Samiti Pub. pp. 4–5. [September issue]

Mukhopadhyay S N (ed.). 2005. ***Advanced Process Biotechnology***. New Delhi: Viva Books. 382 pp.

Mukhopadhyay S N. 2006. **Eco-friendly products through process biotechnology in the provision of biotechnology economy: recent advances.** *Technorama* **55**(T): 10–20

Mukhopadhyay S N. 2006. **Process technology products development.** [Invited NRDC talk delivered, 5 October 2006, Ashok Hotel, New Delhi]

Mukhopadhyay S N. 2007. ***Experimental Process Biotechnology Protocols***. New Delhi: Viva Books. 214 pp.

Pimentel D, McNair S, Janecka J, Wightman J, Simmonds C, O'Connell C, Wong E, Russel L, Zern J, Aquino T, Tsomondo T. 2001. **Economic and environmental threats of alien plant, animal, and microbe invasions.** *Agriculture, Ecosystems, and Environment* **84**(1): 1–20

Piszczek E A. 1965. ***Today's Health Guide: a manual of health information and guidance for the American family***, edited by W W Bauer. Chicago: American Medical Association. 624 pp.

Schroder J, *et al*. 1983. **The use of Ti plasmids for the genetic engineering of plants**. In *Basic Biology of New Development in Biotechnology*, edited by A Hollander *et al*, pp. 173–193. New York: Plenum Press.

Sharmashanker F C, Sharmashanker S, Miler S D, Vance G F, Zhang R. 2001. **Water and biofertilizer use efficiencies for rice plants**. *Agricultural Water Management* **46**: 241

Short J M. 1995. **Enzymes development via combinatorial natural and non-natural evolution**. In *Applied Molecular Evolution: international business communications*, edited by M A Southborough, D E Robertson, *et al*., pp. 3–7

Subba Rao N S. 1979. ***Recent Advances in Biological Nitrogen Fixation.*** New Delhi: Oxford and IBH Pub. Co. Pvt. Ltd

USBC (US Bureau of the Censor). 1998. **Statistical Abstract of the United States**. Washington, DC: US Bureau of the Censor, US Govt Printing Office

Volkenstein M V. 1983. ***Biophysics***. Moscow: Mir Publishers

2

Biotic Chemistry

2.1 INTRODUCTION

All chemicals of life are formed from innumerable elements. All elements on earth are composed of atoms that comprise electrons, protons, and neutrons. These atoms mostly exist as molecules. Atoms in a molecule are bound by chemical bonds. Therefore, chemical bonds play a significant role in the formation of chemicals of life. In dealing with biotic chemistry, a fundamental knowledge of chemical bonds is a prerequisite. Chemical bonds like hydrogen bonds, covalent bonds, and ionic bonds play important roles in biotic chemistry. This section briefly reviews chemical bonds.

Atomic and molecular structures and the nature of the participating chemical bonding are of great significance in manifesting the properties of all molecules in chemical as well as biological systems. Electrons, protons, and neutrons are key components in all atoms in a molecule bound by molecular bonds.

An electron in its ground state occupies the orbital with the lowest energy. There may be some electrons in the same atom, which are located in orbitals with even lower energy. The further the electron from the atom's nucleus, the greater the energy associated with that electron. An electron in the ground state may absorb energy from an external source (for example, energy in the form of heat) and get promoted to a higher energy orbital. The electron is now considered to be in an excited state. In a similar fashion, an excited electron may release some amount of energy and occupy a lower energy orbital (not necessarily the ground state). This energy is released in the form of a photon of light.

The electrons of an atom are distributed so that they occupy concentric atomic orbitals, with each orbital containing two electrons. With increasing distance from the nucleus, there is a corresponding increase in the energy levels of the orbital. Each atomic orbital has a characteristic energy level; hence, they are said to be quantized.

The quantum theory describes the characteristics of an electron by assigning four quantum numbers to it: principal quantum number, angular momentum quantum number, magnetic quantum number, and spin quantum number.

The principal quantum number n denotes the main energy level, or energy shell, in which the electron is present. The main energy levels of an atom are assigned integral values. The shell with integer value 1 is closest to the nucleus. The value of a shell increases as the energy level becomes more distant from the nucleus. Hence, $n = 1, 2, 3, \ldots$ Upper case letters, K ($n = 1$), L ($n = 2$), M ($n = 3$), N ($n = 4$), and so on, are sometimes used to denote the electronic shells.

The angular quantum number, or the Azimuthal quantum number, l determines the shape of the volume in which the electron is most likely to be found. The shape of this volume can be illustrated with electron cloud diagrams and depends upon the magnetic and electrostatic interactions of the electron with the nucleus and other electrons in neighbouring orbitals. The interaction between the electron and the nucleus will vary with the distance of the orbital from the nucleus. Consequently, the angular momentum of the electron depends upon the main energy level in which it resides. It has been determined that the quantum numbers l and n are related as shown in Equation 2.1.

$$l = n - 1 \tag{2.1}$$

Therefore, l takes the values of 0 to $n - 1$; for each value of n, there are n values of l. The various sub-shells are labelled *s*, *p*, *d*, and *f*. When $l = 0$, the sub-shell is called *s* orbital; when $l = 1$, it is called *p* orbital; when $l = 2$, it is called *d* orbital; and when $l = 3$, it is called *f* orbital.

A moving charge generates a magnetic field. Since an electron is an elementary particle with a negative charge and moves within some volume described by l, it propagates a magnetic field. This is accounted for in the quantum theory by the magnetic quantum number m, which depends upon l (Equation 2.2).

$$m = -l, 0, 1 \tag{2.2}$$

Thus, when $l = 0$, $m = 0$; when $l = 1$, $m = -1, 0, 1$; and when $l = 2$, $m = -2, -1, 0, 1, 2$. The number of values of m will determine the number of orbitals contained in each of the main energy levels. For example, when $n = 2$, $l = 1$ and $m = -1, 0, 1$. Thus, there are three orbitals in the *p* sub-shell. It is similarly found that there are five orbitals in the *d* sub-shell and seven orbitals in the *f* sub-shell.

The fourth quantum number describes the rotational motion of an electron. The spin of an electron on its own axis is either clockwise or counterclockwise. The spin quantum number *s* accounts for this rotational motion. The values +½ and −½ are used to describe the spin of the electron. The spin of the electron is indicated by half arrows. The clockwise spin is represented by an arrow pointing upwards, while the counterclockwise spin is represented by an arrow pointing downwards. Each orbital can accommodate a maximum of two electrons of opposite spins.

The Pauli Exclusion Principle states that no two paired electrons in an atom may have identical sets of quantum numbers. Consequently, paired electrons in the same atomic orbital cannot have the same spin quantum number, even if they have the same principal, angular momentum, and magnetic quantum numbers.

With this basic knowledge of quantum theory, one can assign an electron configuration to an atom. Electron configurations show how electrons within an atom are distributed. From the electron configuration, one can deduce the valence of the atom and hence determine its bonding capacity.

The Aufbau process of filling atomic orbitals is used in determining the electron configuration of an atom. By this process, the electrons fill the orbitals in order of increasing energy. When there is more than one orbital in an energy sub-level (for example, there are three orbitals in the *p* sub-shell), the electrons fill the orbitals in such a way that each has one electron before an orbital receives a pair of electrons. The relative energies of the orbitals are given as $s < p < d < f$.

One can now assign the electron configuration for oxygen and deduce its bonding capacity from its valence. There are eight electrons in an oxygen atom; that is, its atomic number is eight. In the first main energy level, $n = 1$, $l = 0$, and $m = 0$. There is just one orbital in this sub-shell. Since an orbital can accommodate two electrons, the first main energy level will hold two electrons, leaving six electrons unaccounted. In the second main energy level, $n = 2$, $l = 0, 1$, and $m = -1, 0, 1$. There is one orbital in the *s* orbital ($l = 0$) and three orbitals in the *p* orbitals ($l = 1$) in this level. Since *s* orbitals are lower in energy than *p* orbitals, it will be filled first, leaving four electrons unaccounted. These electrons fill the three orbitals in the *p* orbitals one at a time such that each orbital has one electron. The last electron then completes one of the orbitals in the *p* sub-shell. Hence, the three orbitals in the *p* orbitals are filled such that one has a pair of electrons and the other two have one electron each. The electron configuration of oxygen is represented in Figure 2.1.

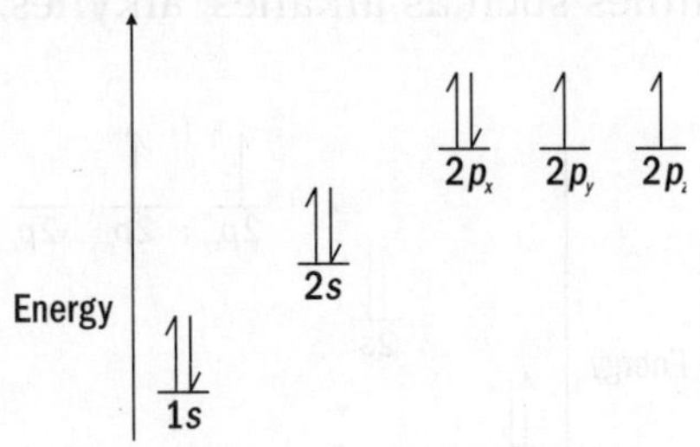

Note The numbers 1, 2, and 3 represent the principal quantum number *n*. The three *p* orbitals are labelled p_x, p_y, and p_z for differentiation.

Figure 2.1 Electron configuration of oxygen

The valence of an atom is the number of electrons that take part in the bonding of one atom with another. The octet rule states that all atoms lose, gain, or share electrons so as to have eight electrons in their valence shell (outermost shell). By looking at the electron configuration of oxygen, one can see that it requires two electrons to fill the valence shell. Hence, the bonding capacity (or the number of unpaired electrons available for bonding) of oxygen is two.

One can look at the electron configuration of carbon, which is one of the most biologically significant elements. There are six electrons in a carbon atom (atomic number = 6). As in oxygen, two electrons will occupy the lone *s* orbital in the first main energy level. If the four remaining electrons filled the orbitals in the second main energy level just like an oxygen atom, the electron configuration for carbon will be as shown in Figure 2.2.

In this case, the valence of carbon is four, and the bonding capacity is two. This is because there are two unpaired electrons in the 2*p* orbitals. It should be noted, however, that this representation of carbon's electron configuration does not comply with the octet rule. In this case, a fully bonded carbon atom will contain only six electrons in its valence shell and has one empty orbital ($2p_z$). This is an energetically unstable condition. If it is at all possible for carbon to complete its valence shell, it will do so. It so happens that carbon does comply with the octet rule and has all its orbitals in the first two main energy levels filled to capacity when it bonds with other atoms. This is facilitated by the hybridization of atomic orbitals. The electrons in the second main energy level occupy an orbital that is intermediate in energy between the 2*s* and 2*p* orbitals. The hybrid orbitals, in this case, are specifically called sp^3 hybrid orbitals because they are formed by the hybridization of one *s* and three *p* orbitals. The four sp^3 orbitals formed each contain an unpaired electron. The electron configuration of carbon after hybridization is shown in Figure 2.3.

Now it can be seen that carbon has a valence of four and a bonding capacity of four. In this case the octet rule is satisfied.

The diversity of organic compounds is so vast that these organic compounds have been divided into families such as alkanes, alkynes, and aromatic compounds,

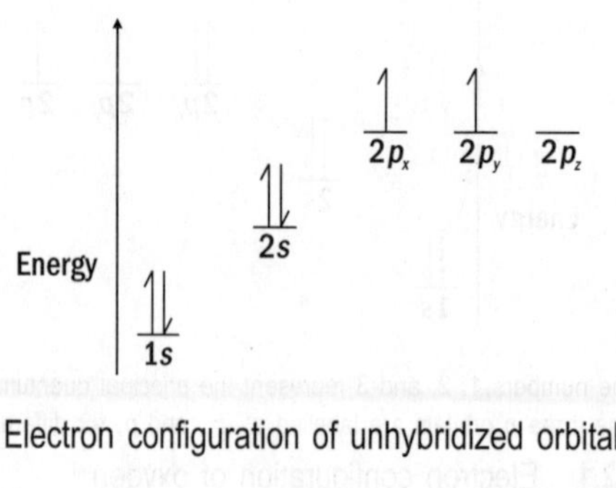

Figure 2.2 Electron configuration of unhybridized orbitals of carbon

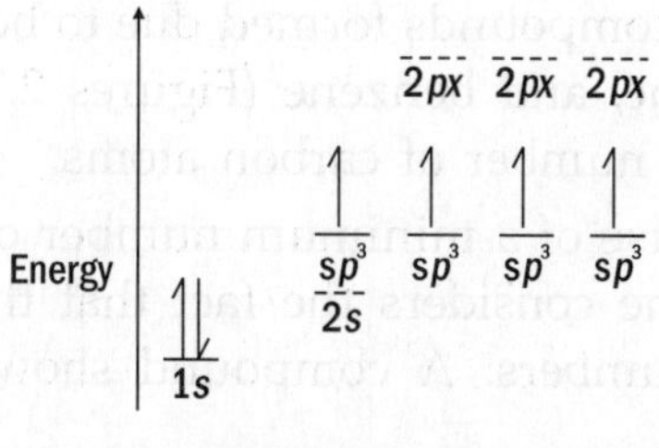

Figure 2.3 Electron configuration of hybridized orbitals of carbon

which have no counterparts among inorganic compounds. The tremendous variety of these compounds is made possible by the unique properties of carbon. To be able to understand how carbon can form such a huge number of compounds with a great variety of properties, the way in which the atoms in these molecules are bonded together must be looked at.

Carbon has a valence number (the number of bonds, an atom of an element can form) of four. This means that each carbon atom always has four bonds, which can be bonded to four other atoms, as in methane; to three other atoms, as in formaldehyde; or to two other atoms, as in hydrogen cyanide. In other words, carbon is capable of forming single, double, and triple bonds (Figures 2.4–2.6).

Apart from bonding with atoms of other elements, carbon atoms are capable of bonding with each other to form unique chain and ring structures. This bonding of atoms of the same element to form long-chain structures is called catenation. This process accounts for the tremendous variety of organic compounds.

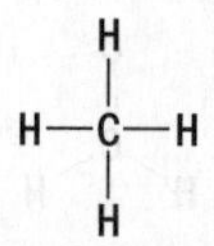

Figure 2.4 Methane

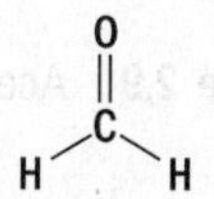

Figure 2.5 Formaldehyde

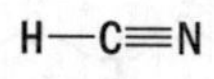

Figure 2.6 Hydrogen cyanide

Some examples of organic compounds formed due to bonding between carbon atoms are ethane, ethylene, acetylene, and benzene (Figures 2.7–2.10). These examples show linkage between a minimum number of carbon atoms.

These examples show the use of a minimum number of carbon atoms. The diversity possible is apparent when one considers the fact that these different types of bonds can be formed in infinite numbers. A compound shown in Figure 2.11 exemplifies this fact.

A multitude of compounds with different structures can be drawn similar to the one shown in Figure 2.11; the only constraining factor is that carbon must have four bonds. The freedom of bonding for carbon is apparent when contrasted with halogens (iodine, chlorine, bromine, and fluorine) as these are monovalent and can form only one bond (Figures 2.12 and 2.13).

Another unique feature of carbon compounds is the existence of isomers (compounds having the same molecular formulae but different spatial arrangements), for example, two compounds of molecular formulae C_4H_{10} and C_5H_{10}. The former can exist in two forms, *n*-butane and 2-methyl propane (Figures 2.14 and 2.15), while the latter can exist as trans 2-pentene and cis 2-pentene (Figures 2.16 and 2.17).

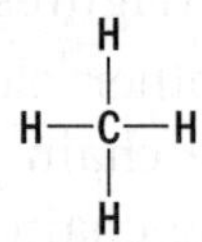

Figure 2.7 Ethane

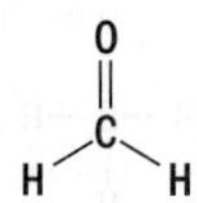

Figure 2.8 Ethylene

H—C≡N

Figure 2.9 Acetylene

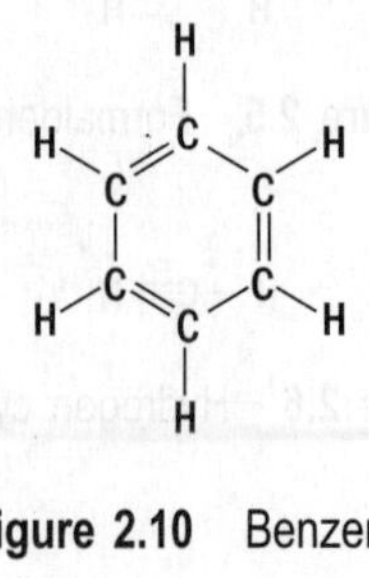

Figure 2.10 Benzene

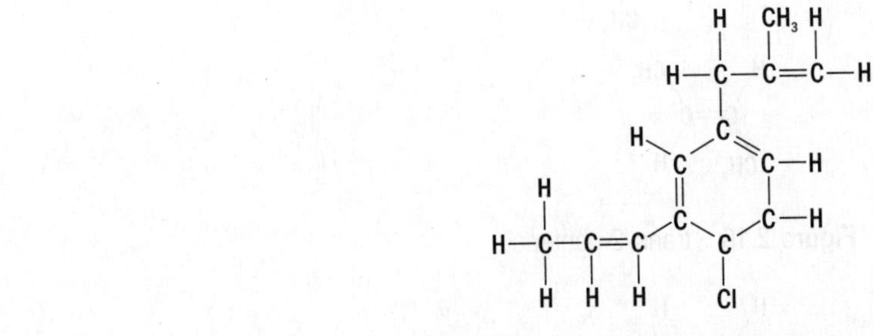

Figure 2.11 A long-chain organic compound

H—Cl

Figure 2.12 Hydrochloric acid

Br—Br

Figure 2.13 Bromine

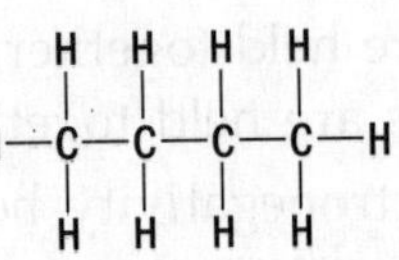

Figure 2.14 *n*-butane

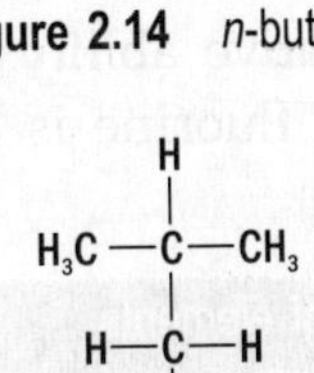

Figure 2.15 2-methyl propane

Although these compounds have the same chemical formulae, they have different chemical, physical, and biological properties due to the difference in their arrangement in space.

In a covalent bond, two atoms are held together by a shared pair of electrons. In an ionic bond, oppositely charged ions are held together by electrical attraction.

In general, the difference in electronegativity between two elements influences the character of their bond (Table 2.1).

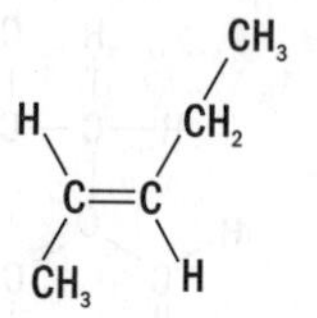

Figure 2.16 trans 2-pentene

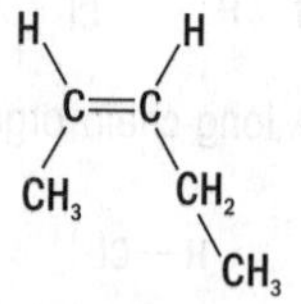

Figure 2.17 cis 2-pentene

Although these compounds have the same chemical formulae, they have different chemical, physical, and biological properties due to the difference in their arrangement in space.

In a covalent bond, two atoms are held together by a shared pair of electrons. In an ionic bond, oppositely charged ions are held together by electrical attraction.

In general, the difference in electronegativity between two elements influences the character of their bond (Table 2.1).

Electronegativity measures the relative ability of an atom to attract electrons in a covalent bond. Using Pauling's scale, fluorine is arbitrarily given the value 4.0 units.

Table 2.1 Electronegativities of main groups of elements

I A	II A	III A	IV A	V A	VI A	VII A
H						
2.1						
Li	Be	B	C	N	O	F
1.0	1.5	2.0	2.5	3.0	3.5	4.0
Na	Mg	Al	Si	P	S	Cl
0.9	1.2	1.5	1.8	2.1	2.5	3.0
K	Ca		Ge	As	Se	Br
0.8	1.0		1.8	2.0	2.4	2.8
Rb	Sr		Sn	Sb	Te	I
0.8	1.0		1.8	1.9	1.8	2.5
Cs	Ba		Pb	Bi		
0.7	0.9		1.7	1.8		

Other elements are assigned electronegativity values relative to that of fluorine. A bond exhibits 50% ionic and 50% covalent character if the difference in the electronegativity of the atoms forming it is 1.7 units. A bond between two atoms with an electronegativity difference greater than 1.7 units is mostly ionic in character. If the difference is less than 1.7, the bond is predominantly covalent.

The hydrogen bond is one of the most significant bonds in biological systems. Hydrogen bonds determine the structure of large macromolecules like proteins. The following section discusses the hydrogen bond.

2.2 HYDROGEN BOND

A hydrogen bond is a molecular force in which a hydrogen atom is shared between two atoms. Hydrogen bonds occur as a result of uneven distribution of electrons in a polar bond, such as an O–H bond. Here, the bonding electrons are attracted towards the more electronegative oxygen atom, resulting in a slight positive charge (δ^+) on the hydrogen and a slight negative charge (δ^-) on the oxygen. The slightly positive hydrogen atom gets attracted to a relatively negative atom of some other polarized bond, thus forming a weak hydrogen bond (Figure 2.18).

The electronegative atom that participates in the formation of the polar covalent bond is called the hydrogen donor, while the electronegative atom that takes part in the formation of the hydrogen bond is called the hydrogen acceptor. In the given example, oxygen is the hydrogen donor and nitrogen is the hydrogen acceptor. A hydrogen bond can also be thought of as an intermediate type of acid–base reaction. It is to be noted, however, that the electronegative bond is an electrostatic one; no electrons are shared or exchanged between atoms in such a bond.

Hydrogen bonds are highly directional (note the arrows in Figure 2.18) and are strongest when all three atoms are collinear. The bond energy of a hydrogen bond is in the range of about 3 kCal/mol to 7 kCal/mol. This bond energy is intermediate between the energy of a covalent bond and a van der Waals bond. The energy of the bond is strongest when the hydrogen acceptor is fluorine, oxygen, or nitrogen. Hydrogen bonds formed with any of the three halogens are significant.

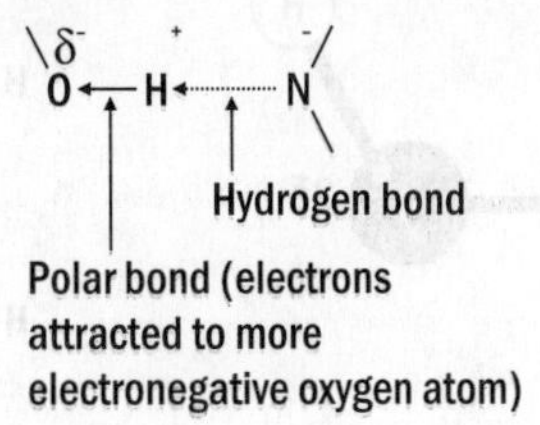

Figure 2.18 Formation of hydrogen bond

Hydrogen bonds are responsible for the structure of water and its special properties as a biological solvent. There is extensive hydrogen bonding between water molecules, forming the water matrix. This structure has significant effects on the freezing and boiling points of water and its solubility properties. Any molecule capable of forming a hydrogen bond usually does so when mixed with water. Thus, a variety of compounds dissociate and are soluble in water.

Hydrogen bonds are also responsible for maintaining the shape of proteins. Since shape is crucial to their function (as both enzymes and structural components), this bonding is extremely important. For example, hydrogen bonds maintain the helical shape of keratin and collagen molecules and give them their characteristic strength and flexibility.

More importantly, hydrogen bonds hold together the two helices of a deoxyribonucleic acid (DNA). The bonding occurs between the base pairs: adenine (A) with thymine (T) and guanine (G) with cytosine (C). The intermediate bond strength of the hydrogen bond is ideal for the function of DNA; it is strong enough to give the molecule stability, yet weak enough to be broken with sufficient ease for replication and ribonucleic acid (RNA) synthesis.

The formation of a hydrogen bond between two adjacent water molecules is shown in Figure 2.19. The figure indicates the lengths of the hydrogen and the covalent bonds.

2.3 WATER

2.3.1 Structure

In order to understand the function of water, one needs to be thoroughly acquainted with its structure first. Its state (solid, liquid, or gas) when present in a material, particularly a biological system, needs to be identified in order to understand its consequent behaviour. Structurally, the six valence electrons of oxygen in a water molecule are hybridized to four sp^3 orbitals. These orbitals are elongated to the corners of a tetrahedron. Of the four hybrid orbitals, two form O–H covalent bonds with a

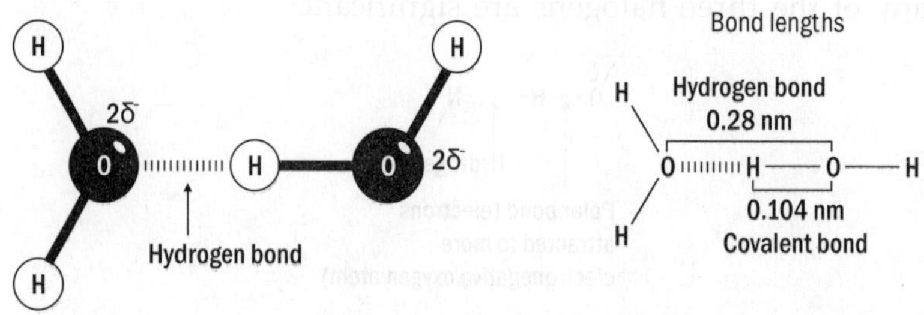

Figure 2.19 Feature of hydrogen bond formed between two water molecules

bond angle of 105° for H–O–H. The other two hybrid orbitals hold two non-bonding electron pairs. The O–H covalent bonds, due to the highly electronegative oxygen, have a partial (40%) ionic character. Figures 2.20(a)–2.20(d) illustrate the molecular structure of water.

Two atoms of different elements connected by a covalent bond exert different forces of attraction on the electrons. In such cases, the bond is dipolar, with one end slightly negatively charged (δ^-) and the other slightly positively charged (δ^+). A covalent bond in which both the donor and acceptor atoms are of the same element is called non-polar.

(a) Tetrahedral structure

(b) Aggregation property

(c) Structural geometry (C is the centre of gravity)

(d) Orbital model

Although a water molecule has an overall neutral charge, the electrons within it are asymmetrically distributed, which makes the molecule polar. The nucleus of the oxygen

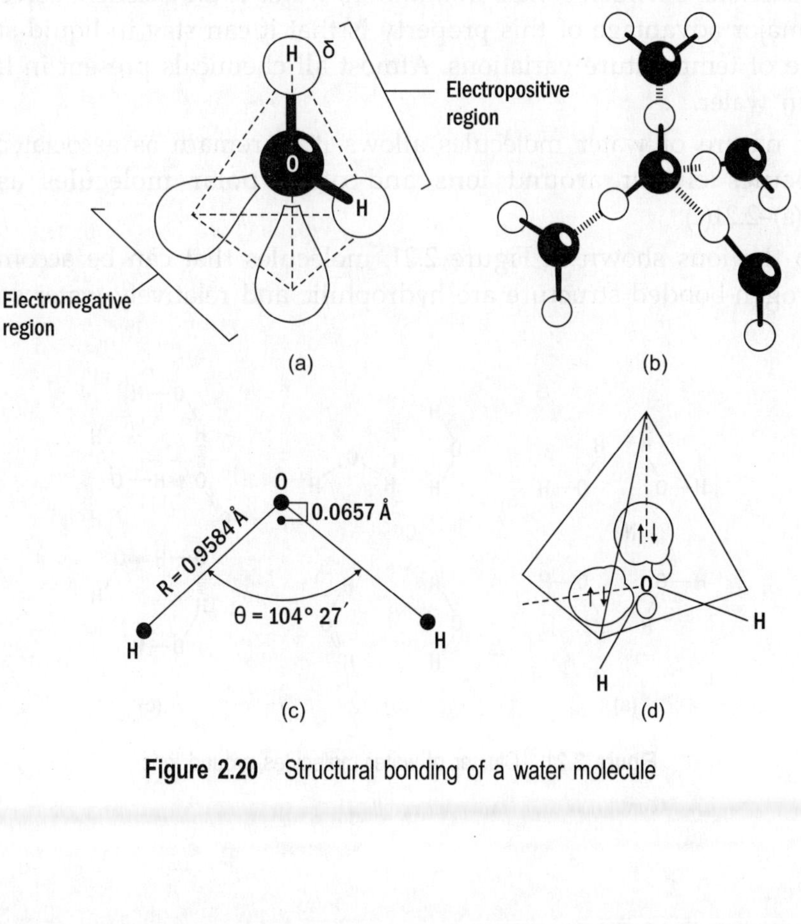

Figure 2.20 Structural bonding of a water molecule

draws electrons away from the nuclei of the hydrogen atoms, leaving these nuclei with a small net positive charge. Thus, an excess of electron density on the oxygen atom creates weak negative regions at the other two corners of an imaginary tetrahedron (Figure 2.20a).

Molecules of water join together transitionally in a mhydrogen-bonded lattice (Figure 2.20b). The property of aggregation of water $(H_2O)_n$ molecules has great atmospheric implications. Even at 37°C, 15% of water molecules are joined by four others in a short-lived assembly known as a flickering cluster.

The cohesive nature of water is responsible for many of its unusual properties such as high surface tension, specific heat, and heat of vaporization.

2.3.2 Properties of Water

The chemistry of life on this planet is essentially the chemistry of water and carbon. Water is the most abundant molecule in a living cell as well as on earth. In fact, it makes up 70%–90% of the weight of most life forms.

The properties of water shaped the chemistry of living organisms when life began in oceans. Life easily developed in the liquid phase as reactions in solutions are more rapid than reactions between solids. In addition, water is an excellent solvent for living systems. A major advantage of this property is that it can stay in liquid state through a wide range of temperature variations. Almost all chemicals present in living matter are soluble in water.

The polar nature of water molecules allows it to remain as associated molecules. Water molecules cluster around ions and other polar molecules as shown in Figures 2.21(a)–2.21(c).

Similar to the ions shown in Figure 2.21, molecules that can be accommodated in water's hydrogen-bonded structure are hydrophilic and relatively water soluble.

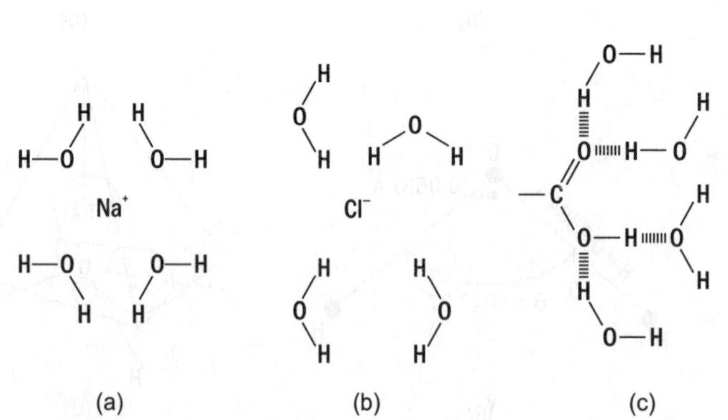

Figure 2.21 Cluster of water molecules around ions

Scientifically, water exists in monomeric, dimeric, trimeric, and tetrameric forms in nature. The presence of water in natural hypercycles is crucial. For both biological and non-biological activities, water is the most important substance. Recent discovery of liquid water beneath the surface of Mars has further confirmed that it is a universal liquid. This discovery may shed some light on the existence of extraterrestrial life.

The properties of water can be divided into chemo-physical and biogenic properties. These are discussed in subsequent sections.

2.3.2.1 *Chemo-physical properties*

It is important to consider that for biological systems, water acts as a strong hydrogen-bonding agent. Intramolecular bonds may be formed even in strong polar solvents such as water. For better understanding of the role of water in biology, knowledge of physical biochemistry, computational chemistry, and biochemistry in relation to aqueous solvent systems is very important. This is because water contains significant numbers of aggregates.

Water in liquid form has a unique structure, which is responsible for its chemo-physical properties in relation to temperature. The increase in density of water that occurs on warming from 0°C to 4°C may be attributed to a decrease in the concentration of aggregates and increase in the number of complexes with other denser structures. The structural features of liquid water reveal a crystal structure with a hydrogen-bonded framework of 46 water molecules per cubic unit, with eight water molecules per cubic unit in the centre of the polyhydra formed without hydrogen bonds. The number of hydrogen bonds in liquid form of water is 85% more than that in ice, as indicated by the enthalpy of fusion. The density of a water crystal is 1 g/cm^3, which is the same as that of liquid water. Furthermore, it is found on analysis that the pentagonal dodecahedra structure may be arranged relative to one another in a number of ways such that random structures for liquid water may be based upon aggregates of water molecules bonded to one another in different ways. The radial distribution function for water as calculated from X-ray diffraction pattern, the dispersion of dielectric constant, and some other chemo-physical properties of water is found to be compatible with this structure. An advantage the structure provides water is the availability of spaces between intermolecular spaces in associated/aggregated liquid water molecules, which is responsible for its solvent property. The hydrogen atoms, or protons, in water have a tendency to "wrap" themselves with electrons from whichever source in which electrons are available. In aqueous solution, the protons combine with water and form H_3O^+ (hydronium) ion complex (Equation 2.3).

$$H^+ + H_2O \rightleftharpoons H_3O^+ \tag{2.3}$$

If water is used as a solvent, the above ionization shifts to the right so that strengths of strong acids in aqueous solutions cannot be estimated. On the other hand, the strength of weak acids like acetic, propionic, and phosphoric acids can be estimated as they do not ionize as easily in aqueous solutions.

Water is a weak base, but its conjugate acid, H_3O^+, is a strong base. The influence of water as a solvent on the ionization and apparent strength of bases is also similar to that of acids. The strength of weak bases can be estimated in aqueous solutions because water is a weak proton acceptor. In biological systems and in many industrial processes, constancy of pH is an essential prerequisite. Different types of solvents are known to keep the pH of solutions at a constant level. Such solutions are known as buffer solutions. The amount of base added to change the pH of a buffer solution by unity is called its buffer capacity (Equation 2.4).

$$\beta = \frac{db}{d(pH)} \tag{2.4}$$

where β is the buffer capacity and b is the amount of base added.

Based upon electroneutrality, Equation 2.4 may be written as shown in Equation 2.5.

$$b + C_{H^+} = C_{OH^-} \tag{2.5}$$

From Equations 2.4 and 2.5, it is possible to arrive at an expression of buffer capacity of water (Equation 2.6).

$$\beta = 2.303\ [C_{H^+} + C_{OH^-}] \tag{2.6}$$

The calculated value of β for water (from Equation 2.6) shows an extremely small value when water is neutral (pH 7). This value rises to a certain extent when the solution is either acidic or alkaline (below 2.4 and above 11.6 as experimentally determined in biosystems and non-biosytems).

In general, water in cells and cellular functions exhibits properties similar to those of water in bulk aqueous solutions. However, alternative biogenic descriptions of properties of cell water with regard to various cellular activities are also available. Functions such as water-powered bio-pumping in cellular activities have been studied. Ions can make and break water molecules in vitro. Water pricing based upon chemo-physical property has industrial implications.

Table 2.2 lists the different chemo-physical properties of water with regard to bio-aqueous solvent systems.

2.3.2.2 Biogenic properties

The existence of life forms may not have been possible if water did not possess biogenic properties. Factors affecting the biogenic characteristics of water are listed in Table 2.3. Moreover, water has atmospheric weather implications for sustenance of life forms on

Table 2.2 Chemo-physical properties of water

Properties	Value	Unit
Colour/odour (ideal)	Colourless/odourless	
Density (at 4°C)	1.00	g/cm^3
Density (at boiling point)	0.958	g/cm^3
Boiling point	100.0	°C
Freezing point (at 1 atm pressure)	0.0023	°C
Triple point (at 4.58 Hg)	+0.0098	°C
pH	7.00	–
Ion product (at 25°C) constant	1.008×10^{-14}	mol^2/dm^6
Viscosity (at 20°C)	0.001	Pas
Permittivity (at 18°C)	81.7	–
Specific conductance (at 18°C)	4×10^{-6}	per ohm per cm
Thermal conductivity	0.537	$kCal/h/m^2/°C$
Heat of fusion	1.434	kCal/mol
Heat capacity (liquid)	0.9986	cal/g/°C
Heat capacity (vapour)	0.484	cal/g/°C
Heat of formation (20°C)	68.35	kCal/mol
Heat of vaporization	9.719	kCal/mol
Latent heat of vaporization	576.6	cal/g
Critical temperature	374.1	°C
Surface tension (at boiling point)	58.8	dyn/cm
Electrolytic dissociation constant (at 18°C)		
K_1 ($H_2O \rightarrow H^+ + OH^-$)	1.07×10^{16}	–
K_2 ($OH^- \rightarrow H^+ + O^{-2}$)	$< 10^{-36}$	
Dipole moment of vapour	1.85	cgs units $\times 10^{48}$
Strength of a hydrogen bond	5.0	kCal/mol
H-bond length in water	0.26–0.31	nm
H–O–H valence angle	104.5	degree
Redox potential (E_h)	2.13	volt

Table 2.3 Factors affecting the biogenic characteristics of water

Items	Per cent involved (%)
Quantum of active radiation consumed in photosynthesis, with respect to total incident radiation on the earth's surface	43
Maximum absorption of active radiation for consumption in photosynthesis (20% is lost)	80
Conversion efficiency of absorbed light	23–30
Loss of biomass by respiration at night	20–40

earth. For such sustenance, it is well known that the earth's temperature, weather, and climate are determined by absorption of solar radiation, with water playing a dominant role in the process.

Scientists have developed an "etch canal" method to assess the distinction between chemical and biogenic weathering of micrometeorites. Materials composed of olivine and interstitial glass bars have been found etched out in deep-sea spherules due to chemical etching in water. However, in spherules extracted from Greenland sediments (made of cocoons of siderobacteria), olivine is etched out through biogenic weathering.

The densities and partial pressures of hydrogen-bonded water complexes $-(H_2O)_n-$indicate that they absorb a part of solar radiation. Their spectra shift and broaden from those of water vapour monomers. Therefore, the role of $(H_2O)_n$ when n is greater than 2 requires a study of the kinetics of $nH_2O \rightarrow (H_2O)_n$ thoroughly. Recent studies inferred that dimers and trimers of water influence the amount of solar radiation striking the earth's surface. However, higher oligomers of mole fractions less than 10^{-11} are unlikely to absorb appreciable amounts of radiation. This finding relates to the fact that variation in the amount of solar radiation striking the earth's surface affects global biomass production potential. In addition, the amount of solar radiation reaching the earth's surface depends upon climatic conditions and varies with geographical location. The level of dependence of the corresponding country on agricultural production will thus change accordingly. Further, several physical and biological constraints are involved in it. Maximum efficiency for conservation of solar energy into biomass may be computed considering the items listed in Table 2.3.

The different biogenic properties of water with regard to bio-aqueous solvent systems are briefly listed below.

- Non-toxic to living beings
- Required as an aqueous solvent in a human body bioreactor
- Essential in biogenic weathering
- Multiphasically heterogeneous
- Supports ionic interactions between the chemicals of life in aqueous media
- Aids in developing multiphase systems separated by membranes
- Required in water-powered bio-pump operations
- By virtue of the biogenic weathering property of water, ice crystal nucleation or formation control can be stimulated by bacteria
- Biogenic ions can make and break water molecules in vivo
- Ions can make and break water molecules in vivo
- In protein folding-unfolding in cells, entropy release of water plays a critical role

Water pricing based upon biogenic property has cultural values and value judgement implications as it is a basic human and animal need.

From Table 2.3, it is clear that the maximum overall efficiency, $E = 0.43 \times 0.8 \times$ (0.23 to 0.30) × (0.60 to 0.80) = 5% to 8%. In India, E = (0.05 to 0.08) × 4950 kCal/m^2/day ≈ 250–400 kCal/m^2/day.

This amount of solar energy incidence, in coordination with India's temperate weather, has made it an agricultural country, that is, most suitable for outdoor photosynthetic food production.

In photosynthesis, Ribulose-1, 5 diP is the key compound in plants that fix CO_2 through reverse Embden–Meyerhof–Parnas (EMP) pathway as shown in Figures 2.22 and 2.23.

2.3.3 Biological Functions of Water

The various properties of water assist in assuming the particular state in which it exists. Depending upon the pressure and temperature conditions, the transitional states are water (liquid), ice (solid), and vapour (gas) (Figure 2.24). Variation in the states of water has made it a multipurpose prime resource on earth, which plays a significant role in ecological balance. This property also makes it a vital material for industrial growth, especially in the beverage industry (Mukhopadhyay 1995; Mukhopadhyay and Baisya 1997) and in the preservation of food. Thus, in ecological engineering, the biogenic character of water serves as an essential component of living matter and agricultural produce as it shapes the chemistry of all living organisms. By biogenic properties, water dissolves waste products of metabolism in body reactors (Figure 2.25) and assists in their removal from organisms to the environment.

Water functions as an insulator in body heat regulations. Due to its high specific heat capacity, water has a good capacity of absorbing heat with only minimal changes

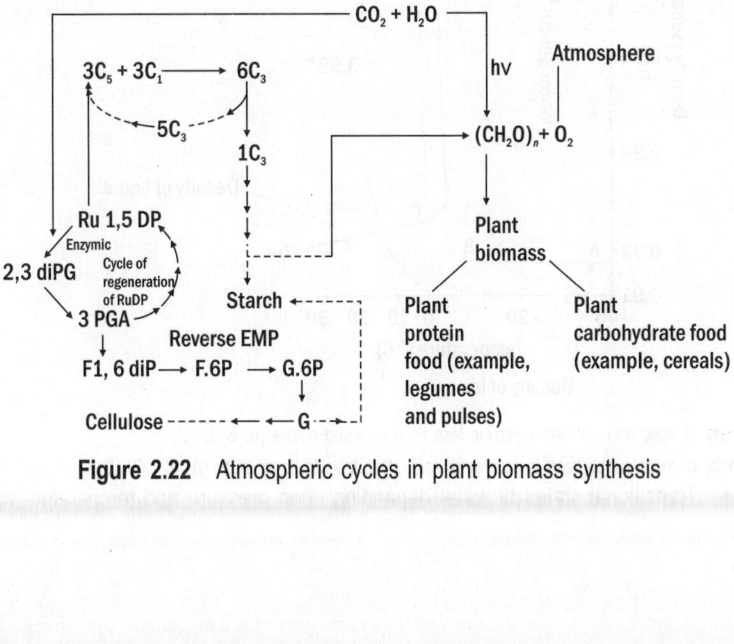

Figure 2.22 Atmospheric cycles in plant biomass synthesis

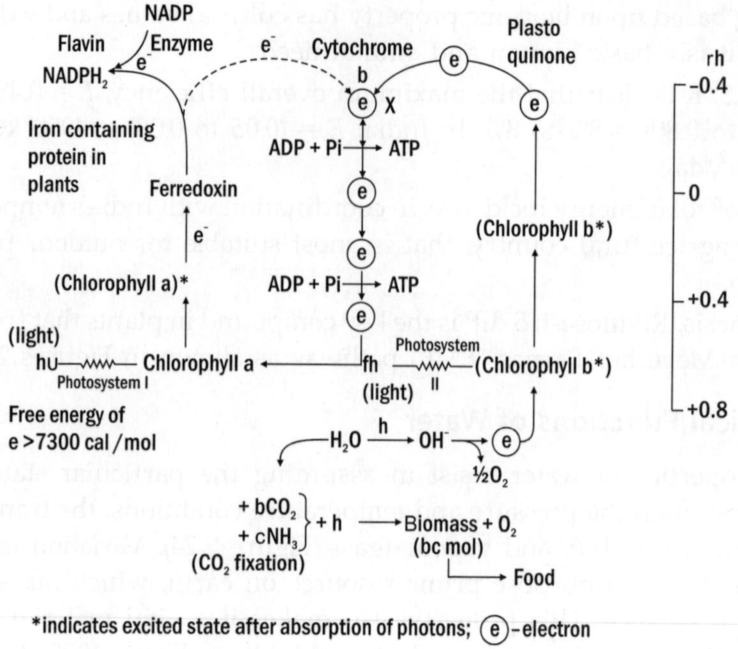

Figure 2.23 Biological redox systems involvement in biomass production

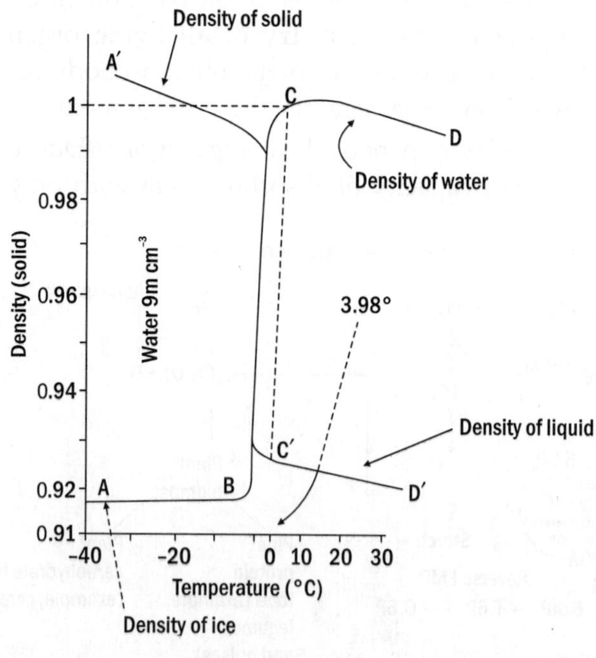

Notes 1. Density of solid state of water (ice) is less than its liquid phase (A, B, C, D)

2. Density of most substances is greater in solid state than in liquid state (A′, B′, C′, D′).

Figure 2.24 Transitional states of water depending upon pressure and temperature conditions

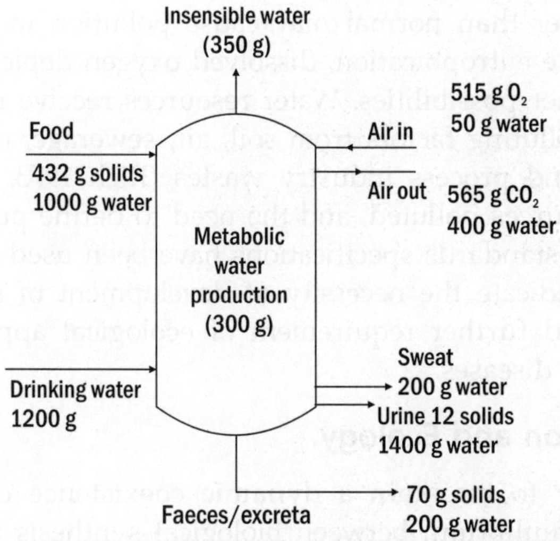

Note Correlations of physiological variables in human body bioreactor and steady state mass balances on the body for an average person at rest per day (neglecting solid content in sweat less than 2 g).

Figure 2.25 Fate of drinking water via food metabolism in human body bioreactor

in its own temperature. Excess heat energy in the body is dissipated by breaking the hydrogen bonds of water molecules. In this way, living organisms are protected against sudden thermal changes. Further, the conductivity property of water makes it possible for heat to be distributed evenly throughout bodily tissues. Water is also present in bodily fluids that serve as lubricants wherever one organ or bone rubs against another. There are many other significant functions of water that make it a unique fluid in biological systems.

A general mass balance can be written as shown in Equation 2.7.

$$\Sigma m_{in} - \Sigma m_{out} = dm_t/dt \text{ (zero at steady state)} \tag{2.7}$$

where m is the mass flow rate and m_t is the total body mass at any time t.

2.3.4 Water as a Driver of Life Forms

All living beings daily use water from natural resources like rainwater, surface water (such as oceans, seas, rivers, and lakes), stored water (such as ponds), and groundwater for various purposes. This ultimately leads to pollution in these waterbodies and generates another type of water known as wastewater. Human activities for domestic, municipal or industrial purposes mostly produce this wastewater. Nearly 97% of water on earth is surface water, which is extensively used by humans as aquatic transport avenues or waterways. In such uses, the presence of dissolved or suspended foreign matters (like silt, chemicals, faeces, metallic elements, organic materials/nutrients or oil spills) in concentrations higher than normal may cause pollution in natural waterways by various phenomena like eutrophication, dissolved oxygen depletion followed by death of aquatic life, and other possibilities. Water resources receive microbial cells, organic nitrogen, and other polluting factors from soil, air, sewerage, municipal wastes, dead plants and animals, and process industry wastes (Rich 1973; Mukhopadhyay 1995). This makes water resources polluted, and the need to define purity of water standard arises. Different water standards specifications have been used for different purposes. These specifications indicate the necessity of development of strategies to break the transmission chain and further requirement of ecological approaches to reduce the burden of water-borne diseases.

2.3.5 Water Pollution and Ecology

All living matters try to maintain a dynamic coexistence on earth. It is helpful towards keeping an equilibrium between biological synthesis and cleavage reactions (Equations 2.8 and 2.10) in water ecology. If none of these ecosystem processes is disturbed, a peaceful earth will be established for living systems with their environments. This is called ecology. Thus, in biospheric and abiospheric worlds, ecology plays an important role. The balance of these two worlds is essential. However, ecological imbalance is caused by pollution in water, soil, and air. Among them, soil and water pollution are the major storehouses of various types of micro and macro-organisms, which are the principal agents for redesigning one biological form to another and for maintaining their food cycle.

Biocellular enzymes

$$\underset{\text{Carbohydrate}}{C_6H_{12}O_6} + \underset{\text{Nutrients in water}}{NH_3 + O_2} \xrightarrow{\text{Enzyme}} \underset{\text{Cell mass}}{(C_5H_7NO_2)_n} + 4H_2O + CO_2 \quad (2.8)$$

Biodegradations

$$C_5H_7NO_2 + 12H^+ \xrightarrow[\text{(Anoxic)}]{\text{Enzyme}} \underset{\text{Fuel}}{4CH_4} + CO_2 + NH_3 \quad (2.9)$$

$$C_5H_7NO_2 + O_2 \xrightarrow[\text{(Oxic)}]{\text{Enzyme}} 5CO_2 + 2H_2O + NH_3 \quad (2.10)$$

2.3.6 Water Treatment, Management, and Pollution Abatement

Abatement of water pollution has now become an essential mission for all nations. For pollution abatement/management, water is treated by various methods (Table 2.4). These methods are broadly divided into four categories: (1) biological treatment, (2) chemical treatment, (3) electrochemical treatment, and (4) physical separation/purification.

Table 2.4 Ecology balancing water pollution management by treatment methods

Treatment method	Specifics
(A) Physical/chemical	
Aeration	Water fall aeration
Bubble aeration	
Sedimentation/flocculation	Discrete particle
Flocculent	
Hindered	
Compression	
Filtration	Carbon
Carbon	
Sand	
Membrane	
Disinfection	Chlorine
Demineralization	Ion exchange
(B) Biological	
Aerobic	
(a) Activated sludge process (ASP)	Conventional
(b) Step aeration	High rate
(c) Contact stabilization	Trickling filter
(d) Extended aeration	
(e) Aerated lagoons	
Anaerobic	
(a) Anoxic digestion	Mesophilic
(b) Biotrophic methanation	Thermophilic

Biological treatment is still the most widely used procedure in managing polluted water. As it cannot cope with many specific industrial pollutants in water, many other methods of treating such wastewaters have been adopted. Solvent treatment, electrodialysis, membrane separation (such as reverse osmosis and ultrafiltration), and chemical oxidation are some methods used. Biological treatment/management makes use of microbial biomass to break down/consume organic matter, resulting in stabilization of pollutants.

The methodology of water purification or treatment/management may be considered for drinking water and waste/effluent water.

In the management of drinking water purification, filter media systems have now been designed for high filtering speed with greatest possible reduction of pollutants. They have also been customized corresponding to the customer's requirements. The selection of filter media has been oriented by the volume of water to be purified and the types and quantity of impurities. The filter media vary to a great extent depending upon the application properties of drinking water. These filter media have been accordingly graded. For example, a filter medium of grade 1579 is a pure cellulose paper filter without adsorbents and is the best filter for the purpose of eliminating chlorine from treated water. Also, the presence of activated carbon is absolutely necessary for reducing pollutants like atrazene (a pesticide used in Europe and the USA) and similar compounds (such as lindan, a pesticide used in the USA) in drinking water (Binzer 1997). Potable water must conform to biological and tolerable standards. For biological standards of potable water, the levels of *Escherichia coli* count in any 10 ml sample, coliform count in 95% of 100 ml samples in any year, and coliform organisms in a second sample should be zero or coliform organisms in any 100 ml sample should be at a maximum of 10. In case of total coliform count, 1/100 ml is the acceptable standard. Tolerable standards (highest limit) for toxic chemical components to be taken care of for ecological acceptability are as follows—As: 0.05, Cd: 0.005, NO_3^-: 45.00 (all expressed in parts per million [PPM]), polynuclear aromatic hydrocarbons: 0.20 mg/L, radioactive substances: gross-activity 3.0 picoCuries per liter (pCi/L). The standard set of components affecting the ecological acceptability of water is given in Table 2.5.

In the management of drinking water technology, the most conventional method of treatment of water is chlorination.

Table 2.5 Components affecting the ecological acceptability of water

Component	Highest limit
Colouring (on platinum and cobalt scale)	5–50 units
Odour	Non-objectionable
Taste	Non-objectionable
Turbidity	5 in NTU
pH	7–7.5
TS	500 mg/L
Total hardness	2 m Eq/L
Fe	0.1 mg/L
Mn	0.25 mg/L
Cu	0.05 mg/L
Zn	5.00 mg/L
Ca	75.00 mg/L
Mg	30.00 mg/L
SO_4	200.00 mg/L
C1	200.00 mg/L
Phenolics	0.001 mg/L
Mineral oil	0.01–0.30 mg/L
Anionic detergents (as MBAs)	0.2–1.0 mg/L

NTU — number of turbidity unit; MBA — monoclonal antibody

Disinfective chlorine demand (Figure 2.26) indicates that its overdose may pose biohazard possibility (Figure 2.27) and, therefore, influence ecology (Mukhopadhyay and Baisya 1997). In addition to chlorination of water for rendering it safe from a health hazard, the industrial processor often further chlorinates water in the plant (such as food processing plants) for use in disinfecting products and machinery. Since

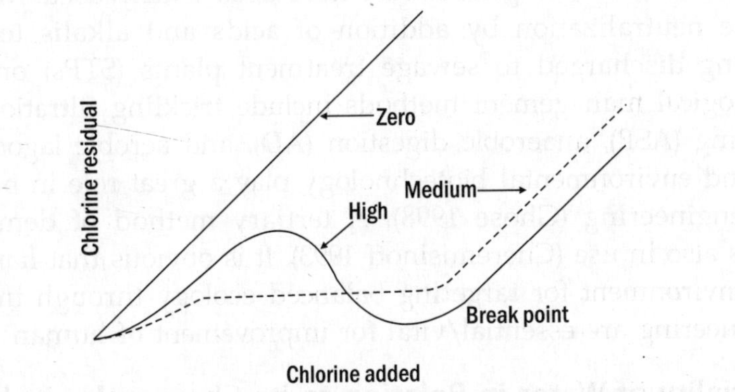

Figure 2.26 Disinfective chlorine demand characteristics of water

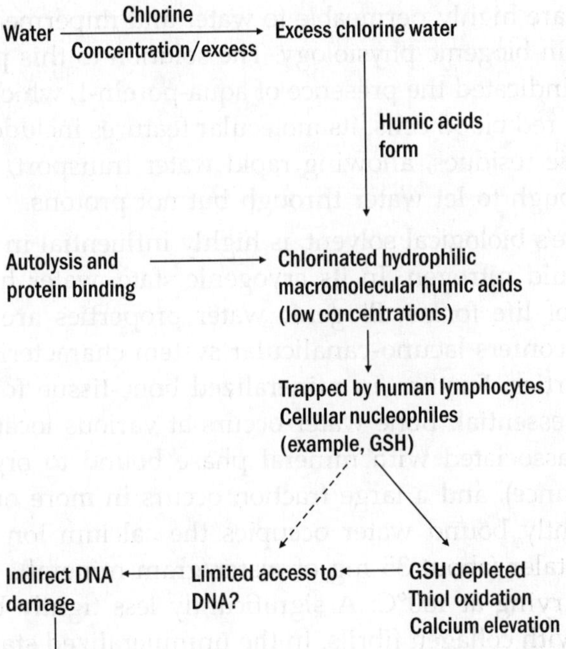

Figure 2.27 Overdose chlorination of water on biohazard possibility

wastewaters may contain a wide variety of materials, their compositions and pollution loads vary greatly. Nevertheless, wastewaters can be broadly classified according to the nature of their impurities and pollution potentials, as stated earlier. These will determine the methodology suitable for taking care of ecological balance. Primary methods of physical nature (like screening out of gross particulates) and secondary methods of chemical and biological nature have been widely used. Many wastewaters usually require neutralization by addition of acids and alkalis to a pH range of 6–9 before being discharged to sewage treatment plants (STPs) or natural waters. Secondary biological management methods include trickling filtration (TF), activated sludge processing (ASP), anaerobic digestion (AD), and aerobic lagooning. Therefore, environment and environmental biotechnology play a great role in balancing domain in ecological engineering (Ghose 1998). A tertiary method of demineralization by ion-exchange is also in use (Cheremisinoff 1993). It is obvious that handling, care, and protection of environment for targeting balanced ecology through implementation of ecological engineering are essential/vital for improvement of human lives.

2.3.7 Essentiality of Water in Relation to its Chemo-physical and Biogenic Properties

Cellular membranes are highly permeable to water and impermeable to protons, which has been a dilemma in biogenic physiology. The solution to this problem was provided by bioscientists who indicated the presence of aqua-porein-1, which is the water channel found in mammalian red blood cells. Its molecular features include an aqueous pathway line with hydrophobic residues, allowing rapid water transport, and a pore spanning just 3 Å—narrow enough to let water through but not protons.

Water, being nature's biological solvent, is highly influential in determining survival after exposure to liquid nitrogen. In its cryogenic state, water has served extensively in cryopreservation of life forms. Biogenic water properties are essential in forming calcified matrix that confers lacuno-canalicular system characteristics in bone through its viscoelastic properties. Further, in mineralized bone-tissue formation, the biogenic property of water is essential. Bone water occurs at various locations and in different binding states. It is associated with mineral phase bound to organic phase (collagen and cementing substance), and a large fraction occurs in more or less free form (bulk water). The most tightly bound water occupies the calcium ion coordination sites in the appetite-like crystaler (about 35 mg of water/gram mineral). This water cannot be replaced by simple drying at 100°C. A significantly less tightly bound reaction is the association of water with collagen fibrils. In the unmineralized state, as the bone matrix is laid down by obsteoblasts, the collagen fibres produced contain a large fraction of water (about 60%), which is essentially needed.

2.4 BIOCHEMISTRY AND IMMUNOCHEMISTRY: A CONCISE TREATISE

In the growth and functions of biological systems, metabolic reactions manifest activities of living cells. Also, the capacity of living cells to adapt to changes in the environment appears to be a universal attribute in their adaptive regulatory reactions. A concise treatise of biochemistry and immunochemistry will be useful.

2.4.1 Biochemical Pathways in Life Forms

In different life forms, catabolic, anabolic, and amphibolic reactions occur. Remarkable similarities are observed in the ways in which cells of different life forms obtain energy and form the intermediate compounds they need for biosynthesis.

2.4.1.1 Anoxic/anaerobic pathways

It is from the studies of fermentation that the concept of chemical changes in microorganisms/life forms emerged. Furthermore, the concept of enzymes also emerged from the work on fermentation. In life forms, the transformations or chemical changes of materials that take place have been termed bioconversions. In animal bodies, the bioconversion of sugar to lactic acid through sequential reactions is called glycolysis and the reactions pathway that takes place has been called EMP pathway. On the other hand, the bioconversion of sugar to ethyl alcohol and carbon dioxide, caused by yeasts, microorganisms, and other factors, is called alcoholic fermentation. Scientists have concluded from researches that biocatalytic/enzymic reactions in glycolysis and fermentations are identical in many steps. Thus, the EMP pathway and alcoholic fermentation pathway have many similarities and a few dissimilarities.

The processes of EMP pathway and alcoholic fermentation are collectively called metabolism, from the ancient Greek word "metabole" meaning change (here, it specifically means the change that takes place in life forms). The sequential reactions of change of material from one form to another within life forms have been called metabolic pathway or metabolic network. The substrates used for sustenance of life forms come from the consumption of food/nutrients. Among the substrates, carbohydrates/sugars are most commonly/abundantly used as a source of carbon. The other components in food/nutrients include proteins, fats/lipids, and minerals. When food/nutrient is consumed by life forms, the components in them are first degraded. The purpose of carbohydrate/sugar degradation in life forms is unlocking of the energy enclosed in carbon bonds and its simultaneous capture in special energy-rich phosphate bonds present in the form of adenosine triphosphate (ATP).

Sugar degradation leads to the fragmentation of complex sugar molecules into its simpler mono-molecular form that goes into the synthesis of a number of usable compounds such as amino acids, Lipids, purines, and pyrimidines. A flow chart of fermentation and glycolytic biochemical pathway to show carbohydrate metabolism is given in Figure 2.28, and the overall pathway is chemically represented by Equation 2.11.

$$G + 2ATP + 2NAD^{+} \rightarrow 2\ \text{Pyruvate} + 4ATP + 2NADH_2 \quad (2.11)$$

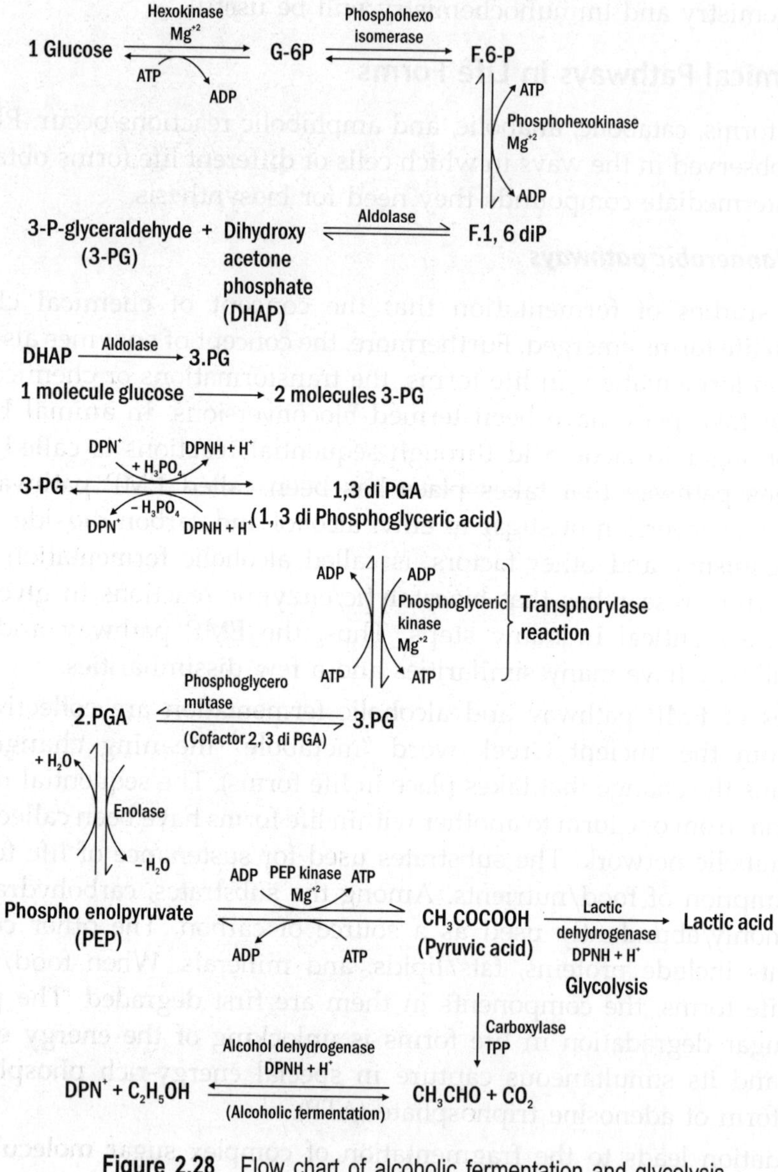

Figure 2.28 Flow chart of alcoholic fermentation and glycolysis

In animal life forms, no acetaldehyde/ethanol (CH_3CHO) is formed from pyruvate. Lactic dehydrogenase is specifically present in animal tissues and certain bacteria like lactic acid bacteria. At pH 7, the reaction is far towards the formation of lactate.

Alcoholic fermentation may proceed in the presence of air as well as in its absence. However, the velocity of the reaction is decreased by aeration. In air, the growth of yeast is accelerated, while the consumption of sugar is decreased. In the absence of air/oxygen, the reverse is true. Pasteur showed that 1 g of yeast was produced from 4 g to 10 g of glucose in the presence of air and from 60 g to 80 g of glucose in the absence of air/oxygen. This effect is, sometimes, referred to as the Pasteur effect. Since there are possibilities of side reactions in the metabolic pathway, concentrations of many cofactors in the environment may fluctuate, resulting in fluctuations of micro-environmental conditions. This causes auto-oscillations in the behaviour of life forms in product formations, manifesting non-linearities in the kinetics. Typical examples of such auto-oscillations exist in many organisms. Periodic, auto-oscillatory phenomena are characteristic of the process of glycolysis. Glycolysis is the anoxic/anaerobic bioconversion of hexose sugars to tricarboxylic acids, leading to synthesis of ATP through the participation of enzymes and cofactors in appropriate steps as described in the following sections.

2.4.1.1.1 Oscillations in biochemical/microbial reactions

Oscillation in cell reaction kinetics is one of the reasons for requirements in pathway/metabolic regulation. Oscillation/non-linearity in the kinetics of cellular and enzymatic reactions has been observed in a number of cases. Periodicity is observed in many bioreaction systems. Some examples of such cases include the following.

- Periodic oscillations in glycolysis/ethanol fermentation by yeasts
- Periodic oscillations in cell behaviour in enzyme synthesis by fungal system
- Prey–predator oscillations
- Dynamics of a chemostat

Let us consider the glycolytic reaction pathway in biological systems.

The first oscillatory behaviour in the glycolytic reaction pathway was observed in anoxic yeast cells. The anoxic bioconversion of hexose sugar to tricarboxylic acid (TCA), leading to biosynthesis of ATP molecules or $NADH_2$ molecules, is an example (Figure 2.29).

2.4.1.1.2 Periodic auto-oscillations in glycolysis/ethanol fermentation

Periodic changes in glycolytic metabolism were observed in microbial cells. Time dependence of metabolites could be seen under differing conditions such as pH temperature fluctuation. Also, various effectors of the enzymes of glycolysis could influence on the type and form of oscillations.

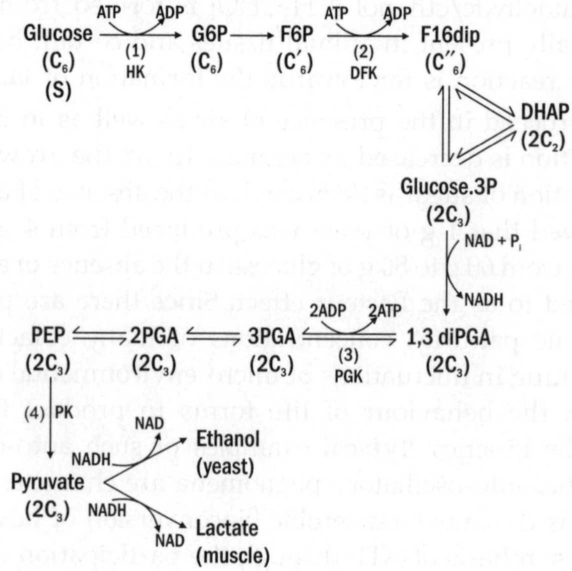

HK – hexokinase; PFK – phospho fructokinase; PK – pyruvate kinase; PGK – phosphoglyceric kinase

Figure 2.29 Flow chart of anoxic bioconversion of hexose sugar to tricarboxylic acid

Origin of oscillation

The factors that influence the origin of oscillation are: (1) autobiocatalysis, (2) interactions between the elements of a multi-enzymes system, (3) interactions of multi-enzymes systems with each other, and (4) force process parameters.

Typical auto-oscillatory reaction

Reactions inhibited by an excess of substrate(s) or coenzyme (Y') provide typical non-linear process. Equations 2.13–2.16 show the reaction steps that determine the kinetics of the process as a whole.

Evidence of oscillation

Oscillations in the concentration of one of the products of glycolysis and $NADH_2$, which are almost sinusoidal in shape, could be observed as shown in Figure 2.30.

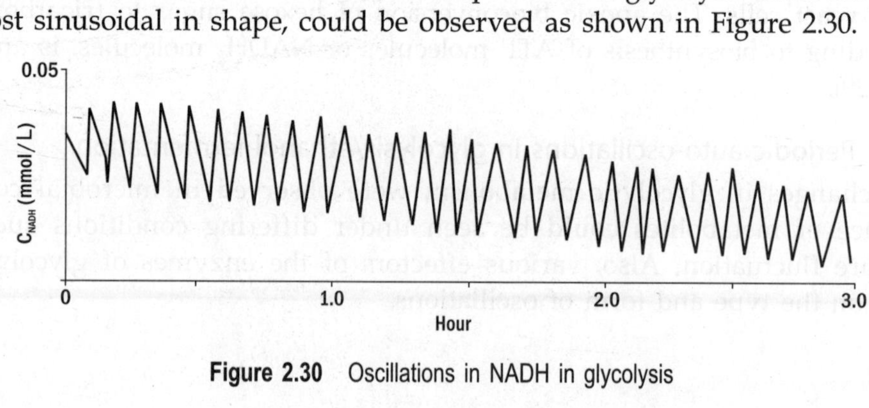

Figure 2.30 Oscillations in NADH in glycolysis

In the above reactions (Figure 2.29), the first damped concentration oscillations and then non-damped auto-oscillations were discovered by Chance and Hess in 1964.

Proof of auto-oscillation

There are three requirements for auto-oscillation. These are (1) kinetics, (2) regime characteristics, and (3) phase-plane equation. The kinetics of the glycolytic reactions process is determined on the whole by several "narrowsites" (Equations 2.13–2.16).

Kinetics

Let us consider υ_0 = rate of reaction (Equation 2.11), which is catalysed by hexokinase (HK). Also, let the step $C_6' \rightarrow C_6''$ be fast and irreversible.

Concentration (S) of any substances, C_6' and C_6'', increases at a rate of υ_0 due to reaction shown in Equation 2.5, which is catalysed by phosphofructokinase (PFK).

This is followed by several rapidly reversible reactions (up to the reaction shown in Equation 2.15).

Therefore, any intermediate concentration is proportional to each other. Let this concentration be S_1.

Since PFK is activated by S_1, the increase in S_1 due to reaction shown in Equation 2.14 is $\int$(S and S_1). That is,

$$\frac{dS_1}{dt} = \frac{dS}{dt} \tag{2.12}$$

In the reaction shown in Equation 2.15, catalysed by pyruvic kinase (PK), the concentration of S_1 is reduced.

$$\frac{dS}{dt} = \upsilon_0 - k_1 \frac{S}{ks + S} \cdot \frac{S_1}{ks_1 + S_1} \tag{2.13}$$

$$\frac{dS_1}{dt} = k_1 \frac{S}{ks + S} \cdot \frac{S_1}{ks_1 + S_1} - k_2 \frac{S_1}{k + S_1} \tag{2.14}$$

For reaction regime

$K_S \gg S,\ K_{S1} \gg S_1$

Here,

$\upsilon_0 \propto C_{HK}$, where C_{HK} is the concentration of HK,

$k_1' \propto C_{PFK}$, where C_{PFK} is the concentration of PFK,

$k_2' \propto C_{PK}$, where C_{PK} is the concentration of PK,

$k_1' = k_1 k_S k_{S1}$, and

$$k_2' = \frac{k_2}{K}$$

$$\frac{dS}{dt} = \upsilon_0 - k_1' S \cdot S_1 \tag{2.15}$$

$$\frac{dS_1}{dt} = k_1' S \cdot S_1 - k_2' \frac{S_1}{1 + \frac{S_1}{K}} \tag{2.16}$$

$$\frac{dx}{dt} = 1 - xy \tag{2.17}$$

$$\frac{dy}{dt} = qy\left[x - \frac{1+r}{1+ry}\right] \tag{2.18}$$

Here,

$$q = \frac{\left(k_2' - \frac{\upsilon_0}{K}\right)^2}{\upsilon_0 k_1} \tag{2.19}$$

$$r = \frac{S_1^0}{K} = \frac{\upsilon_0/K}{k_2' - \frac{\upsilon_0}{K}} \tag{2.20}$$

Thus,

$$\frac{dy}{dx} = \frac{qy\left[x - \frac{1+r}{1+ry}\right]}{1 - xy} \tag{2.21}$$

Dynamic property of Equation 2.21

Together with Equations 2.15 and 2.19, Equation 2.21 defines on the phase plane a certain vector field with singular points. The phase pattern of the equation in the $x - y$ plane indicates isoclines as shown in Figure 2.31. At steady state,

$$S_0 = \frac{k_2' - \frac{\upsilon_0}{K}}{k_1'} \tag{2.22}$$

$$S_1^0 = \frac{\upsilon_0}{k_2' - \frac{\upsilon_0}{K}} \tag{2.23}$$

Normalizing $x = \frac{S}{S_0}$ and $y = \frac{S_1}{S_1^{0}}$,

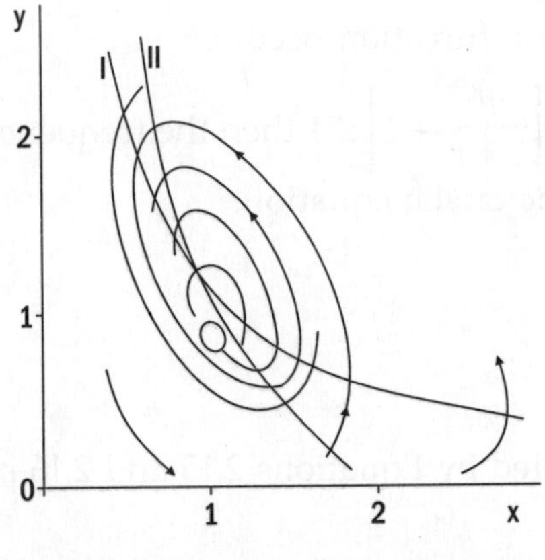

Figure 2.31 Phase–plane plot of Equation 2.21

$$\tau = k_1' S_1^0 t = \frac{k_1' \upsilon_0}{k_2' - \frac{\upsilon_0}{K}} t \qquad (2.24)$$

and introducing in Equations 2.15 and 2.16, we can derive a phase–plane plot of Equation 2.21.

The properties of the isoclines can be observed from three principal lines.

1. The isocline of the vertical tangent line (I) is a hyperbola.

$$y = \frac{1}{x} \qquad (2.25)$$

The isocline of the horizontal tangent line (II) is a hyperbola.

$$x = \frac{1 + r}{1 + ry} \qquad (2.26)$$

2. The straight line $y = 0$.

The character of the singular point can be determined by linearizing Equations 2.17 and 2.18. The roots of the characteristic equation are given by Equation 2.27.

$$R_{12} = \frac{1}{2}\left(1 - \frac{qr}{1 + r}\right) \pm \frac{1}{2}\sqrt{\left(1 - \frac{qr}{1 + r}\right)^2 - \frac{4q}{1 + r}} \qquad (2.27)$$

This may exhibit several characteristics, as discussed in subsequent sections.

Root characteristics

Case I If $\frac{4q}{1 + r} > \left(1 - \frac{qr}{1 + r}\right)^2$, then the point 0 is a focus.

Case II If $\frac{qr}{1 + r} < 0$, then the focus is stable.

Case III If $\frac{qr}{1 + r} > 0$, then the focus is unstable and there appears a limit cycle.

Case IV If $\frac{qr}{1+r} = 1$, then bifurcation occurs.

Case V Near the point 0, if $\left[\frac{qr}{1+r} - 1\right] \leq 1$ then the frequency of oscillation ≈ imaginary part of the root of the characteristic equation.

That is,

$$\omega \approx \sqrt{\frac{q}{1+\tau}} \tag{2.28}$$

Thus, the system represented by Equations 2.15 and 2.16 proves to be auto-oscillatory under certain conditions.

The characteristics parameter q contributes to the amplification of auto-oscillation when it increases and contributes to damping when it decreases. Equation 2.14 provides information as given in Table 2.6.

2.4.1.1.3 Distinction between glycolysis and fermentation

Glycolysis is the anaerobic degradation of carbohydrates in animals. The end product of degradation is lactic acid, and the degradation pathway is called the EMP pathway. So glycolysis is the anaerobic break down of sugar to lactic acid through the EMP pathway in animal bodies. On the other hand, alcoholic fermentation occurs in most microorganisms (for example, yeast and bacteria). It is the anaerobic break down of carbohydration microbial cells. The end product of degradation in this process is ethyl alcohol and CO_2.

2.4.1.1.4 Anoxic fermentation by prokaryote

Alcoholic fermentation of carbohydrates by yeasts is known. The only exception to yeast is the fermentation of carbohydrates to ethanol by a bacterium/prokaryote named *Zymomonas* sp. Like yeast, the advantage of using *Zymomonas* for fermentation process is that there are no by-products. In addition, *Zymomonas* grows faster than yeast and, therefore, produces ethanol at a faster rate. Pathway engineering has been used to

Table 2.6 Observed properties of *q*

Effect	Observation	Remark
Decrease in k_1' and υ_0 (rate of influx)	Self-excitation of oscillation in pathway	Conforms to experimental results of Selkov
Increase in k_2' (addition of PK)	Self-excitation of oscillation in pathway	—
Resulting oscillations in bioreactions in cells	Self-excitation of oscillation in pathway	Online diagnosis is not available

PK — pyruvate kinase

redesign *Zymomonas* that can break down carbohydrates/polysaccharide polymers releasing large amounts of xylose. This xylose was then efficiently fermented to ethanol. This process has been done in two stages. In the first stage, pathway engineers introduced genes for metabolism of xylose in *Zymomonas*. The engineers found that xylose A (*xyl A*) and xylose B (*xyl B*) genes encode enzymes xylose isomerase and xylulose kinase, respectively, which converts xylose into xylulose and then to xylulose S-phosphate as per pathways shown in Figures 2.32(a) and 2.32(b).

In the second stage, the accumulated C_5 compound undergoes interconversions. In the interconversion, transketolase (*tkt A*) transforms two C_5 molecules of sugar into one C_3 and one C_7 molecule. These molecules are converted by the presence of transaldolase

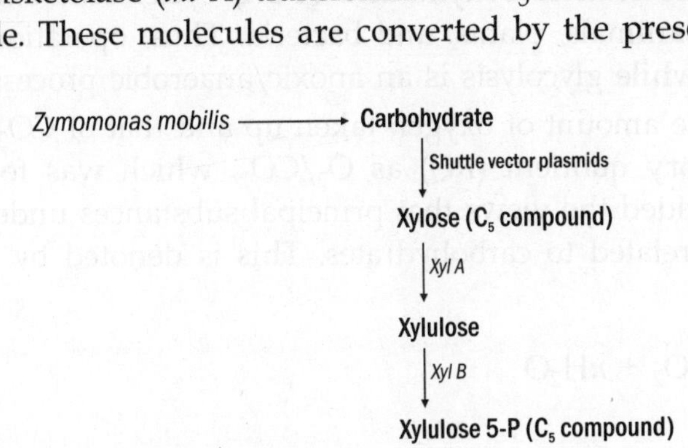

Figure 2.32(a) Xylose metabolic pathway

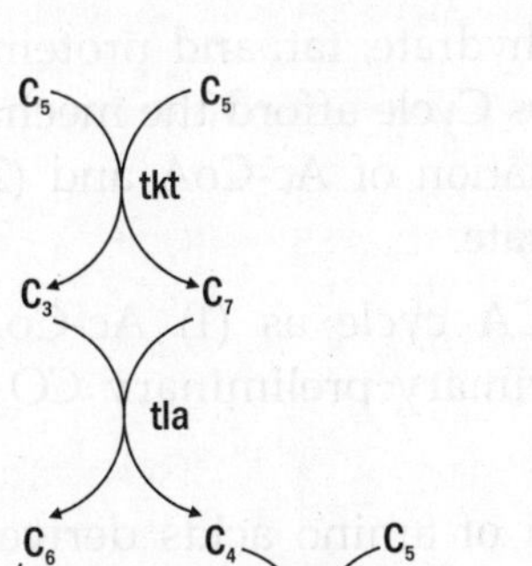

Figure 2.32(b) Accumulated C_5 sugar metabolic pathway

(*tal*) gene into a C_4 sugar and fructose 6-phosphate (F-6P)—a C_6 sugar molecule. In turn, the F-6P entering into the glycolysis process forms ethanol. The C_4 and another C_5-5P are transformed by *tkt A* into C_6 and C_3 sugars that feed into the glycolytic pathway to make ethanol as shown in Figure 2.32(b).

2.4.1.2 Oxic/aerobic pathways

The pyruvic acid from the EMP pathway is rapidly oxidized to CO_2 and water following a cyclic pathway. This pathway was first proposed by H A Kreb in 1937 and is, therefore, called the Krebs Cycle or tricarboxylic acid (TCA) cycle. This oxic/aerobic metabolic pathway operates in animals, plants, and bacteria. Thus, operation wise, this cycle is purely oxic/aerobic, while glycolysis is an anoxic/anaerobic process.

Measurement of the amount of oxygen taken up and that of CO_2 produced provides the ratio of respiratory quotient (RQ) as O_2/CO_2, which was found to be close to unity. This ratio provided the views that principal substances undergoing oxidation in animal bodies were related to carbohydrates. This is denoted by the reaction shown in Equation 2.29.

$$\underset{\text{Carbohydrate}}{(CH_2O)_n} + nO_2 \rightarrow nCO_2 + nH_2O \qquad (2.29)$$

A schematic of the Krebs Cycle is shown in Figure 2.33.

2.4.1.2.1 Metabolic significance of Krebs Cycle

It is possible to integrate carbohydrate, fat, and protein metabolism through the Krebs Cycle. The reactions of the Krebs Cycle afford the mechanisms for (1) the final oxidation of fatty acids through the oxidation of Ac-CoA and (2) the oxidation of carbohydrate through the oxidation of pyruvate.

Pyruvate may enter the TCA cycle as (1) Ac-CoA after decarboxylation or (2) oxaloacetic acid (OAA) after primary preliminary CO_2 fixation reactions catalysed by OAA decarboxylase.

The final stages of oxidation of amino acids derived from protein metabolism are affected through the reactions of the Krebs Cycle. A number of amino acids give rise to Ac-CoA in the course of metabolic reactions, while others may give rise to OAA or α-KG (two members of the Krebs Cycle) either by transamination or by deamination. Figure 2.34 shows the metabolism of amino acid.

2.4.1.2.2 Hexose monophosphate shunt pathway

There is evidence that enzyme systems present in many biological systems, including animal tissues, plant tissues, and some microorganisms, may metabolize glucose through reactions that bypass anaerobic glycolysis down to glyceraldehyde-3 phosphate.

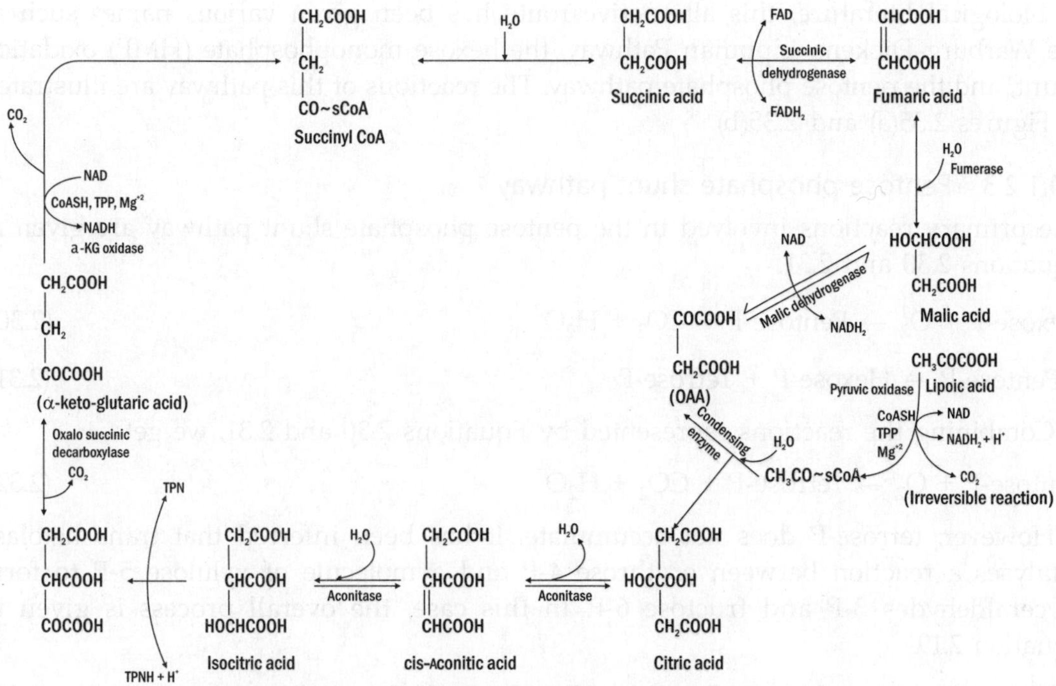

OAA – oxaloacetic acid; TPN - triphosphopyridine nucleotide

Note This cycle occurs in the respiring tissues of animals ranging from protozoa to mammals.

Figure 2.33 Krebs Cycle scheme

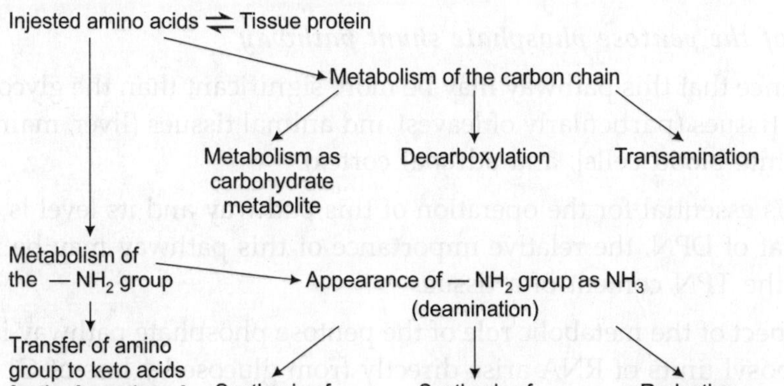

Figure 2.34 Amino acid metabolism

In biological literature, this alternative route has been given various names such as the Warburg–Dickens–Lippman Pathway, the hexose monophosphate (HMP) oxidation shunt, and the pentose phosphate pathway. The reactions of this pathway are illustrated in Figures 2.35(a) and 2.35(b).

2.4.1.2.3 Pentose phosphate shunt pathway

The primary reactions involved in the pentose phosphate shunt pathway are given in Equations 2.30 and 2.31.

$$\text{Hexose-P} + O_2^{1} \rightarrow \text{Pentose-P} + CO_2 + H_2O^{2} \tag{2.30}$$

$$2\ \text{Pentose-P} \rightarrow \text{Hexose-P} + \text{Tetrose-P} \tag{2.31}$$

Combining the reactions represented by Equations 2.30 and 2.31, we get

$$\text{Pentose-P} + O_2 \rightarrow \text{Tetrose-P} + CO_2 + H_2O \tag{2.32}$$

However, tetrose-P does not accumulate. It has been inferred that trans ketolase catalyses a reaction between erythrose 4-P and a molecule of xylulose 5-P to form glyceraldehydes 3-P and fructose 6-P. In this case, the overall process is given in Equation 2.19.

$$2\ \text{Pentose-P} + O_2 \rightarrow \text{Hexose-P} + \text{Triose-P} + CO_2 + H_2O \tag{2.33}$$

In the sequence of reactions in this pathway, the C-1 of glucose gets converted to CO_2. The most important intermediate is ribulose 5-P and the most important enzymatic step involves trans ketolase.

Significance of the pentose phosphate shunt pathway

There is evidence that this pathway may be more significant than the glycolytic pathway in some plant tissues (particularly of leaves) and animal tissues (liver, mammary glands, leucocytes [white blood cells], and adrenal cortex).

Since TPN is essential for the operation of this pathway and its level is usually much lower than that of DPN, the relative importance of this pathway may be influenced by variations in the TPN content of a tissue.

Another aspect of the metabolic role of the pentose phosphate pathway is the question whether D-ribosyl units of RNA arise directly from glucose by loss of C1 of the hexose and utilization of the pentose phosphate without further cleavage of the C-chain.

---

[1] The molecules of triphosphopyridine nucleotide (TPN) have been reduced. Therefore, one molecule of oxygen is required.

[2] Reoxidation of two molecules of TPNH forms two molecules of water. However, since one molecule of water was required for opening the lactone ring of 6-P glucono-δ-lactone by lactonase reaction, the net gain of water molecule is 1.

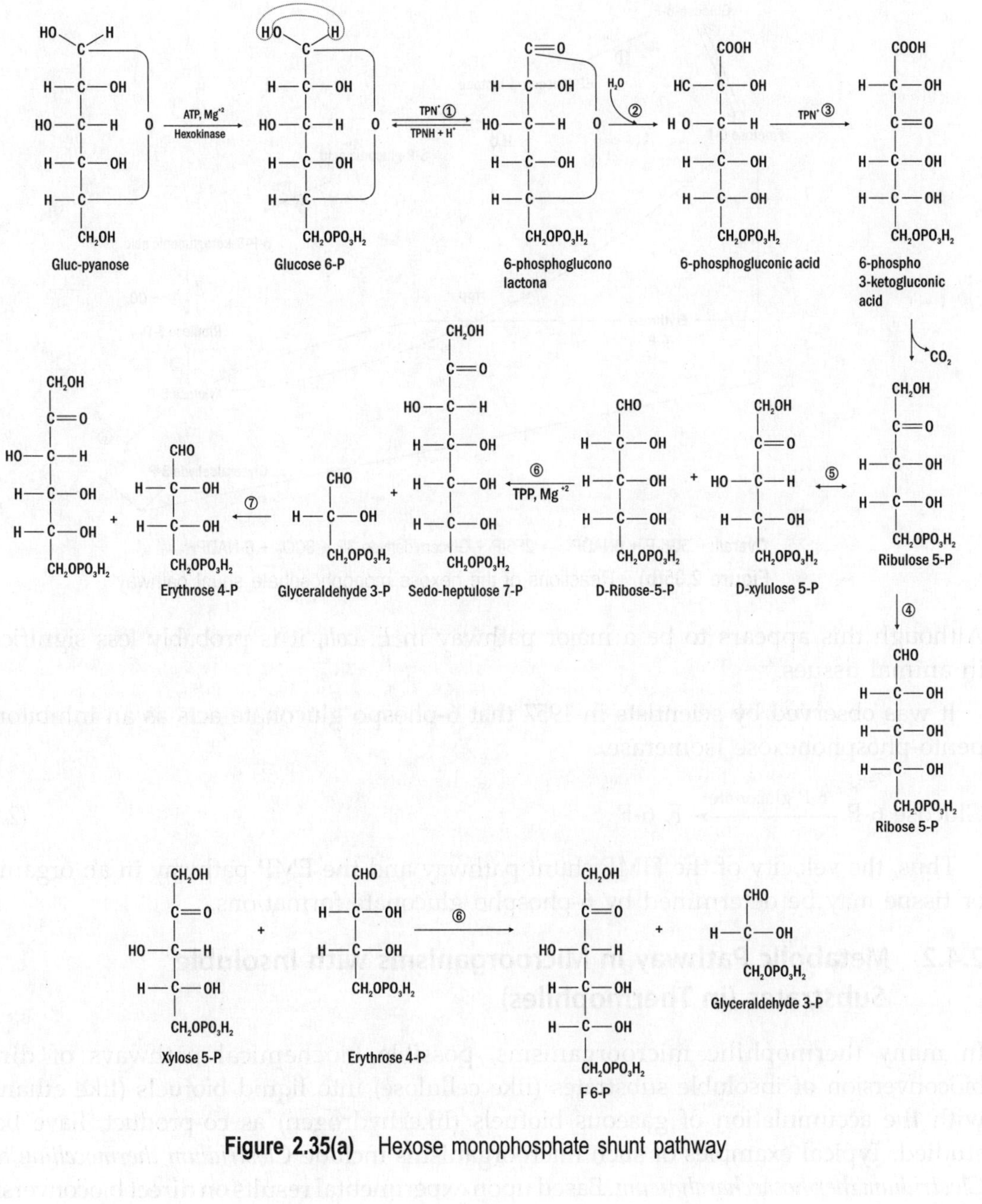

Figure 2.35(a) Hexose monophosphate shunt pathway

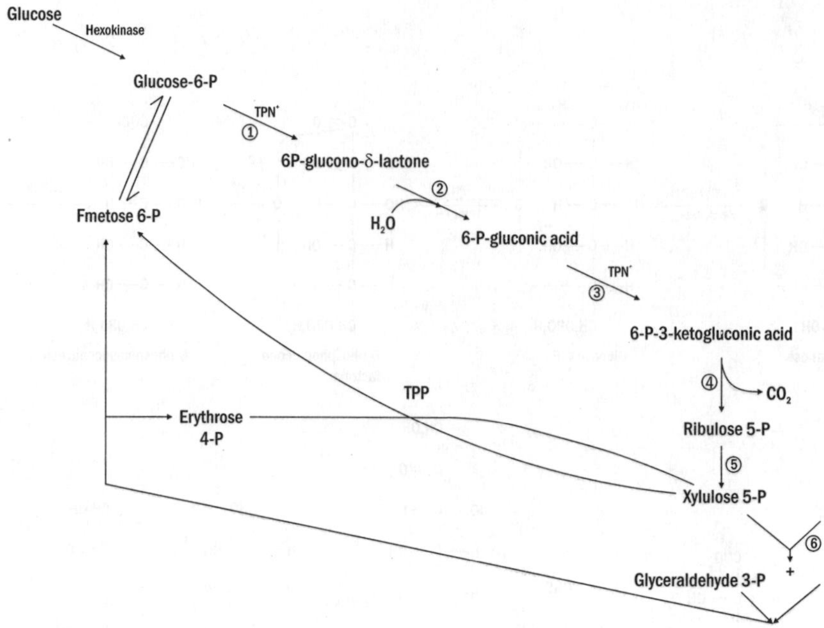

Overall: $3G6\text{-}P + 6NADP^+ \rightarrow 2F6\text{-}P + \text{Glyceraldehyde } 3P + 3CO_2 + 6\ NADPH_2$

Figure 2.35(b) Reactions of the hexose monophosphate shunt pathway

Although this appears to be a major pathway in *E. coli,* it is probably less significant in animal tissues.

It was observed by scientists in 1957 that 6-phospo gluconate acts as an inhibitor to pento-phosphohexose isomerase.

$$\text{Glucose 6-P} \xrightarrow{\text{6-P gluconate}} \text{F, 6-P} \tag{2.34}$$

Thus, the velocity of the HMP shunt pathway and the EMP pathway in an organism or tissue may be determined by 6-phospho gluconate formations.

2.4.2 Metabolic Pathway in Microorganisms with Insoluble Substrates (in Thermophiles)

In many thermophilic microorganisms, possible biochemical pathways of direct bioconversion of insoluble substrates (like cellulose) into liquid biofuels (like ethanol), with the accumulation of gaseous biofuels (like hydrogen) as co-product, have been studied. Typical examples of such microorganisms include *Clostridium thermocellum* and *Clostridium thermosaccharalyticum*. Based upon experimental results on direct bioconversion of cellulose by thermophilic *C. thermocellum*, a proposed schematic and metabolic pathway in the cell, including enzymes inhibition, is shown in Figures 2.36 and 2.37. Direct bioconversion of cellulose to ethanol by thermophiles has many advantages (Table 2.7).

Relatively little is known about the entire spectrum of thermophilic bacteria that may be of interest in bioethanol production. Thermophilic ethanologens are easily obtained from a variety of self-heating environments (for example, manure, composts or soils) or volcanic features (for example, thermal springs or decomposing bacterial algal mats).

C. thermocellum (ATCC 27405) is described as a Gram-negative rod 0.6 × 4 μm in size, forming terminally spherical spores on solid but not in liquid medium. The organism has been reported to grow on cellulose, cellobiose, glucose, fructose, and mannitol. Two recent literature reports, however, indicate that *C. thermocellum* was only capable of growth on cellulose and cellobiose. Other investigators have also reported growth on xylose, but this capability usually coincides with the simultaneous ability to use a variety of carbohydrates, including starch, which is uncharacteristic of this organism. A defined medium has been developed by Fleming for *C. thermocellum*, showing the requirements of vitamins and amino acids for this organism. However, more recent studies have shown that these requirements can be reduced to only biotin and pyridoxine in media containing minimal salts.

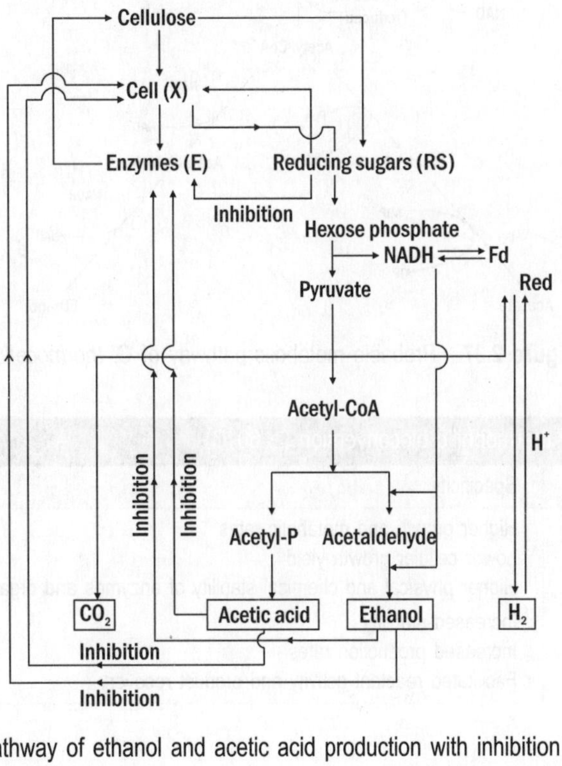

Figure 2.36 Schematic pathway of ethanol and acetic acid production with inhibition of enzyme and cell growth

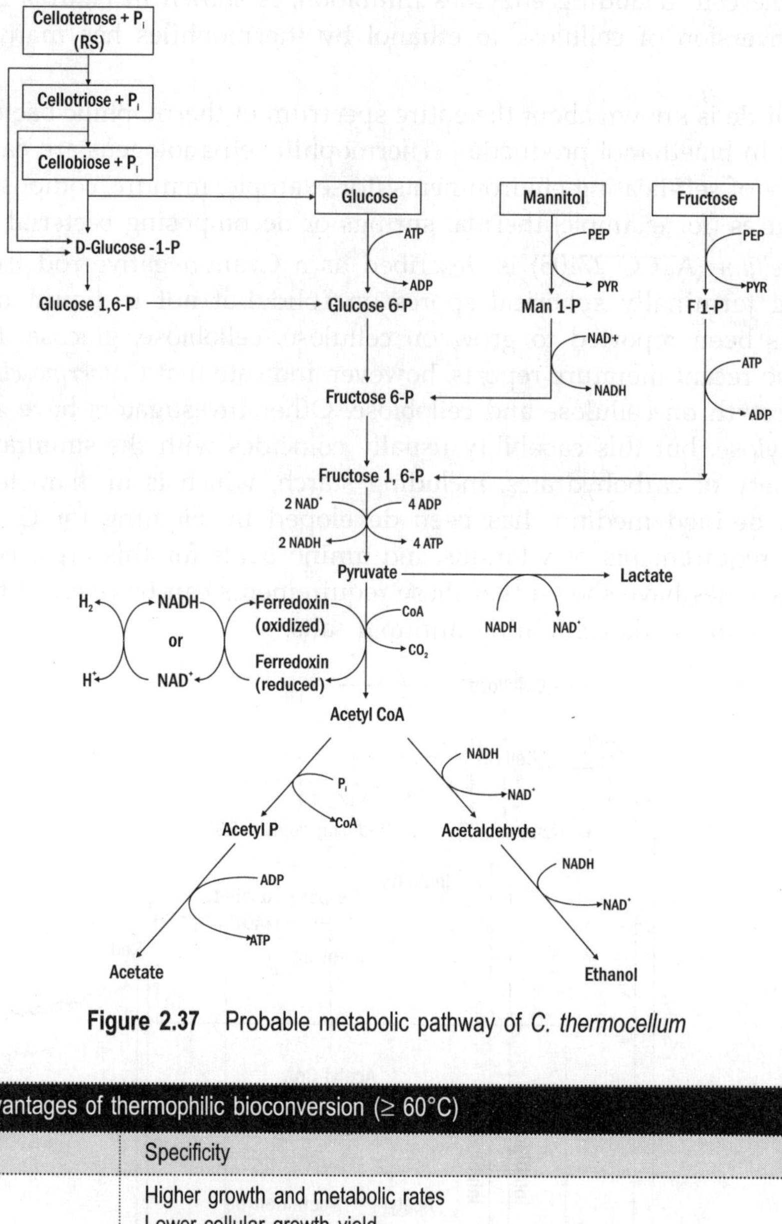

Figure 2.37 Probable metabolic pathway of *C. thermocellum*

Table 2.7 Advantages of thermophilic bioconversion ($\geq 60°C$)

Advantage	Specificity
Microbial	Higher growth and metabolic rates Lower cellular growth yield Higher physical and chemical stability of enzymes and organisms
Process	Increased stability Increased production rates Facilitated reactant activity and product recovery

The major bioconversion products of *C. thermocellum* are ethanol, acetic acid, lactic acid, CO_2, and hydrogen, along with smaller amounts of butyric and formic acids. The metabolic pathway proposed by Weimer and Zeikus is shown in Figure 2.36. However, a strain has been developed through a programme of adaptation by serial transfer in media containing increasing concentrations of ethanol, mutation and selection for ethanol tolerance, and reduced lactate production.

2.4.3 Fundamentals of Immunochemistry

For a long time, research-based process development of vaccines for the prevention of infectious diseases in humans is being carried out. In the past century, many vaccines have been developed against infectious diseases that disabled or killed millions of children and adults worldwide, such as smallpox and polio. Extensive experience-based information has now been accumulated in the development of vaccines for the prevention of viral respiratory tract diseases. In the literature, general considerations that influence the development of a vaccine strategy against viral pathogens have been taken into account.

First, it is essential to define the protective antigens of the pathogen and to determine whether antibodies or T-cells play a predominant role in protection. Second, the goal of immunization must be well defined. Third, the age of the subject in whom the vaccine is to be used must be defined/known since pre-existing passively transferred maternal antibody can interfere with the response of very young infants to a vaccine. The choice of vaccine type (for example, live or non-alive) and the dose or the route of administration of the vaccine may need to be modified to overcome the effects of passively acquired antibody. Last, the site to be protected by the vaccine needs to be properly defined.

At present, vaccines have been categorized as given in Figure 2.38. Newer advances in molecular biology/rDNA technology and immunology have created avenues and

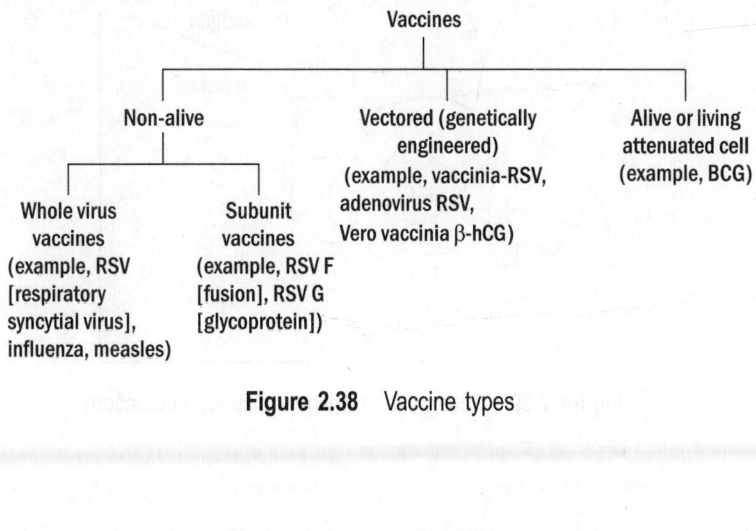

Figure 2.38 Vaccine types

opportunities for improving the safety, efficacy, and efficiency of existing vaccines and for enhancing the development of newer ones.

2.4.3.1 Vaccines

Vaccines are therapeutic agents (consisting of living attenuated strains of virus) that are derived (directly or indirectly) from pathogenic organisms or are effective because of an immunological reaction. Vaccines are most effective against viral types of infection. Immunological reactions occur when antigens are introduced into a body. Antigens are high molecular weight compounds (4000 onwards), generally produced from biological origin. When introduced into the internal organs, antigens change the metabolic balance to produce specific antagonistic compounds called antibodies.

Vaccination is a special term for immunization procedures in which an antigen (in this case referred to as vaccine) is given (either orally or internally) to the subject for establishing resistance to an infectious disease (that is, immunity in literal sense) by producing antibodies. The vaccine is a suspension of infectious agent(s) or some part of them. Thus, in reality, vaccination involves antibody–antigen reaction system or antibody-ligand reaction system.

Therefore, antibodies or antagonistic compounds can be defined as proteins produced by cells in the body's immune system in response to the presence of specific antigens. These organisms have unique traits that recognize the shape of particular composition on the surface of the antigens (Figure 2.39). They bind with only those antigens that have such surface compositions. This specificity makes antigens potentially useful for a number of applications on the diagnosis and treatment of various diseases.

Antigen is any substance that activates this immune response. Monoclonal antibodies are glycosylated proteins (other examples include vaccines, growth factors, and liver hormones). These proteins are produced by most feasible approach of animal cell culture

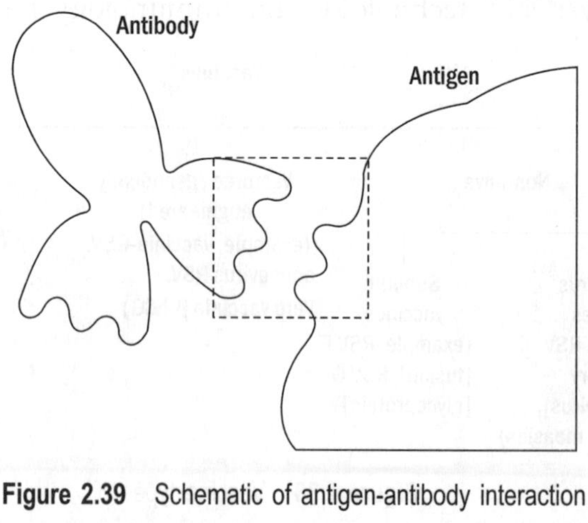

Figure 2.39 Schematic of antigen-antibody interaction

bioprocessing. Their production involves the creation of a hybrid cell, hybridoma, by combining an individual specific antibody-producing cell with a cell that will reproduce indefinitely. The hybridoma can, therefore, continually produce a single type of antibody. These are truly grown separately as a single type of antibody by overcoming all practical unfeasibilities.

2.4.3.2 *Valence and affinity of antibodies*

Ligands may be univalent (or monovalent) and multivalent in nature. A simple reaction between a univalent ligand (L) is shown in Equation 2.35.

$$Ab + L \underset{k_1'}{\overset{k_1}{\rightleftharpoons}} AbL \tag{2.35}$$

$$AbL + L \underset{k_2'}{\overset{k_2}{\rightleftharpoons}} AbL_2$$

$$\vdots \quad \vdots \quad \vdots$$

$$AbL_{n\text{-}1} + L \underset{k_n}{\overset{k_n}{\rightleftharpoons}} AbL_n$$

where k_n is the rate constant, and n is the number of ligand binding sites per antibody molecule or antibody valence.

This reaction is called intrinsic binding reaction if (1) the binding sites on antibody act independently, (2) the forward reaction rate constants are equivalent (that is, $k_1 = k_2 = \cdots = k_n$), and (3) the backward rate constants are also equivalent. The reaction is then expressed as shown in Equation 2.36.

$$S + L \underset{k'}{\overset{k}{\rightleftharpoons}} SL \tag{2.36}$$

where S is the special binding site on antibody and k and k' are the rate constants for association and dissociation.

The intrinsic association constant is given as shown in Equation 2.37.

$$k = \frac{k}{k'} = \frac{C_{SL}}{C_S \cdot C_L} \tag{2.37}$$

It is still not known where and by which mechanism the production of antibodies takes place. In fact, these may vary from one animal to another and with the site of application. Cell grouping involved in this reaction are thymus (T-cell) gland and the reticuloendothelial system, particularly lymph nodes, the momentum, spleen, and bone marrow; however, mobile cells are also known to play a part. The reticuloendothelial system includes a range of cells of similar morphological characteristics that appear in every organ. Lymphocytes and/or plasma cells derived from lymph nodes may also be involved. Human lymphocytes comprise about 30% of the total number of leucocytes in the body.

Antigens like toxins, bacteria, and foreign blood cells proliferate in the gut and release their modified toxins, which in turn are absorbed into the blood stream and induce the formation of a specific type of antibody.

2.4.3.3 Duration of immunity

The most puzzling problem in immunology is the duration of protection obtained either by being affected by a specific disease or by vaccination. The duration of immunity is variable for different diseases. For example, cold and influenza provide immunity for only four weeks, but measles and many serious viral attacks provide immunity for life.

In most cases, the antigen is completely destroyed within a few weeks. On the other hand, at the outset, the antibody reaches high levels of concentration that gradually declines over months (or years) to a low or even undetectable value. It is known that all antibodies are protein in nature and usually of fairly high molecular weight. As a group, they are more resistant to embryonic hydrolysis than other proteins. After immunization, this resistant quality seems to increase with time.

2.4.3.4 Antibodies in blood serum

There are two ways of ensuring that blood contains a required antibody.

1. ***Passive immunization:*** Some blood containing the required antibody is prepared in another animal or human, and the preformed antibody is injected into the patient. This is known as "passive immunization", and the material is known as antiserum.

   Passive immunization has limited success as a therapeutic agent because antiserum has a relatively short lifespan.

   Passive immunization is useful for dealing with emergency cases and can, if necessary, be combined with the second method to give temporary cover until the patient's own antibodies are developed. These preformed antibodies have a half-life of approximately not more than two weeks.

2. ***Active immunization:*** An antigen is injected or otherwise administered so that the patient produces a specific antibody. This is known as "active immunization" and forms the basis of most current treatments aimed at protecting patients from getting a specific disease.

Active immunization is generally not effective once the disease has attacked as it duplicates the mechanism already triggered by the disease agent, but there are exceptions to this.

The earliest example of active immunization was demonstrated by British physician Edward Jenner, where he used the exudates from cowpox lesions as a prophylactic against human smallpox.

Figure 2.40 illustrates the typical response to an antigen administered subcutaneously and intravenously in equal doses, 15 days apart. The titre is measured on separated blood serum by mutual precipitation of antigen and specific antibody. It is shown on a logarithmic scale as this demonstrates a tendency for exponential development following the second dose.

The points illustrated by the graph (Figure 2.40) and similar experiments are discussed below.

- Subcutaneous injection is usually more effective as primary dose, but there is negligible difference at the secondary stage. The reason for this can be explained as follows. The effect is initially concentrated on adjacent lymph nodes instead of being dispersed throughout the body. However, at later stages, the entire endothelial system is prepared to react quickly.
- Although initial titres of antibody are influenced by the size of the dose, the final levels are influenced to a limited extent. On a quantitative scale, it has been shown, for example, that three doses (50 μg each) of purified capsule polysaccharide from strains of Pneumococcus produced a total of 50 g active antibody in human patients over a period of four years.

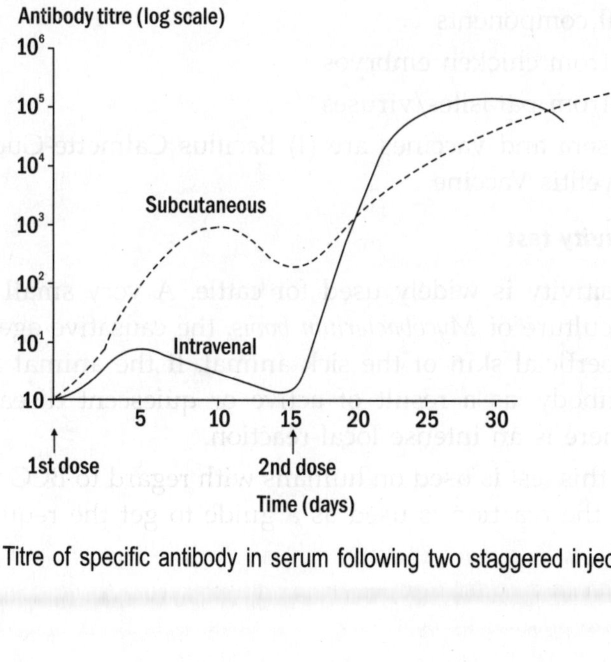

Figure 2.40 Titre of specific antibody in serum following two staggered injections of antigen

- The rate of fall in the titre of the antibody primarily depends on the antigen. There is a tendency for particulate ones, including active or inactivated bacteria, virus or rickettsiac, to be retained longer than those derived from small molecules. Toxoids and preformed antibodies decline quite rapidly according to bimolecular reactions.

$$\frac{dx}{dt} = k(a - x)^n \tag{2.38}$$

where a is the maximum titre of antibody and n is the power factor, varying from 1.2 to 3.5.

It should be noted that many of these antigens are complex in nature. Thus, even a simple virus contains at least a protein and a nucleic acid. However, the antibody always appears to be a γ-globulin, sometimes conjugated with lipid. Regardless of the antigen and route, humans and rabbits produce antibodies of molecular weight approximately 160 000 amu, whereas horses produce antibodies of molecular weight approximately 900 000 amu. Cows and pigs give variable results.

2.4.3.5 Commercial sera and vaccines

There are many varieties of sera and vaccines. These can be broadly classified under eight headings, either by type or by manufacturing method.

1. Preferred antibody
2. Whole animal organs
3. Dermal culture
4. Living but attenuated organisms
5. Toxoids
6. Insulated bacterial components
7. Vaccines derived from chicken embryos
8. Vaccines derived from parasites/viruses

Two examples of sera and vaccines are (1) Bacillus Calmette-Guerin (BCG) Vaccine and (2) Oral Poliomyelitis Vaccine.

2.4.3.6 Hypersensitivity test

A test for hypersensitivity is widely used for cattle. A very small amount of extract from an attenuated culture of *Mycobacterium bovis,* the causative agent of bovine TB, is injected into the superficial skin of the sick animal. If the animal already has a high level of specific antibody, as a result of active or quiescent disease or has recently recovered from it, there is an intense local reaction.

A modification of this test is used on humans with regard to BCG vaccination against TB. The intensity of the reaction is used as a guide to get the required dose. There is a high incidence of severe reaction to certain vaccines, including anthrax. This has led to the suggestion that the first dose of the vaccine should be accompanied by a tattoo code as a warning against further administration at a later date.

2.4.3.7 Viruses

Viruses, or phages, are the smallest microbes. Viruses cannot synthesize protein because they do not have ribosomes, which are the structures needed for protein synthesis. They are a unique class of sub-cellular infectious parasites and can multiply only inside the cells of a living host. They are, therefore, obligate intracellular parasites found in animals, plants, insects, fungi, algae or bacteria. Viruses do not contain water and have little or no synthetic or metabolic activity in themselves. Their nomenclature is based on the name of the host in which they multiply. For example, a virus growing on bacteria is called bacteriophage and one growing on an actinomycete is called actinophage. Growth and multiplication of viruses take place intracellularly, where the virus directs the host cell to synthesize new virus population. This frequently results in damage and death of host cells.

Viruses use the energy sources, biosynthetic machinery, and even much of the replicative functions of the host cells. They provide only the genetic information for their own replication through reconstitution of entire genome by transfectional recombinations of overlapping fragments. The genetic material of viruses may be RNA or DNA. In the host, their genes code for the following.

- Proteins that replicate the viral genome
- Properties that have structural or catalytic roles in assembling new infectious particles
- Regulatory proteins that direct the overall process of viral reproduction

2.4.3.7.1 Examples of viruses

Many viruses having different ultra-microstructural geometric figures are known. Bacteriophage 12, MS2, poliovirus, and λ phage are known to possess icosahedral (20-sided) geometric figures. Phage M13 and tobacco mosaic virus (TMV) are known to have filamentous structure. Vaccinia, polioma, corynebacterium phage β, T4 phage, and P2 phages are also known to have ultra-microstructural geometric figures. Many other plant and animal viruses consist of a single molecule of RNA associated with protein subunits arranged in a polyhedron showing cubic symmetry. Thus, viruses have unique naturally acquired design engineering (NADE) structures.

The myxoviruses show helical symmetry, with the nucleoprotein helix enclosed by a lipoprotein sheath.

Viruses vary in diameter, from 10 μm (for example, foot and mouth virus) to 300 μm (for example, vaccinia virus). Some viruses are shaped like a tadpole, with a hexagonal head (90 × 60 μm) and a tail (100 μm × 25 μm). A few phages have no tail and many are highly specific to the strain of host they attack.

The nucleic acid in a virus particle is surrounded by a protein coat called a capsid. The capsid consists of protein subunits called capsomers. A complete virus particle is called a virion. It may be covered by an envelope. A typical geometric NADE structure of phage P2 of *E. coli* is shown in Figure 2.41. Another virus, vaccinia, of animal cell Vero has NADE as shown in Figure 2.42. This virus is also an essential fundamental agent for the transfectional fermentation process. Its electron microscopic structural design appearance also gives some principal dimensions. Viruses also play an important role in protein kinesis engineering (PKE).

2.4.3.7.2 Industrial and commercial importance of viruses

Phages can produce industrially important materials/chemicals by transfectional fermentation biotechnology (TFB). The cultivation of animal viruses to test antiviral drugs and for vaccine production is an important undertaking all over the world. Vaccines are being developed against viruses such as influenza, polio, herpes, AIDS, and HIV using TFB/tDNA (recombinant DNA) techniques.

2.4.3.7.3 Disadvantages of viruses in industrial process biotechnology

Viruses may cause or jeopardize the productivity of various industrial and commercial bioproducts. A few examples include (1) *Streptococci* cultures used in cheese making,

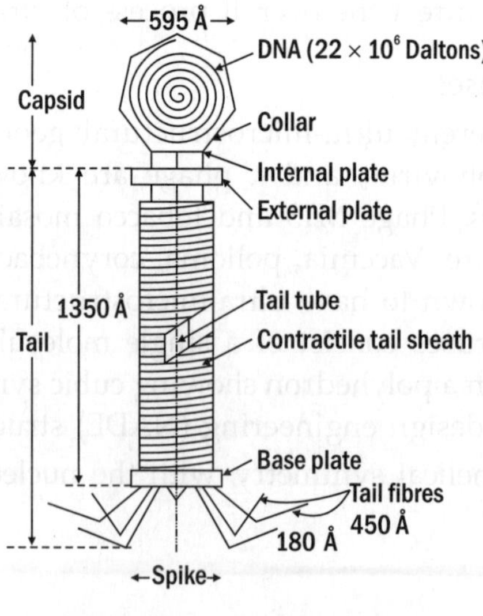

Figure 2.41 Phage P2 of *E. coli*

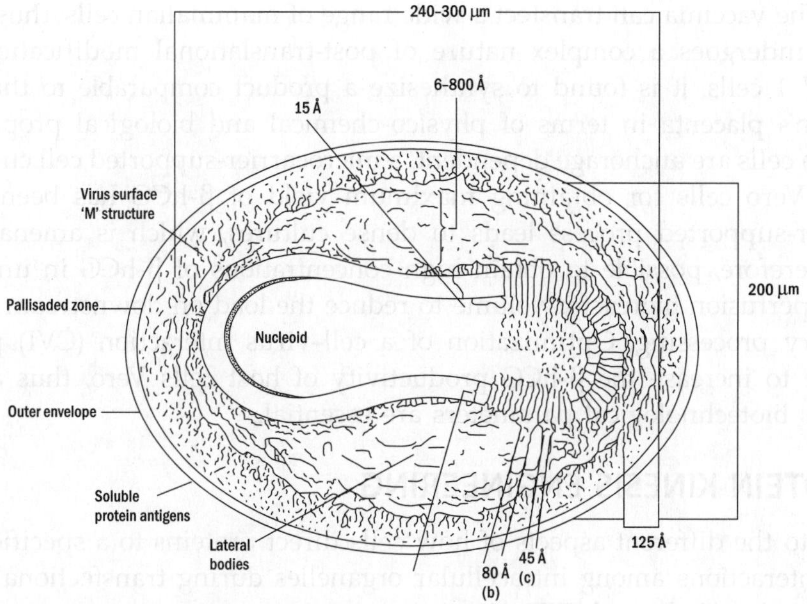

Figure 2.42 The structure of vaccinia virus, showing various layers identifiable with their suggested structure and principal dimensions

(2) *Clostridia* sp. for acetone butanol fermentation, and (3) *Pseudomonas* sp. in organic acid production.

2.4.3.7.4 Viral vaccine production

In biomedical engineering science, prevention of diseases and control of life processes are of important concern. For a long time, research-based process biotechnology has carried out the development of vaccines for regulation and control of birth in humans and prevention of infectious diseases. Newer vaccines are being invented and developed for the same purposes. In recent years, in countries that are facing the problem of population explosion, one of the major research emphases has been to develop birth control vaccines. India is one such country, and researches by Indian scientists have revealed that human chorionic gonadotropin (hCG) is a birth control hormone. It is extensively used as the promoter of both male and female fertility.

The birth control vaccine currently in clinical trials employs β-hCG as a birth control vaccine. β-hCG belongs to a family of glycoprotein hormones. The production of urinary β-hCG of pregnant woman involves processes that are expensive and time consuming. Moreover, each step of its production involves some loss of material, resulting in poor yield. For enhancing the production, scientists have utilized a vaccinia recombinant construct, in which the gene for β-hCG is cloned under vaccinia P 7.5

promoter. The vaccinia can transfect a wide range of mammalian cells; thus the protein expressed undergoes a complex nature of post-translational modifications through PKE. In CV 1 cells, it is found to synthesize a product comparable to that produced in a woman's placenta in terms of physico-chemical and biological properties. Since mammalian cells are anchorage dependent, a microcarrier-supported cell culture system employing Vero cells for obtaining maximum yield of β-hCG has been developed. Microcarrier-supported process leads to dense cultures, which is amenable to scale up. It is, therefore, possible to obtain high concentrations of β-hCG in unit upstream processing perfusion bioreactor volume to reduce the load on downstream purification and recovery processing. Optimization of a cell–virus interaction (CVI) process is a prerequisite to increase the β-hCG productivity of host cells Vero, thus ascertaining that process biotechnological parameters are essential.

2.5 PROTEIN KINESIS ENGINEERING

PKE refers to the different aspects of how cells direct proteins to a specific target and modulate interactions among intracellular organelles during transfectional processing to deliver product(s) through TFB.

2.5.1 PKE Relevant to β-hCG

Protein kinesis involves sizing, shaping, and modulation of protein molecules, using transfectional tools, and designing molecules *in vivo* by non-classical engineering termed PKE, which plays a great role in biomedical sciences.

2.5.2 Transfectional Fermentation

Transfectional fermentation is the fermentation of a medium by cellular organisms that are genetically engineered by the introduction of a viral DNA for molecular blending into the host organism.

The production of β-hCG is a physico-heterogenic Vero–vaccinia transfectional fermentation example.

2.5.3 PKE-participated TFB

PKE promises to provide birth control that cannot be done by conventional vaccines. In new vaccines, β-hCG production that control birth vaccinia virus is not employed, but some of its proteins transfected in host Vero cells, eliminating the risk of contaminating the vaccine with live virus. Thus, recombinant TFB is important in β-hCG production.

2.5.4 Vero Cell Origin and Characters

Vero cell, which is used for producing β-hCG by PKE, has been obtained from its origin as shown in Table 2.8. The characteristics of a typical Vero cell are also given in Table 2.8.

2.5.5 Typical Macromolecular Components in Vero Cell and *E. coli* cells

Vero cell is eukaryotic type and *E. coli* is prokaryotic type. Their typical macromolecular components vary quantitatively as given in Table 2.9.

2.5.6 Vero Cell Line

Vero cell line is an established diploid cell line. The attachment of cells to microcarrier surface is an affinity-based multi-step adhesion process, which occurs on specific surfaces. This cell line has high cell-substrate interaction. In producing viral vaccine using diploid cell line, high cell affinity cell-microcarrier system is preferred. Vero cell line serves as a suitable cell for use as a substrate for viral vaccine.

Vero cells are robust cells in terms of their physical dimensions as given in Table 2.10.

2.5.7 Vaccinia Virus

Vaccinia virus is a very important fundamental agent for transfectional fermentation process. The electron microscopic structural appearance is shown in Figure 2.42.

Table 2.8 Origin and characteristics of a typical Vero cell

Origin	Characters Physical	Chemical	Biological
Kidney of a normal adult African green monkey	Anchorage-dependent fibroblast (solid–solid interaction systems)	Have complex requirements for growth	Support propagation of numerous viruses, including vaccinia

Source Mukhopadhyay (1995)

Table 2.9 Comparative properties of eukaryotic and prokaryotic cells

Components	Vero (animal cell)	*E. coli* (microbial cell)
Total DNA	~10–18 pg	0.017 pg
Total RNA	~20 pg	0.10 pg
Total protein	200–300 pg (5×10^9 molecules of average molecular weight 40 000)	0.20 pg (3×10^6 molecules of average molecular weight 40 000)
Cytoplasmic ribosome	~10^6	3×10^4
Cytoplasmic tRNA	~10^7	4×10^5
Cytoplasmic mRNA	~10^5	4×10^3
Total dry weight	~300–400 pg	0.4 pg

DNA – deoxyribonucleic acid; RNA –ribonucleic acid; pg – picogram; tRNA – transfer RNA; mRNA – messenger RNA

Table 2.10 Dimensional parameters of Vero cell

Parameter	Value
Diameter (μ)	12.0
Surface (S, μm^2)	430.0
Volume (V, μm^3)	6600.0
S/V (μ/m)	0.7

2.5.7.1 Morphogenesis

Vaccinia has two morphologically distinguished fundamental particle forms like mulberry (M) and capsule (C). The proposed structure for the threads of the M form (Figure 2.43) of virus represents the points at which the strands forming the double helix turn into the axis of the electron beam and may show as a series of dots. Structurally, vaccinia contains phospholipids, neutral fat, carbohydrate, flavin, and biotin.

The M form is similar to bacteria in proportion. It is a threaded structure DNA that comprises less than 5% of the dry weight of the vaccinia and over 90% consists of protein. It plays an important role in transfection.

The conversion of vaccinia virus from the M form to the C form is effected by fat solvents. Such a conversion leads to impermeable layer to PTA, impermeability due to incorporation of lipid in the layer, and confer resistance to the action of photolytic enzymes. Figure 2.44 shows the infection and life cycle of vaccinia virus.

A schematic of Vero cell processing and engineering concerns is given in Figure 2.45. The materials and methods used are listed in Tables 2.11 and 2.12.

2.5.8 Fundamentals of β-hCG Protein Kinesis Engineering

Based on earlier reports, the upstream design of β-hCG hormonal protein kinesis involves the base items given in Table 2.13. The bioprocessing base parameters used are given in Table 2.14. These indicate the need of a bioreactor for PKE information, which was computed in a stirred tank bioreactor (STBR) having design base items as

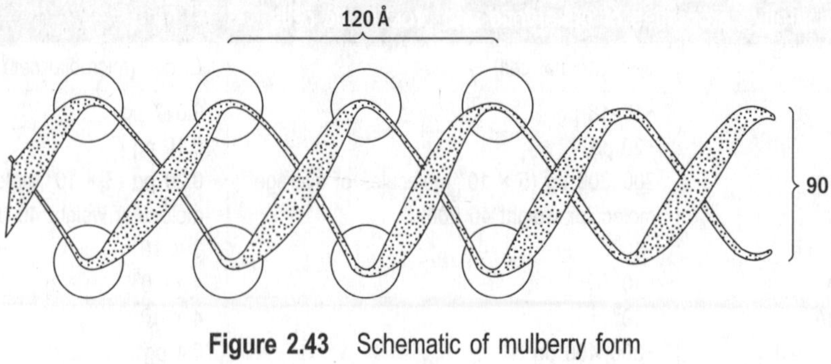

Figure 2.43 Schematic of mulberry form

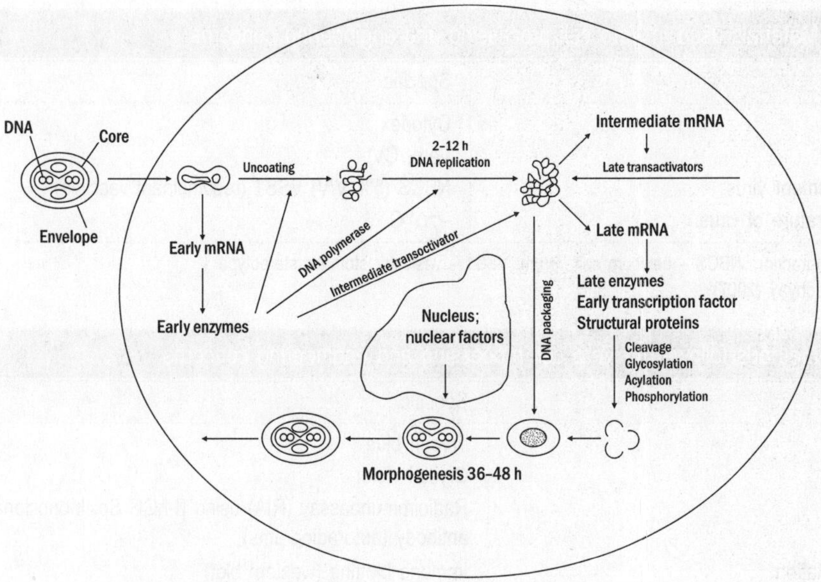

Figure 2.44 Infection and life cycle of vaccinia virus
Source Mukhopadhyay (1995)

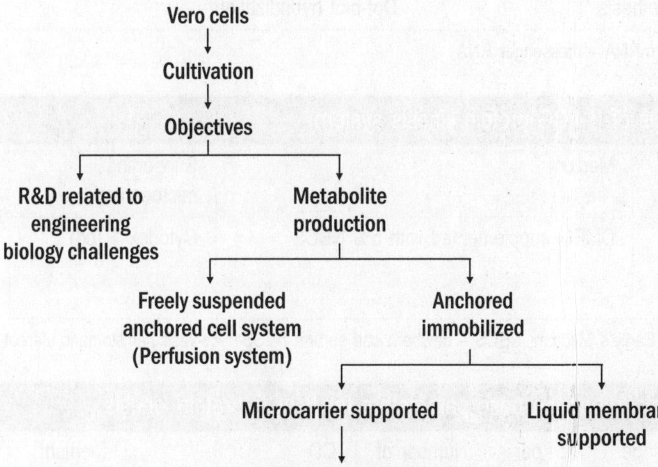

Figure 2.45 Vero cell processing and engineering concerns

Table 2.11 Materials used in Vero cell processing and engineering concerns

Material	Specific
Microcarrier	Cytodex
Cells	Vero, CVI
Medium supplement virus	NBCS (5% V/V) VSS1 (recombinant vaccinia)
Storage temperature of virus	−70°C

CVI – cell–virus interaction; NBCS – newborn calf serum; VSS1 – vesicular stomatic stereotype 1
Source Mukhopadhyay (2007)

Table 2.12 Methods followed in Vero cell processing and engineering concerns

Parameter	Specific
Viable cell	Trypan blue
Total cells	Crystal violet
β-hCG	Radioimmunoassay (RIA) using β-hCG Sp. Monoclonal antibody (autoradiograms)
β-hCG glycosylation	Immuno blotting (western blot)
Infection level	MOI
Metabolic labelling of β-hCG in post-infected cells	Pulse chase
β-hCG specific mRNA synthesis	Dot-blot hybridization

MOI – multiplicity of infection; mRNA – messenger RNA

Table 2.13 Base items of β-hCG protein kinesis system

Item	Cell line	Medium	Supporting microcarrier	Transfectional virus
System	Vero	DMEM supplemented with 5% NBCS	Cytodex 1 (cd)	r-VSS1 expressing β-hCG hormonal protein

DMEM – Dulbecco's Modified Eagle's Medium; NBCS – newborn calf serum; r-VSS1 – r-vesicular stomatic stereotype 1

Table 2.14 Items of bioprocessing base parameters for β-hCG protein kinesis

Item	Cell diameter (μm)	Passage number	MOI per cells (pfu)	Number of cells/bead	CD		Medium		
					Bead diameter (μm)	Density in medium (g/L)	ρ (kg/m^3)	η (kg/m/s)	γ (m^2/s)
Value	50	90–125	2.5	7	105	5	10^3	1.5×10^{-3}	1.5×10^6

CD – cytodex; ρ – density; η – viscosity; γ – kinematics viscosity; MOI – multiplicity of infection; pfu – plaque forming unit

given in Table 2.15. In addition, the ratio of vessel diameter to height diameter was 3, and the ratio of impeller diameter to vessel diameter was 1/3.

2.5.9 PKE Process Operation Factors

2.5.9.1 *Hydrodynamic effect*

Taking into consideration the operation in an STBR, one may assume that when shear flow occurs over a Vero cell anchored to the cytodex surface, the microscopic flow field is likely to be affected by the protrusion of the cell system and influence the kinesis engineering operation. This hydrodynamic effect on the transected Vero cell system may be pronounced. This is because the cell system experiences a net torque created by flow around its circumstances, which is outered by the adhesive force between the microcarrier and cell system surface. Thus, the process needed evaluation and checking of the possibility of hydrodynamic effects and hydrodynamic cell death. The hydrodynamic effect analysis was carried out as per literature. The computed value of hydrodynamic death of cells ($5.57 \times 10^{-9}\ h^{-1}$) indicated its negligibility. Therefore, scale-up considerations with the given parameters (Tables 2.13–2.16) are safe to be used in the β-hCG PKE bioreactor (STBR).

In addition, cell tech parameters of bioprocessing, including oxygen supply requirements, productivity, and yield considering intracellular and extracellular products of the kinesis reaction, were required to be examined in terms of development of its process biotechnology.

2.5.9.2 *Physiological status of cell*

Innoculum physiological suitability for PK by Vero cells growing on microcarrier is essential. Cells growing in log phase of physiological state are found to adhere and grow rapidly on microcarriers than cells of other phases. From earlier reports, inoculation has to be done only once by using Vero cells in its exponential phase of growth. Also, the time gap for the recovery process of Vero cells has to be maintained for 1–10 min. As the final yield of β-hCG is a function of multiplicity of infection (MOI), it should provide the most suitable protein kinesis condition (found to be at 800 ng/mg cells) at a vaccine virus plaque forming unit (pfu) of 25–50 per cell. The yield of this protein hormone is reduced on either side of this range of MOI.

Table 2.15 Design base items of the stirred tank bioreactor (14 L, working volume 10 L)

Item	STBR diameter (m)	Agitator/impeller type	Diameter (m)	Width (m)	Tip speed (m/s)	Speed (rpm)
Value/type	0.118	R turbine blade	0.06	0.0	0.126	40

STBR – stirred tank bioreactor

Table 2.16 Hydrodynamic effect analysis

Parameter	Value	Maximum value	Remark
Integrated shear factor	2.1 (per second)	21 (per second)	Acceptable
Time-averaged shear rate	2.04 (per second)	2.5 (per second)	Acceptable
Maximum time-averaged shear rate (in trailing vortex)	5.02 (per second)	45 (per second)	Acceptable
Average shear stress	0.0068 N/m^2	0.65 N/m^2	Acceptable
τ_{rc} (Reynolds shear stress)	1.7 N/m^2	0.65 N/m^2	Acceptable because the exposure time and probability are very low
Boundary layer shear stress (on impeller blade)	1.03 N/m^2	0.65 N/m^2	Acceptable because the exposure time and probability are very low
Kolmogorov length of smallest eddy	618.9 μm	>>cell diameter	No hydrodynamic effect on cell
τ_{max} (due to eddy energy dissipation)	0.031 N/m^2	0.65 N/m^2	Acceptable
τ_{max} (due to highest localized energy dissipation)	3.88×10^{-5} N/m^2	0.65 N/m^2	Acceptable
q_1	1.74×10^{-7} (per hour)	Specific growth rate (0.031 [per hour])	Effect can be neglected
q_2C_m	5.57×10^{-9} (per hour)		

q_1 – death rate due to eddy cell interaction; q_2 – collision death rate; C_m – microcarrier concentration; τ_{max} – maximum stress due to highest localized energy dissipation

2.5.9.3 *Substrate-transient response*

In transfected Vero–vaccinia system, the effect of glucose and glutamine (substrates) on metabolic hormonal protein production can be examined by the pulse-fed response analysis of transfected Vero cells behaviour. From the results (Figure 2.46), it was computed that in this PK, the specific ATP level in Vero cells infected with vesicular stomatic stereotype 1 (VSS1) was increased by 70% after 30–35 h of infection as compared to uninfected cells under identical conditions. Also, during this period, the measured ratio of ATP/adenosine diphosphate (ADP) declined from 10 to 2.5, indicating an increase in ATP utilization despite its comparatively greater generation. Glucose and glutamine substrates could support the synthesis of β-hCG and virus (VSS1); however, the yield is higher in the presence of both carbon sources.

2.5.9.4 *Oxygenation*

As published in earlier reports, oxygenation in vaccinia-transfected Vero cells showed that transfected cells consumed more oxygen in β-hCG hormonal protein kinesis than uninfected cells. Increase in ATP yield and consumption of oxygen appeared to be due to higher conversion of carbon through the TCA cycle and subsequent oxidation via oxidative phosphorylation.

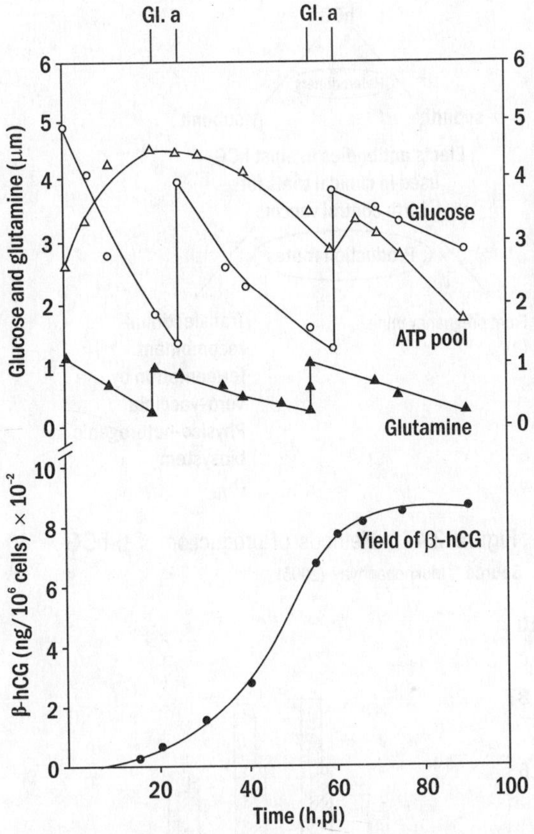

Figure 2.46 Synthesis of β-hCG hormonal protein in the presence of pulses of glucose and glutamine as carbon sources

Obtaining α-subunit hCG is a laborious, time-consuming, and more expensive process than obtaining β-subunit hCG (Figure 2.47).

Figure 2.48 indicates that the optimum level of infection in the Vero–vaccinia transfectional expression system occurs at 25 pfu/cell.

Figures 2.49(a)–(c) show the photomicrographs of Vero cells at different post-infection time (100 ×).

Figure 2.50 shows the β-hCG expression vector development with vaccinia, while Figure 2.51 shows the construction of recombinant plasmid (PSS1) and Figure 2.52 shows the TFB process flow chart of β-hCG production.

Table 2.17 shows the cell tech parameters of Vero–vaccinia transfected culture (β-hCG production).

Figure 2.53 shows the downstream processing for β-hCG recovery and purification.

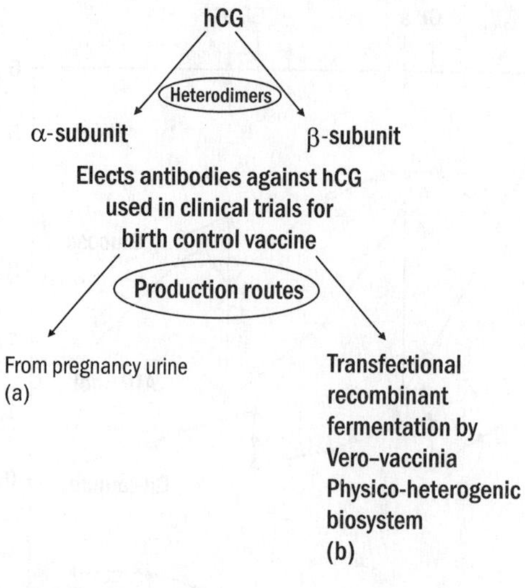

Figure 2.47 Methods of production of β-hCG
Source Mukhopadhyay (2005)

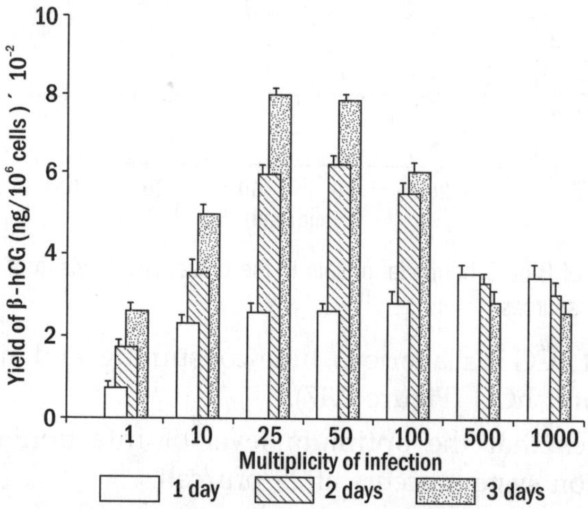

Figure 2.48 Yield of mean ± sem based on cells at different multiplicity of infection

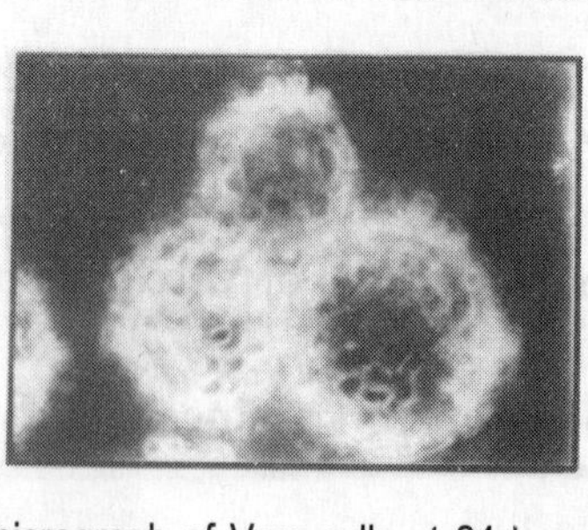

Figure 2.49(a) Photomicrograph of Vero cells at 24 h post-infection time (100 ×)

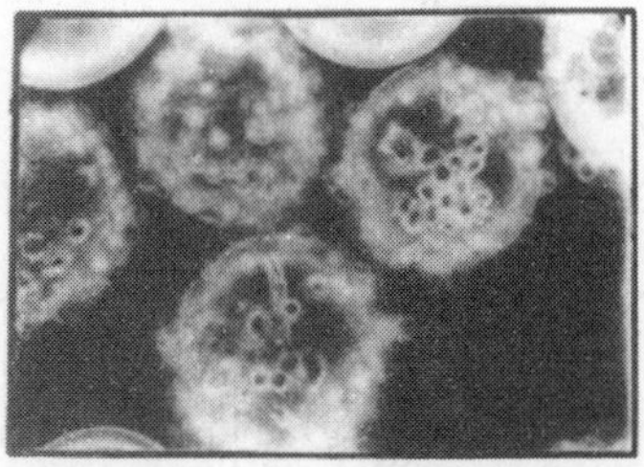

Figure 2.49(b) Photomicrograph of Vero cells at 48 h post-infection time (100 ×)

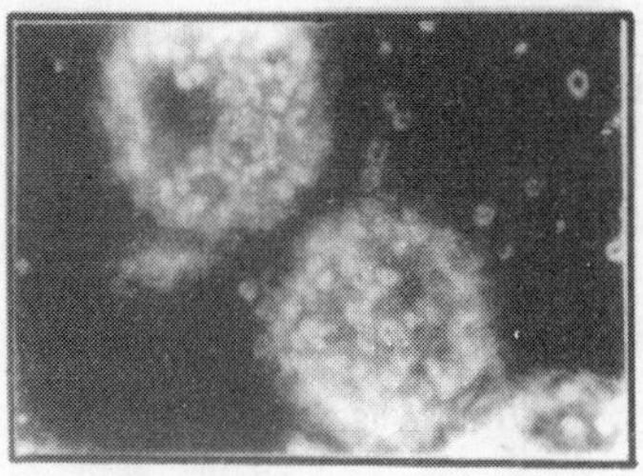

Figure 2.49(c) Photomicrograph of Vero cells at 72 h post-infection time (100 ×)

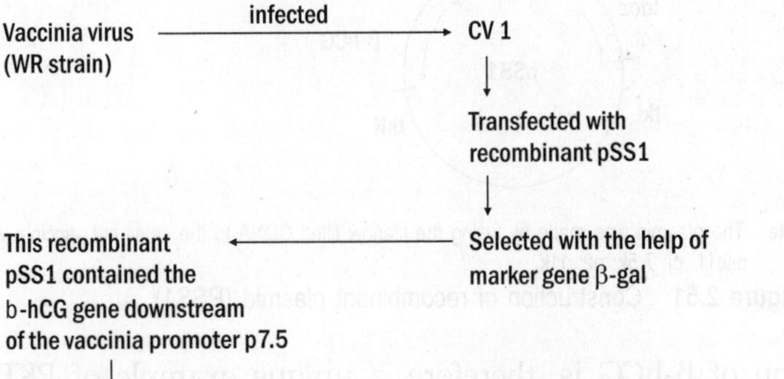

Figure 2.50 β-hCG expression vector development with vaccinia
Source Mukhopadhyay (2005)

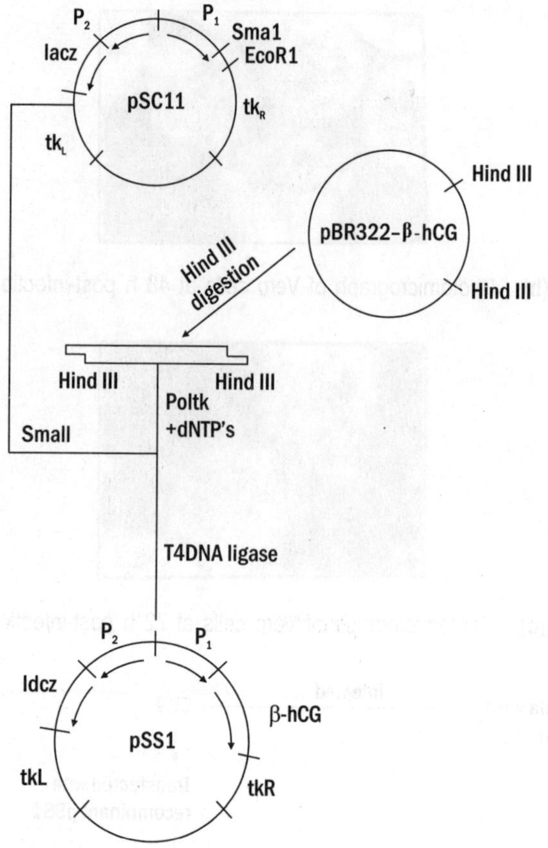

Note The plasmid was made by fusing the klenow filled CDNA to the small cut vaccinia vector psc11. pj: 7.5k: p2: 11k.

Figure 2.51 Construction of recombinant plasmid (PSS1)

The production of β-hCG is, therefore, a unique example of PKE. In the process, necessary biotechnological design parameters have been described. Also, the supply and uptake rates of oxygen by transfected Vero cell system should be assessed systematically. This will provide useful information for design of recombinant Vero cell culture system, especially when it is propagated on a microcarrier.

2.6 TUTORIAL PROBLEM

Antibodies attached to porous polyacrylamide beads are used to purify precious biochemicals (antigens) from biological fluids such as blood serum or cell culture broths. When the broth is passed over a column containing the beads, the antigen binds with the antibody, while all other unrecognized molecules pass through undisturbed.

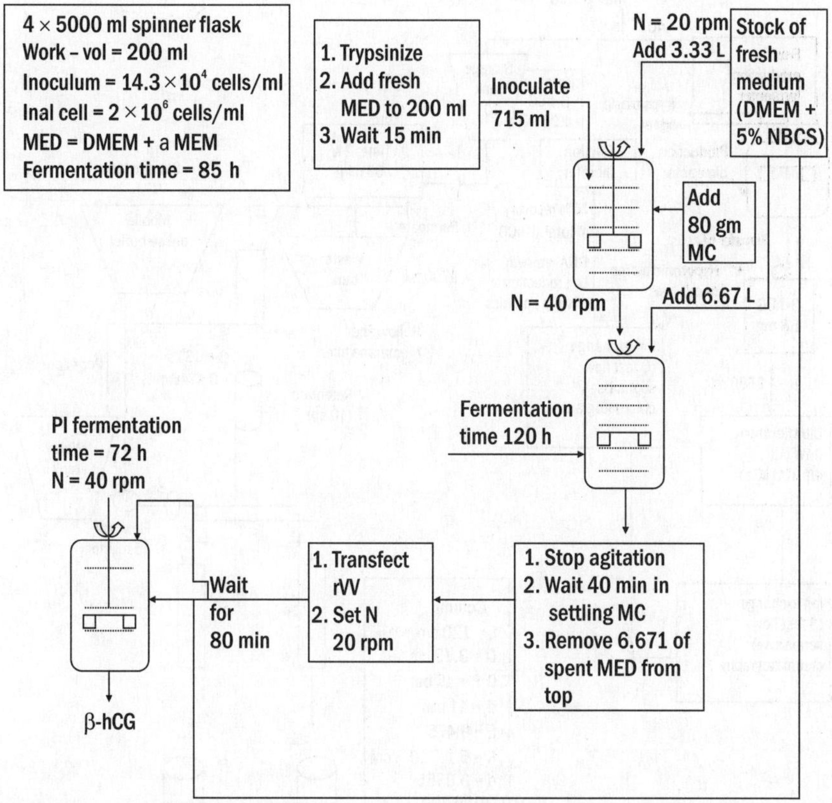

Figure 2.52 Transfectional fermentation biotechnology process flow chart of β-hCG production
Source Ghosh and Mukhopadhyay (2000)

Table 2.17 Cell tech parameters of Vero–vaccinia transfected culture (β-hCG production)

Parameter	Value
Doubling time (h)	22–37
Cell density (cells/ml)	6.0×10^6
[s]SP. glucose uptake rate (μ mol/10^6 cells/h)	110–168
	0.40
Product concentration (g/L)	4.8×10^{-3}
Productivity (μg/L/h)	66.0
Approximate medium cost	
$/L	12.5
₹/L	390.0

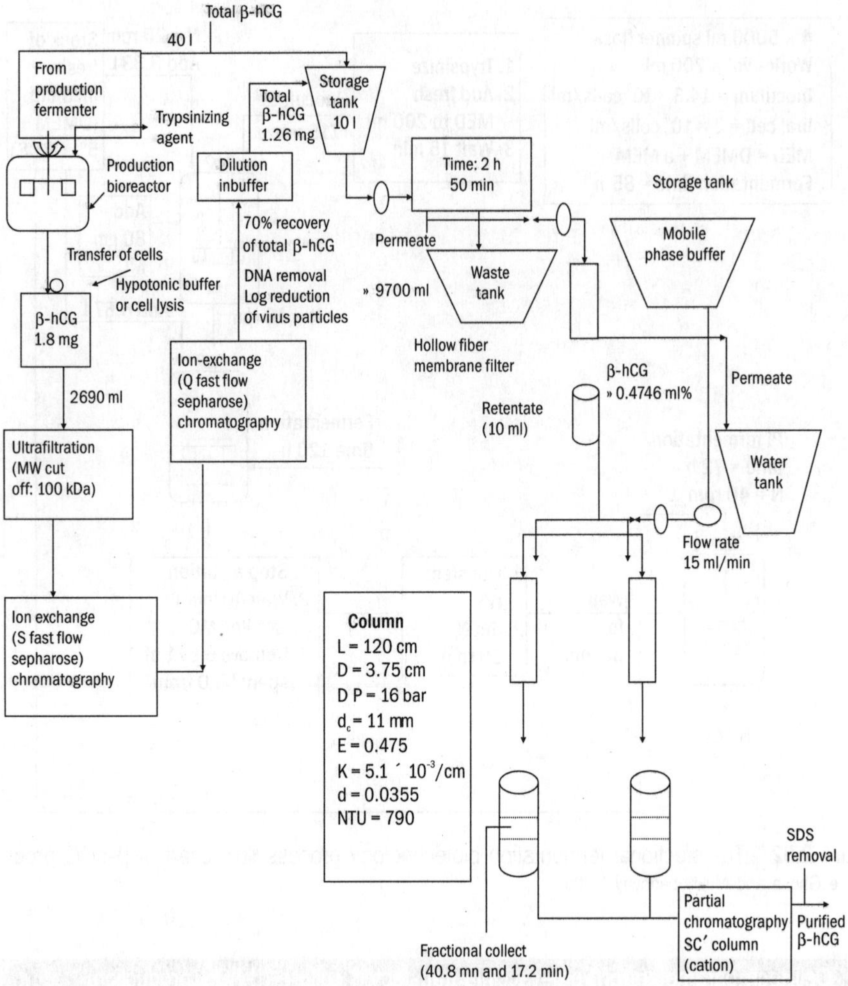

Figure 2.53 Downstream processing for β-hCG recovery and purification
Source Ghosh and Mukhopadhyay (2000)

The retention of antigens on the column is proportional to the strength of binding, as described in Equation 2.39 for the volume of fluid required to displace the antigen from the column.

$$V_e = \{\varepsilon + (1 - \varepsilon)\, K\}\, V \tag{2.39}$$

where V_e is the elution volume, ε is the column void fraction, $(1 - \varepsilon)$ is the volume fraction of the column taken up by the beads, V is the total volume of the column, and K is the apparent equilibrium constant for adsorption.

$$V_e = \left.\frac{C_{\text{surface}}}{C_{\text{bulk fluid}}}\right|_{\text{equilibrium}} \tag{2.40}$$

$$V_e = \left.\frac{\text{Bound antigen}}{\text{Antigen in bulk fluid}}\right|_{\text{equilibrium}} \tag{2.41}$$

1. If the interaction between the antibody and the antigen is described as shown in Equation 2.42, then

$$Ab + Ag \underset{k_A'}{\overset{k_A}{\rightleftharpoons}} Ab \cdot Ag \tag{2.42}$$

Derive the expression for the solution volume of the antigen in terms of antigen concentration in the bulk liquid and the maximum capacity for antigen binding, $C_{s, \max}$.

2. In the cell culture broth, there is a molecule (I) that competes with antigen for antibody-binding sites. This inhibitor interacts with the antibody in a manner similar to the antigen and inhibits binding between antibody and antigen.

$$Ab + I \underset{k_I'}{\overset{k_I}{\rightleftharpoons}} Ab \cdot I \tag{2.43}$$

$$Ab + Ag \underset{k'_A}{\overset{k_A}{\rightleftharpoons}} Ab \cdot Ag \tag{2.44}$$

Derive an expression for the antigen elution time as a function of the inhibitor concentration.

REFERENCES

Binzer J C. 1997. ***Drinking water filter media separation technology.*** pp. 28–31 [Paper 16 (97E)]

Cheremisinoff P N. 1993. ***Water Management and Supply.*** Englewood Cliffs, New Jersey: Prentice Hall Inc.

Ghose T K. 1998. ***Environmental and Biotechnology.*** Second A N Roy Memorial Oration Lecture, NRC, I.I.Ch.E. IIT Delhi

Ghosh A and Mukhopadhyay S N. 2000. ***Indian Chemical Engineering Congress Proceedings,*** pp. 133–136, Calcutta: Indian Chemical Engineering Association

Mukhopadhyay S N. 1995. **Water resources, pollution, and purification for environment and hazard protection**, pp. 189–200. [Proceedings of the *Environmental Development of Process Industries*, organized by the Indian National Academy of Engineering, New Delhi]

Mukhopadhyay S N (ed.). 2005. ***Advanced Process Biotechnology.*** New Delhi: Viva Books. 382 pp.

Mukhopadhyay S N and Baisya R K. 1997. **Role of water composition in relation to food biotechnology**. In *Proceedings of the Fourth APBioCheC '97*, vol. 2, edited by Z Y Shen, F Oyang, J T Yu, and Z A Cao. Beijing: Tsinghua University Press. pp. 1354–1357

Rich L G. 1973. ***Environmental System Engineering.*** Tokyo: McGraw Hill-Kogakusha Ltd.

BIBLIOGRAPHY

Alberts B, Bray D, Lewis J, Raff M, Roberts K, Watson J D. 1994. ***Molecular Biology of the Cell,*** 3rd edn. New York: Garland Publishing Inc. 26 pp.

Bailey J E and Ollis D F. 1986. ***Biochemical Engineering Fundamentals,*** 2nd edition. New York: McGraw-Hill Book Company. 918 pp.

Clark D P and Pazdernik N J. 2009. ***Biotechnology applying the Genetic Revolution.*** Amsterdam: Academic Press

Day E D. 1990. ***Advanced Immunochemistry,*** 2nd edn. New York: Wiley-Liss. 296 pp.

Garcia L E. 2005. **Water pricing: an outsider's perspective.** *Water Resources Development* **21**(1): 9–17

Ho C S and Wang D I C. 1991. ***Animal Cell Bioreactor.*** Boston: Butterworth-Heinemann.

Kantawala D. 1998. **Abatement of water pollution**. In *Chemical Technology for Better Environment*, edited by T K Roy. New Delhi: INAE (Indian National Academy of Engineering). pp. 76–115

Mukhopadhyay S N. 2007. ***Experimental Process Biotechnology Protocols.*** New Delhi: Viva Books. 214 pp.

Mukhopadhyay S N. 1998. **The dimension of biotechnology in transportation system studies**. In *Proceedings of the International Conference on Transportation Systems Studies*, edited by P S Satsangi and A L Agarwal. New Delhi: TMH Publishing Co. Ltd. pp. 350–354

Mukhopadhyay S N and Das D K. 1994. ***Oxygen Responses, Reactivities, and Measurements in Biosystems.*** Boca Raton, Florida: CRC Press

Mukhopadhyay S N and Mukhopadhyay A. 1998. ***Proceedings of the APBioChEC '97,*** vol. 1, edited by Z Y Shen, F Ougyang, J T Yu, and Z A Cao, Beijing: Tsinghua University Press. pp. 124–129.

Mukhopadhyay S N and Shrivastava A. 1997. **Aspects of cellulose: synthesis, conversion, fabric bioprocessing, and applications.** In *Indian Institute of Chemical Engineers Proceedings of the Golden Jubilee Congress*, vol. 1, edited by S Nath. New Delhi: Replica Press. pp. 209–222

Reed B M (ed). 2008. ***Plant Cryopreservation: a practical guide.*** New York: Springer. pp. 15–17

Talwar G P, Singh O, Pal R, Chatterjee N, Sahai P, Dhall K, Kaur J, Das S K, Suri S, Buckshee K. 1994. **A vaccine that prevents pregnancy in women.** *Proceedings of the National Academy of Sciences* **91** (18): 8532–8536

3

Cell Biology and Microbiology

3.1 DEFINITION OF A BIOCELL

A living cell is a chemical-manufacturing machine with soul imbibed in it. A biocell is a fundamental unit of life (definition of life is given in Chapter 1). It is a super micro-bioreactor, having a unique design that originated under the influence of nature. Biocells are, therefore, naturally acquired design engineering (NADE) materials, and non-classical engineering principles and tools can redesign/modify them. Redesigned cells are, therefore, recombinatorially acquired design engineering (RADE) materials. Biocell extracts are enzymes, fission-promoting metabolites, hormones, plasmids, immunobiologicals, cell organelles, and so on.

3.2 BIOCELL TYPES, POSITIONING, CHARACTERISTICS, AND IDENTIFICATION TESTS

3.2.1 Introduction

Till today, a majority of bioprocesses use living organisms/biocells to produce materials of commercial importance. In recent years, attempts have been made to use animal and plant cells in the production of useful materials for research, development, and other purposes.

The main microbial cells associated with industrial biotechnological processes are bacteria and fungi. Microscopic forms of fungi are referred to as yeasts and moulds to distinguish them from macroscopic or higher fungi such as mushrooms. The metabolic and economic capacities of microbes in terms of bioprocessing requirements, scale-up, control for industrial production of metabolites, and their use are now well known. Presently, the number of species of microbes is nearly 50000. Among them, nearly 2000 are bacteria and 48 000 are fungi, including yeasts and moulds (Salle 1961; Buchman and Gibbons 1974; Lodder 1970; Ainsworth 1971).

From early days of investigations on microbial cells, bacteria and fungi are regarded as chlorophyll-lacking plants, and their systematic positions have been traditionally held in the divisions of protophyta and thalophyta.

These microorganisms, including others, take part in biological cycles of matter on earth. All regions of earth containing living cells may be collectively called biosphere. The activities of living organisms have profoundly affected the composition and structure of the surface layer of the planet and influenced the global atmosphere. Living microbial cells are, therefore, one of the powerful agents in the geochemical changes of the hypercycles of nature. In a biomaterial cycle, the maintenance of microorganisms of the biosphere in a steady state mostly depends on (1) the cycle turnover of elements essential for life, particularly carbon, nitrogen, oxygen, hydrogen, sulphur, and phosphorus and (2) the steady input of energy from the universal solar energy source. Through conversion of solar energy by photosynthesis, many elements are withdrawn as inorganic compounds from the environment and are accumulated in the organic constituents of cells and tissues of living organisms. These organic materials provide the energy source for other forms of life, either directly or indirectly, through the use of a biomaterial cycle (Figure 3.1).

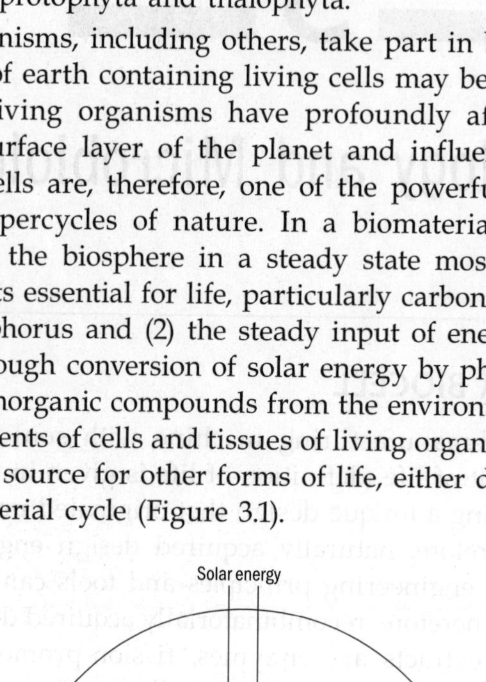

Figure 3.1 Biomaterial cycle in nature

For purposeful utilization of microorganisms and effective participation of a biomaterial cycle, some basic knowledge of different common groups of microorganisms, their organelles, behavioural functions, and actions on materials is required.

3.2.2 Basics of Living Cells

As mentioned, a cell is the fundamental organizational unit of life; therefore, all living things are made up of cells. According to the cell theory, all cells arise from other living cells. Living things can, thus, be defined as chemical organizations of cells that are capable of reproducing themselves.

3.2.3 Thumb Rule of Classification

The binomial method of nomenclature and classification of microbial cells has been universally accepted and used (Ainsworth 1971). According to this method, an organism has two parts to its name: (1) name of genus and (2) name of the specific epithet (species).

This method has been adopted from the conventional way of classifying living beings (Figure 3.2). In principle, this method of naming living organisms is simple, straightforward, and followed universally in identifying microorganisms. The present position of microorganisms in the universe belongs to biological or phylogenetic

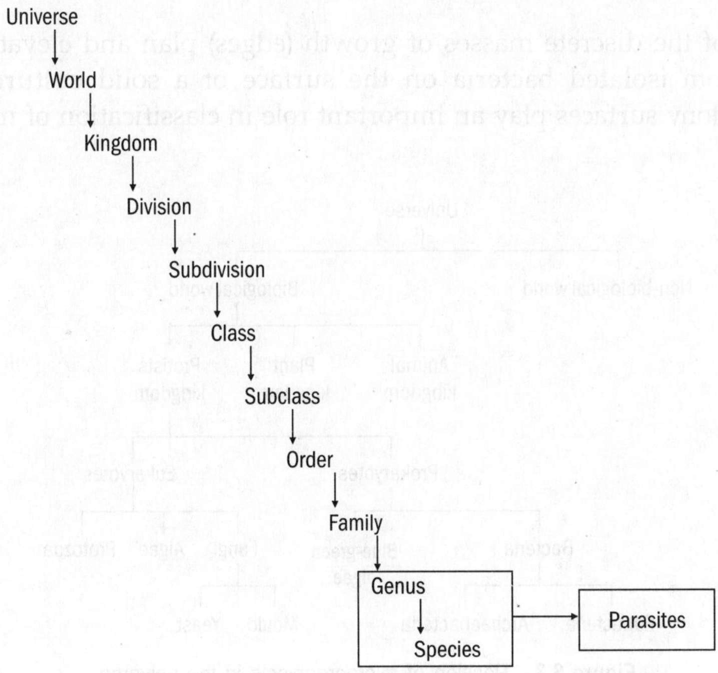

Figure 3.2 Thumb rule of positioning living cells in the universe

world, as shown in Figure 3.3 (Lapage, Clark, Lessel, *et al.* 1973; Aiba, Hamphrey, and Millis 1973; Bailey and Ollis 1986), which shows that parasites are the smallest biocells in this universe.

3.2.4 Common Biocells of the Universe

3.2.4.1 Microbes and their organelles

Among the microbes, eubacterium is a single-cell microorganism found practically everywhere in nature—from high mountains and oceanic depths to polar ice and hot springs. Bacteria are microorganisms that are much smaller than other microbial cells. A bacterium is of microscopic size, about 0.4–0.5 μ (1 μ = 10^{-3} mm), and their width is about 4 μ.

The ability of microbes to withstand extreme physical conditions may be attributed to their minute size, simplicity of structure, and the protective mechanisms and structures that most of them possess. Examples of the existence of various shapes of bacteria are shown in Figure 3.4. Among the major characteristics of bacterial cells are their shape, arrangement, and size. Existence of different shapes (Hawker, Linton, Folks, *et al.* 1980; Davis, Dulbbecco, Ginsberg, *et al.* 1968; Cruickshank 1970; Gunsalus and Sterier 1960) and morphological colony forms (pleomorphism) such as cocci, rods, spiral, and filaments depend largely on the growth phase and the environment of the microbial cells.

Appearance of the discrete masses of growth (edges) plan and elevation of colonies can be seen from isolated bacteria on the surface of a solid culture media. Such nutrient agar colony surfaces play an important role in classification of microorganisms (Figure 3.5).

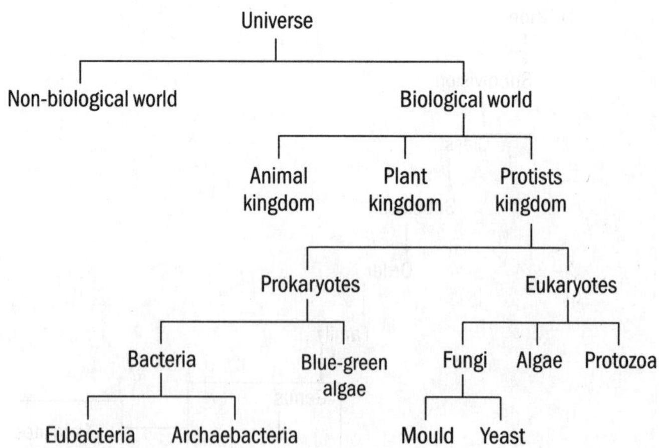

Figure 3.3 Position of microorganisms in the universe

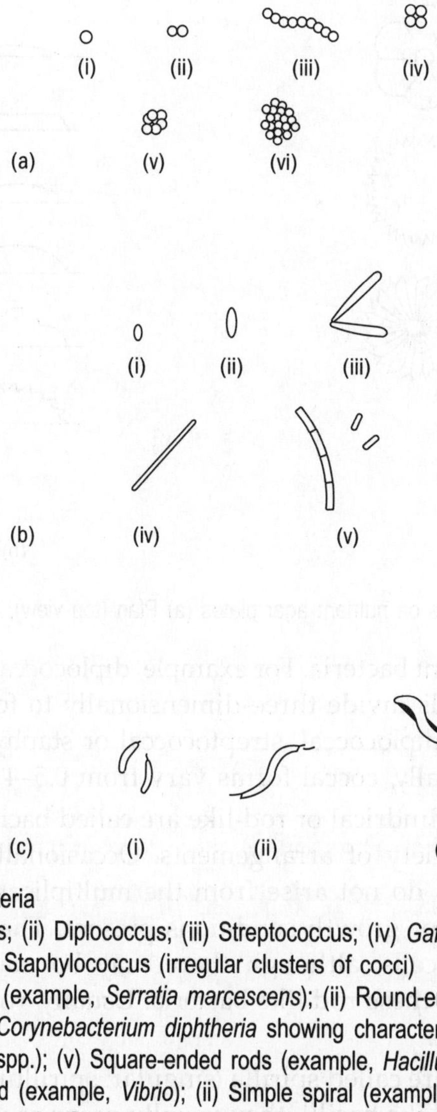

Figure 3.4 Shapes of bacteria

(a) Coccus: (i) Single coccus; (ii) Diplococcus; (iii) Streptococcus; (iv) *Gaffkya tetragena* (tetrads of cocci); *Sarcina* (packets of 8–16 cocci); Staphylococcus (irregular clusters of cocci)

(b) Rod: (i) Small ovoid rod (example, *Serratia marcescens*); (ii) Round-ended rod (example, *Escherichia coli*); (iii) Club-shaped (example, *Corynebacterium diphtheria* showing characteristic "hinge" cells); (iv) Long slender rod (example, *Lactobacillus* spp.); (v) Square-ended rods (example, *Hacillus antharis*)

(c) Spiral: (i) Comma shaped (example, *Vibrio*); (ii) Simple spiral (example, *Spirillum*); (iii) Long spiral surrounding central axial filament (example, *Spirochacta*); (iv) Slender spiral (example, *Treponema*); (v) Slender coiled spirals (example, *Leptospira*)

Although there exists thousands of different species of bacteria, the cells of most bacteria have one of the three fundamental shapes: (1) the coccus—spherical or ellipsoidal, (2) the rod—cylindrical, and (3) the spiral—helically coiled. The spherical bacterial cells have been termed cocci (singular: coccus). They may exist in a wide variety of arrangements. Many of these bacteria form patterns of arrangement useful for identification. These patterns have been explained by the peculiarities in the

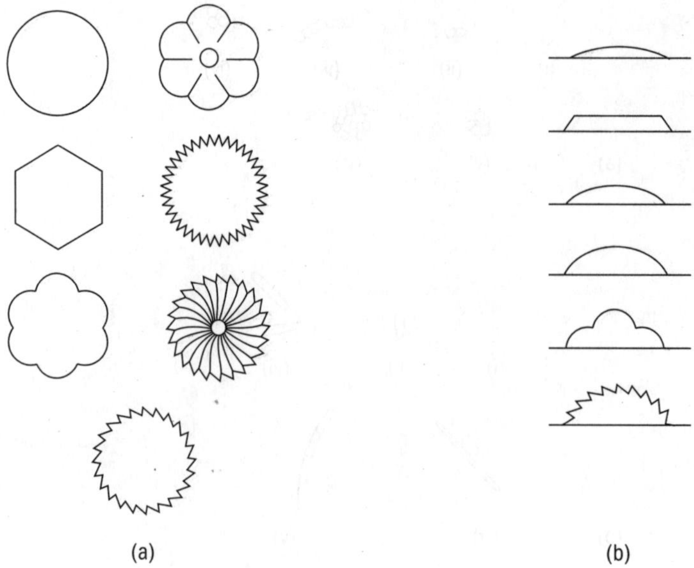

Figure 3.5 Microbial colony surfaces on nutrient agar plates (a) Plan (top view), (b) Elevation (side view)

multiplication process of different bacteria. For example, diplococcal bacterial cells divide to form pairs. Streptococcal cells divide three-dimensionally to form irregular clusters of cocci. Each of the cells in a diplococcal, streptococcal or staphylococcal aggregate is an independent organism. Usually, coccal forms vary from 0.5–4 μ in diameter.

The bacterial cells that are cylindrical or rod-like are called bacilli (singular: bacillus). These do not have a wide variety of arrangements. Occasionally, they are found in pairs or chains. These patterns do not arise from the multiplication process but from the particular stage of growth or growth conditions present. Therefore, bacilli usually appear as single, unattached cells, although there may be many variations in the thickness (width) and length of these rod-like bacteria. Usually, rod-shaped bacteria are 0.5–20 μ long and 0.5–4 μ wide.

Spiral-shaped bacterial cells are called spirilla (singular: spirillum). They may be more than 10 μ long and 0.5 μ wide. Like bacilli, they usually occur as unattached individual cells. Besides length, they may exhibit considerable difference in the amplitude and frequency of spirals. An important consequence of the smallness of bacterial cells, weighing 10^{-12} g on an average, is the high ratio of surface area to volume (Oginsky and Umbreit 1959). The bioconversion significance of this ratio in bioprocessing is that it allows a large portion of the cells to be in contact with its environment and has a great relevance of rapid ingestion of nutrients and growth factors with corresponding high metabolic rate. This high metabolic rate enables bacteria to adjust and to introduce changes in their environment in short time periods. These changes may be beneficial to the bacteria producing them or to other species of bacteria. For instance, the release of CO_2 by certain bacteria increases the acidity of the growth medium and favours the growth of bacteria requiring a low pH environment. Another important factor involved in the capability of bacteria to alter their environment is their ability to multiply rapidly. Within the multitudes of bacteria produced by a newly settled bacterium, there will be a handful of mutants. Natural selection may allow one of these mutants to be most fit for survival in the changing environment and numerous offspring will be derived from it. This cycle may be repeated demonstrating the capability of the bacterial species to survive under changing environmental conditions. Many bacteria are mobile, and their movement in a liquid is affected by hair-like projections called flagella, which arise from the cytoplasm present just inside the cell wall (Figure 3.6). Although some bacteria do not possess flagella, they can move by flexing and gliding their bodies over solid surfaces (for example, *Microscilia* and *Cytophaga*) or swim through liquids (for example, *Borrellia*). Most bacteria, whether motile or non-motile, can be carried to other habitats by air current, water, aerosols, surfaces of animals, foods, and so on. Various parts, sizes, and compositional characteristics of bacteria as presented by Aiba, Hamphrey, and Millis (1973) indicate that any organism can be characterized morphologically, to a great extent, by their electron micrograph.

Bacteria multiply by simple binary fissions (Oginsky and Umbreit 1959). When a cell has reached the optimum size, it splits into two daughter cells, which then continue to grow and divide to form four cells and so on, following a geometrical progression series. Under favourable conditions, a single bacterial cell can divide every few minutes (usually, 15–20 min) to give rise to millions of progeny within a day.

By dispersing a known volume of aqueous cell suspension in a nutrient medium containing solidifying agents (like agar-agar and gelatin) and incubating the culture at a suitable temperature, each viable cell will proliferate in the medium to produce a colony containing millions of bacteria. Colony characteristics such as size, shape, and colour often help identify bacteria since many form highly individual colonies on specific media. Microscopic examinations of this colony smear are used as criteria in their classification. An important criterion used in classifying bacteria is the ability of some to retain certain dyes when stained, according to the method introduced by Hans Christian Gram in 1834. The bacteria that retain the dye and appear purple under the microscope are called Gram-positive bacteria. On the other hand, Gram-negative bacteria fail to retain the purple dye and are coloured red by a counter-stain like safranin and carbolfuchsin. A fundamental difference in the two groups of bacteria is based upon the presence or absence of magnesium ribonucleate on the surface layers of the bacterial cell. However, many bacteria lose their Gram-positive character with age, which restricts carrying out this particular test only on young actively growing cultures.

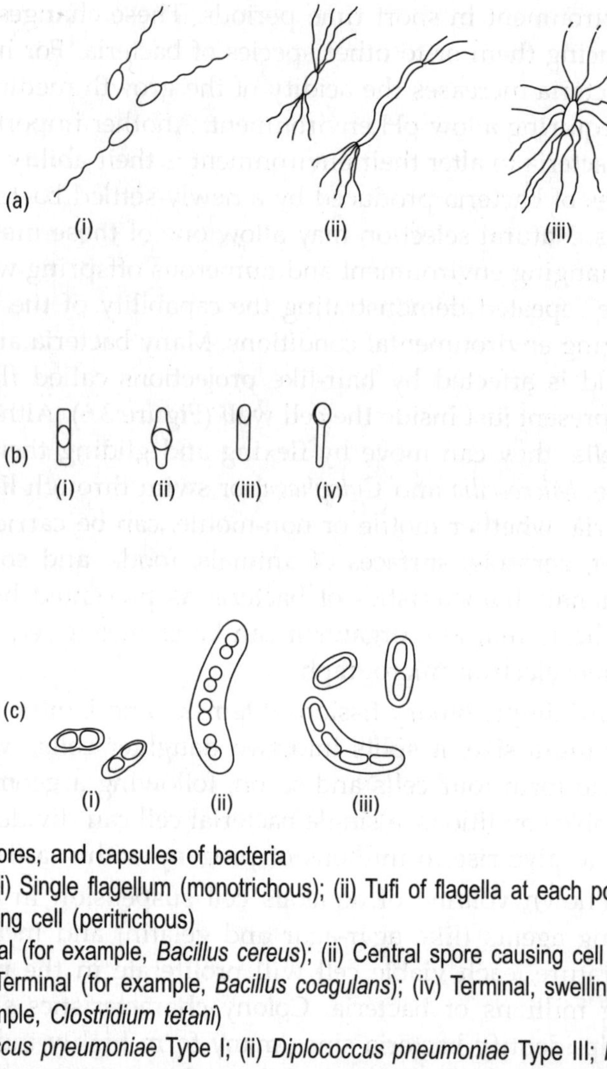

Figure 3.6 Flagella, spores, and capsules of bacteria

(a) Flagellated bacteria: (i) Single flagellum (monotrichous); (ii) Tufi of flagella at each pole (lophotrichous); (iii) Flagella surrounding cell (peritrichous)

(b) Endospores: (i) Central (for example, *Bacillus cereus*); (ii) Central spore causing cell to swell (for example, *Clostridium* sp.); (iii) Terminal (for example, *Bacillus coagulans*); (iv) Terminal, swelling the cell to form "drumsticks" (for example, *Clostridium tetani*)

(c) Capsules: (i) *Diplococcus pneumoniae* Type I; (ii) *Diplococcus pneumoniae* Type III; *Klebsiella pneumonia*

In Gram staining, a smear of bacterial cell is fixed by a mild flame to a microscope glass slide. The slide is treated with different solutions in the following order: (1) crystal violet (1 ml), (2) iodine solution (1 ml), (3) ethyl alcohol, and (4) safranin counter stain (1%) or basic fuchsin (1%) for a few seconds. Gram-positive bacteria retain the crystal violet iodine complex and are dyed a deep violet colour. On the other hand, Gram-negative bacteria lose the crystal violet iodine complex when treated with ethyl alcohol and are stained by safranin, which imparts a red colour. The reason for this difference is due to the composition of the cell walls of the bacteria.

The cell walls of bacteria are composed of peptidoglycan. This layer contains much more Gram-positive cells than Gram-negative cells. In Gram-negative bacterial cells, the lipopolysaccharide layer is present surrounding the thin peptidoglycan layer. The difference in staining is due to the high lipid content (about 20%) of the cell walls of these bacteria. These lipids include fats, oils, steroids, and certain other large organic molecules. During the staining procedure of Gram-negative cells, the alcohol treatment extracts the lipids from the cell wall, resulting in its increased permeability. The crystal violet iodine complex is thus leached from the cell during the alcohol wash. This decolourized cell then absorbs the red safranin dye. On the other hand, the cell walls of Gram-positive bacteria have lower lipid content and become dehydrated during the treatment with alcohol, which decreases permeability of the wall. Therefore, the crystal violet iodine complex cannot leave the cell wall and the cell is thus coloured violet. In addition, the thick peptidoglycan layer of the Gram-positive cell wall prevents decolourization by alcohol. Most bacteria fall into one of these two staining groups, and Gram staining thus serves as an important means of bacterial classification. The constitutional differences between Gram-positive and Gram-negative bacteria are given in Table 3.1.

Generally, Gram-positive bacteria tend to react or respond to various agents in different ways and the opposite response is exhibited by Gram-negative bacteria. Some of these differences are listed in Table 3.2.

Table 3.1 Constitutional distinctions between Gram-positive bacteria and Gram-negative bacteria

Constituent	Gram-positive	Gram-negative
Aromatic amino acids, arginine, protein	Absent	Present
Diaminopyrimidine	Present	Absent
Polysaccharides, lipids	Present in less quantity	Present in excess
Hexosamin	Present in greater quantity	Present in smaller amount

Table 3.2 Differences in properties and reactions or responses of Gram-positive and Gram-negative bacteria

Properties/sensitivity	Gram-positive bacteria	Gram-negative bacteria
Isoelectric range	pH 2–3	pH 5
Sensitivity to triphenyl methane dyes	Pronounced	Less
Inhibition of growth in acidic media	Pronounced	Less
Acid fastness	Probable	Never
Solubility in 1% KOH	Negative	Positive
Digestion by trypsin or pepsin	Negative	Positive
Sensitivity to streptomycin	Negative	Positive
Sensitivity to penicillin	Positive	Negative
Presence of Mg-ribonucleate	Positive	Negative

Acid-fastness is another property associated with bacteria. It refers to the resistance of some bacteria to stain under normal procedures. This is due to the presence of waxy substances present in the cell wall, which make the entry of the stain difficult. Thus, more drastic treatments are resorted to for staining bacteria. In acid-fastness, more powerful stains and application of heat to the preparation serves in staining. Once the cells have been stained by this method, they cannot be decolourized by the action of acid (usually 20% H_2SO_4). For instance, mycobacteria have high lipid content, from which an unsaponifiable wax can be isolated. This unsaponifiable wax is known as ingolic acid; it does have acid-fast properties.

In acid-fast staining (Ziehl–Neelsen staining), the smear of microorganisms is dried on a slide. Carbolfuchsin solution is poured on it and heated carefully until steam is expelled. Staining is allowed for 3–5 min, taking care that the smear is not dried. It is then washed with water and decolourized with acid for 10–20 s. It is then counterstained with methylene blue for a few minutes.

The ability of most microorganisms to exist and even to multiply in apparently unfavourable conditions is mainly due to one or a combination of factors that are both morphological or biochemical. These factors are associated with the formation of organelles like spores, capsule, and (depending on this) enzyme systems. Presence of these organelles can be examined under a microscope after staining. In spore staining, a smear of the spores is prepared on a slide and is fixed by passing it occasionally over a flame. The smear is then flooded with 5% malachite green dye solution and heated with steam for 3–4 min. The slide is then washed with water and covered with 0.5% safranin for 30–45 s, after which it is washed and dried. The capsules of bacteria are often clearly stained by treating with common stains like basic fuchsin, polychrome methylene blue, Leishman's stain, and Gram stain.

The best method for staining capsules on bacteria is the wet-film Indian ink. A microscope slide is wiped free from grit particles and loopful Indian ink is placed on it. A small portion of solid bacterial capsule bearing culture is emulsified in the ink, after which a clear cover slip is placed in the emulsion and pressed down. Next, a blot soaking of India ink is carried out so that the ink film becomes thin and the gripped capsule is unable to move. It is then examined with an oil immersion objective, and the highly refractile outline of the bacterium is observed.

Due to their internal metabolism, bacteria are able to withstand adverse conditions like absence of air and presence of unwanted elements like sulphur, iron, and salt in the surroundings. Some of these habitats and their nutritional requirements are listed in Table 3.3. Modern molecular biology has revealed that cells contain certain special constitutive enzymes that are accountable for their self-defence mechanism. These enzymes are found to have restrictions of cleavage of the foreign deoxyribonucleic acid (DNA) double helix at definite or specific sites and are collectively called restriction

Table 3.3 Some parameters required for the existence of certain bacteria

Parameter	Example	Characterization
A. Physical		
1. *Temperature*		
(a) Optimum below 20°C (b) Optimum between 20–45°C (c) Optimum above 45°C (d) Capable of withstanding high temperatures	• *Pseudomonas* spp. • *Athromobacter* spp. • Most bacteria, including pathogens • *Bacillus stearothermophilus* • *Clostridium thermosaccharolyticum* • Spore-bearing bacteria	• Psychrophiles • Mesophiles • Thermophiles • Thermoduric organisms
2. *Acidity*		
(a) Preference for pH < 4.5 (b) Preference for medium pH range (c) Preference for alkaline conditions (pH > 8)	• *Lactobacillus* spp. • *Acetobacter* spp. • Most bacteria • *Alcaligenes*	Aciduric
3. *Light*		
Illumination is essential	• *Rhodospirillum* • *Chromatium*	Photosynthetic bacteria
4. *Osmotic concentration*		
(a) Tolerant of high sugar concentrations (b) Require 12%–30% salt for growth* (c) Tolerant of high salt concentrations*	• *Leuconostoc* • *Mesenteroides* • *Halobacterium* • *Staphylococcus aureus*	• Osmophiles • Obligate halophiles • Facultative halophiles
B. Chemical		
1. *Oxygen requirements*		
(a) Oxygen necessary for growth (b) Oxygen prevents growth (c) Growth with or without oxygen (d) Require low oxygen tensions	• Most bacteria • *Clostridia* • *Escherichia coli* • Many lactic acid bacteria	• Aerobes • Obligate anaerobes • Facultative anaerobes • Microaerophiles
2. *Other gaseous requirements*		
(a) Nitrogen (b) Hydrogen (c) Carbon dioxide (d) Methane (e) Carbon monoxide	• *Azotobacter*, *Klebsiella* • *Aerogenes*# • *Hydrogenomonas chlorobium* • *Thiobacillus* • *Methanomonas* • *Carboxydomonas*	• Nitrogen-fixing bacteria • Photosynthetic autotrops • Chemosynthetic autotrops

Contd.

Table 3.3 *Contd...*

Parameter	Example	Characterization
3. *Other inorganic sources*		
(a) Ammonia (b) Nitrites (c) Hydrogen sulphide, sulphur, thiosulphates (d) Iron and manganese	• *Nitrosomonas* • *Nitrobacter* • *Thiobacillus* spp. • *Gallionella* • *Sphaerotilus* • *Ferrobacillus*	• Nitrifying bacteria • Colourless sulphur bacteria • Iron and manganese bacteria
4. *Organic compounds*	• Majority of bacteria	Heterotrophs

* Not entirely an osmotic phenomenon

\# *Klebsiella aerogenes* fixes nitrogen under neutral or alkaline conditions

Source Mukhopadhyay (2005)

enzymes. However, the restriction enzyme of any organism is not capable of cutting its own DNA. These enzymes are neither overproduced in the cell nor inducible. Most recognition sites of the enzymes contain a dyad axis of symmetry and, in most cases, all bases within the site are uniquely specified. The symmetrical recognition sequence of these enzymes has been termed palindromes. Restriction endonucleases that recognize identical sequences are called isoschizomers.

As shown in Figure 3.6, flagellum is another organelle present in many bacterial cells and is responsible for its locomotion. Bacteria that do not have flagella are called atrichous, whereas those that have only one flagellum are called monotrichous. Bacteria that have many flagella are called peritrichous. After suitable staining, flagella appear long (3–12 μ), slender, filamentous, and helicoidal in shape under a microscope. They are extremely thin (100–200 Å) appendages. Some believe that flagella originate from the cell wall, while others believe that they traverse the cell wall into the cytoplasm. These delicate organs can get easily detached from the cells. Although flagella primarily consist of proteinous material, they are not essential to the life of the bacteria. This protein is called flagellin and has a molecular weight of about 20 000. Unlike celia, which protect a cell by lashing movements, the flagella protect a bacteria by rapid contractions, which pass through the flagella from one end to the other.

Identification of flagella is carried out following the Bailey's method. For staining, the smear on the slide is dried and allowed to react with stain solution A for 3.5 min. Solution A is a mixture of 18 ml of 10% tannic acid with 6 ml of 8% ferric chloride ($FeCl_3$). Stain is poured off and, without washing, covered with solution B and allowed to react for 7 min. Solution B is a mixture of 3.5 ml of solution A, 0.5 ml of basic fuchsin, in 0.5% ethyl alcohol, 8.5 ml concentrated HCl, and 2 ml of formalin. The slide was then washed and covered with Ziehl Neelsen's carbolfuchsin, steamed for 1 min. The smear is then dried and examined under a microscope.

Cytological investigations also reveal the presence of highly resistant endospores (Figure 3.6) in the bacterial cells. Endospores are organelles of variable shapes and are formed as part of the defensive mechanism of bacteria. Endospores enable bacteria to withstand unfavourable environmental conditions like freezing, boiling, exposure to disinfectant, desiccation, osmosis, chemicals, and exhaustion of nutrients for varying periods of time, often in years. They are formed in well-nourished cells where conditions for vegetative growth gradually become unfavourable.

Formation of spores takes place in several stages. In the first stage, some granules aggregate at the site of formation and, at the same time, the nuclei of the cell arrange themselves in two groups in this space. In the second stage, an elliptical envelop similar to the cytoplasmic membrane develops enclosing the nuclei, which merge to give rise to the forespore. Within a few minutes, the forespore develops into a mature dense endospore. The remainder of the cell either disintegrates or turns into a sac. When conditions are favourable for growth, each spore germinates into a separate vegetative cell. In bacteria, the spores are not reproductive in function and do not serve to in-species multiplication. They are basically resting forms, one phase in the right cycle of a sporangeous bacterium. Sometimes, they remain dormant for years. Spores contain much less water than a vegetative cell.

Pili (fimbriae) are completely proteinaceous organelles in bacterial cells. These finer and filamentous appendages are shorter and straighter than flagella. In the same cell, they usually vary in thickness (75–100 Å) and length (up to several microns). Like flagella, they appear to arise from the basal bodies in the cytoplasmic membrane. There is evidence that in bacterial conjugation, pili apparently form a bridge between the male and female cells, through which DNA is transferred.

Ribosomes are somewhat spherical organelles present in the cytoplasm of a cell, with a diameter of approximately 180 Å and a molecular weight of 2.8×10^6. They are readily recovered from the homogenate by high-speed centrifugation (100 000 g). Ribosomes are composed of 60% DNA and 40% protein, and they constitute up to 40% of the cells dry weight and 95% of its RNA. They are active components in protein synthesis in cells, where they are grouped in chains called polysomes. In certain bacteria (usually Gram-positive sp.), one or more large, irregular envoluted invaginations of the cytoplasmic membrane have been observed. These are termed mesosomes. In some bacterial species (like *Bacillus subtilis*), mesosomes are vesicular, whereas in others, they appear as concentric *lamellar.* In either form, they increase the total membrane surface of the cell. Their main function is connected to wall synthesis and in segregation of nuclei to the daughter cells during cell division. Capsular organelles in most bacteria are mucilaginous envelops completely surrounding the cell (Figure 3.6). These are composed of long-chain polysaccharides and/or polypeptides. Capsules protect bacteria by making the entry of the harmful substances difficult, if not impossible. They are also associated with causing diseases within a host. The cells that lack capsules are avirulent; they produce rough-textured colonies that are distinct from the smooth colonies of pathogenic forms.

The internal biochemistry of cells having various organelles may be such that adverse external conditions are counter-balanced. Several common organisms found in the gut of humans and animals, if placed in an acidic environment, may form certain enzymes like carboxylase. These enzymes may catalyse reactions, leading to the removal of CO_2 from amino acids with a consequent increase in the internal pH of the cell. On the other hand, if the cell is placed in an alkaline medium or environment, other enzymes, like deaminase, may be formed instead of decarboxylase.

Besides the parameters listed in Table 3.3, bacteria have certain properties that separate them from other microbial groups. Unlike fungi, yeasts, protozoa, and most algae, bacteria do not have a nuclear structure surrounded by a membrane. This type of nucleus is called prokaryon (means before the nucleus). On the other hand, the nuclei of fungi, yeasts, moulds, and algae are surrounded by a membrane and are called eukaryon (means true nucleus).

Bacteria belonging to the prokaryotic group multiply by simple fission; in such cases, mitosis and meiosis are not observed. Unlike eukaryotic organisms, bacteria usually do not have highly specialized organelles for photosynthesis and respiratory process. They are not complex like eukaryotic organisms; for example, their flagella are smaller in diameter than those of eukaryotic organisms. Similarly, the bacterial cell wall differs from that of eukaryotic organisms. In identifying a bacterial cell, it is subjected to several tests and then referred to the system of classification. These tests relate to general taxonomy, chemotaxonomy, and numerical taxonomy. An outline of the general taxonomic characteristics provides important information, and a number of cellular organelles and materials have been identified this way. Their cytotechnological dimensions (Table 3.4), properties, and functions are helpful in the classification of bacteria. Their concerns are related to cellular and genetic materials dealing with DNA base composition, DNA–DNA hybridization, cell wall composition, cellular fatty acids, phospholipids, quinines, among others.

3.2.4.1.1 Tests commonly used to identify microbes

In order to examine and identify a bacterial cell, a large number of microscopic (phase electron), biochemical, and other tests are performed. The outline of some of these methods is given in Tables 3.4 and 3.5.

Various organelles of bacteria (Table 3.6) can be studied under an electron microscope/micrograph, which indicate their cytological minuteness. Among these organelles, some permeable cell membranes play a significant role in cellular growth because it is the

Table 3.4 Cytotechnological dimensions of bacterial organelles

Organelle	*Dimension*
1. Slime layer	Microcapsule (5–10 μm) Capsule (0.5–2 μm) Slime (Indefinite)
2. Cell wall (i) Gram-positive sp. (ii) Gram-negative sp.	 10–20 μm 10–20 μm
3. Cell membrane (thickness)	5–10 μm
4. Flagellum/celium	10–20 μm by 4 × 12 μm
5. Pili (fimbriae)	5–10 μm × 0.5–2 μm
6. Inclusions (i) Spore (ii) Storage granules (iii) Chromatophores (iv) Ribosomes (v) Volutin (vi) Plamids	 1–1.5 μm × 1.6–2 μm 0.5–2 μm 50–100 μm 10–30 μm 0.5–1 μm 2–5 μm
7. Nuclear material	About half the cell volume
8. Overall size of a bacteria	Rod-shaped bacteria: 0.5–20 μm long × 0.5–4 μm wide Cocci-shaped bacteria: 0.5–4 μm in diameter Spiral-shaped bacteria: >10 μm long × 0.5 μm wide

Source Aiba, Hamphrey, and Millis (1973); Bailey and Ollis (1986)

site of differential permeability for substrates and other materials. The cell membrane exhibits osmotic property also, even though the extent of its porosity is still uncertain. However, there are two means of transport across a two-sided cell membrane, which has led to speculation regarding the pore theory of transport by the membrane. The outer surface of the membrane has been termed the "cis" side and the inner surface, the "trans" side. In some instances, the process of transport appears to be simple pore diffusion and is dependent on the nature of the membrane, the relative distribution of solutes and air particles on the two sides of the membrane, and the particles or solutes attendant charges, if there are any. This type of transport across a cell membrane is termed passive transport. However, in other instances, material is transported across the membrane only during active metabolism in the cell. The metabolic energy is responsible for transport of material across the membrane. Such transport is termed active transport.

There are many hypothetical models explaining the transportation of dissolved materials across a differentially permeable membrane, which have been illustrated in Figure 3.7.

Table 3.5 Most probable numbers per 100 ml

Tubes positive			MPN	Tubes positive			MPN	Tubes positive			MPN
10 ml	1.0 ml	0.1 ml		10 ml	1.0 ml	0.1 ml		10 ml	1.0 ml	0.1 ml	
0	0	1	3	1	2	0	11	2	3	3	53
0	0	2	6	1	2	1	15	3	0	0	23
0	0	3	9	1	2	2	20	3	0	1	39
0	1	0	3	1	2	3	24	3	0	2	64
0	1	1	6	1	3	0	16	3	0	3	95
0	1	2	9	1	3	1	20	3	1	0	43
0	1	3	12	1	3	2	24	3	1	1	75
0	2	0	6	1	3	3	29	3	1	2	120
0	2	1	9	2	0	0	9	3	1	3	160
0	2	2	12	2	0	1	14	3	2	0	93
0	2	3	16	2	0	2	20	3	2	1	150
0	3	0	9	2	0	3	26	3	2	2	210
0	3	1	13	2	1	0	15	3	2	3	290
0	3	2	16	2	1	1	20	3	3	0	240
0	3	3	19	2	1	2	27	3	3	1	460
1	0	0	4	2	1	3	34	3	3	2	1100
1	0	1	7	2	2	0	21	3	3	3	1100
1	0	2	11	2	2	1	28	—	—	—	—
1	0	3	15	2	2	2	35	—	—	—	—
1	1	0	7	2	2	3	42	—	—	—	—
1	1	1	11	2	3	0	29	—	—	—	—
1	1	2	15	2	3	1	36	—	—	—	—
1	1	3	19	2	3	2	44	—	—	—	—

Note Using three tubes each and innoculated with 10, 1.0, and 0.1 ml of sample
Source Jacobs and Gerstein (1960)

1. Hydrogen sulphide production

In the H_2S production test, strips of filter paper are soaked in basic lead acetate solution (10%) and then dried. They are then placed on top of a nutrient broth culture and incubated. Blackening of the strip of filter paper indicates H_2S production.

2. Lipolytic activity

In the lipolytic activity test, a homogeneous suspension of tributyrin agar containing the culture is incubated. A clear zone developed around the colonies indicates the lipolytic activity of the organism.

3. MRVP tests

Methyl red reaction test

Table 3.6 Difficulties in developing methanogen-derived stable rDNA molecules (*Methanobacterium sarcina*)

Envisaged difficulty in genetic analysis of methanogens	Remark
Methanogens are neither truly prokaryotic nor eukaryotic	Characterization limitation
Special newer class of bacteria	Belongs to archaebacteria whose characteristics are not well understood
Macromolecules of their cell envelope • do not contain muramic acid or D-amino acids • contain analogous sugar peptide polymer that plays a role of bacterial peptidoglycan and their lipids have either linkages	Pose problems in application of rDNA techniques
Antibiotics that are used as markers are inactive against methanogens	Problems in identifying specificity
Translation system is sensitive to some antibiotics that are known to affect eukaryotic ribosomes and to ADP ribosylation	

In the methyl red reaction test, the culture is incubated for two days in dextrose phosphate medium (pH 7). If the final pH is 4.5 or less, it is considered that "mixed acid" type fermentation has taken place. A few drops of methyl red solution are then added to the culture. If the colour is red, the test is positive; if yellow, it is negative; and if orange, it is uncertain.

Voges–Proskauer reaction test

The Voges–Proskauer (VP) reaction test is meant for testing the formation of acetyl methyl carbonyl (acetoin) from dextrose. Dextrose is oxidized by the reagent to diacetyl, which produces red colour with a constituent of the peptone medium. In this test, the culture in the dextrose phosphate medium is incubated for 2–5 days. An equal volume of 40% KOH and a pinch of creatin are then added with vigorous shaking.

$$\underset{\text{(Acetoin)}}{CH_3COCHOHCH_3} \longrightarrow CH_3CHOHCHOHCH_3 \longrightarrow \underset{\text{(Diacetyl)}}{CH_3COCOCH_3}$$

4. Nitrate reduction

In the nitrate reduction test, the culture is first incubated in 0.1% KNO_3 broth for 2–5 days. A solution of sulphanilic acid (1 ml) is added, followed by a few drops of α-naphthylamine solution. Development of pink or red or maroon colour indicates the presence of nitrites. Some organisms further reduce nitrites. Thus, if the above test is negative, a few milligrams of Zn dust is added to test for unreduced nitrites, which is confirmed by the development of red colour.

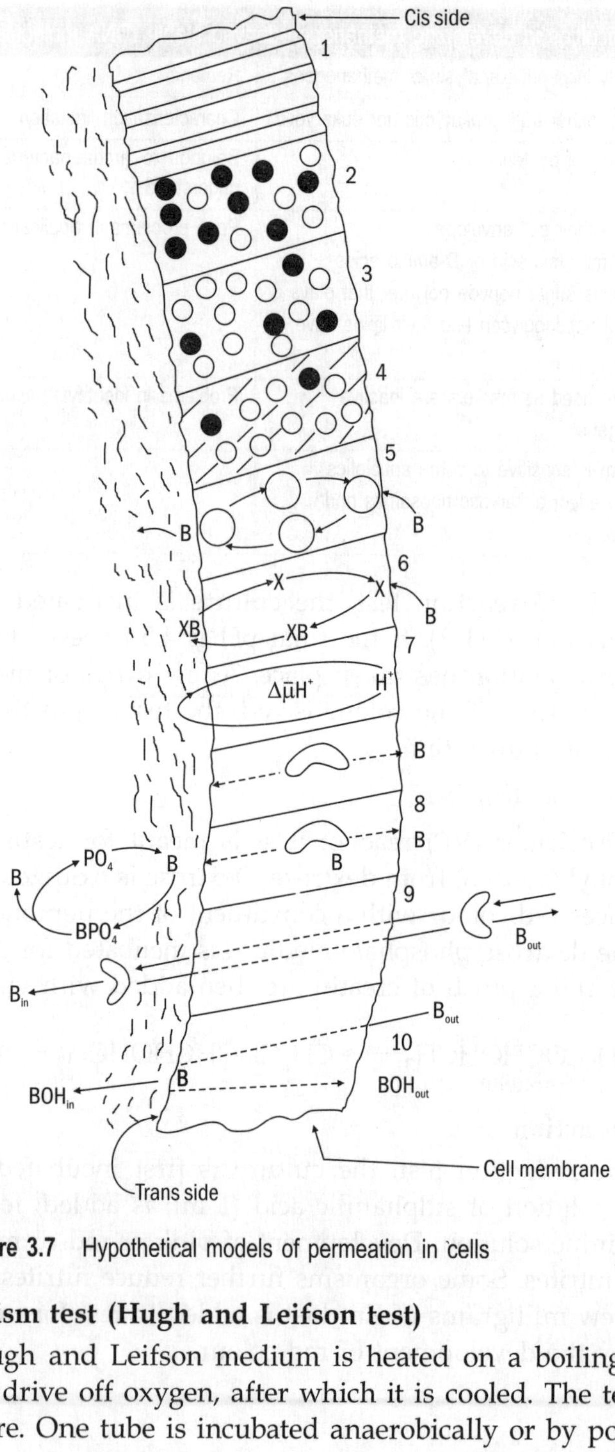

Figure 3.7 Hypothetical models of permeation in cells

5. Carbohydrate metabolism test (Hugh and Leifson test)

In two test tubes, the Hugh and Leifson medium is heated on a boiling water bath for 10 min to release and drive off oxygen, after which it is cooled. The test tubes are inoculated with the culture. One tube is incubated anaerobically or by pouring liquid paraffin to create an anaerobic condition. The other tube is incubated aerobically. This test shows that after incubation, only oxidative acid is produced in the aerobic tube while fermentation acids are produced in the anaerobic tube.

The Hugh and Leifson medium contains peptone (2 g), NaCl (5 g), K_2HPO_4 (0.3 g), agar (3 g), and water (1000 ml). These constituents are dissolved by heating, and the pH of the medium is adjusted to 7.1 and 15 ml of 0.2% bromothylmol blue. About 8–10 ml is taken in the tube, sterilized, and sterile glucose is added to give final 1% concentration.

6. Urease test

In Christensen's medium containing the culture, urea is split to form ammonia, which changes the indicator of the medium from yellow to pink.

7. Eijkman test

The Eijkman test is performed in differentiation of coliform bacilli like *E. coli* and *Aerobacter aerogenes*. This test is carried out by directly sub-culturing from the primary test tubes (MacConkey's broth), where showing acid and gas changes into green lactone bile broth on incubation at 44°C for 24 h. It indicates whether there is a false positive result due to other organisms that may give positive reaction in the MacConkey's broth at 37°C. This test also shows positive results for the formation of indole at 44°C. Results as expected are given in Table 3.7.

8. Cholera red test

The cholera red test is used to identify vibrios. In this test, a few drops of concentrated sulphuric acid are added to a four-day peptone-water culture. Development of red colour due to the formation of nitrosoindol indicates the presence of vibrios.

9. Most probable number test

The most probable number (MPN) test estimates coliform bacilli using the MacConkey broth. It can also be used to test for other organisms in liquid samples, if the growth is easily observable (for example, by turbidity or acid production).

In this test, the samples are mixed in nine different test tubes by shaking vigorously. About 10 ml of MacConkey's medium (double strength) is added to three of the test tubes, 1 ml of the medium (single strength) is added to three other tubes, and 0.1 ml of

Table 3.7 Eijkman's test results

Type	Gas at 44°C	Indole formation at 44°C	Inference
1	+	+	Typical coliform (*Escherichia coli*)
2	+	–	Atypical coliform (irregular type 1)
3	–	+	Other coliform
4	–	–	Other coliform

the medium (single strength) is added to the rest. The double strength broth is added to tubes containing larger volumes; otherwise, the medium would be too dilute. The tubes are incubated at 37°C for 24 h and examined for acid and gas production. The numbers of positive tubes in each set of three are tabulated and matched with the appropriate data given in Table 3.5.

10. Viability test

Ringer's solution, saline or 0.1% peptone are suitable diluents for the viability test. About 9 ml of diluent is taken in sterile test tubes, which act as dilution blanks. The samples are mixed by shaking under sterile conditions. About 1 ml of the sample is transferred to the first dilution blank and mixed. Next, 1 ml of the sample is transferred to another dilution blank. This procedure is continued for a required number of dilutions. An example of the test tabulation is shown in Table 3.8.

One millilitre of the above dilutions are put in sterile tubes containing sterile nutrient agar or other suitable media (14 ml) at 45°C, well mixed, and the tubes are labelled with the dilution number. From each tube, 1 ml of each dilution is pipetted in the centre of appropriate sterile Petri dishes. The Petri dishes are shook gently by a rotary motion by hand, and next the medium is allowed to set one by one. After the medium is set, the Petri dishes are incubated at 30°C for 24 h. Next, the colonies showing viable cells are counted with the help of a colony counter.

11. Pigmentation or chromogenic test

In the pigmentation or chromogenic test, in glucose agar slant slopes, the sample containing the organism is inoculated and incubated at 25–30°C until growth has occurred. Formation of pigment and its colour are examined after incubation. Water solubility of the pigment is also examined. Some examples of bacteria that produce pigments are *Pseudomonas* sp., *Sarcina* sp., and *Rhodospirillum* sp.

12. Photic test

In the photic test, the test organism produces fluorescence in the medium used for growth.

By employing the above-mentioned tests and examinations, identification of a pure microbial culture isolated by any method may be carried out. The methods of isolation of a pure microbial culture are discussed below.

Table 3.8 Viability test

Tube number	Sample	1	2	3	4	5
Dilution factor	—	1/10	1/102	1/103	1/104	1/105
Volume of original fluid per ml	1	10–1	10–2	10–3	10–4	10–5

3.2.4.1.2 Methods of isolation of a pure microbial culture

A pure culture consists of a nutrient medium containing the growth of a single species of organism. The following methods are employed to isolate a pure microbial culture.

1. Pour plate method

The pour plate method is used for isolating a pure bacterial culture. The nutrient agar tubes are melted in a boiling water bath and cooled to 43–45°C. The melted nutrient agar tube is then inoculated with a loopful of mixed bacterial cultures. The contents of the tube are mixed thoroughly by rotating the tube between the palms of the hands. One loopful of this mixed culture tube is transferred to a second molten nutrient agar tube and mixed thoroughly.

Again, one loopful from the second inoculation tube is inoculated to a third tube as before. The inoculated agar of the three tubes is poured with three sterile Petri plates under aseptic conditions. The agar in the plates is allowed to set and then correctly labelled. The Petri plates/Petri dishes are then inverted and incubated at 37°C for 24–48 h. After incubation, the plates are examined for the presence of different kinds of colonies, from which a well-isolated colony is selected. The purity of this colony is examined by Gram staining. After this, a loopful of this pure colony is transferred to sterile nutrient agar slants and incubated at 37°C for 24–48 h.

2. Streak plate method

Like in the power plate method, the streak plate method involves the usage of agar tubes. Agar is poured carefully in two sterile Petri plates/Petri dishes and allowed to set until the agar becomes firm. A loopful of the mixed culture is aseptically transferred in it. The loop is streaked back and forth carefully on the surface of the agar in the plate to about 0.65 cm apart. The first streak should contain more culture than the second; the second streak should contain more culture than the third; and so on. The last streak should thin out the culture sufficiently to give isolated colonies. Without re-inoculating the wireloop, if the second plate is streaked as carefully as before, the certainty of securing a well-isolated pure colony of a culture increases.

Several difficulties may arise while isolating a pure microbial culture. Rapidly growing organisms may suppress the growth of desirable slow-growing organisms. Sometimes, the use of a synthetic medium at a pH of 5–5.5 for the growth of one type of microorganism may encourage the growth of other types of microorganisms, which interferes with the isolation of a pure culture. Similarly, an incubation temperature of 25–30°C for growth of bacteria may also support the growth of certain fungi. In order to overcome these difficulties, optimum temperature and pH, proper medium for growth of the desirable culture, and suitable inhibitors that hinder the growth of undesirable organisms are employed.

For example, in the isolation of actinomycetes in pure state, the medium is supplemented with 0.4% sodium propionate, which inhibits the growth of all other types of microorganisms. Also, to isolate fungi from other organisms, the medium is supplemented with dyes like rose benzol (1:13 000) or crystal violet (1:30 000) and streptomycin (30 γ/ml).

3.3 BIOCELLULAR DESIGN AND ACTIVITY

Bacterial, Plant, and Animal Cell Structural Designs

The uniqueness of the bacterial, plant, and animal cells relates to their morphological and other cellular characteristics (Figures 3.8, 3.9, and 3.10).

3.4 ARCHAEBACTERIA: THE NEWER BIOCELLS

In the late 1970s, the concept and proposal of Woese and his associates reclassified all living organisms into three primary kingdoms—eukaryotes, eubacteria, and archaebacteria (Fox, Stackebrandt, Hespell, *et al.* 1980). According to this classification, prokaryotic organisms were considered to comprise two phylogenetically distinct groups. The first is the eubacteria that consist of the traditional bacterial groups discussed earlier, including photosynthetic bacteria, blue-green bacteria, endospore-forming bacteria, actinomycetes, and spirochaetes. The second is the archaebacteria that consist of methanogenic bacteria (extremely halophilic bacteria). *Thermoplasma* species

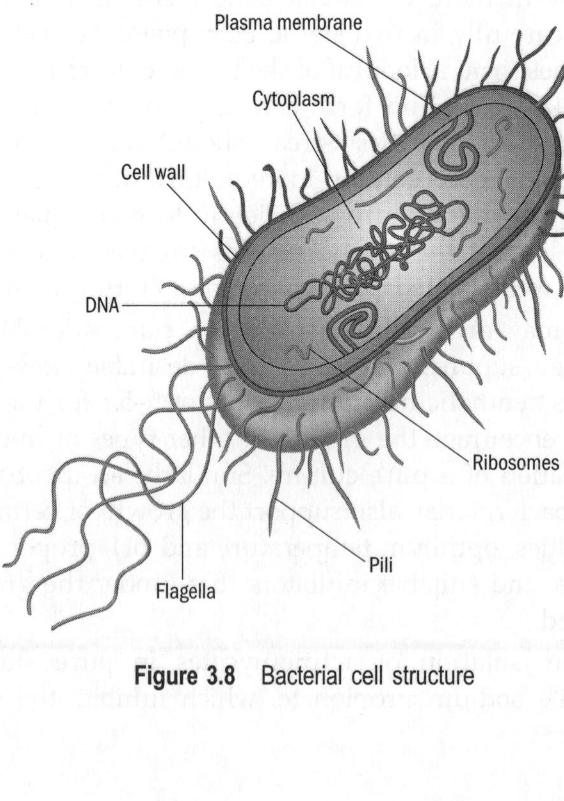

Figure 3.8 Bacterial cell structure

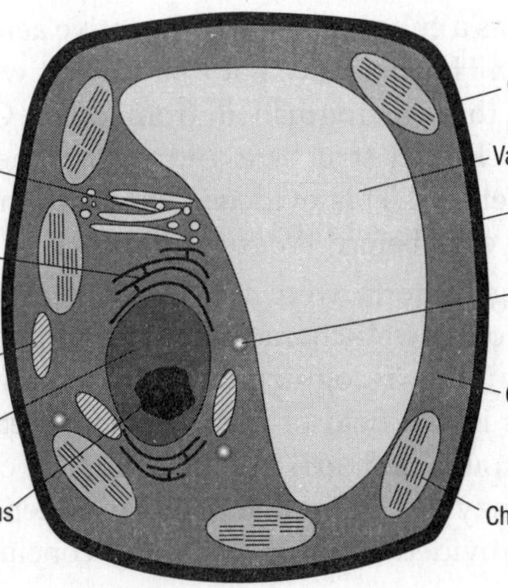

Figure 3.9 Plant cell

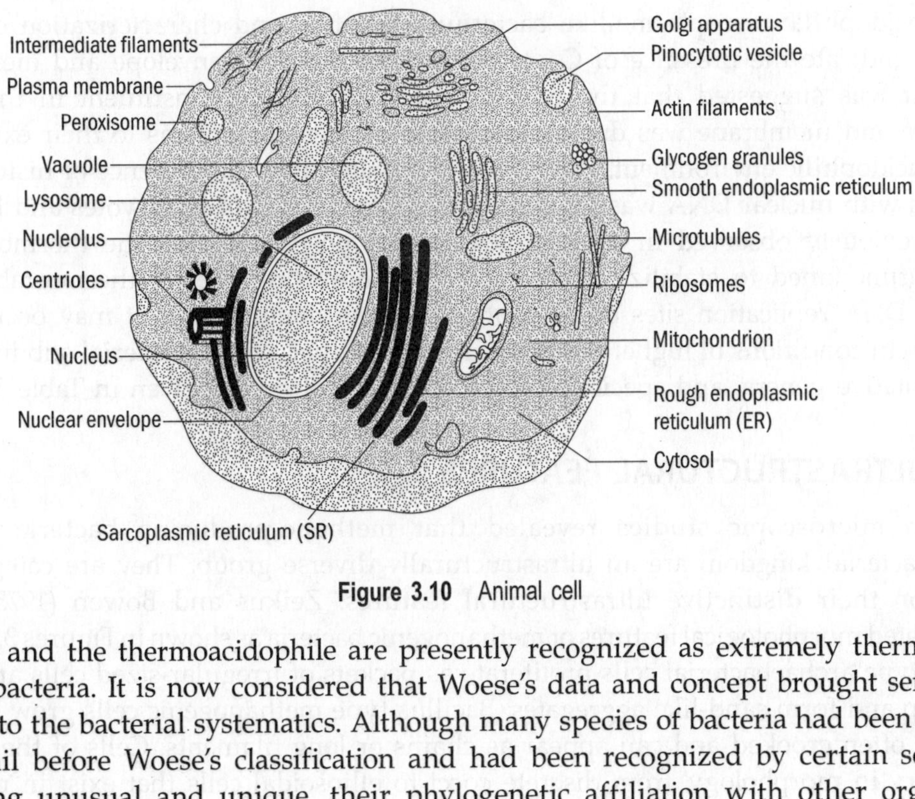

Figure 3.10 Animal cell

(or sp.) and the thermoacidophile are presently recognized as extremely thermophilic archaebacteria. It is now considered that Woese's data and concept brought sense and clarity to the bacterial systematics. Although many species of bacteria had been studied in detail before Woese's classification and had been recognized by certain scientists as being unusual and unique, their phylogenetic affiliation with other organisms was uncertain. For example, *Halobacterium* (now classified as an archaebacteria) had previously been classified as a eubacterial Gram-negative aerobic rod like *Psuedomonas* sp., *Azotobacter*, and *Acetobacter*. However, even at that time, it was known that *Halobacterium* possessed characteristics that distinguish it from other Gram-negative aerobic rods, including differences in cell wall structure and membrane lipids, which composed of C_{20} isoprenoid glycerol diethers. This evidence supports the concept of archaebacteria, which existed almost 20 years before Woese's proposal.

The methane-producing bacteria were the first extensively studied archaebacteria. Previously classed among other well-characterized bacteria on the basis of morphological criteria, the methanogens were recognized as a taxonomic group by Baker in 1956. Being one of the pioneers in the field of methanogen research, he first concluded that methanogenic bacterial groups had striking physiological characteristics. Subsequently, the presence of novel coenzymes in methanogens was demonstrated by Wolfe and his associates. This further provided support to Baker's conclusion.

3.4.1 Isolation and Characterization of *Sulfolobus* and *Thermoplasma*

Sulfolobus is an aerobe found in a Yellowstone hotspring, and *Thermoplasma* is a thermoacidophilic mycoplasma-like bacterium. Isolation and characterization of these bacteria indicate the presence of C_{40} tetraethers in their cell envelope and membrane lipids. It was suggested that the presence of this unusual constituent in their cell envelope and membrane was due to adaptation of these organisms to their extremely thermoacidophilic environment. Later, Searcy revealed that the presence of histone-like proteins with nuclear DNA was thought to be characteristic of eukaryotes and had not been previously observed in prokaryotes. Searcy postulated that the thermoplasma histone functioned to stabilize DNA, either by conferring greater thermostability at specific DNA replication sites or by protection from depurination. It may occur at *in situ* growth conditions of higher temperature and low pH. Archaebacterial subdivisions, representative genera, and species as reported in literature are given in Table 3.9.

3.5 ULTRASTRUCTURAL FEATURES

Electron microscopic studies revealed that methane-producing bacteria in the archaebacterial kingdom are an ultrastructurally diverse group. They are categorized based on their distinctive ultrastructural features. Zeikus and Bowen (1975) have enumerated morphological features of methanogenic bacteria as shown in Figures 3.11–3.18. *Sarcina*-type archaebacterial cells proliferate as packets of irregular-sized cells and tend to clump and form sand-like aggregates. Bacillus type methanogenic cells grow as rods that are often crooked and can appear as chains or long filaments. Cells of the coccus type vary in morphology, from discrete cocci to ellipsoidal cells that exist in pairs or as chains. *Methanobacterium ruminantium* (Figure 3.11c) is morphologically similar to

Table 3.9 Archaebacterial subdivisions, representative genera, and species

Subdivision	Genus	Species
Methanogenic bacteria	*Methanococcus* group	*Methanococcus vannielii* *Methanococcus voltae* *Methanococcus thermolithotrophicus* *Methanococcus jannaschii*
	Methanobacterium group	*Methanobacterium formicicum* *Methanobacterium bryantii* *Methanobacterium thermoautotrophicum* *Methanobacterium smithii* *Methanobacterium arboriphilus* *Methanobacterium ruminantium* *Methanosphera stadtmanae* *Methanothermus fervidus*
	Methanomicrobium group	*Methanosarcina barkerii* *Methanococcoids methylutans* *Methanothias sochangenii* Methanospirillum hungatii *Methanomicrobium paynteri* *Methanogenium cariaci* *Methanomicrobium methisnigri* *Methanoplamus limicola*
Halobacteria	*Halobacterium* group	*Halobacterium volcanii* *Halobacterium cutirubrum* *Halobacterium sodomense* *Halobacterium trypanicum* *Halococcus morrhuae*
Thermoplasma	*Thermoplasma* group	*Thermoplasma acidophilum*
Thermococcus	*Thermococcus* group	*Thermococcus celer*
Extreme thermophiles	*Sulfolobus*	*Sulfolobus solfotaricus*

"streptococci" and should be considered a coccus-type cell. Spirillum-type cells grow as regularly curved rods that form continuous spiral filaments.

The general appearance of *Methanosarcina barkeri* is shown in Figures 3.12–3.15. The cell envelope structure is of Gram-positive type, although atypical in appearance. The amorphous outer layers of the wall often appear as laminations (Figures 3.13 and 3.14). It is diffusely attached to a more electron-dense layer closely apposed to the plasma membrane (Figure 3.15). The thick-layered wall of *M. barkeri* resembles that of *Sarcina ventriculi*. The cytoplasm of *Methanosarcina* is densely packed with ribosomes, and the DNA appears evenly distributed throughout the cell. Internal membranes of

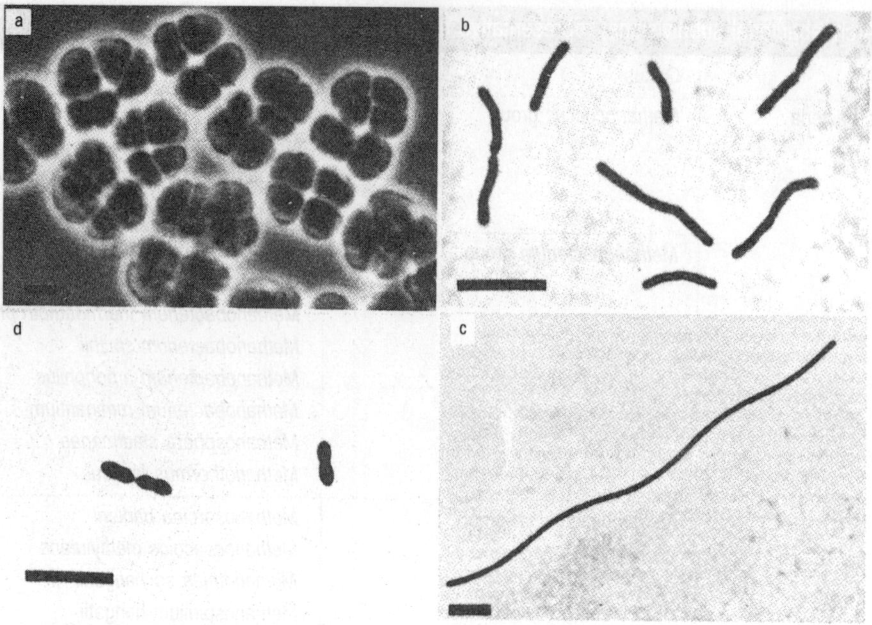

Figure 3.11 Phase-contrast photomicrograph of (a) *Methanosarcina barkeri*, (b) *Methanobacteriumc thermoautotrophicum*, (c) *Methanobacterium ruminantium*, (d) *Methanospirillum* strain 3P3 [Bar indicates µm]

Source Zeikus and Bowen (1975)

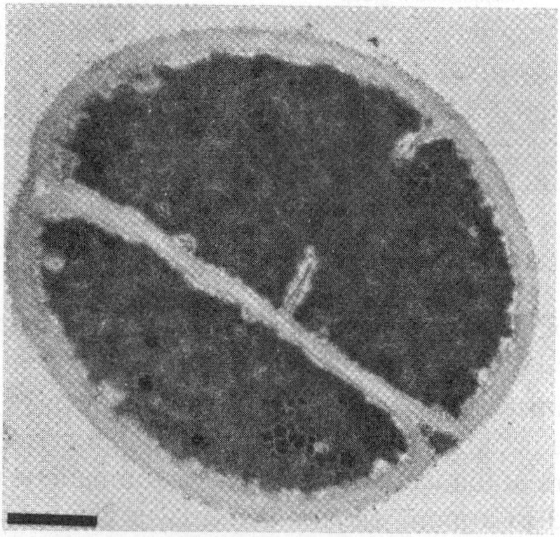

Figure 3.12 Thin section of *Methanosarcina barkeri* revealing cell segmentation and cross-wall formation [Bar indicates 0.52 µm]

Source Zeikus and Bowen (1975)

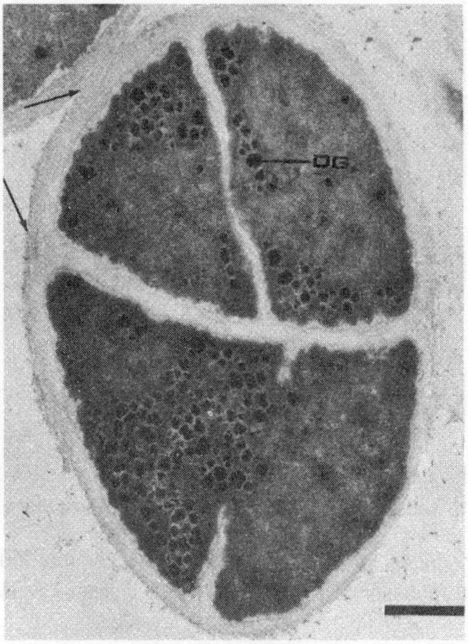

Figure 3.13 Section through *Methanosarcina barkeri* illustrating a multilayered outer wall (arrows) and numerous dense cytoplasmic granules [Bar indicates 0.49 μm]

Source Zeikus and Bowen (1975)

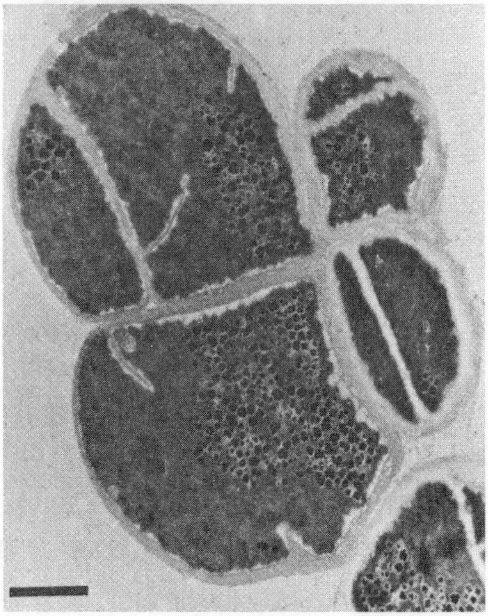

Figure 3.14 Micrograph of *Methanosarcina barkeri* showing the variation in size of segmented daughter cells [Bar indicates 0.71 μm]

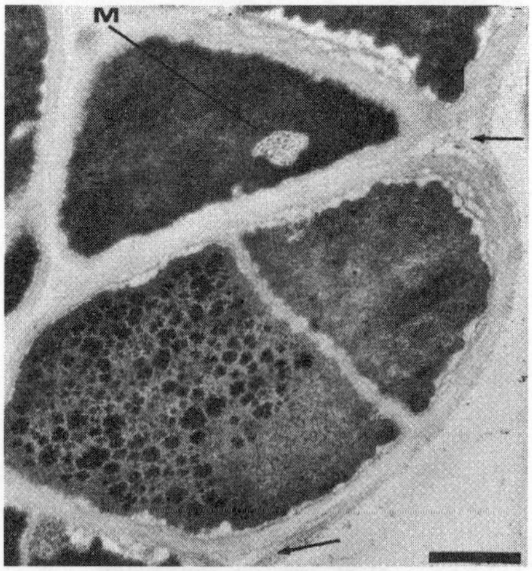

Figure 3.15 Thin section of *Methanosarcina barkeri* revealing the common outer wall layer shared by daughter cells (arrows) and a mesosomal structure (M) [Bar indicates 0.54 μm]

Source Zeikus and Bowen (1975)

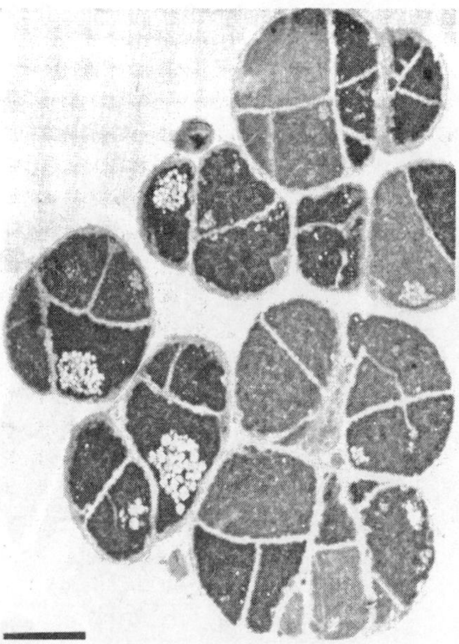

Figure 3.16 Low-power survey micrograph showing a mature *Methanosarcina* aggregate [Bar indicates 1.26 μm]

Source Zeikus and Bowen (1975)

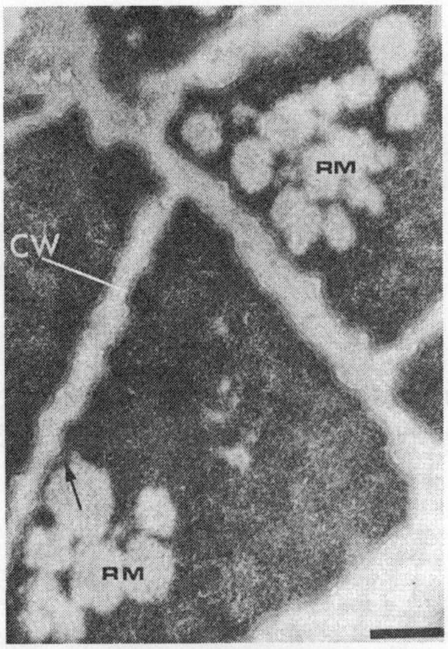

Figure 3.17 High-power micrograph of *Methanosarcina barkeri* illustrating an electron-dense cell wall layer (CW) attached to the cytoplasmic membrane and regions of low electron density that probably contain reserve materials (RM) [Bar indicates 0.20 μm]

Source Zeikus and Bowen (1975)

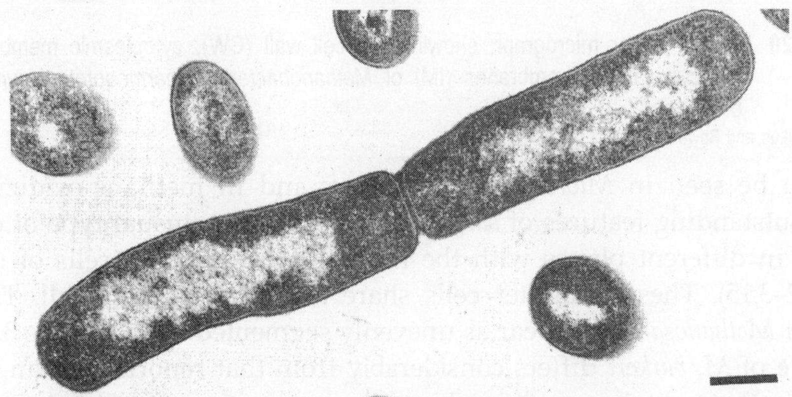

Figure 3.18 Grazing sections revealing the general ultrastructural appearance of *Methanobacterium* strain MoH [Bar indicates 0.19 μm]

Source Zeikus and Bowen (1975)

the vesicular, mesosomal type are often present. Cells at log phase cultures have been shown to contain numerous electron-dense granules of unknown composition (Figures 3.12–3.15).

Cytoplasmic regions of low electron density with circular to oval profiles are often present in some older cells (Figure 3.19). These inclusion bodies do not appear to have limiting membranes (Figure 3.20) and probably contain reserve materials. Similar

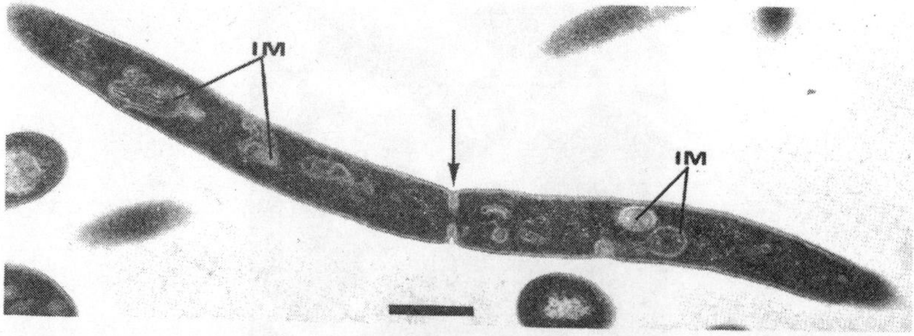

Figure 3.19 Thin section of *Methanobacterium formicicum* illustrating the mechanism of cell division (arrow) and numerous intracytoplasmic membranous bodies (IM) [Bar indicates 0.36 μm]
Source Zeikus and Bowen (1975)

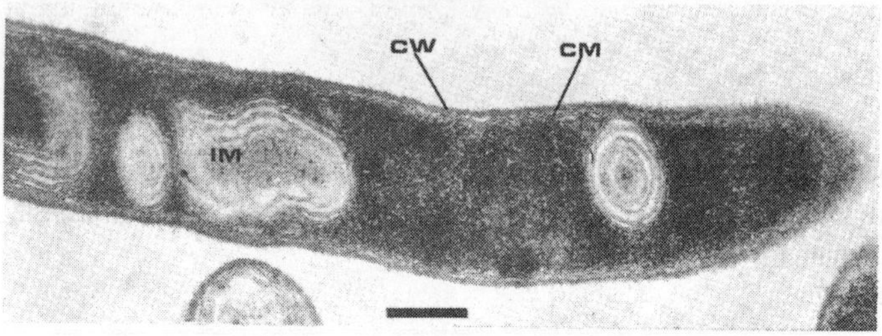

Figure 3.20 A high-power micrograph showing the cell wall (CW), cytoplasmic membrane (CM), and intracytoplasmic membranes (IM) of *Methanobacterium thermoautotophicum* [Bar indicates 0.15 μm]
Source Zeikus and Bowen (1975)

granules can be seen in *Micrococcus radiodurans* and in methane-oxidizing bacteria. One of the outstanding features of *sarcina*-type cells is the unusual type of cell division. Cells divide in different planes with the formation of daughter cells of unequal size (Figures 3.12–3.15). These daughter cells share a common outer wall. Thus, mature aggregates of *Methanosarcina* appear as unevenly segmented cocci (Figure 3.12). The fine ultrastructure of *M. barkeri* differs considerably from that reported for an undescribed *Methanosarcina* that contains gas vacuoles.

The ultrastructure of *Methanobacterium* strain MoH, *Methanobacterium thermoautotrophicum*, and *Methanobacterium formicicum* as shown in Figures 3.13–3.15. It can be seen that *Methanobacterium* species have a sharply defined, smooth, Gram-positive wall, along with a well-delineated cytoplasmic membrane, a disperse granular ribosome region, and a centrally located nuclear material with DNA fluids. The cell division mechanism (Figure 3.14) of *Methanobacterium* species resembles that of a typical Gram-positive bacterium like *Bacillus subtilis*. Many *Methanobacterium* species have been found to contain numerous intracytoplasmic membranes (Figures 3.14 and 3.15). These membranous inclusions are most numerous in *M. formicicum* and *M. thermoautotrophicum*.

These organisms have been shown to consist of closely apposed unit membranes that are formed from invagination of the plasma membrane. Cytoplasmic reserve materials are not observed in *Methanobacterium* cells.

The sections of *Methanobacterium ruminantium,* shown in Figures 3.19–3.21, are representative of coccus type of cells. The Gram-positive cell envelope of this organism has a distinctive triple-layered appearance. The wall consists of an inner electron-dense layer closely adjoined to the plasma membrane, followed by a thicker and more electron-transparent middle layer and a rough irregular outer layer. The cytoplasm is highly granular, and the DNA is obscured by ribosomes. Membranous cytoplasmic bodies resembling mesosomes are also present and often associated with cell division. Also, cytoplasmic inclusion bodies of low-electron density or high-electron density are not present. Cells of *M. ruminantium* appear to be undergoing constant cell division; before one cross-wall is completed, a second division is initiated. The fine structure of *M. ruminantium* is very similar to that reported for streptococci.

The general appearance of *Methanospirillum* cells is shown in Figures 3.19 and 3.20. The ultrastructure of *Methanospirillum* strain 3P3 differs dramatically from other methanogenic cell types.

Methanogenic bacteria are the best characterized group among the archaebacteria. Although they exhibit a wide variety of morphologies (Figures 3.17–3.24) and are found in many habitats, they are homogeneous with regard to metabolism. All methanogenic bacteria are anaerobes that have the unique ability to synthesize CH_4 from H_2 and various C_1 compounds such as CO_2, formate, methyl amines, methanol or acetate, depending on the organism. Regardless of the substrate, genes or species, methanogenesis appears to proceed through a common step involving the enzymatic reductive demethylation

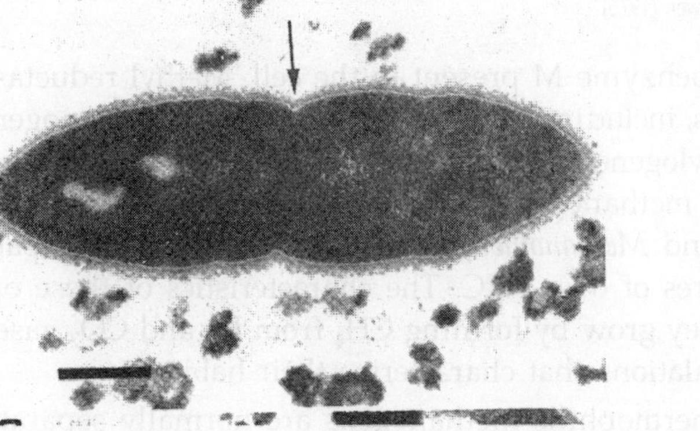

Figure 3.21 Longitudinal section illustrating general appearance of *Methanobacterium ruminantium.* Arrow shows developing cross-wall. [Bar indicates 0.37 μm]

Source Zeikus and Bowen (1975)

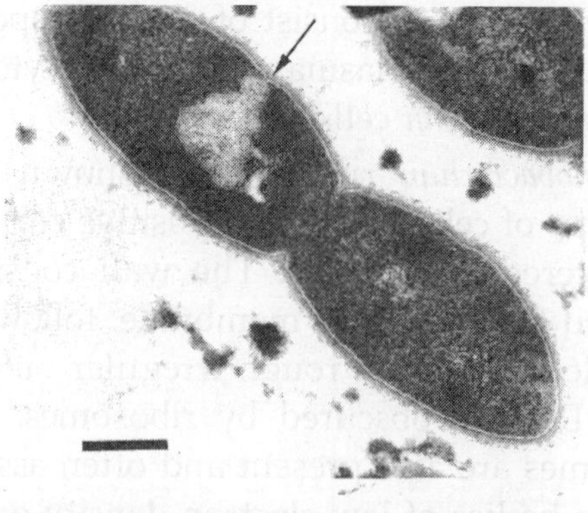

Figure 3.22 Thin section of *Methanobacterium ruminantium* undergoing cell division. Arrow shows membrane association with the second division site. [Bar indicates 0.32 μm]

Source Zeikus and Bowen (1975)

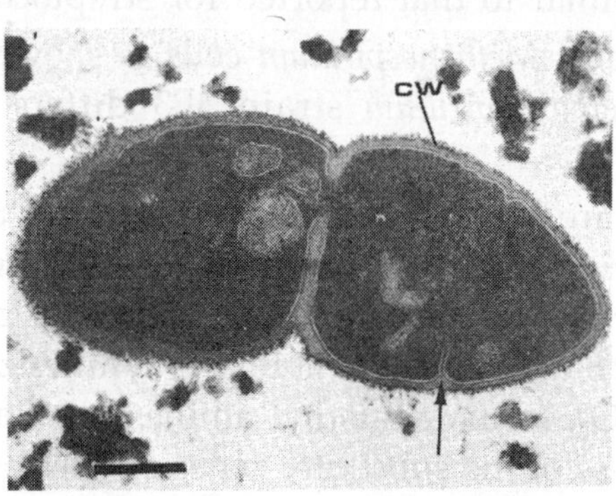

Figure 3.23 Longitudinal section of *Methanobacterium ruminantium* illustrating triple-layered cell wall (CW) and cross-wall development (arrow) in the second division plane [Bar indicates 0.21 μm]

Source Zeikus and Bowen (1975)

of the methyl coenzyme M present in the cell. Methyl reductase enzyme is present in all methanogens, including the highly thermophilic methanogens; it has been proposed as a simple phylogenetic marker for the four families of methanogens. Among the four families of methanogens, only two are extremely thermophilic microorganisms—*Methanococcus* and *Methanothermus*. These organisms are capable of growing even in high temperatures of over 80°C. The characteristics of these organisms are presented in Table 3.10. They grow by forming CH_4 from H_2 and CO_2 gases present in abundance in volcanic exhalations that characterize their habitats.

Moderately thermophilic methanogens are normally separated from sewage sludge and thermally heated sediments and wastes. Reactions in these anaerobic, thermophilic microorganisms in degradation of waste biomaterials and associated production of

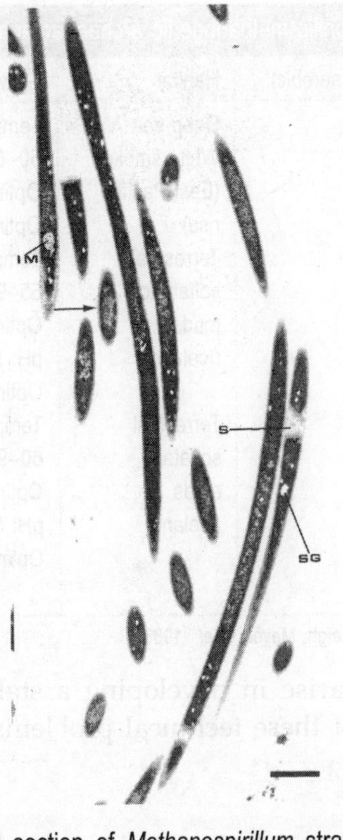

Figure 3.24 Grazing and longitudinal section of *Methanospirillum* strain 3P3 revealing double track cell wall (arrow), intracytoplasmic membranes (IM), storage granules (SG), and structured cell spacer (S) (connected spacer [CS] and the inner [IW] and outer [OW] layers of the cell wall). [Bar indicates 0.31 μm and 0.13 μm, respectively]

Source Zeikus and Bowen (1975)

CH_4 as a clean fuel are biotechnologically feasible, although efficiency and economics still pose a problem. In recent years, approaches and techniques of molecular biology and genetic engineering are being used in order to enhance the efficiency of these microbial cells. However, the established techniques of modern molecular biology, as applied to eubacteria, may not be applicable to archaebacteria. More research information about gene regulation in archaebacteria is required for understanding and development of successful approaches to the molecular cloning and expression of the genes. These approaches require overcoming technical problems involved in constructing the recombinant DNA molecule of the bacteria. One of the major problems in using the methane-producing bacteria in biomethanation is the slow growth rate. In order to overcome this problem, recombinant DNA technology has been used to obtain methane-producing bacteria with high specific growth rates for higher biological efficiency.

Table 3.10 Characteristics of thermophilic methanogenic archaebacterial cells

Organism DNSI number	Metabolism (anaerobic)	Habitat	Growth condition	Miscellaneous character
Methanococcus jannaschii 266i	Obligate autotroph $H_2/CO_2 \rightarrow CH_4$	Deep sea White smoker (East-Pacific rise)	Temperature: 50–86°C Optimum: 85°C Optimum pH: 6	Irregular cocci, 0.8–1.5 μm, stimulates growth, G+C 31%
Methanothermus fervidus (3537, 3538)	Obligate autotroph $H_2/CO_2 \rightarrow CH_4$	Terrestrial solfataric muds (Iceland)	Temperature: 55–97°C Optimum: 80°C pH: 5.5–7.5 Optimum: 6.5	0.3–0.4 μm, by 1–4 μm growth, G+C 33%
Methanotherus sociabilis (3496)	Obligate antroph H_2/CO_2 CH_4	Terrestrial solfataric muds (Iceland)	Temperature: 60–97°C Optimum: 80°C pH: 5.5–7.5 Optimum: 6.5	Rod shaped (0.3–0.4 μm, by 3–5 μm) forms aggregates, G+C 33% G+C – Guanine + Cystosine

Sources Wiegel and Ljungdahl (1986); Jones, Leigh, Mayer, *et al.* (1983)

Several technical problems arise in developing a stable γ-DNA molecule from the *Methanosarcina* species. Some of these technical problems are presented in Table 3.6.

3.6 YEASTS

Yeasts are microorganisms belonging to a group of fungi and are widely used in bioprocessing for production of useful materials of industrial and commercial importance. This group of microorganisms usually does not have true mycelium and are predominantly unicellular. However, some of them may develop in forms similar to true mycelium. Yeasts are larger in size than bacteria and require a slightly more acidic environment than bacteria to survive. A typical yeast cell (Figures 3.25a–3.25e) may be ovoid, spherical or sausage shaped, with inclusion bodies scattered throughout the granular cytoplasm. The energy reserve carbohydrate material stored in this organism is glycogen. Oil droplets are also found in older cells and in certain species of yeasts. Volutin granules or mitochondria are found in the cytoplasm and are considered to be storehouses of the enzyme system, which acts as the catalytic fuel cell in the organism.

Yeasts are mostly saprophytic in character. Taxonomically, most yeasts belong to the family Saccharomycetaceae of the order Endomycetales and class Ascomycetes. Generally, most yeast cells reproduce asexually under balanced and good growth conditions. The proliferation of yeasts takes place by budding or binary fission. In general, yeasts may be found on a wide variety of substrates, including soil, water, plant, fruit, nectar of flower, and exudates of tree and animals.

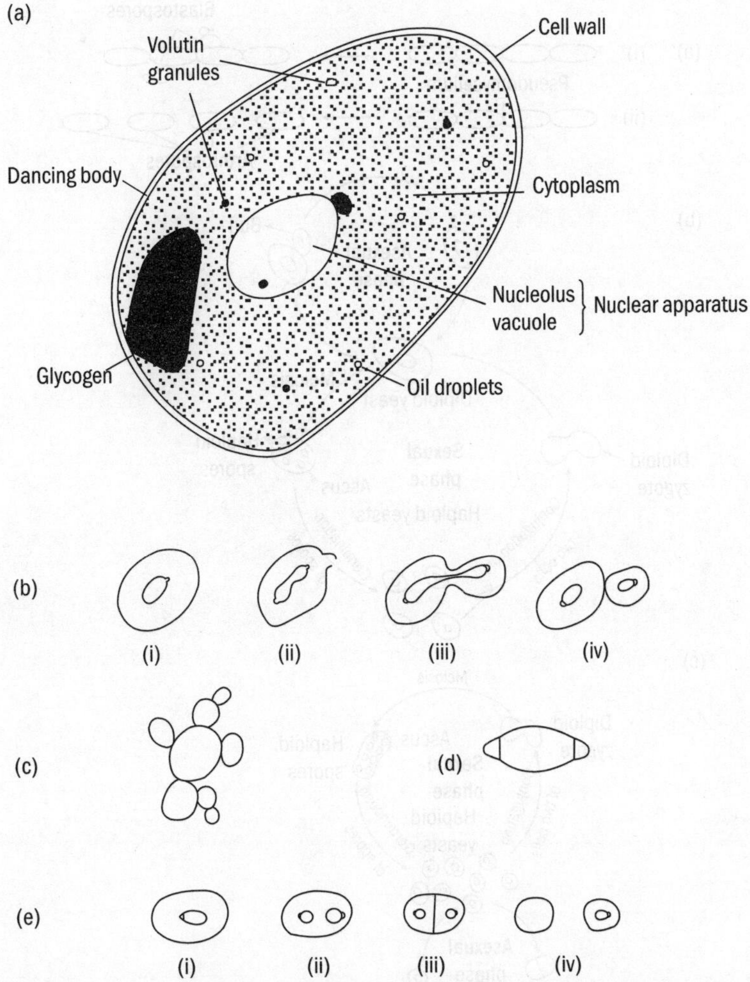

Figure 3.25 (a) Diagram of a "typical" yeast cell; (b) (i–iv) Stages in the formation of a "bud"; (c) Multipolar budding resulting in a mass of adhering yeast cells; (d) Polar budding in an apiculate yeast cell; (e) (i–iv) Cell division of a "fission" yeast

Figures 3.26a(i) and 3.26a(ii) show pseudomycelial yeast producing blastospores and arthrospores, respectively. Figures 3.26b and 3.26c show the life cycles of a diplobiontic and haplobiontic yeast. Figure 3.26d shows the shapes of ascospores in various yeasts.

3.7 CELL AND TISSUE SYSTEM BIOLOGY

It is possible to grow plant and animal tissues and cells as a culture. Such cultures can be grown in a suitably designed bioreactor using proper strategy. Both rotating and

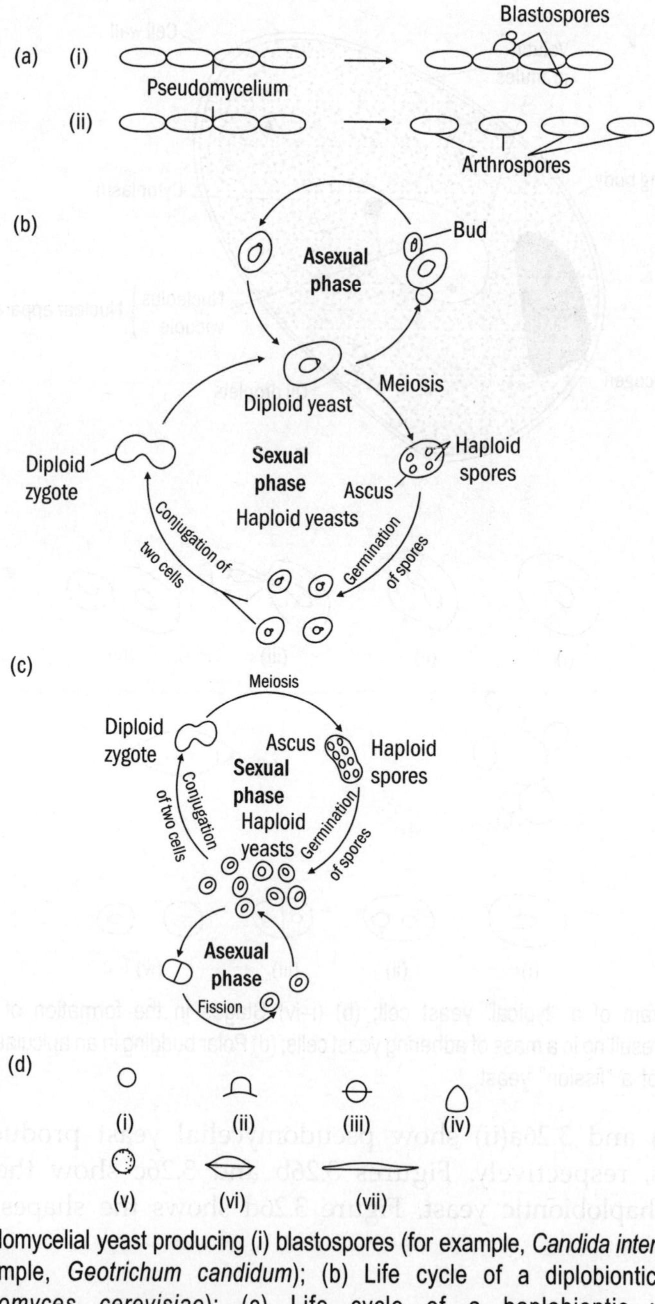

Figure 3.26 (a) Pseudomycelial yeast producing (i) blastospores (for example, *Candida intermedia*); (ii) arthrospores (for example, *Geotrichum candidum*); (b) Life cycle of a diplobiontic yeast (for example, *Saccharomyces cerevisiae*); (c) Life cycle of a haplobiontic yeast (for example, *Schizosaccharomyces octosporus*); (d) Shapes of ascospores in various yeasts: (i) Spherical and smooth (for example, *Saccharomyces*); (ii) Bowler hat-shaped (for example, *Hansenula*); (iii) Saturn-shaped (for example, *Hansenula saturnus*); (iv) Helmet-shaped (for example, *Zygopichia*); (v) Spherical and warty (for example, *Debaryomyces*); (vi) Walnut-shaped and warty with equatorial rim (for example, *Schwanniomyces*); (vii) Needle-shaped (for example, *Monosporella*)

non-rotating types of bioreactors have been used for different types of tissues. With appropriate stress agents, a bioreactor can synthesize useful stress protein products (Figures 3.27 and 3.28).

The tissue culture of gall cells of plants could regenerate species having higher productivity using suitable bioprocessing and biomolecular separation strategy (Figure 3.29). A foreign gene is inserted into the Ti plasmid, and this is used to infect plants, producing a crown gall. The plant cells in the gall contain the Ti plasmid with its piece of foreign DNA. Gall cells are then grown in culture to produce plantlets, which can be transferred to soil where they grow into mature plants. Since each plant is derived from a single cell carrying the foreign gene, all the cells in the fully grown plant contain that gene.

3.7.1 Plant Tissue Culture

Plant cell cultures have been grown in various conventional and semi-conventional stirred tank bioreactors (STB) of 20 000–75 000 L. Reports from many investigators indicated that the STB design, as well as the airlift bioreactor design, may not be suitable for high density (200–400 g/L wet biomass or 10–20 g/L dried biomass) plant cell suspension cultures. The major reason for this is associated with shear sensitivity and the rapid settling character of cells by the solid–solid interfacial phenomenon for plant cell aggregation, as well as stickiness, foaming, and cell flotation tendencies, and viscous shear timing behaviour of culture fluids, whose homogeneous mixing at low shear should not be underestimated. For proper oxygenation, this may also become problematic in the conventional bioreactor as well as in an airlift pneumatic oxygenator. However, production of secondary metabolite-like steroidal alkaloids (for example, conessine) in a suspension culture of *Holarrheria antidysenterica* in Murashige and Skoog

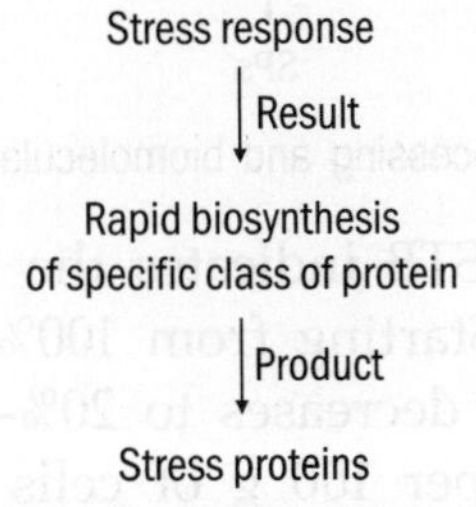

Figure 3.27 Typical strategy of stress protein formation

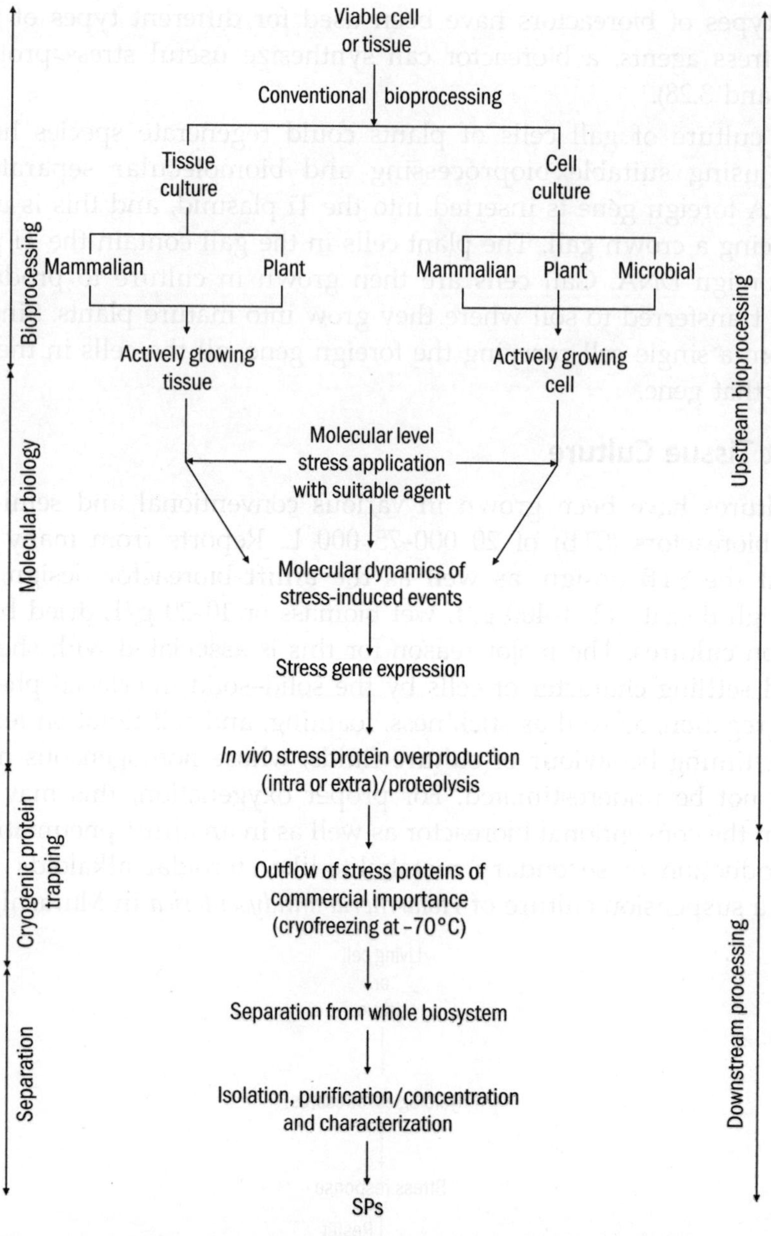

Figure 3.28 Bioprocessing and biomolecular separation strategy

(MS) medium in a conventional STB indicates the feasibility of this design in plant cell cultivation for oxygenation. Starting from 100% oxygen saturation set point, the percentage of O_2 saturation level decreases to 20%–30% during cell cultivation, thus resulting in 110 mg of alkaloid per 100 g of cells per day. Also, since commercial

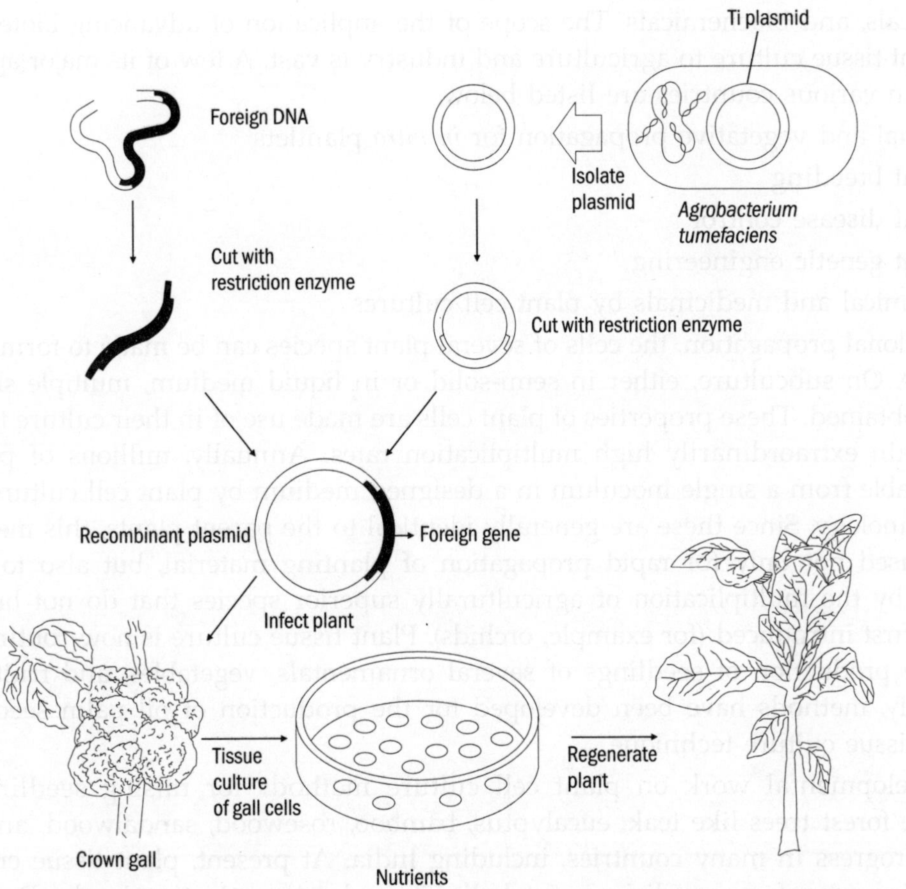

Figure 3.29 Plant genetic engineering with Ti plasmids from *Agrobacterium tumefaciens*

exploitation of plant cell culture for the production of embryos, artificial seed, and valuable phytochemicals lags behind microbial, mammalian, and insect cell culture processes, many recently developed mixer design bioreactors for high-density plant cell suspension cultures have been developed in recent years. Unconventional bioreactors like the horizontal rotary thin-film bioreactor (HRTFB), which have been used in microbial cell culture, have also been tested for their suitability as oxygenators in a plant cell suspension culture system. The membrane stirrer bioreactor is another type that has been used in plant cell culture. Both of these unconventional bioreactors have been developed to perform better in improving oxygenation efficiency without imparting shear stress.

Rapid progress has been made in medium design for plant tissue culture biotechnology in recent years. This biotechnology has been applied to agriculture, horticulture, chemicals, and biochemicals. The scope of the application of advancing biotechnology in plant tissue culture to agriculture and industry is vast. A few of its major application areas in various countries are listed below.

- Clonal and vegetative propagation for *in vitro* plantlets
- Plant breeding
- Plant disease control
- Plant genetic engineering
- Chemical and medicinals by plant cell cultures

By clonal propagation, the cells of several plant species can be made to form plantlets *in vitro*. On subculture, either in semi-solid or in liquid medium, multiple shoots are often obtained. These properties of plant cells are made use of in their culture technique to obtain extraordinarily high multiplication rates. Annually, millions of plants are obtainable from a single inoculum in a designed medium by plant cell culture process biotechnology. Since these are generally identical to the parent plants, this method has been used not only for rapid propagation of planting material, but also to increase yields by the multiplication of agriculturally superior species that do not breed true when first introduced (for example, orchids). Plant tissue culture is now routinely used for the production of seedlings of several ornamentals, vegetables, and fruit species. Recently, methods have been developed for the production of oil palm seedlings by plant tissue culture technique.

Developmental work on plant cell culture methods for raising seedlings from mature forest trees like teak, eucalyptus, bamboo, rosewood, sandalwood, and poplar is in progress in many countries, including India. At present, plant tissue cultivation in a bioreactor for a medicinal metabolite is under investigation in the Department of Biochemical Engineering and Biotechnology, Indian Institute of Technology Delhi, India.

Many recent developments in plant cell culture in designed medium are due to studies related to an excrescence called crown gall, which is a type of tumour that afflicts many flowering plants. The lumpy crown gall itself consists of a mass of plant cells that proliferate rapidly. The growth and proliferation of cells in a crown gall is similar to that in mammalian tumours. The cause of this rapid growth is *Agrobacterium tumefaciens*, a bacterium that contains small circular pieces of DNA known as Ti plasmids, which induce tumour. In plant genetic engineering, Ti plasmids have opened up a new green revolution, providing a technique to regenerate plants. A crown gall is developed when Ti plasmids are transferred from the bacteria into the chromosomes of the infected plant. After infection, the resultant change in the genetic constitution of the plant cells induces them to grow and divide rapidly. Thus, Ti plasmid is a potential vector for introducing bacterial genes into plants. Through this technology, one can attempt to insert genes from one plant species into another. A schematic of plant genetic engineering with Ti plasmids from *A. tumefaciens* is shown in Figure 3.26.

3.7.2 Stem Cell System

Stem cells are non-specialized cells that have the dual capacity to divide indefinitely (self-renew) and to differentiate into more mature cells with specialized functions. They may be obtained from a cell bank, which is a storage facility for stem cell lines.

3.7.2.1 Stem cell criteria

Cells or populations of cells that fulfil certain criteria at a given instance are termed actual stem cells. On the other hand, cells that do not actually fulfil these criteria or capabilities are termed potential stem cells. Stem cells criteria have been listed in Table 3.11.

Deterministic and stochastic concepts of cell/population growth used in the application of stem cell/tissue engineering are given below.

- **Deterministic concept of population growth**

  Figure 3.30 shows the kinetic formation of stem cell proliferation. During proliferation, the function of postmeiotic stem cells remaining in the stem cell pool at steady conditions is 0.5.

Table 3.11 Stem cell criteria

Criteria	Stem cell	Transient cell	Matured cells
Differentiation marker	No	Onset	Yes
Proliferation	Yes	Yes	No
Self-maintenance	Yes ($P_{sm} \geq 0.5$ possible)	No ($0.5 > P_{sm} \geq 0$ possible)	No
Several programme cells	Yes	Limited	No
Regeneration of tissues after injury	Yes (long term)	Temporary	No
Flexibility in options	(2)–(5)	(2), (4)	No

P_{sm} – function of post-meiotic stem cells remaining in the stem cell pool

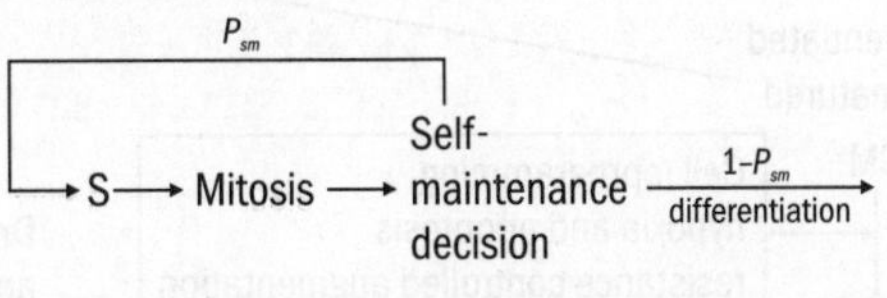

P_{sm} – function of postmeiotic stem cells remaining in the stem cell pool

Figure 3.30 Kinetic formation of stem cell proliferation

- **Stochastic concept of single-cell growth**
  During single-cell growth, the probability of symmetric cell division generating one stem cell at steady state is unity. If the probability of symmetric cell division generating two stem cells is equal to the probability of stem cell loss by symmetric cell division or extinction and if both are not unity, then long-term random extinction for finite populations is a probability.

  Figure 3.31 shows the growth of a single cell and Figure 3.32 shows the potential applications of human embryonic stem cell line.

3.8 CELLULAR MATERIALS AND METABOLISM IN BIOTECHNOLOGY

When nutrients reach the relatively enormous catalytic power inside the cell, microorganisms contribute to the key role they play in chemical/biochemical transformations, which has been termed cellular metabolism. This metabolism takes place through a biosystematics pathway in the presence of oxygen/air, as shown in

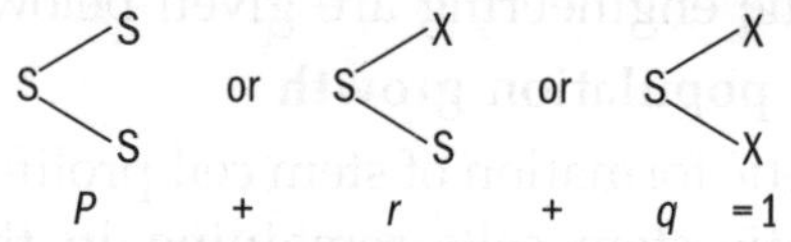

p – probability of symmetric cell division generating two stem cells; *r* – probability of symmetric cell division generating one stem cell; *q* –probability of stem cell loss by symmetric cell division or extinction

Figure 3.31 Growth of a single cell

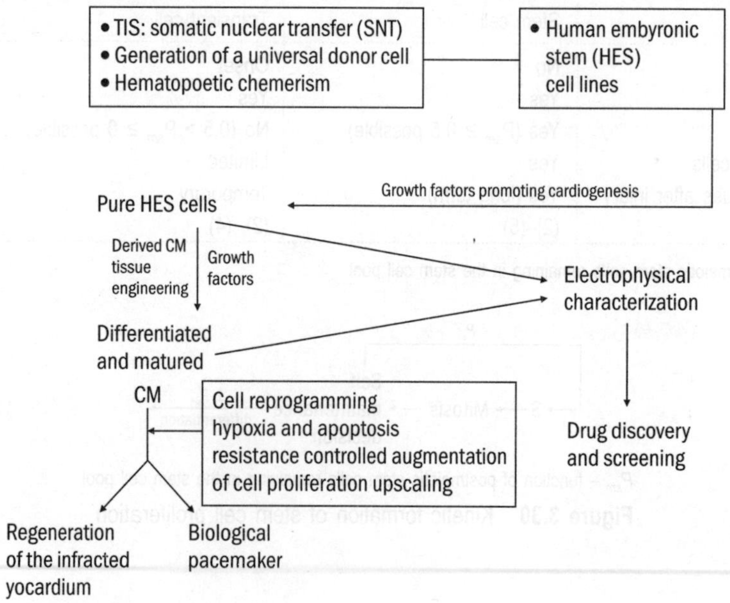

Figure 3.32 Potential applications of human embryonic stem cell line

Figure 3.33. Depending on the organism type, the carbon source in the nutrient medium may vary widely.

Utilization of alkanes as a carbon source for microbial production of useful substances has been extensively studied. Many important successes have been achieved, although increased price of alkanes due to the so-called "petroleum crisis" has discouraged such activities to an extent.

The microbial products from alkanes are classified into three groups (Table 3.12). The first group involves compounds produced by conventional fermentation from carbohydrate substrates, although metabolic pathways of alkenes and conventional carbon sources are different.

3.9 MICROBES AND MICROBIOLOGY

3.9.1 Biocell/microbial Cell Growth Principles

The features in the growth of a biocell are listed below (Figure 3.34).

• A biocell does not divide until it matures

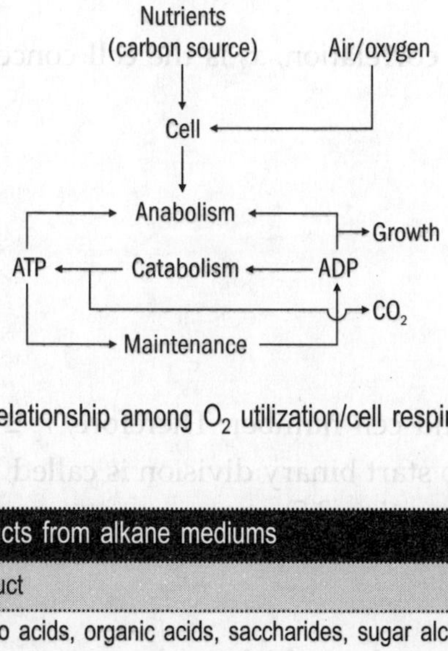

Figure 3.33 Biospheric pathway relationship among O_2 utilization/cell respiration, energy production (ATP), and cell growth

Table 3.12 Classification of products from alkane mediums

Product type	Product
Common to those from carbohydrate substrates	Amino acids, organic acids, saccharides, sugar alcohols, nucleosides and nucleotides, vitamins and coenzymes, antibiotics, enzymes, cell mass
Owing to physiological and metabolic features in alkane utilization	Lipopolysaccharides, lipoproteins, enzymes vitamins and coenzymes (FAD, FMN, CoA, CoQ, cytochromes, biotin vitamers, isoprenoids, and so on)
Specific to alkane substrates	Higher monocarboxylic and dicarboxylic acids, ω-hydroxy fatty acids, fatty alcohols, wax esters, methylcitrate, methylisocitrate, and so on

FAD – flavin adenine dinucleotide; FMN – flavin mononucleotide; CoA – coenzyme A; CoQ – coenzyme Q

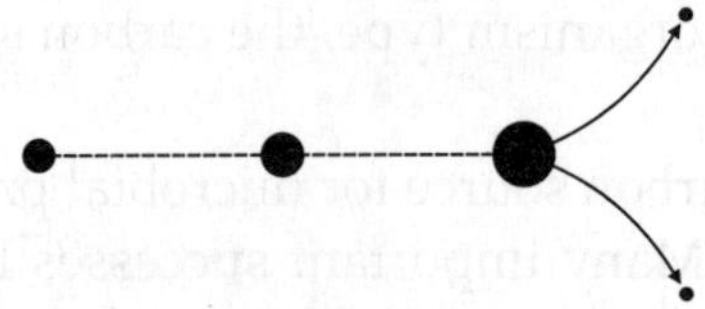

Figure 3.34 Simple representation of biocell growth

- When a biocell is placed in a medium, a few important manifestations are usually observed.
  - The size of the cell increases to a threshold limit suitable for cell division
  - Increase in size takes place due to swelling from osmotic intake of water
  - This is not considered growth because no synthesis of macromolecules of cells occurs
  - For growth to occur, a bivariant property of cell must be exhibited

The bivariational correlation is shown in Equation 3.1.

$$r_c = \frac{\Sigma x_1 x_2}{\sqrt{\Sigma x_1^2 \, \Sigma x_2^2}} \tag{3.1}$$

where r_c is the coefficient of correlation, x_1 is the cell concentration (g/L), and x_2 is the cell count.

$$\Sigma x_1 x_2 = \Sigma x_1 x_2 - \frac{\Sigma x_1 \, \Sigma x_2}{n} \tag{3.2}$$

$$\Sigma x_1^2 = \Sigma x_1^2 - \frac{\Sigma x_2^1}{n} \tag{3.3}$$

$$\Sigma x_1 x_2 = \Sigma x_1 x_2 - \frac{\Sigma x_1 \, \Sigma x_2}{n} \tag{3.4}$$

Thus, r_c relates cell size and cell number. Therefore, $r_c \leq 1$.

The time taken by a cell to start binary division is called lag time (t_l) and is measured by the following relation (Equation 3.5).

$$t_l = \frac{c' - \alpha v}{\beta x_o + v} \tag{3.5}$$

where c' is the critical value of c, which is the concentration of fission promoting metabolite (FPM). This can be verified from experimental data.

$$c = \alpha v + \beta x_o t + vt \tag{3.6}$$

where v is the culture filtrate volume (CFV) added in inoculum, x_o is the cell concentration at the time of inoculation, α is the concentration of critical substance, β is the average increase in critical substance per unit time per unit cell mass, and γ is the increase in critical substance due to internal cell material production.

Growth phases and growth rates in relation to segregation of mechanism of attachment to bacterial DNA to the membrane of cell in batch culture growth are shown in Figure 3.35.

3.9.1.1 Lag phase

The lag phase appears as a shock of rapid switch over to a new environment that needs adaptation. The time the cell takes to adapt to the new environment and start multiplying is termed lag time or germination lag. This time is a physiological distance between initial and steady state. Several reactions may determine the rate of this building process, depending on the cell's physiological/morphological form, that is, spore or vegetative.

Equation 3.6 can be used to assess FPM or CFU concentration. This equation relates FPM concentration (c), culture filtrate volume (v), and initial cell concentration (x_o). For multiplication to occur, $c = c'$ Again, c' is the critical concentration of FPM at $t = t_l$, which is the end of lag time.

When nutrient is provided to a cell inoculum, it is assumed that reproduction will take place. If in a suitable nutrient, one cell in the inoculum (say, X) is taken

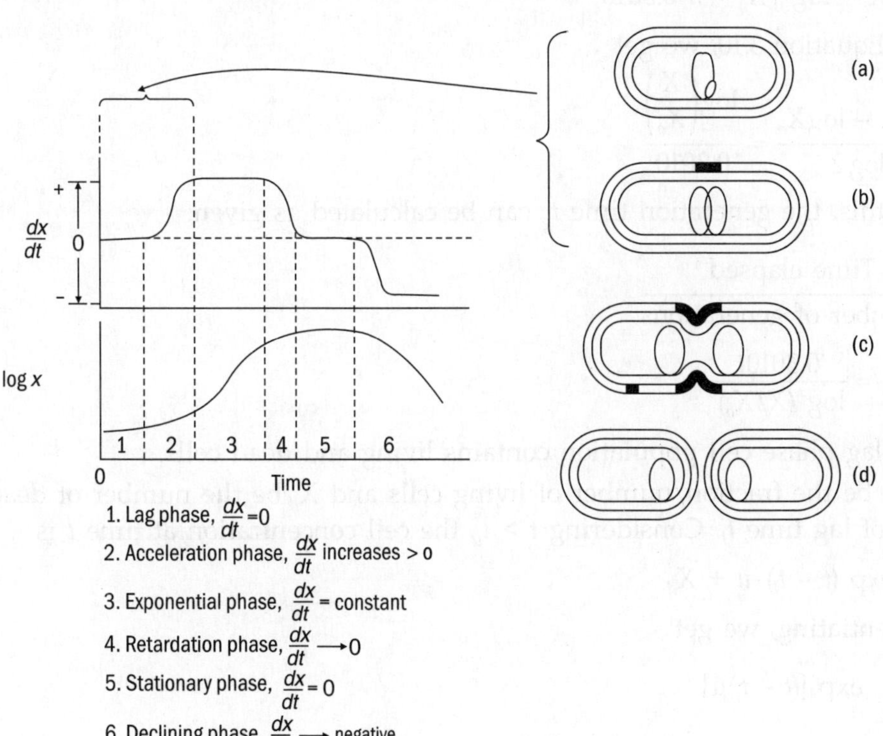

Figure 3.35 Growth phases and growth rates

into consideration, then the number of cells in any succeeding generation will be as shown.

1st generation: $X = 1 \times 2 = 2$

2nd generation: $X = 1 \times 2 \times 2 = 2^2 = 4$

3rd generation: $X = 1 \times 2 \times 2 \times 2 = 2^3 = 8$

.

.

nth generation: $X = 1 \times 2 \times \ldots = 2^n$

Thus, the following can be deduced from the above relations.

$$X = X_o \times 2^n \tag{3.7}$$

where X_o is the number of cells in the inoculum.

Taking $\log_2$ of both sides, we get the following relation.

$$\log_2 X = \log_2 X_o + n \tag{3.8}$$

$$\text{or } \log_{10} X = \log_{10} X_o + n \log_{10} 2 \tag{3.9}$$

$$\text{or } \log_{10} X = \log_{10} X_o + n\ 0.3010 \tag{3.10}$$

From Equation 3.10, we get

$$n = \frac{\log X - \log X_o}{\log 2} = \frac{\log\left(\frac{X}{X_o}\right)}{0.3010} \tag{3.11}$$

From this, the generation time t_g can be calculated as given.

$$t_g = \frac{\text{Time elapsed}}{\text{Number of generations}}$$

$$\text{or } t_g = \frac{t}{n} = \frac{0.3010t}{\log (X/X_o)} \tag{3.12}$$

Thus, lag phase cell population contains living and dead cells.

Let X_o be the fraction/number of living cells and X_d be the number of dead cells at the end of lag time t_l. Considering $t > t_l$, the cell concentration at time t is

$$X = X_o \exp (t - t_l) \cdot u + X_d \tag{3.13}$$

Differentiating, we get

$$\frac{dX}{dt} = \mu X_o \exp [(t - t_l)\mu] \tag{3.14}$$

$$\text{or } \frac{dX}{dt} = \mu(X - X_d) \tag{3.15}$$

Since the cells start growing/proliferating exponentially,

$$\frac{dX}{dt} = \mu' x \tag{3.16}$$

where μ′ is the apparent value of μ.

Thus, from Equations 3.15 and 3.16, we get

$$\mu(X - X_d) = \mu' X \tag{3.17}$$

$$\text{or } \mu' = \frac{\mu(X - X_d)}{X} = \mu\left(1 - \frac{X_d}{X}\right) \tag{3.18}$$

$$\text{So, } \frac{X_d}{X} = (\mu - \mu')/\mu = \left(1 - \frac{\mu'}{\mu}\right) \tag{3.19}$$

This gives the ratio of non-viable to viable cells. Substituting Equation 3.10 in Equation 3.19, we get

$$\frac{X_d}{X_o \exp(t - t_l)\mu + X_d} = \left(1 - \frac{\mu'}{\mu}\right) \tag{3.20}$$

That is,

$$\frac{X_o}{X_d} \exp\,(t - t_l)\mu = \frac{\mu'}{\mu - \mu'} \tag{3.21}$$

$$\text{or } \frac{X_d}{X_o} = \frac{\mu - \mu'}{\mu'} \exp\,(t - t_l)\mu = \left(\frac{\mu}{\mu'} - 1\right) \exp\,(t - t_l)\mu \tag{3.22}$$

In lag phase, if the death rate is proportional to the number of living cells with a rate constant K_d, the cell loss during the this period is given by

$$-\frac{1}{X}\frac{dX}{dt} = K_d \tag{3.23}$$

Thus, at the end of lag time t_l, the cell number is

$$X = X_d + X_o \exp\,(-K_d\, t_l) \exp\,[\mu(t - t_l)] \tag{3.24}$$

Thus, the non-viable to viable fraction of cells is

$$\frac{X_d}{X_o} = \left(\frac{\mu}{\mu'} - 1\right) \exp\,[\mu\,(t - t_l)] \exp\,(-\,K_d t_l) \tag{3.25}$$

3.9.1.2 Log phase

1. Growth

In this phase, the viable cells are well adapted in the new culture medium conditions and grow exponentially. Thus,

$$\frac{dX}{dt} = \mu X \tag{3.26}$$

Taking $X = X_0$ at $t = t_l$ and integrating Equation 3.26, we get

$$\int_{X_0}^{X} \frac{dX}{X} = \mu \int_{t=t_l}^{t} dt$$

$$\text{or } \ln \frac{X}{X_0} = \mu(t - t_l) \tag{3.28}$$

$$\text{or } X = X_0\, e^{\mu(t - t_l)} \tag{3.29}$$

where $t > t_l$

2. Growth and substrate uptake rate relation

In case of cell growth without producing any product, it is observed that cell growth rate is proportional to the substrate (carbon source) uptake rate by the cells. In log phase, Equation 3.29 may also be written in terms of cell weight (X) as the cell number is proportional to cell mass (weight). Thus

$$\frac{dX}{dt} \propto \left(-\frac{ds}{dt}\right) = Y_G\left(-\frac{ds}{dt}\right) \tag{3.30}$$

where Y_G is the proportionality factor known as growth yield coefficient and S is the substrate concentration (Figure 3.36).

The consumption rate of carbon source $\left(-\frac{ds}{dt}\right)$ is

$$\left(-\frac{ds}{dt}\right) = \left(-\frac{ds}{dt}\right)_m + \left(-\frac{ds}{dt}\right)_G \tag{3.31}$$

where $\left(-\frac{ds}{dt}\right)_m$ is the consumption rate of carbon source for maintenance and $\left(-\frac{ds}{dt}\right)_G$ is the consumption rate for true growth.

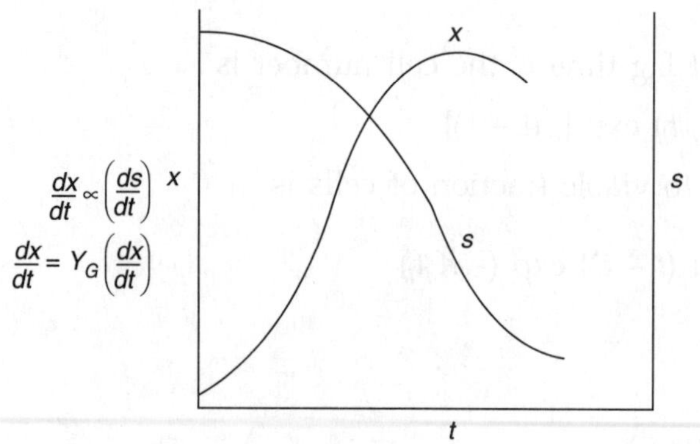

Figure 3.36 Time profile of cell growth and substrate uptake

Thus,

$$\left(-\frac{ds}{dt}\right) = \left(-\frac{ds}{dt}\right)_m + \left(-\frac{ds}{dt}\right)_G \tag{3.32}$$

In energy-coupled growth (Figure 3.37), Equations 3.33–3.36 may hold good.

$$\left(-\frac{ds}{dt}\right)_m = mX \tag{3.33}$$

and $$\left(-\frac{ds}{dt}\right)_G = \frac{1}{Y_G}\mu X \tag{3.34}$$

In these relations (Equations 3.33 and 3.34), m is the maintenance coefficient (mole substrate/g cell/h) and Y_G is the growth yield constant (g cell/mole substrate).

From Equation 3.32, we get

$$\left(-\frac{ds}{dt}\right) = -\left[mX + \frac{1}{Y_G}\mu X\right] \tag{3.35}$$

Therefore,

$$\frac{1}{X}\frac{ds}{dt} = m + \frac{1}{Y_G}\mu \tag{3.36}$$

or $$q = m + \frac{1}{Y_G}\mu$$

where q is the specific limiting nutrient/substrate uptake rate (Figure 3.38).

Substituting Equation 3.36 in Equation 3.30, we get

$$\left(\frac{dX}{dt}\right) = Y_G\left(-\frac{ds}{dt}\right) - mY_GX \tag{3.37}$$

In Equation 3.37, if the first term on the right-hand side is high compared to the second term, then

$$Y_G\left(-\frac{ds}{dt}\right) >> mY_GX \tag{3.38}$$

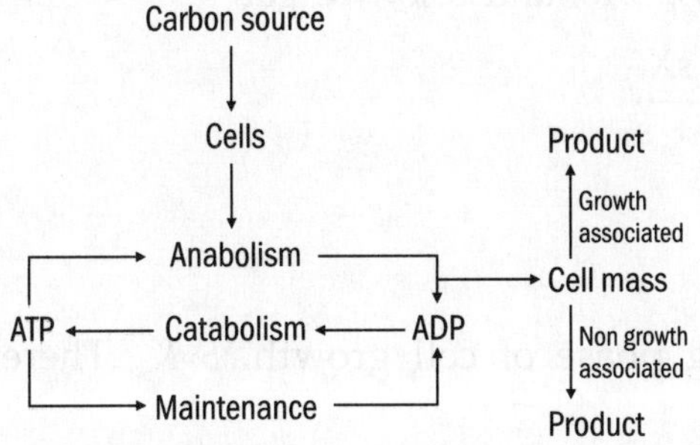

Figure 3.37 Anabolism and catabolism with energy couple

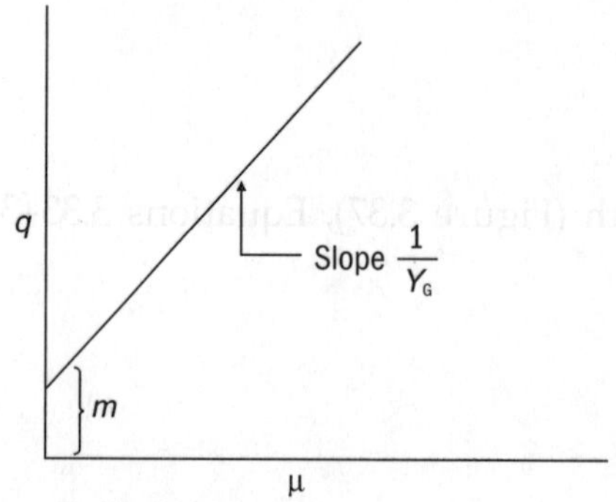

Figure 3.38 Specific substrate uptake rates as a formation of specific growth rate

$$\text{and } \frac{dX}{dt} = Y_G\left(-\frac{ds}{dt}\right) \quad (3.39)$$

If the concentration of limiting nutrient in the medium varies and the change in the concentrations of all other nutrients is negligible, then the growth rate changes hyperbolically. This variation of rate limiting nutrient concentration with respect to specific growth rate (μ) was proposed by Monod as shown in Equations 3.40 and 3.41.

$$\frac{1}{X}\frac{dX}{dt} = \mu = \frac{\mu_m s}{K_s + s} \quad (3.40)$$

$$\text{or } \frac{dX}{dt} = \frac{\mu_m sX}{K_s + s} \quad (3.41)$$

Also, in this cell growth phase, the specific limiting nutrient/substrate uptake rate (q) follows similar relation with respect to the Michaelis–Menten law. Thus,

$$\frac{1}{X}\left(-\frac{ds}{dt}\right) = q = q_m \frac{s}{K_m + s} \quad (3.42)$$

$$\text{or } -\frac{ds}{dt} = q_m \frac{s}{K_m + s} X \quad (3.43)$$

So, from Equations 3.39, 3.40, and 3.43, we get

$$\frac{dX}{dt} = Y_G\, q_m \frac{sX}{K_m + s} = \frac{\mu_m sX}{K_s + s} \quad (3.44)$$

That is,

$$\mu_m = Y_G \times q_m \quad (3.45)$$

In the exponential/log phase of cell growth, $S \gg K_s$. Therefore, from Equation 3.45, we get

$$\mu_m = \frac{1}{X}\frac{dX}{dt}$$

(since K_s is negligible) (3.46)

Thus, in this phase

$$\frac{dX}{dt} = \mu_m X \quad (3.47)$$

From Equation 3.47, we get

$$\int \frac{dX}{X} = \mu_m \int dt \quad (3.48)$$

$$\int_{X_o}^{X} \frac{dX}{X} = \mu_m \int_{0}^{t} dt \quad (3.49)$$

$$\text{or } \ln \frac{X}{X_0} = \mu_m^t \quad (3.50)$$

$$\text{or } X = X_0 e^{\mu_m^t} \quad (3.51)$$

When $X = 2X_0$ and $t = t_g$, the doubling or generation time. From Equation 3.50, we get

$$\ln \frac{2X_0}{X_0} = \mu_m t_g$$

$$\text{or } t_g = \frac{\ln 2}{\mu_m} = \frac{0.063}{\mu_m} \quad (3.52)$$

If the cell population/mass was to grow, without change, at a rate μ in the base biomass, a longer time (T) would be required for the population/mass to double, which is simply the reciprocal specific growth rate (Equation 3.53).

$$\text{or } \tau = \frac{1}{\mu} = 1.45 = t_g \quad (3.53)$$

where τ is the birth time or the time at which an organism must exist before dividing when present in a population having a particular growth constant μ.

μ_m as a function of temperature

In log phase, growth proceeds at a rate (maximum) that increases with temperature until cell functions are impaired. The change in μ_{max} with temperature is expressed by a temperature quotient (Q_{10}), which is the temperature rise that causes the growth rate to double. It is also expressed in terms of activation energy (E_a) as given in Equation 3.54.

$$\mu_m = A \exp (E_a/RT) \quad (3.54)$$

3. Cell quota

In some organisms, it was noticed that the amount of limiting nutrient per cell decreased with growth rate as it becomes limiting. These observations led to the development of the concept of cell quota (Q'). As inter-species and chemical comparisons are impossible without knowledge of cell size, the value of Q' is sometimes defined as the "nutrient per gram of cells". Cell quota approaches the minimum value K_Q.

$$\mu = \mu'_m \left[1 - \frac{K_Q}{Q'}\right] \quad (3.55)$$

where μ'_m is a constant. It is considered to be inflated over maximal specific growth rate μ.

In the unique case where all of the substrate collected is laid down and retained as cell material, then

$$Q' = \frac{1}{Y_G} \quad (3.56)$$

The substrate uptake is then given as

$$Q = \mu Q'X \quad (3.57)$$

This is appropriate for uptake of substrate when excretion is negligible.

3.9.1.3 *Transient and stationary phase*

- The appearance of the phases mentioned above has been explained and attributed due to the following factors.
- Lack of nutrients or growth factors
- Lack of oxygen
- Accumulation of inhibitory or toxic metabolites
- Lack of biological space
- Changes in ion concentration in medium, especially pH
- Factors for programmed cell death (PCD)

Considering toxic/inhibitory product accumulation in this phase (Figure 3.39), we get

$$\frac{dX}{dt} = kX\,[1 - f(\text{toxin concentration})] \quad (3.58)$$

If the toxin linearly decreases, the specific growth rate is

$$\mu = k\,[1 - bC_t] \quad (3.59)$$

where C_t is the toxin concentration, and b and k are constants.

When $C_t \propto f\,(X)$, (3.60)

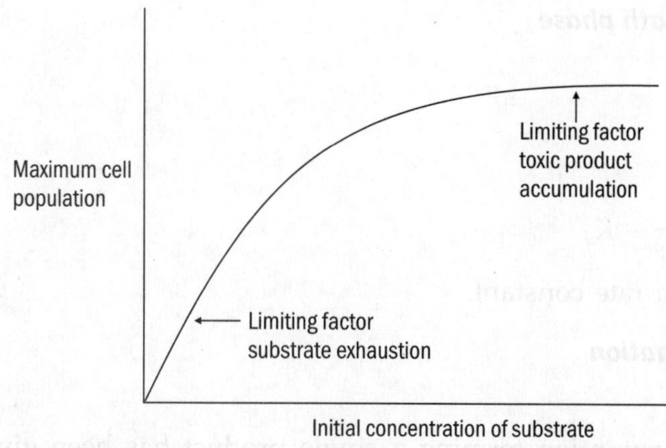

Figure 3.39 Inhibitor accummulation with relation to growth of microbes

$$\frac{dC_t}{dt} = qX \tag{3.61}$$

$$\text{or} \int_0^{C_t} dC_t = q \int_0^t Xdt \tag{3.62}$$

Assuming $C_t = 0$, at $t = 0$

$$C_t = q \int_0^t Xdt \tag{3.63}$$

From Equation 3.58, we get

$$\mu = \frac{1}{X}\frac{dX}{dt} = k\left[1 - bq\int_0^t Xdt\right] \tag{3.64}$$

or the effective specific growth rate is

$$\mu_{eff} = \frac{1}{X}\frac{dX}{dt} = k\left[1 - bq\int_0^t Xdt\right] \tag{3.65}$$

μ_{eff} diminishes rapidly with time in these phases. Thus,

$$\frac{d}{dt}(\mu_{eff}) = -kbqX < 0 \tag{3.66}$$

$$\text{and} \frac{d^2}{dt^2}(\mu_{eff}) = -kbqX\frac{dX}{dt} < 0 \tag{3.67}$$

The growth stops when

$$\frac{1}{bq} = \int_0^t Xdt \tag{3.68}$$

3.9.1.4 *Declining/death phase*

In this phase,

$$\frac{dX}{dt} = \mu X - K_d X \tag{3.69}$$

$$\text{or } \frac{1}{X}\frac{dX}{dt} = \mu = \frac{\mu_m s}{K_s + s} - K_d \tag{3.70}$$

where K_d is the death rate constant.

3.9.1.5 *Product formation*

1. Simple models

Simple substrate consumption forming a single product has been given by the three models listed below.

Growth-associated model

The growth-associated model considers a substrate to be stoichiometrically converted into a single product (P). The rate of product formation is related to the rate of growth as given in Equation 3.71.

$$\frac{dP}{dt} = \alpha \frac{dX}{dt} \tag{3.71}$$

provided that α is the stoichiometric constant.

Non-growth-associated model

The non-growth-associated model considers that the rate of product formation is only dependent on cell concentration and that the cell containing constitutive enzymes controls the product formation rate. Thus,

$\frac{dP}{dt} = \beta X$, provided β is the proportionality constant.

Combined model

The combined associated non-growth-associated model takes into consideration the above two models. This model was proposed by Luedeking and Piret, following the productivity behaviour of lactic acid fermentations. In this model,

$$\frac{dP}{dt} = \alpha \frac{dX}{dt} + \beta X \tag{3.72}$$

$$\text{or } \frac{1}{X}\frac{dP}{dt} = \alpha \frac{1}{X}\frac{dX}{dt} + \beta \tag{3.73}$$

$$\text{or } \gamma = \alpha\mu + \beta \tag{3.74}$$

where γ is the specific product formation rate and μ is the specific growth rate.

This model could successfully be applied to some other batches and continuous fermentations.

3.9.2 Role of Oxygen in Cell Systems

The importance of oxygen to microbial activities is well recognized. The response of cells to available oxygen is a complex phenomenon. Like all other living systems, microorganisms utilize oxygen (in molecular form or oxidizing compounds) and eliminate CO_2 through respiration. No organism is known to be indifferent to oxygen, although they differ widely in their sensitivities to free oxygen. Thus, based on oxygen sensitivity, microorganisms are classified as aerobic, anaerobic, and facultative. In aerobic microbial reactions, air is separated into the fermentation liquid and oxygen reaches the microorganisms in the dissolved state.

Depending on the availability of dissolved oxygen, the behaviour of microbial cultures is observed to vary widely. The behavioural variations observed in several microbial systems with the change in dissolved oxygen in the broth have been exemplified in the following section.

3.9.2.1 *Observed effects of dissolved oxygen on microbes*

Usually, in traditional aerobic culture methods, the oxygen supply to the organism seems to be adequate unless the broth is non-Newtonian. In the event of an imbalance between supply and demand of oxygen, a majority of cells are probably starved of oxygen, leading to less product turnover and sometimes accumulation of undesirable products, besides other effects.

Example 1 Shibai, Ishizaki, and Hirose (1973) observed that inosine fermentation by *B. subtilis* was satisfactory. That is, when

$$\frac{\text{Cell respiration rate } (r_{ab})}{\text{Maximum oxygen demand rate } (K_r M)} = 1$$

inosine accumulated in the cells exclusively. However, when $\frac{r_{ab}}{K_r M} < 1$, that is, under limited oxygen supply, the effect was different depending on the degree of limitation. It was observed that when $\frac{r_{ab}}{K_r M} = 0.5 \sim 0.9$, the inosine formation was inhibited and acetoin was the predominant product. When the oxygen supply was further limited, that is, $\frac{r_{ab}}{K_r M} < 0.3$ the cells excreted 2,3-butylene glycol as the main product.

Example 2 The influence of dissolved oxygen concentration on microbial respiration (for a continuous culture of *Klebsiella aerogenes*) has been reported by Harrison and Pirt

(1967). They observed that when the dissolved oxygen tension (DOT) is greater than 10–15 mm Hg, the r_{ab} was independent of DOT. However, when DOT is less than 10 mm Hg, the cells were found to be in an oxygen-limited state. Furthermore, when DOT is less than 5 mm Hg, the r_{ab} immediately increased. An increase in DOT from 5 mm Hg to 10 mm Hg caused the r_{ab} to decrease immediately. The pattern varied with μ over a range of 0.2–0.5 per hour. Therefore, it has been suggested that a decrease in DOT to about 5 mm Hg acts as an uncoupler of oxidative phosphorylation. The stimulation of r_{ab} at low DOT caused a decreased ATP yield that accompanies a change to an alternative electron transport pathway of the uncoupling of oxidative phosphorylation, which results in a rapid electron transfer and oxygen uptake.

***Example* 3** In many aerobic fermentations, cell growth varied with aeration rate. The "Pasteur effect" has been explained in several reports. The glucose uptake rate (r_s) by resting cells under anaerobic conditions is about 15 times greater than that in the presence of oxygen. They explained glycolytic regulation by oxygen at the level of phosphorylation.

Mor and Fiechter (1968) noticed that in continuous cultivation of Baker's yeast under aerobiosis, r_s = 4.15 mmol/g/h, but under anaerobiosis, r_s = 14.9 mmol/g/h. From this observation, they obtained the Pasteur quotient as equal to 0.28.

3.9.2.2 Aerobic-anaerobic growth transitions

These results indicated that when oxygen supply to the cells becomes limiting or exhausted, oxidative phosphorylation is slowed down or even stopped and glycolysis becomes the major source of ATP. The metabolic control is so exercised that as the oxygen supply is shut off, the rate of glycolysis is increased. With the resumption of oxygen supply, the rate of glycolysis is repressed. An outline of the scheme is shown in Figure 3.40.

However, it was observed that the presence of oxygen caused the inhibition of glucose-6-phosphate formation. Based on this, he proposed a model (Figure 3.41) for regulation of carbohydrate transport and metabolism in yeast. According to this model, oxygen regulates citrate formation, which in turn controls hexokinase activity in the glycolytic pathway in yeast. Citrate in glucose-utilizing yeast increases in aerobiosis to levels that are inhibitory to phosphofructokinase. The inhibition appears to be cooperative with that by ATP. These results seemingly confirm that the "Pasteur effect" is due to the regulation of glucose in the cell.

***Example* 4** In a study of citric acid production in a 5.1 stirred-tank fermenter (STF) using *Aspergillus niger* A-588, Khan and Ghose (1973) and Khan (1974) observed that the variation of air flow rate (0.2–0.6 volume per volume per minute [vvm]) and oxygen content of the air caused variation in citric acid production (Table 3.13). It was observed

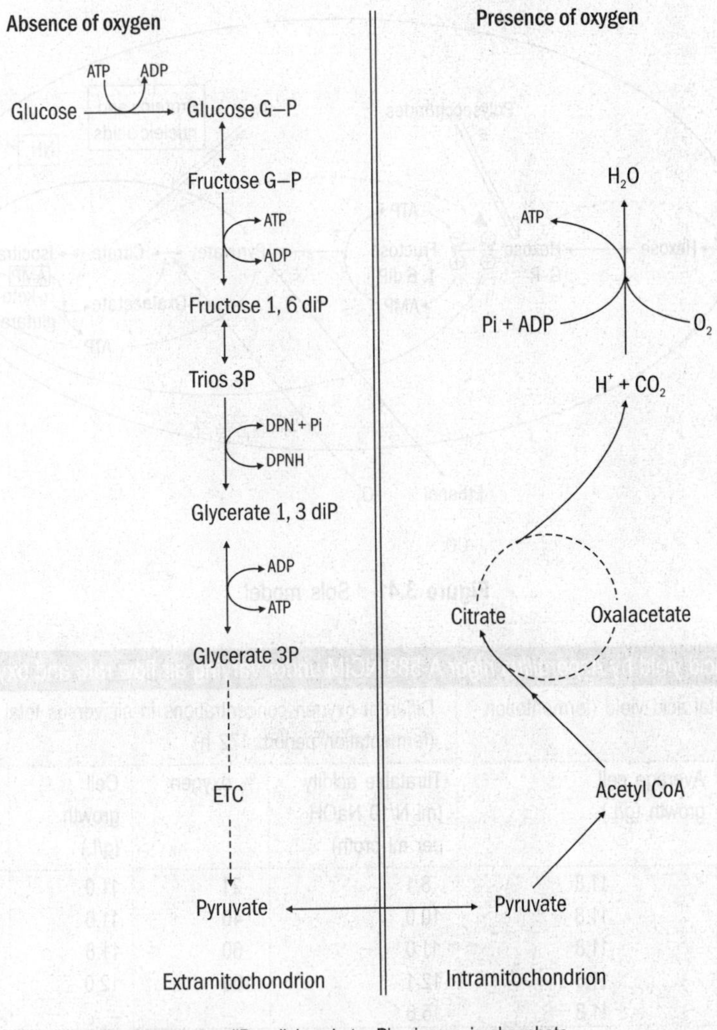

P – phosphate; diP – diphosphate; Pi – inorganic phosphate

Figure 3.40 Glycolysis and pyruvate oxidation in cells
Source Lowenstein (1964)

that a highly aerobic condition is necessary for a good yield of citric acid, although this high aeration rate had little or no effect on growth. In submerged fermentations, oxygen uptake rate by the cell is independent of the level of dissolved oxygen, provided its concentration remains above a critical level. For unicellular organisms, the "critical oxygen" level is very low and typical values are 0.12–1.10 mg/L for bacteria and 0.4–1.5 mg/L for moulds. Strohm, Dale, and Peppler (1959) reported that critical oxygen for growth of yeast is 0.2–0.4 mg/L, although values as high as 2.0 mg/L have also been reported (Table 3.14).

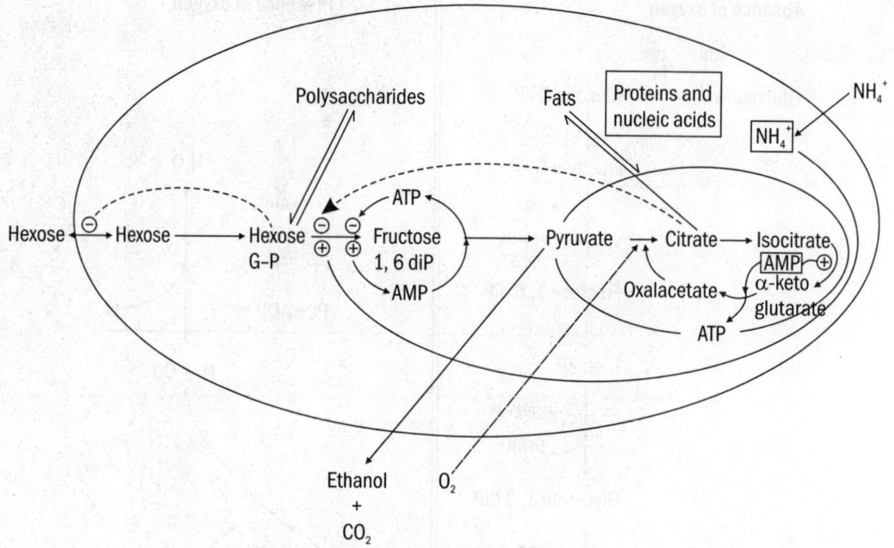

Figure 3.41 Sols model

Table 3.13 Citric acid yield by *Aspergillus niger* A-588 NCIM under varying air flow rate and oxygen concentration

Air flow rate versus total acid yield (fermentation period: 192 h)			Different oxygen concentrations in air versus total acid yield (fermentation period: 172 h)		
Air flow rate (vvm)	Average cell growth (g/L)	Titratable acidity (ml N/10 NaOH per ml broth)	% oxygen	Cell growth (g/L)	Titratable acidity (ml N/10 NaOH per ml broth)
0.2	11.8	8.1	21	11.0	12.1
0.3	11.8	10.0	40	11.6	13.0
0.4	11.8	11.0	60	11.8	13.8
0.5	11.8	12.1	100	12.0	14.3
0.6	11.8	15.6	–	–	–

vvm – volume per volume per minute

Source Khan (1974)

3.9.2.3 Oxygen–microorganism interactions

The varying effects of dissolved oxygen on microbial systems, as shown in the examples above, are indicative that it has different functions or interactions with the microorganism during fermentation processes. Participation of oxygen in microbial metabolism is shown in Figure 3.42.

The various functions that oxygen plays in the life processes of a microorganism are described below.

Table 3.14 Critical oxygen concentrations of various microorganisms.

Organism	Condition	Ccrit (mg/L)
Yeast	Resting + glucose	0.142
	Resting – glucose	0.011
	Batch culture	0.569
Torula utilis	Continuous culture	0.074
	Batch culture	2.0
Penicillium chrysogenum	Batch culture	0.40
Escherichia coli	Batch culture	0.26
Pseudomonas ovalis	Batch culture	1.10
Saccharomyces cerevisiae	Batch culture	0.60
Klebsiella aerogenes	Batch culture	0.455–0.853
	Continuous culture	0.114–0.569
Aspergillus niger	Batch culture	0.320

Ccrit – critical concentration of oxygen

Sources Bennet and Kempe (1964); Brown (1970)

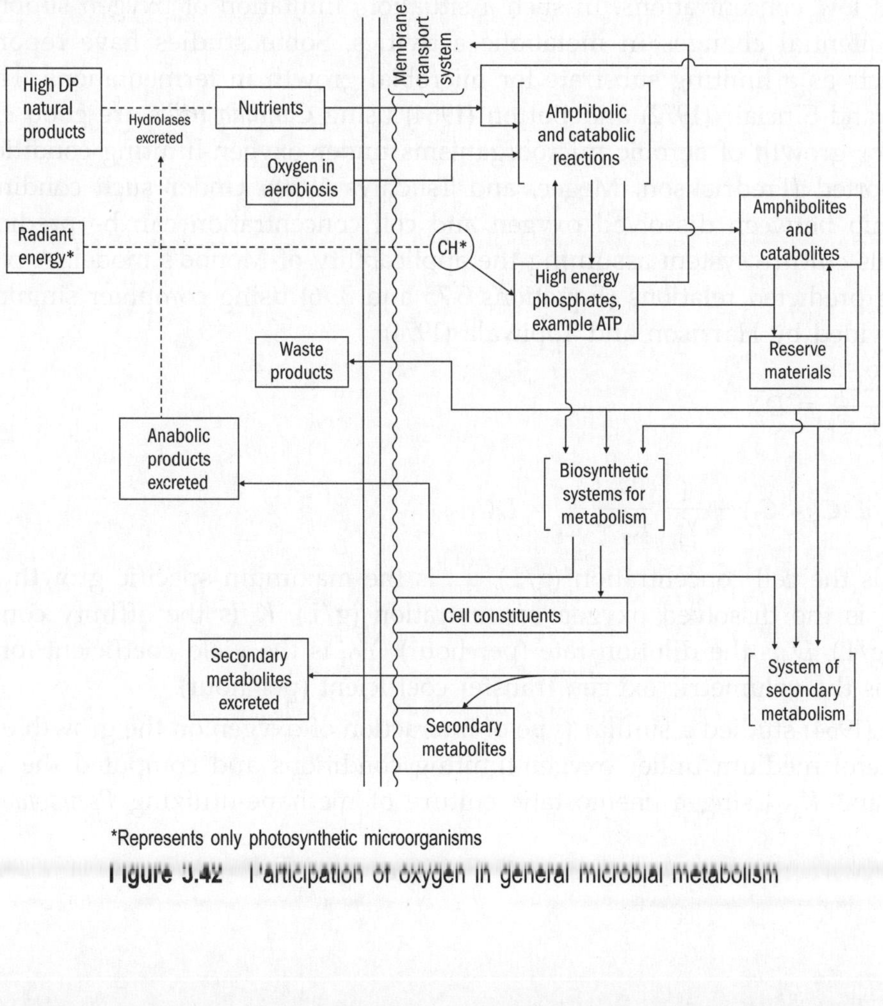

*Represents only photosynthetic microorganisms

Figure 3.42 Participation of oxygen in general microbial metabolism

3.9.2.3.1 Nutritional component

Oxygen acts as a nutrient in methane-utilizing bacteria (like *Pseudomonas methamica* and *Methanomonas methoxidans*) and in some other systems. Most bacteria do not derive oxygen directly from the molecular form or gaseous phase, but when the element is present as dissolved oxygen and in other oxygen-containing compounds (from aqueous phase). In certain hydrocarbon fermentations, oxygen from the gaseous phase is utilized by bacteria as a nutrient via the action of oxygenase enzymes during the initial oxidation of C_1 of the hydrocarbon. In methane-utilizing bacteria, a large proportion of the cell oxygen is derived from molecular oxygen through oxygenase reactions (Higgins and Quayle 1970).

Yeasts and fungi may incorporate molecular oxygen in their systems as unsaturated fatty acids and sterols. Although this oxygen constitutes a small part of the cell mass, it appears to be vital for growth. Inability to produce unsaturated fatty acids in anaerobic environments is related to an essential requirement for trace amounts of oxygen in the medium. Where a significant amount of oxygen is needed for respiratory activities, there is a likelihood of competition between the synthetic and respiratory pathways for oxygen at low concentrations. In such a situation, limitation of oxygen supply results in consequential changes in metabolic activities. Some studies have reported that oxygen acts as a limiting substrate for microbial growth in fermentations. The works of Ryder and Sinclair (1972) and Button (1964) using *Candida utilis* are good examples. Models for growth of aerobic microorganisms under oxygen-limiting conditions have been reported (Fredrickson, Megee, and Tsuchiya 1970). Under such conditions, the relationship between dissolved oxygen and cell concentration can be predicted in a continuous culture system assuming the applicability of Monod's model. A plot of the following predicted relations (Equations 3.75 and 3.76) using computer simulation has been provided by Harrison and Topiwala (1974).

$$\frac{d\overline{X}}{dt} = \frac{N_m \overline{X} C_L}{K_s + C_L} = D\overline{X} \tag{3.75}$$

$$\frac{dC_L}{dt} = K_L\, a\,(C_S - C_l) - \frac{1}{Y_{O_2}} \frac{\mu_m\, C_L}{K_s + C_L} - DC_L \tag{3.76}$$

where $\bar{x}$ is the cell concentration (g/L), μ_m is the maximum specific growth rate (per hour), C_L is the dissolved oxygen concentration (g/L), K_s is the affinity constant for oxygen (g/L), D is the dilution rate (per hour), Y_{O_2} is the yield coefficient for oxygen, and K_La is the volumetric oxygen transfer coefficient (per hour).

Button (1964) studied a similar type of interaction of oxygen on the growth of *C. utilis* in a glycerol medium under oxygen-limiting conditions and computed the values of μ_m, Y_{O2}, and K_s. Using a chemostatic culture of methane-utilizing *Pseudomonas* with dissolved oxygen control, Harrison, MacLennan, and Pritt (1969) found two distinct types of responses in the system, one of which provided the uptake of oxygen for respiratory functions according to the Michaelis–Menten relationship.

$$R' = R'm\ C_L/(K_s + C_L) \tag{3.77}$$

where R' is the rate of oxygen uptake and $R'm$ is its maximum value.

When $C_L >> K_s$, Equation 3.77 reduces to

$$R' = R'm \tag{3.78}$$

This shows that when the concentration of dissolved oxygen is high, the oxygen-utilizing rate reaches a maximum value and becomes independent of the dissolved oxygen concentration.

This means that $R' \to R'm$ asymptotically as C_L increases. The existence of "critical oxygen" concentration (Table 3.14) is thus explained.

When $K_s >> C_L$

$$R' = K\ C_L$$

where K is a constant $= R'm/K_s$ (3.79)

or $R' \propto C_L$

Under this condition, the oxygen uptake rate is directly proportional to the dissolved oxygen level. The other response is concerned with the increased affinity for oxygen at low DOT.

Recently, Soni and Ghose (1974) studied the dynamics of growth of *Saccharomyces cerevisiae* 3132 on cane molasses in a single stage continuous culture to provide additional information on the response of a continuous culture to periodic oxygen inputs. The study is important for the prediction of transient response of the system as well as for providing a kinetic model for the variation of dissolved oxygen concentrations, with time reflecting the nature of the response. The experimental set-up is illustrated in Figure 3.43. The substrate concentration and pH were controlled at constant values, and steady-state conditions were disturbed by replacing air flow with 10 min of oxygen flow (0.1 vvm), followed by 6 min of nitrogen flow (0.5 vvm) periodically. Assuming that the microbial population acts as a servosystem with an input P and outputs Q_0, Q_1, …, Q_n, the oxygen input wave is represented as

$$O_2(t) - \overline{O}_2 = P\cos(\omega t - \theta) \tag{3.80}$$

where $O_2(t)$ is the value of periodic oxygen function at time t, is the mean value of oxygen function over an integral number of cycles, ω is the frequency (rad/s), and θ

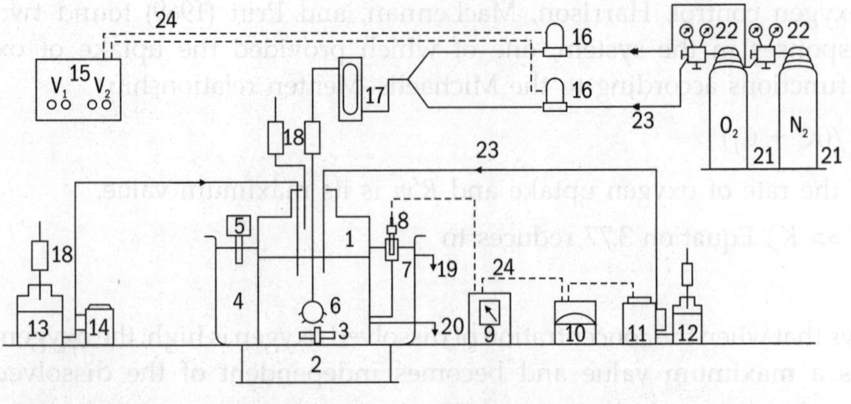

1. Microbial reactor
2. Magnetic stirrer
3. Magnet
4. Water bath
5. Temperature controller
6. Sparger
7. Overflow device
8. pH electrode
9. pH recorder
10. pH controller
11. Micropump
12. NaOH tank
13. Concentration medium tank
14. Micropump
15. Two-way water flow meter
16. Solenoid valve
17. Rotameter
18. Cotton plug filter
19. Continuous outflow
20. Sample outlet
21. Gas cylinder
22. Gas regulator
23. Pressure tubing
24. Electric wiring

Figure 3.43 Experimental set-up

is the phase angle (degrees). Since in the system, pH and sugar concentrations were maintained constant, the throughput of the growth is expressed as

$$N(t) - \overline{N} = Q \cos(\omega t + \psi) + d(t) + e \tag{3.81}$$

where $N(t)$ is the total number of cells at time t, $\overline{N}$ is the mean value of cell number over an integral number of cycles, and e measures the value of the experimental error.

In Equation 3.81, $d(t)$ is the disturbance function that accounts for the random fluctuation of the output due to biological and electrical disturbances of small magnitudes.

Using Bode's plot technique, the values of Q and Ψ can be evaluated. The cyclic behaviour of responses towards cell propagation and dissolved oxgen concentration is illustrated in Figure 3.44. In steady-state conditions, the rate of oxygen transfer from the gas phase to the liquid is equal to the rate of oxygen uptake by the organism. But when a discontinuous periodic input function of oxygen–nitrogen is used, the concentration of dissolved oxygen and C_L varies with time. Under such conditions, the rate of oxygen absorption (r_d) is not equal to the rate of oxygen uptake (R'), but

$$r_d = R' - \frac{d_{C_L}}{dt} \tag{3.82}$$

Expressing the rates in terms of concentration, it is possible to write Equation 3.82 as

$$C_L = C_L(0) + \frac{K'_r Q}{K_1^2 + \omega^2} \left| \frac{\omega \cos(\omega t + \psi)}{K_1 \sin(\omega t + \psi)} \right| - \frac{K_1 P}{(K_1^2 + \omega^2)\, H'} \left| \frac{\omega \cos(\omega t - \theta)}{K_1 \sin(\omega t - \theta)} \right| \tag{3.83}$$

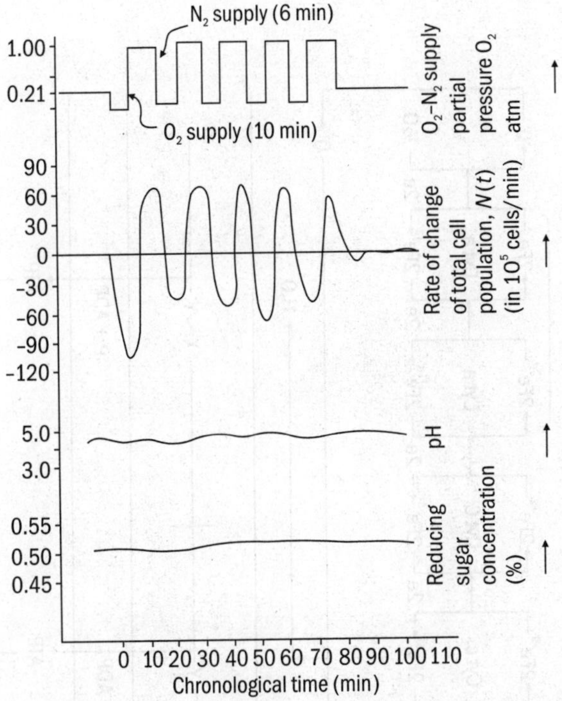

Figure 3.44 Response to periodic oxygen function in a continuous microbial reactor
Source Soni and Ghose (1974)

where $C_L(0)$ is the dissolved oxygen concentration at time $t = 0$; that is, when the broth is saturated with oxygen in air, K_1 and K'_r are the coefficients for rate of oxygen absorption and uptake, respectively. When the dissolved oxygen concentration in the liquid broth falls below the critical level, the oxygen becomes limiting to the metabolism. This subsequently leads to a decline in the total cell population.

Equation 3.83 shows that at "*n*" discrete values of "*t*", the dissolved oxygen concentration falls below the critical level. Transient response at various metabolic levels and over various time periods was also observed with the facultative anaerobe, *K. aerogenes* (Harrison and Topiwala 1974).

3.9.2.3.2 Electron acceptor

The participation of oxygen in respiration is by far its most significant role in aerobic microbes, and this role can be best explained on the basis of the electron acceptor concept.

In fermentation, oxygen is first absorbed in the liquid phase before passing into the cell where it manifests itself as a terminal electron acceptor in the respiratory chain. A sequential mechanism of oxygen participation as an electron acceptor in the respiratory chain is shown in Figure 3.45.

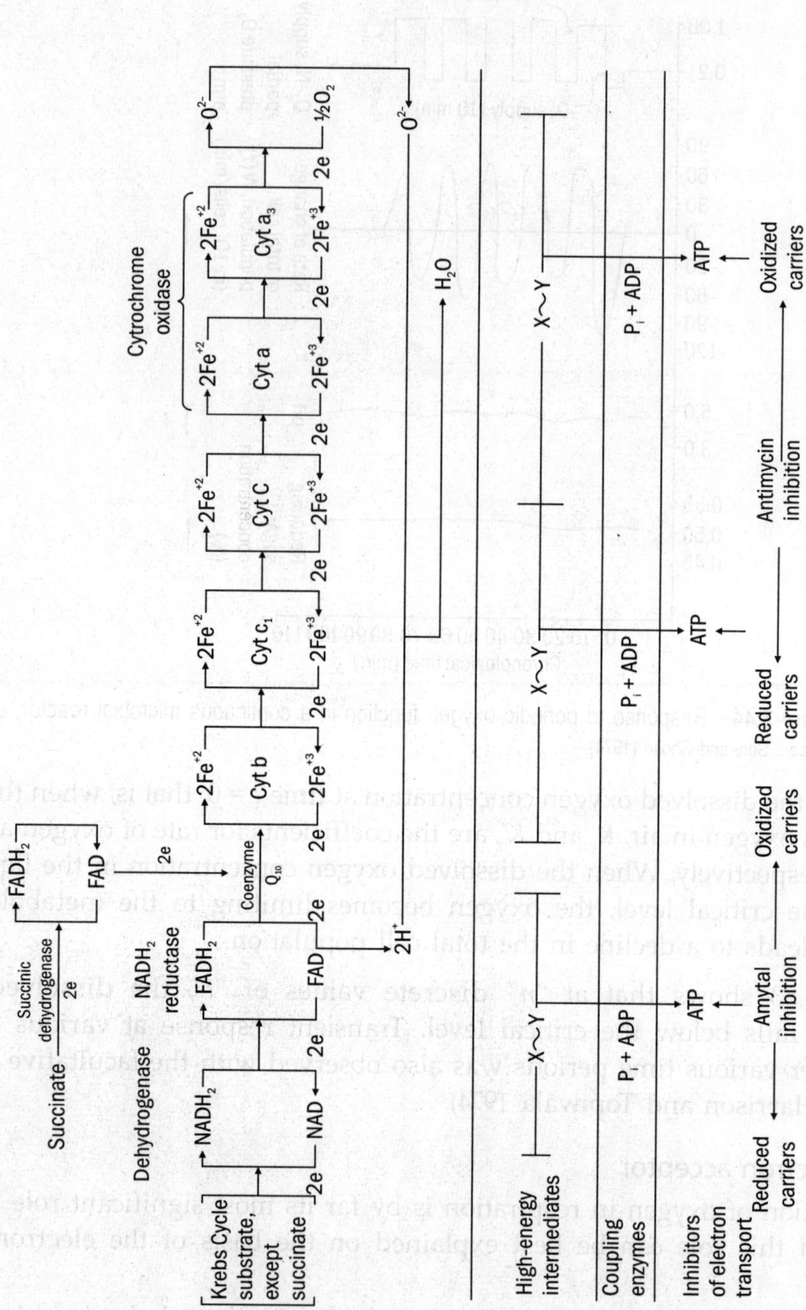

Figure 3.45 Mechanism of oxygen participation in respiratory chain as an electron acceptor

Cytochromes play a significant role in the respiratory mechanism in many facultative aerobes. Oxygen exerts considerable influence on the synthesis of cytochromes, certain enzymes, and thus cofactors in the cells energy mechanisms. However, presence of the cytochrome system does not necessarily indicate that the organism has the ability to grow in the presence of oxygen. Many cytochrome-containing bacteria, particularly obligate anaerobes, die in the presence oxygen (Table 3.15). The importance of critical oxygen concentration can be recognized by the fact that in the culture mediums, it may exercise control over synthesis by exerting influence on the kinetics of respiratory metabolism in growing aerobes and in maintaining proper cellular synthesis. This is particularly true with hydrocarbon substrates. When excess oxygen is supplied, rates of synthesis may become limited by the respiratory activity, and this is manifested in the oxygen uptake rates. In order to ascertain if this activity is a function of the extent of oxygenation of the substrate molecules, Einsele, Fiechter, and Knöpfel (1972) studied the growth of *Candida tropicalis* cells on hexadecane (C_{16}) and glucose (C_6). These carbon compounds were selected for the wide difference in their ability to donate electrons that are supposed to combine with oxygen. The data (Table 3.16) indicate only minor effects on respiration as a result of an enhanced oxygen supply, thus demonstrating the fact that respiratory activity is identical in both types of substrates and, therefore, is the limiting factor in the kinetics of cell synthesis.

3.9.2.3.3 Fermentative metabolism regulator

There are many examples of the ambivalent role of oxygen in enzyme regulation. Several fermentation pathways are either induced or repressed by the presence of

Table 3.15 Development of cytochromes in living systems

Status	Organism	Terminal electron acceptor	Cytochrome type a b c cobinding haemoprotein
Strictly anaerobic	*Clostridium* sp.	Fermentative products	No cytochromes
	Desulphovibrio desulfuricans	Sulphate	– – – +
	Chlorobium thiosulphatophilum	Oxidant generated by photosynthesis	– – + –
Facultative	Chromatium		– + + +
	Rhodospirillum rubum	Oxidant generated by photosynthesis or oxygen	– + + +
	Micrococcus denitrificans	Nitrate or oxygen	– + + +
	Escherichia coli	Fermentation product or nitrate or oxygen	+ + + +
Strictly aerobic	*Azotobacter vinelandii*	Oxygen	+ + + +
	Animals	Oxygen	+ + + –

Source Lascelles (1964)

Table 3.16 Oxygen uptake by *Candida tropicalis* as a function of air flow rate in hexadecane and glucose medium

Substrate	S_0% (w/v)	Aeration (Wm)	O_2 uptake[a] (mmol/h/g)	O_2 transfer[b] (mmol/L/h)
Hexadecane	1	0.4	13.5	650
Hexadecane	1	0.2	14.1	450
Glucose	1	1.6	12.5	380
Glucose	1	0.2	12.0	150

w/v – weight per volume; S_0 – initial concentration of substrate with emulsifying agent
[a] actual oxygen uptake in growth experiment
[b] sulphite method
Note Stirrer speed is 2500 rpm; actual volume is 6.0 L; temperature is 30°C; and pH is 5.0.
Source Einsele, Fiechter, and Knöpfel (1972)

dissolved oxygen in the medium at various levels. In the denitrification process, oxygen completely represses the specific enzyme concerned. The level at which dissolved oxygen exerts an inducing effect on the organism is far below the critical dissolved oxygen level for respiration.

In the evolution of biochemical pathways, microorganisms with the ability to grow either aerobically or anaerobically provide interesting information. Oxygen may inhibit certain enzyme actions causing an imbalance and altering the fermentation pathway. Regulation by oxygen is often caused through repression of enzyme synthesis. In *Pseudomonas saccharophiles*, oxygen at 1 atm inhibited succinate uptake and lipid synthesis but stimulated polysaccharide production. In *Azotobacter vinelandii*, oxygen inhibited nitrogen fixation, although growth in the presence of ammonium ion was unaffected. Cellular synthesis or growth can also be regulated by oxygen. It was found that in *E. coli* colonies on agar, increasing the partial pressure of oxygen (pO_2) from 0.21 atm to 0.955 atm raised the upper limit of glucose concentration (S_o) for Equation 3.84 to be valid from 2.5 g/L to 10 g/L.

$$K_r = K_2 (\sqrt{S_o} - \sqrt{S_i}) \tag{3.84}$$

However, increase in pO_2 also increased the threshold concentration of glucose (S_i) from 0.09 g/L to 1.7 g/L. It indicated the probability that the upper limit of glucose concentration for Equation 3.84 to hold is determined by the formation of toxic products such as alcohols and organic acids produced by oxygen-limited metabolism. Increase in S_o on increasing pO_2 is probably due to oxygen toxicity, since it is surmised that the higher the pO_2, the higher is the glucose diffusion rate necessary to meet the oxidation rate, which lowers the pO_2 below the inhibitory level and thus gives higher radial growth rate (K_r).

3.9.2.3.4 Oxygen induction

Induction by oxygen may occur both at low and high partial pressures. Production of a yellow pigment in *K. aerogenes* has been found to be induced at pO_2 greater than 100 mm Hg (Maclennan and Pirt 1970). The mechanism of induction by oxygen, whether directly with DNA bases or through an intermediate, is still not clear. Molecular oxygen seems to be necessary for the induction of cytochromes in yeasts (Somlo and Fukuhara 1965). Chen and Charalampous (1969) have also reported the induction of cytochromes synthesis by oxygen. Oxygen may also regulate fermentation by regulating cytochrome elaboration, which in turn regulates respiratory activity of the organism. From the responses of *Canida utilis* and *Saccharomyces carlsbergensis* to the differences in oxygen concentration, Rickard *et al.* developed a relationship between the rate of NADH oxidation by yeast extracts and dissolved oxygen concentration. It was predicted that control by oxygen over respiratory activity is due to its ability to regulate the level of one or more components of the respiratory chain.

The influence of dissolved oxygen concentration on the capacity of fatty acid synthesis in cells in a continuous culture of *C. utilis* was studied by Babij, Moss, and Ralph (1969). They showed that the presence of high dissolved oxygen concentration during continuous cultivation of the organism increased the level of polyunsaturated fatty acids.

Growing *S. cerevisiae* and *S. carlsbergensis* in batch culture with and without dissolved oxygen control, Rickard and her associates (1969) also noticed that synthesis of cytochromes in these yeasts was dependent on dissolved oxygen concentration in the growth medium. Similar observations were made when *C. utilis* was grown in a batch culture. Also, the catabolic pattern can be changed by changing the dissolved oxygen concentration in the fermentation liquid. Akbar, Rickard, and Moss (1974) observed that in continuous culture of *S. cerevisiae,* respiration decreased at low dissolved oxygen concentration and high sugar feed rates but cytochromes were glucose repressible at both high and low dissolved oxygen concentrations. It was also demonstrated that in this organism, the fate of pyruvate formed the ingested glucose would depend not only on its rate of formation but also on the dissolved oxygen concentration in the fermentation liquid.

These are few examples of the inducing action of oxygen on microorganisms. Therefore, it appears that the nature of the fermentative metabolism may be regulated by controlling the availability of oxygen at the reaction site.

3.9.2.3.5 Metabolic poison and mutagenic agent

Metabolic changes in glutamic acid fermentation at various levels of dissolved oxygen have been reported (Hirose, Sonoda, Kinoshita, *et al.* 1968). At moderate aeration rates and measured oxygen transfer rate (OTR) (sulphite oxidation) of 10.5×10^{-7} mol/ml/min, the oxygen demand of the cells remained unsatisfied after 10 h of incubation. Substrate uptake and cell growth appeared inhibited, resulting in less product formation. At high aeration rates (68×10^{-7} mol/ml/min), the oxygen growth of cells and sugar utilization were suppressed heavily; no significant amount of glutamate was formed, although the oxygen demand of the cells was met. Thus, cells of the same organism behaved differently when exposed to different aeration rates. In this case, oxygen acted as a toxin even though the culture was aerobic. The mechanism of toxic behaviour of oxygen is not well understood, but there is some evidence that its toxicity to strict anaerobes may be attributed to the formation of H_2O_2, which cannot be removed in the absence of catalase. In other organisms, oxygen acted as a poison through the oxidation of cytochromes (Dolin 1961). Hyperbaric oxygen toxicity has been attributed to several processes, including oxidation of thiol groups, lipid peroxidation, and free radical accumulation. Prolonged exposure to oxygen pressures of 10 atm exercises a mutagenic effect on *E. coli*. It may even repress specific enzyme synthesis. Table 3.17 gives a list of various enzymes repressed by oxygen. It is also possible that, in some cases, oxygen upsets the redox equilibrium inside the cell shifting reducing equivalents from vital biosynthetic reactions to excess oxygen. It is, therefore, unrealistic to offer a single mechanism of toxicity exercised by oxygen.

3.9.2.3.6 Redox potential controller

For aerated cultures, the redox potential follows the Nernst equation (Equation 3.85).

Table 3.17 Oxygen as an enzyme repressor

System	Organism	Specific enzyme repressed
Organic electron acceptor reducing enzymes	*Escherichia coli*	Fumerate reductase
	E. coli	Ethanol dehydrogenase
	Aerobacter aerogenes	Ethanol dehydrogenase
	E. coli	Lactate dehydrogenase
	Staphylococcus aureus	Lactate dehydrogenase
	A. aerogenes	Glyceral dehydrogenase
	E. coli	Formic dehydrogenase
Alternative inorganic electron acceptor reducing enzymes	*A. aerogenes*	Nitrate reductase
	Micrococcus denitrificans	Nitrate reductase
	E. coli	Nitrate reductase
	E. coli	Nitrate reductase
	Proteus vulgaris	Thiosulphate reductase
	E. coli	Hydrogenase
	A. aerogenes	Hydrogenase

Sources Cole and Wimpenny (1966); Cole (1967); Collins and Lascelles (1962); McPhedran, Sommer, and Edmund (1961); Hirsch, Michael, Bernard, *et al.* (1963)

$$E_h = E_o + \frac{RT}{nF} \ln \frac{\text{Oxidant}}{\text{Reductant}} \tag{3.85}$$

where E_h is the measured redox potential, E_o is the standard redox potential, n is the number of electrons taking part in the reaction, F is the Faraday's constant, R is the gas constant, and T is absolute temperature.

Any change in the E_h value in the broth during fermentation sets in changes in the metabolic activities of the culture. It has been suggested that in the regulatory process in facultative anaerobes, the E_h may be used as a control variable for aeration in the oxygen-limited growth condition. On the other hand, the dissolved oxygen concentration may be used as a control variable in the presence of excess oxygen. Although E_h monitoring is not done in most industrial fermentations, knowledge of E_h under oxygen-limited conditions may lead to better control of the system (Lengyel and Nyiri 1965).

In a complex environment, a redox probe can reflect the dissolved oxygen concentration. Here, the redox output is proportional to the logarithm of dissolved oxygen concentration. Thus, at low dissolved oxygen concentration in the broth, large changes in the redox potential will represent small variations in dissolved oxygen. Above 2 mm Hg of DOT, a membrane sensor serves as a more reliable device than a redox probe to measure dissolved oxygen concentrations. However, below this DOT the redox probe will be more reliable. Correlation between metabolism and changes in E_h obtained by Tengerdy (1961) and Daniels, Garcia, and Rosensteel (1970) could be explained by changes in dissolved oxygen. However, it is difficult to distinguish between the effect of oxygen tension and E_h on cells for aerobic cultures.

For anaerobic or facultative cultures, the relationship between metabolic events and E_h of redox probe reading can be obtained. It was reported that in iron-reducing bacterial cultures, a redox probe may be used to follow and control the Fe^{3+}/Fe^{2+} equilibrium. In a continuous culture of a facultative anaerobe, when dissolved oxygen concentration is gradually increased, the potential rises to a high level during oxygen-limiting phases where correct dissolved oxygen concentration cannot be measured using an oxygen electrode. This is seen in the measurement of E_h and pO_2 in glucose-limited cultures of *A. aerogenes*. In this organism, the greatest change in cell physiology occurs in the oxygen-limited phase; this is also true for product formation by *K. aerogenes*.

A number of metabolic processes have their optima for product formation at some point lower than full aerobiosis. These processes could be regulated by E_h. Variations of cytochromes and enzyme content by E_h in batch cultures were observed. A relationship between DOT and E_h value in chemostatic culture of *E. coli* was obtained.

Since the NAD/NADH ratio represents the key E_h to derive cellular oxidation/reduction reactions, it may be possible to regulate E_h values by monitoring NADH. A method of NADH monitoring continuous culture is now available. In K. *aerogenea,* the largest change in NADH occurred when DOTs were very low, but some responses were obtained well above the C_{crit} level of 0.569 mg/L at 10 mm Hg. This would indicate that cells can detect or react to oxygen tensions well above K_S for terminal oxidases. There was a rapid reduction in $NADH^+$ when the pO_2 fell to 0.2–0.5 mm Hg.

3.9.2.3.7 Energy generator

In a microbial system, oxygen usually acts as an energy generator in two ways: (1) involving biothermochemical reaction generating heat energy that can be converted into electrical energy in the biochemical fuel cell, and (2) synthesizing energy-rich cellular chemical, namely ATP. For biochemical fuel cells, the most common bioexothermic reaction is given as shown in Equation 3.86.

$$C_6H_{12}O_6 + 6O_2 \xrightarrow{\text{Organism}} 6CO_2 + 6H_2O + 673 \text{ kCal} \qquad (3.86)$$

The energy produced from 1 kgmol of sugar is equivalent to 780 Wh of electrical energy (4.4×10^{-3} Wh/g). By proper engineering, it is possible to convert this energy into low output power supply.

3.9.3 Cellular Activity Enhancement

3.9.3.1 By mutation

It may be defined as an abrupt, induced or directed molecular change in gene of living cells from one form to another (physiological or morphological) with the aid of some agent(s). By mutation, the nucleotide sequence in a gene molecule is altered, and there may be synthesis of different proteins. According to the classical "one gene one enzyme" hypothesis, a single gene controls the synthesis of one enzyme. Let us consider a biochemical reaction shown in Figure 3.46.

Since enzymes as well as genes are specific in character, the Enzyme I catalyses only reaction $A \rightarrow B$ and Gene I synthesizes only Enzyme I. Similar sequence will occur for other enzymes. If Gene I is altered in some way, it will not synthesize Enzyme I and the reaction $A \rightarrow B$ will be blocked, that is, the organism will lack the capacity to synthesize product *B* from substrate *A*. Also, the organism will fail to grow until and unless *B*, *C*, and *D* are supplied in the medium. In mutation, this blocking may take

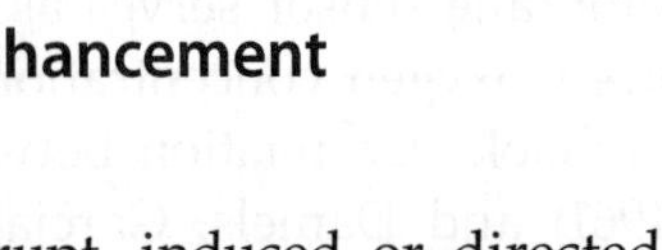

Figure 3.46 Schematic of a biochemical reaction

place reversibly or irreversibly. The substances/agents that cause mutation are called mutagens; these may be grouped under several heads.

3.9.3.1.1 Uses of mutants

1. To increase the productivity of biochemicals

Mutants increase the productivity of biochemicals by induced mutation programme. For example, in the production of penicillin, induced mutations shown in Figure 3.47 have been carried out.

In this way, it is possible to produce mutant cultures giving product as high as 10 000 units/ml by selecting cultures using the natural selection process.

The following examples discuss the possibility of enhanced yields using mutants.

- Let us consider the biosynthesis of a fermentation product (Figure 3.48).

During the biosynthesis of a particular product, the organism simultaneously synthesizes an inhibitor (*I*), which then inhibits the subsequent reaction (say, $B \rightarrow C$). Thus, there will be lower yields of *D* and, therefore, of X. If synthesis of the inhibitor is blocked, then the accumulation of *D* will increase, which increases the production of X. By mutation, the synthesis of inhibitors is blocked, and the resulting mutants will increase the productivity.

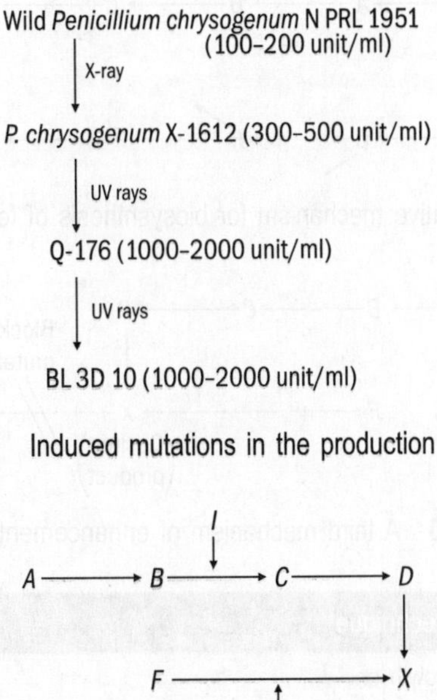

Figure 3.47 Induced mutations in the production of penicillin

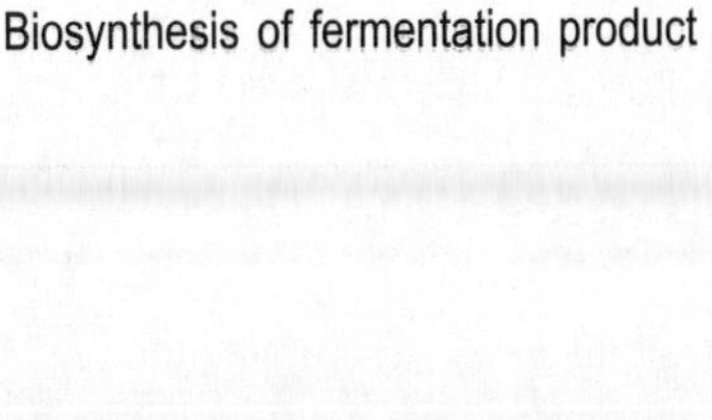

Figure 3.48 Biosynthesis of fermentation product *X* (say penicillin)

- The intermediates formed during the formation of *X* are also involved in some other side reactions. If by some means, these side reactions are blocked at different steps, then more amount of product will be formed, provided the intermediates (*I* and *E*) are in the rate-limiting concentration (Figure 3.49). Mutants block these side reactions and thereby produce more amount of product.
- Due to mutation, there may be blocking of degradation of metabolites formed. This leads to the accumulation of metabolites, which ultimately increases production (Figure 3.50).

Usually, various metabolites are synthesized inside the cell. Mutants cause complete lysis of cells so that more metabolites take part in broth. In this way, the production is increased.

2. Indication of biochemical steps

The use of mutants blocked at different steps in a reaction series has been effective in determining the exact sequence of the reaction. For example, strands of mutants of *E. coli* that require arginine for growth in a minimal medium will grow only if arginine is replaced by citrulline or ornithine. The results of this substitution technique are given in Table 3.18.

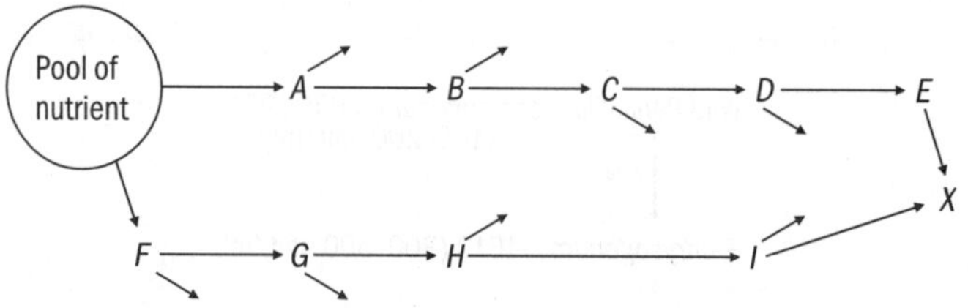

Figure 3.49 Alternative mechanism for biosysnthesis of fermentation product X

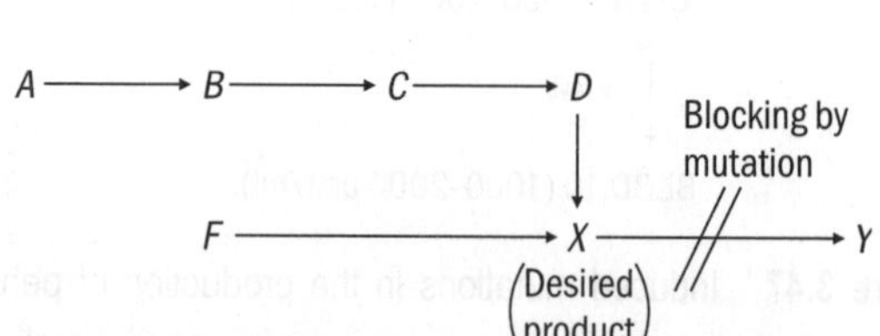

Figure 3.50 A third mechanism of enhancement of product

Table 3.18 Results of substitution technique

Mutant number	Growth on		
	Arginine	Citrulline	Ornithine
1	+	–	–
2	+	+	–
3	+	+	+

On the basis of this data, we can postulate the reaction sequence as shown in Figure 3.51.

Any compounds beyond the blocked step will thus permit growth because it is not further metabolized. Thus, in M_1, citrulline may accumulate since its further metabolism is blocked by the absence of the enzyme required for its conversion to arginine.

Syntropism test for accumulation of products

Experiment Mutants are streaked in pairs on the surface of an agar medium of glucose, ammonium salts, and mineral salts, with sufficient arginine to allow visible growth of mutants.

Observation It is found after incubation that in a pair, the growth of one mutant is more than another (Figure 3.52).

Explanation In the streaked pairs of M_1 and M_2, M_1 is supplied with sufficient arginine for its visible growth. Thus, it produces citrulline and ornithine but not arginine. M_2 utilizes these citrulline and ornithine, and its growth is more than M_1. Similarly, in the streaked pairs of M_2 and M_3, M_2 is unable to convert ornithine into citrulline, leading to an accumulation of ornithine. M_3 utilizes ornithine, which increases its growth as compared to M_2. In case of M_1 and M_3, since there is insufficient arginine for maximum growth of M_1, it cannot convert citrulline to arginine. Thus, there is an accumulation of citrulline and ornithine, leading to an increase in growth of M_3 but not M_1. Thus, increased growth of one mutant in a pair is due to the "feeding" of one mutant by another with compounds piled up behind a genetic block.

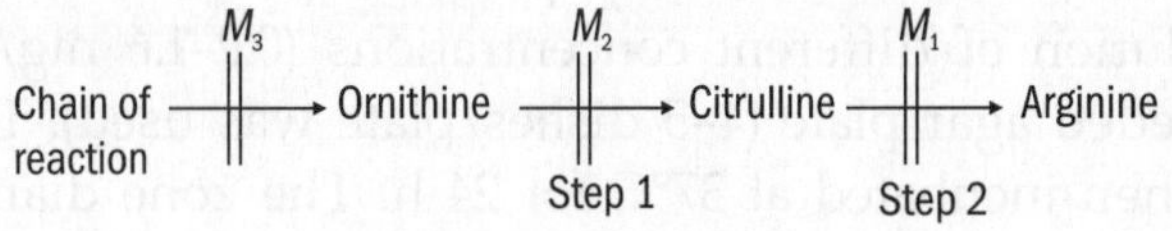

Figure 3.51 Reaction sequence of arginine biosynthesis

Note Mutant 1 (M_1) blocked the reaction immediately before arginine; mutant 2 (M_2) blocked the reaction immediately before citrulline; and mutant 3 (M_3) blocked the reaction immediately before ornithine.

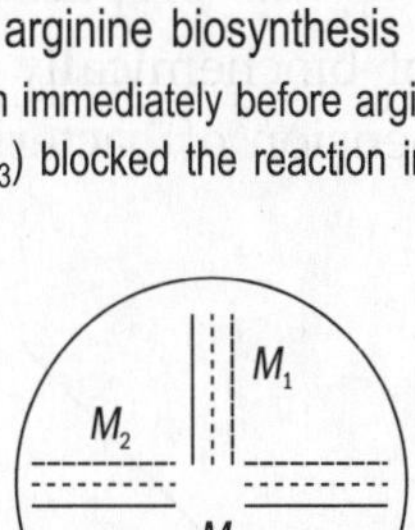

Figure 3.52 Schematic of a typical syntropism test

3. For bioassay of vitamins and amino acids

Mutant strains of microorganisms play a key role in the bioassay of bioproducts. In the bioassay of vitamin B_{12}, a mutant strain of *E. coli* is used. It lacks the capacity to synthesize methionine. Vitamin B_{12} catalyses this synthesis from homocysteine.

$$\text{Homocysteine} \xrightarrow{B_{12}} \text{Methionine}$$

Bioassay of vitamin B_{12} consists of the following steps.

Preparation of inoculum

Twenty-four hours subculture from a slant of a mutant strain of *E. coli* was grown in a medium (peptone: 2.0 g; NaCl: 0.5 g; water: 200 ml; and pH: 7.2 ml) in a 250 ml flask containing 50 ml broth and shaken on a rotary shaker for 24 h at 37°C.

Assay medium preparation

The following components are present in the bioassay medium (pH: 6.8) per titre of distilled water—K_2HPO_4 (7.0 g), KH_2PO_4 (3.0 g), Na-citrate (0.5 g), $MgSO_4 \cdot 7H_2O$ (0.1 g), dextrose (10.0 g), glycine (0.1 g), asparagine (4.0 g), arginine (0.1 g), histidine (0.1 g), glutamic acid (0.1 g), proline (0.1 g), tryptophan (0.1 g), Na-thioglycollate (0.1 g), distilled water (1000 mkl and agar [2.5%]).

Bioassay procedure

One millilitre of bacterial suspension is mixed with 10 ml of molten assay medium in agar (at 45°C) containing all requirements except vitamin B_{12}. The mixture was placed in a Petri dish and allowed to set. Next, filter paper discs (10–12 mm in diameter) saturated with vitamin B_{12} solution of different concentrations (0.2–1.6 mg/ml) were placed on the surface of the seeded agar plate (4–5 dishes/plate was used). The dishes were kept for 1 h at 0°C, and then incubated at 37°C for 24 h. The zone diameter of growth was then measured, and a standard curve was prepared (Figure 3.53).

Figure 3.54 shows the isolation of biochemically deficient mutants. The mutagen is subjected to UV radiation in a suspension of bacterial spore. The mutants can grow in

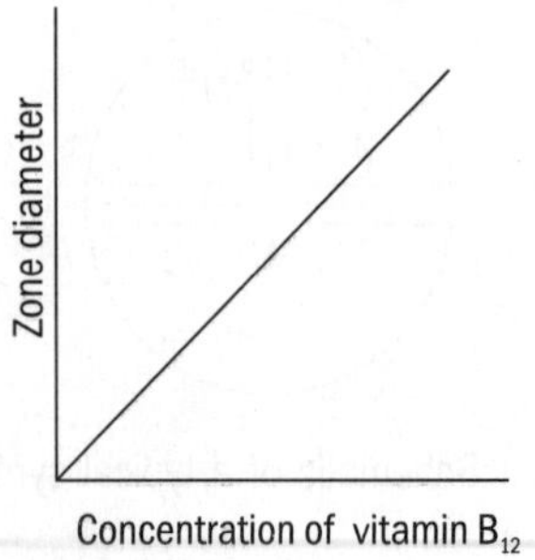

Figure 3.53 Standard curve of zone diameter of growth

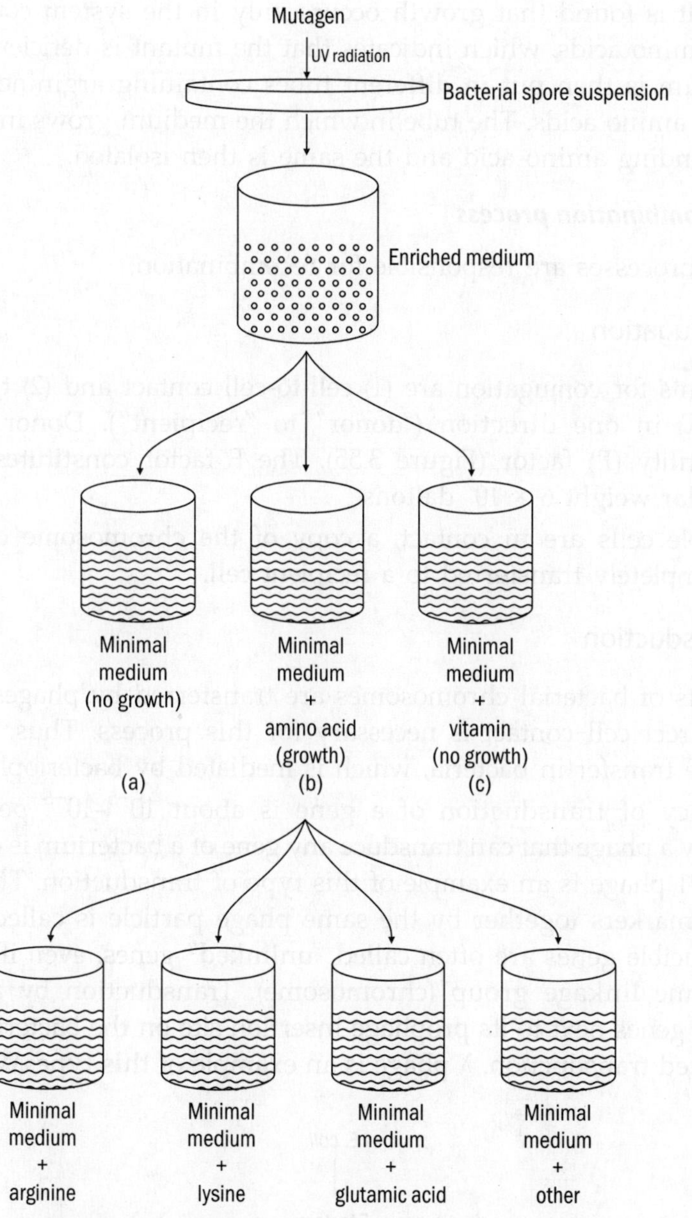

Figure 3.54 Isolation of biochemically deficient mutants

the resulting inoculated enriched medium (minimal medium and complex substances like yeast extract, vitamins, and so on). The culture is then transferred into three tubes containing medium and incubated: the first tube contains minimal medium, the second contains minimal medium and amino acids, and the third contains minimal medium and vitamins. It is found that growth occurs only in the system containing minimal medium and amino acids, which indicates that the mutant is deficient in amino acids. Minimal medium is then put in different tubes containing arginine, lysine, glutamic acid, and other amino acids. The tube in which the medium grows indicates deficiency in the corresponding amino acid and the same is then isolated.

3.9.3.2 By recombination process

The following processes are responsible for recombination.

3.9.3.2.1 Conjugation

The requirements for conjugation are (1) cell-to-cell contact and (2) transfer of genetic material (DNA) in one direction ("donor" to "recipient"). Donor implies bacteria concerning fertility (F) factor (Figure 3.55). The F factor constitutes DNA molecules having molecular weight 6×10^7 daltons.

When suitable cells are in contact, a copy of the chromosome of a donor cell is partially or completely transferred to a recipient cell.

3.9.3.2.2 Transduction

Small fragments of bacterial chromosomes are transferred by phages from one cell to another. No direct cell contact is necessary for this process. Thus, transduction is a process of gene transfer in bacteria, which is mediated by bacteriophage particles.

The frequency of transduction of a gene is about 10^{-5}–10^{-6} per phage particle. Transduction by a phage that can transduce any gene of a bacterium is called generalized transduction. P1 phage is an example of this type of transduction. The transduction of closely linked markers together by the same phage particle is called co-transduction. Non-co-transducible genes are often called "unlinked" genes, even if they are actually part of the same linkage group (chromosome). Transduction by a phage that can transduce only genes next to its prophage insertion site on the bacterial chromosome is called specialized transduction. λ phage is an example of this type. When a transduced

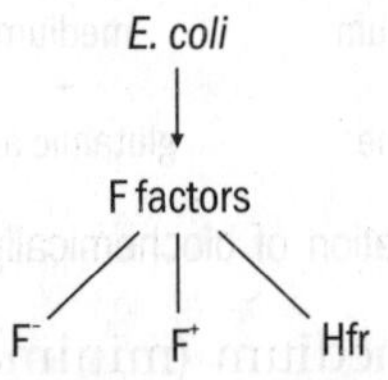

F−: No F factor; function as recipients (female).
F+: Few circular autonomously replicating copies of F factor in cytoplasm.
Hfr: If F+ and F− come in contact, F factor is transferred with a high frequency.

Figure 3.55 Fragments of F factors

chromosome fragment can express its genetic information as a free piece of DNA, it is called abortive transduction (Figure 3.56).

Transduction finds importance in measurement of map distance and serves as one of the ways of gene mapping. It can also sequence a gene on a chromosome.

3.9.3.2.3 Recombinatorial transformation of cells

Cells in a stable state are able to take up DNA that has been isolated from other cells. Cells of certain bacterial strains can be made competent, which means that they can be restructured to be able to take up free DNA.

For successful transformation, the DNA must consist of chromosomal fragments isolated from the same related bacterial strain. This process may occur in yeast cells. An example of recombinatorial transformation of yeast cells is the enhancement of flocculation property of the cells in breweries. Such an enhancement allows the yeast cells to settle quickly, thereby reducing the cost in downstream processing (Figure 3.57).

DNA molecules that can autonomously replicate in the cytoplasm of bacterial cells or yeast cells independent of the cellular chromosome are called plasmids.

3.9.4 Microbial Cell as a Super Microbioreactor

Developments and current status of biomolecular design engineering research in recent advances of biochemical engineering have shown great potential of enhancing, transforming, and/or stabilizing bioproducts through process biotechnology.

Biomolecular design engineering research has gained tremendous momentum in recent years. Uniqueness of DNA function in biocells is bound by its structural engineering features. With the help of molecular engineering tools, DNA structural and functional features are modifiable (Aiba 1984; Arber 1984; Humphrey 1964; Imanaka and Aiba 1981; Mukhopadhyay 1993). Likewise, in recent years, molecular design by molecular blending has provided important tools in developing products related to health care. These modifications relate to structure–function coordination. DNA structural design modification and protein molecule design engineering led to many recent advances in biochemical engineering and process biotechnology. Many molecular engineering tools have been used for technologically redesigning molecules in biocellular machine, proteins of commercial importance like stress proteins, harmful toxin blending for therapeutic

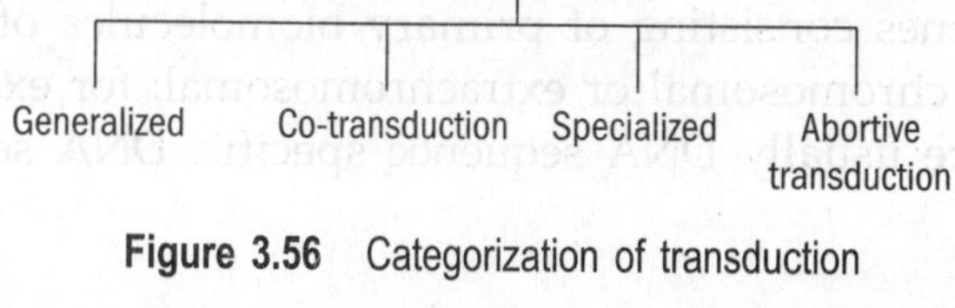

Figure 3.56 Categorization of transduction

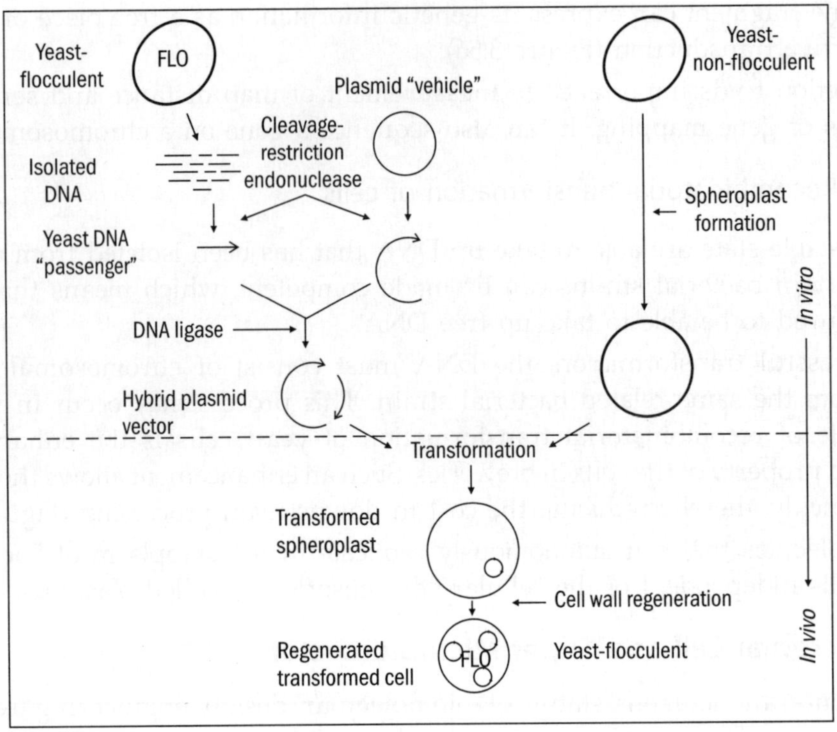

Figure 3.57 *In vitro–in vivo* events in DNA recombination for enhancement of yeast cells flocculation in breweries

use, and so on. These are called restriction endonucleases, chemical nucleases, microbial sword, and so on. A few important purposes in which biomolecular redesigning has been used include stereoselective enzyme catalysed reaction, obtaining hyperproducing transformant of biocells by technologically redesigning microorganisms, enhanced stability of redesigned toxin proteins by molecular blending, among others.

3.9.4.1 Biomolecular systems

3.9.4.1.1 Biocell: a super microbioreactor

A biocell is a super microbioreactor where reactions are carried out in organisms. Physical bioreactor designs are modifiable by machine tools to design the most suitable size and shape with fittings. Similarly, biocell hardware containing living biomolecular systems can be redesigned by using non-classical biomolecular design engineering tools. All biocells contain genes consisting of primary biomolecules of life, called DNA. In biocells, DNA may be chromosomal or extrachromosomal; for example, plasmids. The living cell reactions are usually DNA sequence specific. DNA sequence can be sized through redesigning by modern available techniques; for example, DNA or cloning use biomolecular design engineering tools or hybridome or cell fusion techniques to provide/manufacture products of commercial importance. Biochemical engineers have been successful in reconstructing biotoxin by molecular blending for unusual therapeutic application in hospitals Hambleton (1989).

3.9.4.1.2 Biomolecular design engineering tools

Biomolecular design engineering tools have been used for various purposes depending on their performance; for example, for specific cutting of a molecule, ligating or annealing of one molecular fragment with another fragment. Restriction endonucleases and ligases are examples of biomolecular design engineering tools. Toxin blending for molecular tools has been used to control aberrant host cells.

3.9.4.2 Redesigning/restructuring techniques

Recombinant DNA, cell fusion (hybridoma), molecular blending, and transfection are some of the techniques that have been used in redesigning or restructuring biomolecules. Examples of redesigning by DNA are numerous (Arber 1984; Fuller, Kaguni, and Kornberg 1981; Mukhopadhyay and Ghose 1993). Cell fusion technique of biocellular redesigning for enhanced activity of the cell is well known. Redesigning biocells by hybridoma technique for producing bioproducts, like monoclonal antibodies, is based on the fusion of a single antibody-producing cell to a cell capable of growing indefinitely in a culture (example, a myeloma cell). In molecular blending technique, a toxin is blended with a carrier protein like human albumin and an inert bulking agent like lactose and freeze-dried so that it becomes stable (Hambleton 1989).

3.9.4.2.1 Relevance of biochemical engineering process

By using DNA technique for redesigning biocells, researchers from the Fermentation Research Institute, Tokyo, Japan, have succeeded in redesigning mesophillic *B. subtilis* into a thermophillic bacteria. The aim of the research was to reduce the cooling cost associated with commercial fermentation. This was achieved by increasing the temperature resistance of redesigned microbial biocells through DNA. The plasmids were taken from *Bacillus stearothermophilus*, which thrives at 55–70°C. Resistance to tetrasovoline, an antibiotic, was used as a tracer marker for the two plasmids that were expressed in *B. subtilis*. Oxygen response ambivalence in producing biomolecular design engineering tools like BamH1 is interesting in bioprocessing (Mukhopadhyay and Ghose 1993).

3.9.4.3 In vivo redesigning possibility of newer bioproducts by stressing of biocells

In vivo redesigning of biomolecules (stress proteins), possible interactive stresses, and their retrieval process technology are depicted in Figure 3.58.

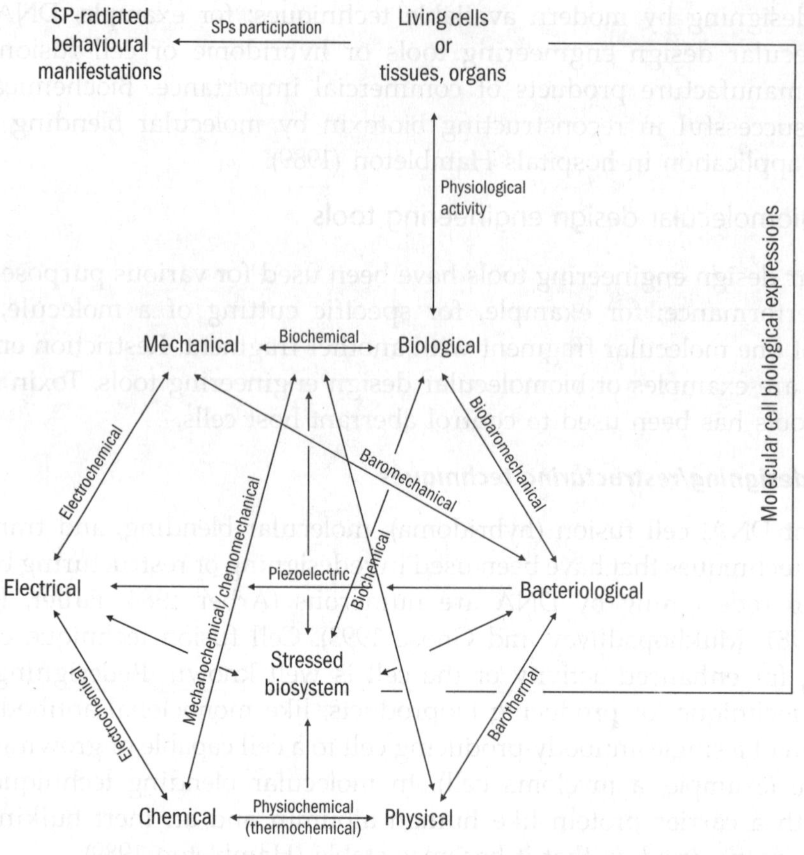

Figure 3.58 Possible interactive stress agents for *in vivo* biomolecular redesigning

Source Mukhopadhyay (2010)

3.9.5 State and Control Variables in Cell Cultivations

The success of all biotechnological processes largely depends on the efficient regulation of its state variables under the influence of control variables. In toxic and anoxic fermentation production systems, various control variables may influence the process state variables (Table 3.19).

The process control variables are environmental parameters that can be regulated through a controller. Under the influence of the control variables, the system manifests its state in terms of the chemical and biological state parameters (Table 3.19).

3.9.6 Cell Cultivation Hardware, Accessories, and Softwares

3.9.6.1 Hardware

In highly metabolic control-requiring bioprocessing and bioreaction systems, high instrumentation and control devices are essential. To achieve meaningful process

Table 3.19 List of process control and state variables of microbial fermentation bioreactor

Process control variable		State variable	
Physical	Chemical	Chemical	Biological
• Temperature • Process • Agitation • $(N_{Re} - N_P)$ • Flow rate (gas and liquid) • Volume • Irradiation	• pH • Dissolved oxygen • Precursor • Sugar/(carbon source) • Substrate(s)	• K_La • CO_2 in exit gas • O_2 in exist gas • Dissolved CO_2 • Protein • Foaming	• X • E_h • RQ • Turbidity • Viscosity (app) • OUR • μ • η • Q • *In vivo* proteolysis • $Y_{x/s}$

N_{Re} – Reynolds number; N_P – Power number; K_La – volumetric O_2 transfer coefficient; X – cell mass; E_h – redox potential; RQ – respiratory quotient; OUR – O_2 uptake rate; μ – specific growth rate; η – specific product rate; Q – specific substrate uptake rate; $Y_{x/s}$ – yield coefficient
Source Mukhopadhyay (2005)

control, many requirements are necessary in the installed equipment hardware, which must consist of the following.

- Bioreactor or fermentation hardware equipment
- Sensors and measuring equipment
- Control devices and interface systems
- Computer hardware and software facilities

These requirements are necessary to (1) carry out biosystemic research on fully monitored environmental needs; (2) correlate environmental parameters with bioreaction regulation and control mechanisms; and (3) reproduce results obtained by assessed environmental conditions through computer monitoring, analysis, and control of bioprocessing.

3.9.6.1.1 Conventional bioreactors

The conventional continuous stirred tank bioreactor (CSTBR) hardware has been used extensively. Laboratory units of a typical CSTBR at the Department of Biochemical Engineering and Biotechnology (DBEB), Indian Institute of Technology (IIT) Delhi and an experimental CSTBR at the Department of Chemical Engineering/Biochemical Engineering Laboratory, California Institute of Technology (Caltech), Pasadena, USA, are shown in Figures 3.59 and 3.60, respectively. Also, a CSTBR hardware, complete with accessories and widely used in microbial engineering processes in industry, is shown in Figure 3.61. Also, a conventional laboratory STEBR is shown in the figure.

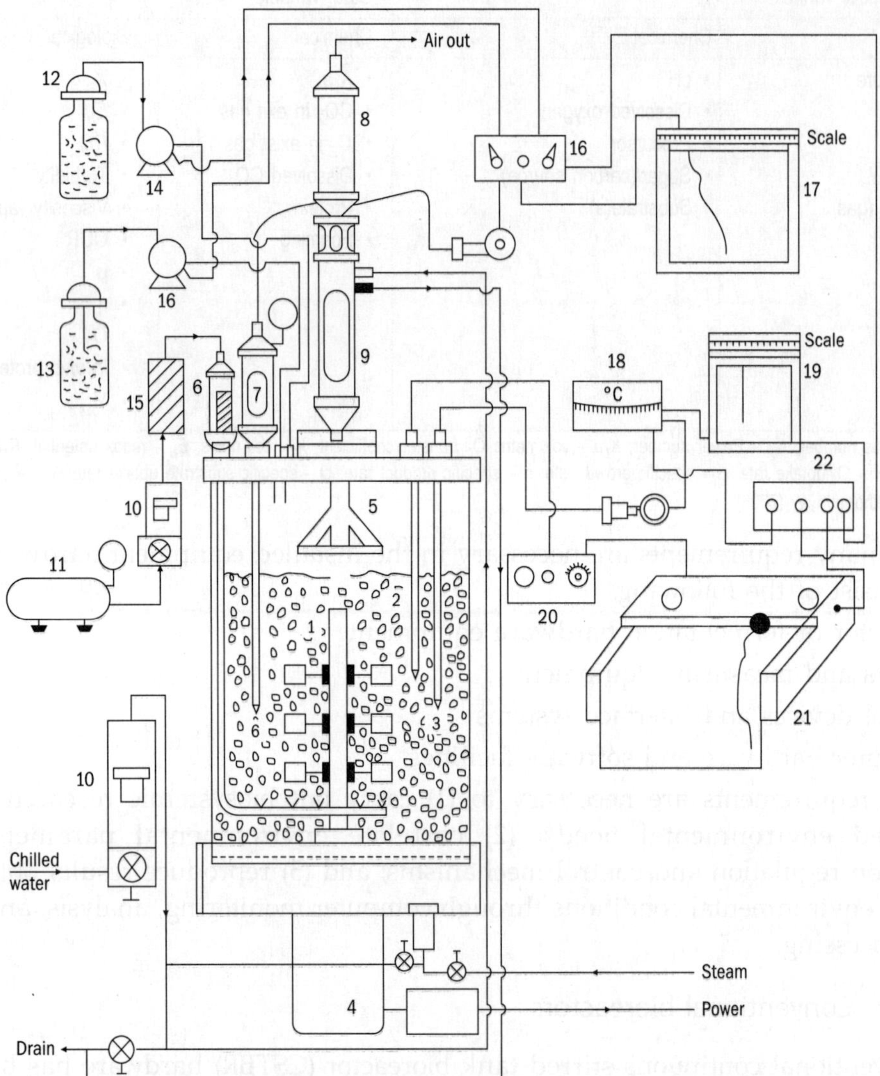

1. Fermenter vessel, 2. Shaft agitator 3. Turbine impeller blade, 4. Agitator motor 5. Funda foam breaker, 6. Ceramic cartridge air filter 7. pH probe fitting, 8. Sterile ceramic candle filter, 9. Condensate housing 10. Rotameter for chilled water flow, 11. Air compressor, 12. Acid bottle, 13. Alkali bottle, 14. Pump, 15. Air filter, 16. Pump and recording switch, 17. Recorder of pH, 18. Temperature indicator, 19. Temperature recorder, 20. Dissolved oxygen indicator, 21. Dissolved oxygen chart recorder, 22. Electrical connection points

Figure 3.59 Conventional continuous stirred tank bioreactor hardware with accessories and fittings of measurement and control of parameters (14I, DBEB IITD, Pilot/Demo Facility)

Figure 3.60 Experimental conventional continuous stirred tank bioreactor laboratory hardware (1 L) (Caltech)

The dimensional and design ratios of a CSTBR design are given as follows.

- Di/Dt = 0.3–0.5
- Li/Di = 0.25
- Wi/Di = 0.2
- H_L/L_t = 0.60
- H_L/Dt = 1.0–2.0
- B/Di = 1.0–1.2
- W/Di = 0.08–0.12
- Hr/Dt = 0.33
- Hr = 1/3 Dt

where Di is the impeller diameter, Dt is the bioreactor diameter, Li is the impeller blade length, Wi is the impeller blade width, W is the baffle width, H_L is the liquid height in the bioreactor, H is the bioreactor height, B is the impeller pitch length, Hr is the sparger to impeller spacing, L_t is the impellar breadth thickness.

3.9.6.1.2 Novel bioreactors

Many novel bioreactor hardware have been designed, fabricated, and developed for laboratory studies and at commercial scales, depending on the biosystemic characteristics, nature, and system based needs.

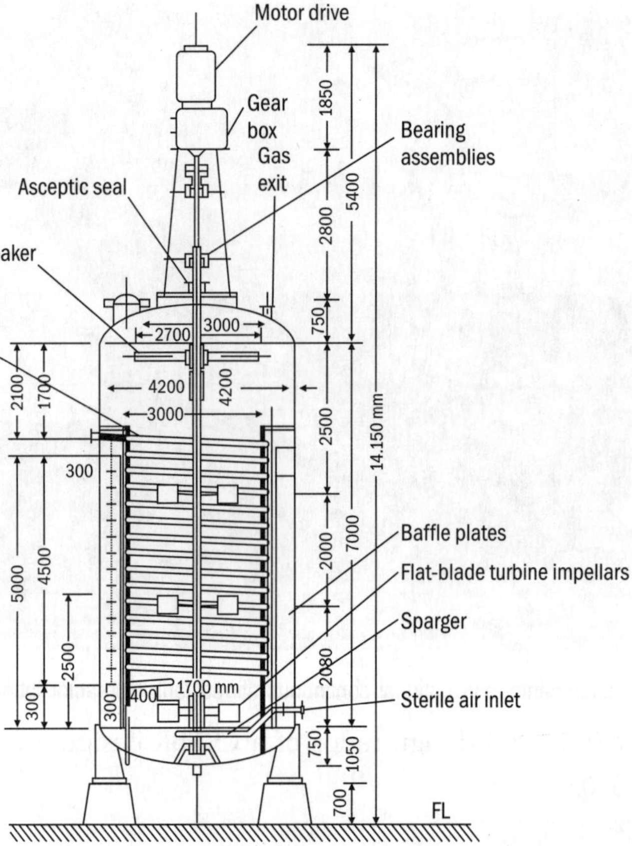

Figure 3.61 An industrial conventional continuous stirred tank bioreactor

Horizontal rotary thin film bioreactor

A horizontal rotatory thin film bioreactor made of standard glass (QVF) has been used in experiments (Figures 3.62a and 3.62b). A typical 60 L capacity stainless steel horizontal rotary fermenter (HRF), having the following components, was designed and fabricated (Figures 3.63 and 3.64).

3.9.6.1.3 New bioreactors

1. Membrane bioreactor

A membrane bioreactor (MBR) has been used in process biotechnology for a long time. An MBR is used to overcome catabolite repression/product level inhibition in process biotechnology (Figure 3.65).

MBR has been successfully used in product separation, chemical plant effluent removal, and agricultural food processing and effluent removal. The main benefit

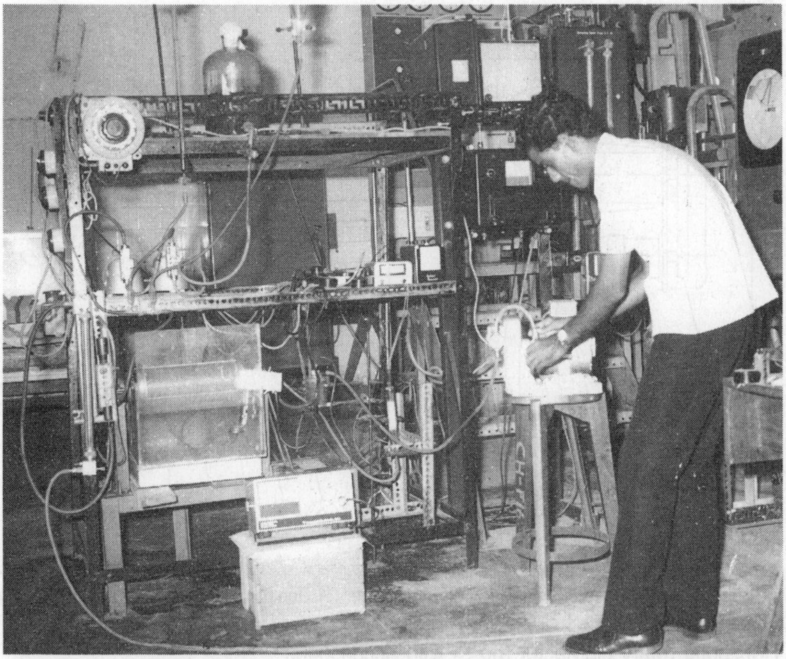

Figure 3.62(a) Horizontal rotatory thin film bioreactor (1 L): front view

Figure 3.62(b) Horizontal rotatory thin film bioreactor (1 L): side view

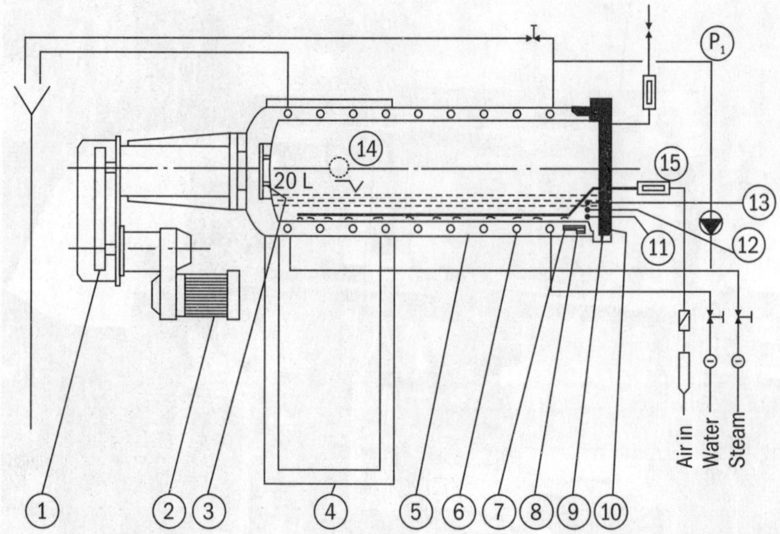

1. Chain drive, 2. Variable speed drive, 3. Mechanical shaft seal, 4. Switch and speed control board, 5. Chamber for condensate, 6. Cooling tube, 7. Sliding bearing, 8. Packing rings, 9. Gland, 10. Cover, 11. Temperature probe, 12. Dissolved oxygen probe, 13. pH probe, 14. Spindle shaft, 15. Air filters

Figure 3.63 Horizontal rotary bioreactor (60 L) with accessories

Figure 3.64 Horizontal rotary thin film bioreactor (60 L) by BERC at Indian Institute of Technology, Delhi

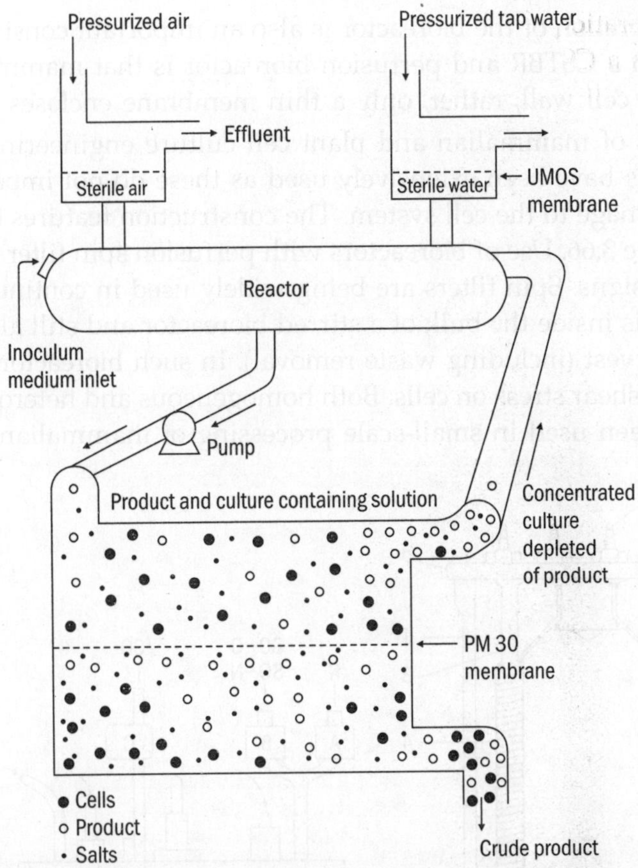

Figure 3.65 Schematic of a continuous microbial membrane reactor using ultrafiltration for product and waste removal

of MBR is the non-accumulation of product, which increases the rate of the forward reactions. Thus, the water quality of the product is significantly higher than that generated by conventional disinfection processes.

2. Perfusion bioreactor

In recent years, mammalian cells have been commonly used to produce complex therapeutic proteins such as monoclonal antibodies (MAbs), erythroprotein (EPO), and interferon beta-1a.

Mammalian and plant cells perfusion bioreactors are preferred as they mostly do not impart sheer force that may cause damage to cell systems. The rate of production is one of the factors that determine the number and size of perfusion bioreactors required to meet the production target. The following relation is used for this purpose.

$$\text{Batch size} = \frac{\text{Production rate, (kg/year)} \times \text{Batch length (days)}}{\text{Titre (g/h)} \times 0.001\ \text{(kg/g)} \times \text{Yearly operation (days/year)}}$$

The mode of operation of the bioreactor is also an important consideration. The main difference between a CSTBR and perfusion bioreactor is that mammalian/animal cells lack a tough outer cell wall; rather, only a thin membrane encloses their cytoplasm.

With the advent of mammalian and plant cell culture engineering, especially these types of bioreactors have been extensively used as these do not impart any shear force that may cause damage to the cell system. The construction features of such bioreactors are shown in Figure 3.66. Use of bioreactors with perfusion spin filter is an advancement of such type of designs. Spin filters are being widely used in continuous perfusion cell culture to keep cells inside the bulk of a stirred bioreactor and still allow for removal of cell-free media harvest (including waste removal). In such bioreactors, a mixing device creates problem of shear stress on cells. Both homogeneous and heterogeneous perfusion bioreactors have been used in small-scale processing of mammalian cell cultures.

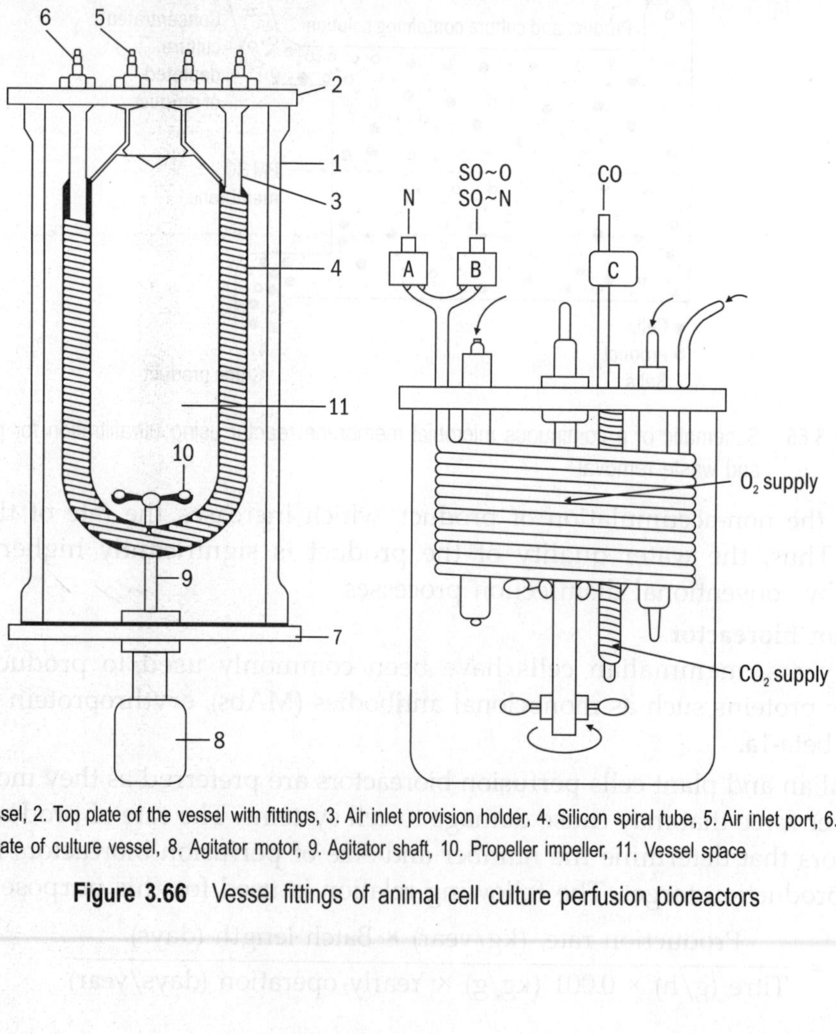

1. Cell culture vessel, 2. Top plate of the vessel with fittings, 3. Air inlet provision holder, 4. Silicon spiral tube, 5. Air inlet port, 6. Air exit provision holder, 7. Base plate of culture vessel, 8. Agitator motor, 9. Agitator shaft, 10. Propeller impeller, 11. Vessel space

Figure 3.66 Vessel fittings of animal cell culture perfusion bioreactors

The perfusion system aims to (1) provide adequate supply of nutrients and remove waste products at high cell densities; (2) perfuse with fresh culture; and (3) achieve steady state ($2^{-5} \times 10^7$ cell/ml).

A cell culture unit consists of a bundle of one or more types of hollow fibres (capillaries) sealed into each end of an 8 mm glass shell by silicone rubber or epoxy resin. Units may be arranged in parallel (as shown) or in series. Nutrient medium stored in a 125 ml reservoir flask was oxygenated and brought to the appropriate pH (Figure 3.67).

Figure 3.68 depicts a microcarrier bioreactor system. It differs from that shown in Figure 3.67 because of the addition of a settling bottle in the recycle circuit between growth and satellite filter vessels. This sequestering of cells and microcarriers outside of the agitated volume of the growth vessel enhances formation of cell-bead aggregates and allows essentially cell carrier free medium to be pumped to the filter.

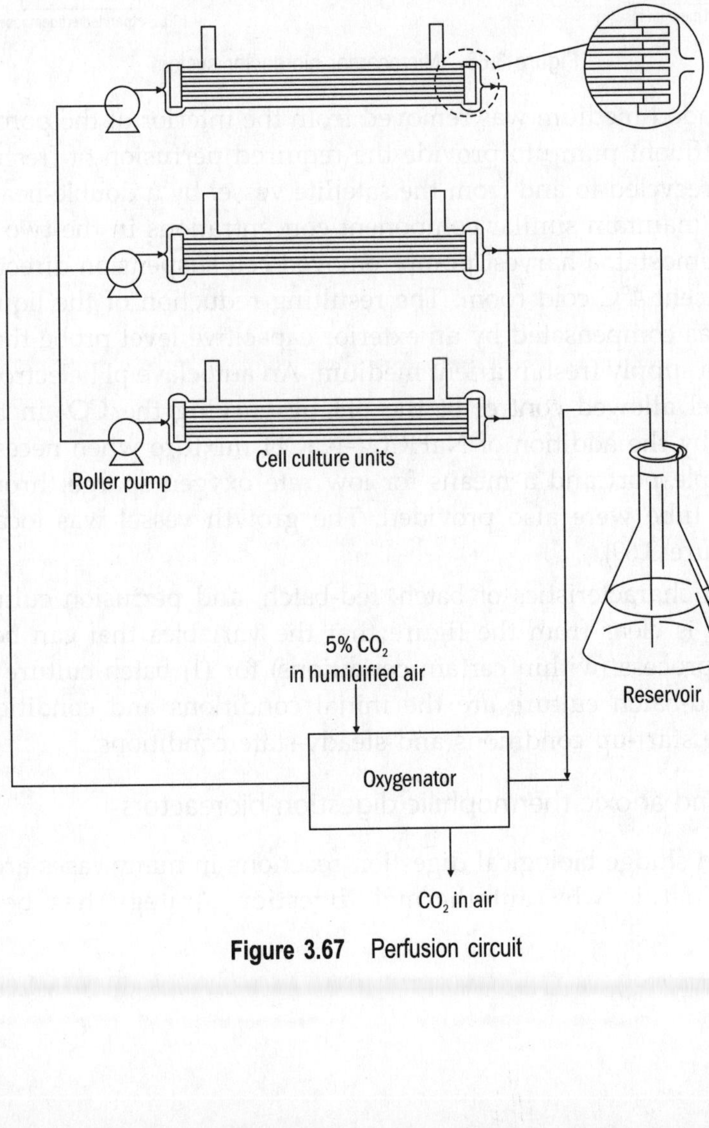

Figure 3.67 Perfusion circuit

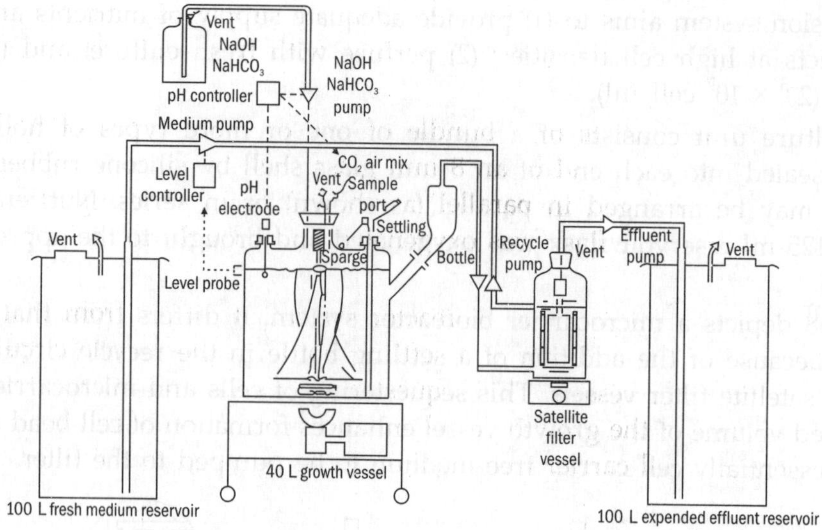

Figure 3.68 Microcarrier bioreactor system

Cell-free expended medium was removed from the interior of the porcelain filter at a rate set on the effluent pump to provide the required perfusion of fresh medium. Cell suspension was recycled to and from the satellite vessel by a double-headed pump at a rate sufficient to maintain similar component concentrations in the two vessels. While operated as a chemostat, a harvest pump removed cell suspension directly to a harvest vessel in an adjacent 4°C cold room. The resulting reduction of the liquid level in the growth vessel was compensated by an exterior capacitive level probe that actuated the medium pump to supply fresh nutrient medium. An autoclave pH electrode penetrating the growth vessel allowed control of the pH by varying the CO_2 in the mixture of overlay gas and by the addition of $NaHCO_3$–NaOH mixture when necessary. A sterile air-shielded sample port and a means for low rate oxygen sparge through a sintered glass dispersion tube were also provided. The growth vessel was located in a 37°C warm room (Figure 3.69).

Environmental characteristics of batch, fed-batch, and perfusion cultures are given in Figure 3.70. It is clear from the figure that the variables that can be manipulated to optimize the process (within certain conditions) for (1) batch culture are the initial conditions; (2) fed-batch culture are the initial conditions and condition f(t); and (3) perfusion are the start-up conditions and steady-state conditions.

3.9.6.1.4 Oxic and anoxic thermophilic digestion bioreactors

Mixed waste food sludge biological digestion reactions in many cases are thermophilic (Zeikus 1979). That is why autothermal digestion strategy has been developed

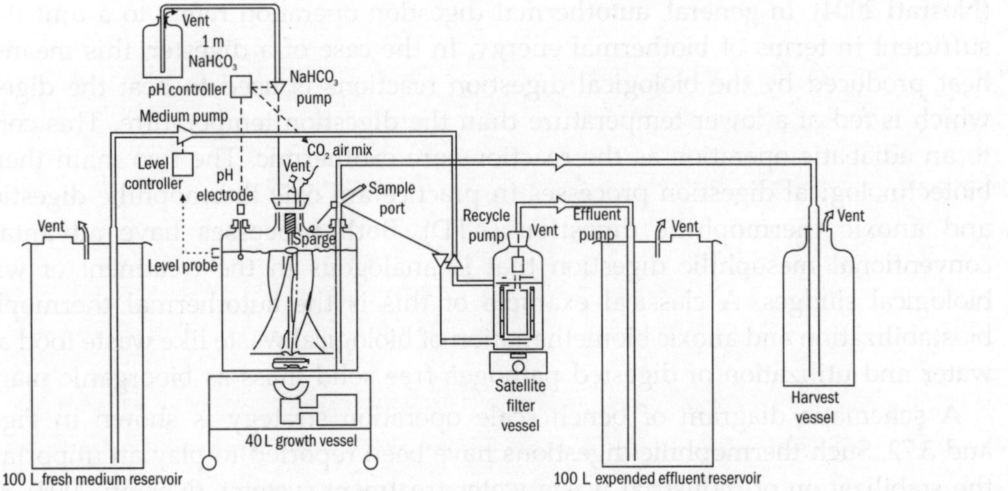

Figure 3.69 Perfusion chemostat system

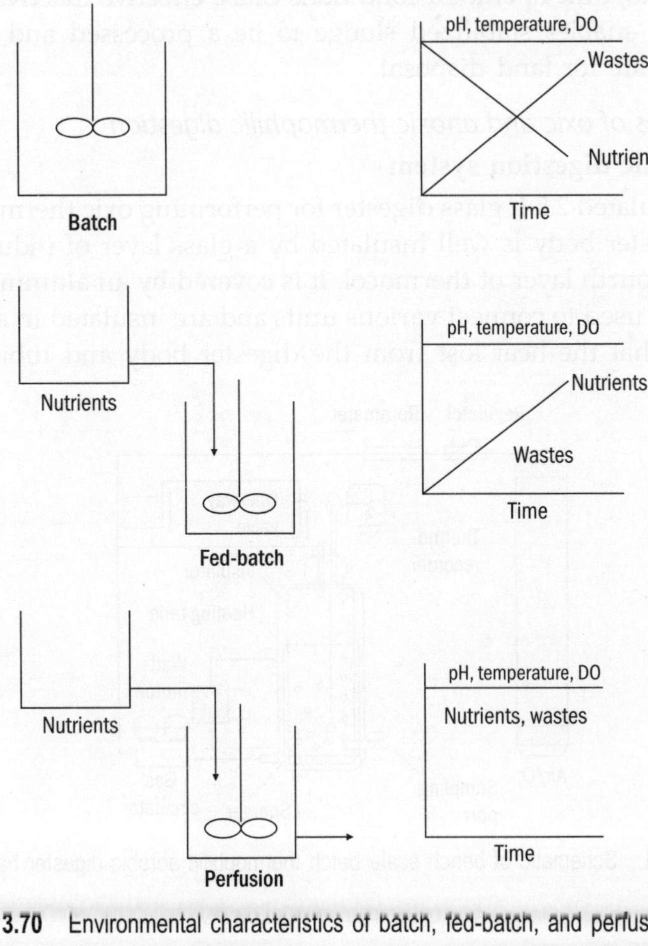

Figure 3.70 Environmental characteristics of batch, fed-batch, and perfusion cultures

(Nosrati 2004). In general, autothermal digestion operation refers to a unit that is self-sufficient in terms of biothermal energy. In the case of a digester, this means that the heat produced by the biological digestion reactions is used to heat the digester feed, which is fed at a lower temperature than the digestion temperature. This corresponds to an adiabatic operation as the reactions are exothermic. The two main thermophilic biotechnological digestion processes in practice are oxic thermophilic digestion (OTD) and anoxic thermophilic digestion (ATD). Both processes have advantages over conventional mesophilic digestion that is analogous in the treatment of waste food biological sludges. A classical example of this is the autothermal thermophilic oxic biostabilization and anoxic biomethanation of biological waste like waste food and waste water and utilization of digested pathogen-free solid mass as bioorganic manure.

A schematic diagram of bench scale operation strategy is shown in Figures 3.71 and 3.72. Such thermophilic digestions have been reported to play an important role in the stabilization of industrial waste water treatment systems (Nosrati 2004). Moreover, more concentrated and more complicated biodegradable organics are decomposed rapidly. Also, thermophilic operation conditions cause effective inactivation of pathogen microorganisms. It enables stabilized sludge to be a processed and pasteurized end product, which is safe for land disposal.

3.9.6.1.4.1 Features of oxic and anoxic thermophilic digestion

a. Oxic thermophilic digestion system

It consists of an insulated 2.5 L glass digester for performing oxic thermophilic microbial reactions. The digester body is well insulated by a glass layer of industrial glass wool followed by a one-fourth layer of thermocol. It is covered by an aluminium foil. Flexible silicone tubings are used to connect various units and are insulated in a similar manner. This is to ensure that the heat lost from the digester body and tubings is as low as

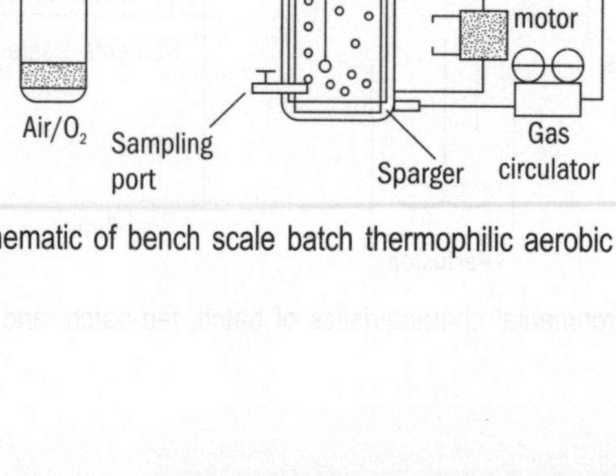

Figure 3.71 Schematic of bench scale batch thermophilic aerobic digester hardware

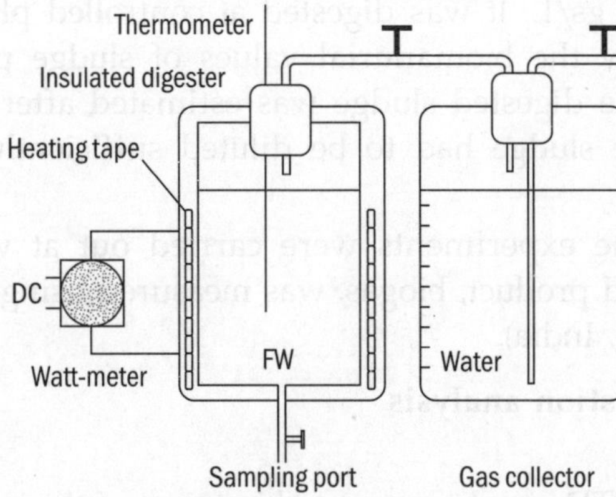

Figure 3.72 Schematic of bench scale batch thermophilic anaerobic digester hardware

possible. The final external area of the digester is 0.09 m^2. A cylindrical aluminium pipe with an internal cross section of 0.1 cm^2 as an air sparging device is installed at the bottom end of the digester. The sparging hole diameter is 1.2 mm.

Operation

Biological sludge obtained from the clarifier bottom (recycle line in the aeration tanks) of the Delhi Jal Board (DJB), containing 8 g/L total solids (TS), 7.4 g/L total suspended solids (TSS), and 6 g/L volatile suspended solids (VSS), was taken in the digester. Air or oxygen was sparged into the digester contents through holes directed downwards to help mixing, avoiding settling of solids when oxic digestion of sludge was carried out.

b. Anoxic thermophilic digestion system

The bench scale batch insulated anoxic digester (2.5 L) that was used (Figure 3.72) in operation has been described in Nosrati (2004).

Waste food from the Indian Institute of Technology Delhi (IITD) hostels was treated for OTD and ATD processes as described in Nosrati (2004). Compositional analysis of the waste food substrate in terms of environmental biotechnological treatment process parameters was also provided.

Materials and methods

For the OTD system, the secondary sludge had a pH of 6.7 and contained solids (g/L) as mixed liquor suspended solids (MLSS) 7.8, TS 9.0, total dissolved solids (TDS) 0.4, total volatile solids (TVS) 6.1, volatile dissolved solids (VDS) 0.1, and VSS 6.0. Also, the sludge had initial cell counts (CFU/L) as total coliform (TC) 4.8 × 10^8, faecal coliform (FC) 1.0 × 10^8, and faecal streptococci (FS) 7 × 10^8. The sludge volume index (SVI) was 54 ml/g, chemical oxygen demand (COD) of 8700 ± 0.5% mg/L, and specific capillary suction time (SCST) of 12 gs/L. It was digested at controlled pH at values of 6.4, 6.7, and 7.0. In order to know the biomanurial values of sludge protein content (SPC), the nitrogen content of the digested sludge was estimated after adequate dilution by the Kjeldahl method. The sludge had to be diluted sufficiently as its colour posed some difficulty.

In the ATD system, the experiments were carried out at various temperatures. During ATD, the generated product, biogas, was measured using a gas chromatograph (AIMILNUCON GC-I 5700, India).

c. Oxic thermophilic digestion analysis

Kinetics

The biokinetic pattern of OTD may be expressed in terms of the increment rate of mixed liquor suspended solids (MLSS), with the removal of VSS (Equation 3.87).

$$\frac{dx}{dt} = \mu X \tag{3.87}$$

where X is the MLSS concentration during digestion (g/L/d), and μ is the specific MLSS increment coefficient. Thus,

$$\mu = \frac{\mu_m s}{K_s + s} \tag{3.88}$$

$$\text{or } 1/\mu = \frac{\mu_m}{K_s + s} + \frac{1}{\mu_m} \tag{3.89}$$

Therefore, a plot of $\frac{1}{\mu}$ versus $\frac{1}{s}$ provides values of μ_m and K_s. In Equation 3.89, μ_m is the maximum value of μ (per day), s is the limiting VSS substrate concentration (*gVSS/L/d*), and K_s is the VSS affinity constant. Since,

$$\mu \propto q \tag{3.90}$$

where q is the specific VSS removal rate

$$\mu = Y \times q \tag{3.91}$$

where Y is the MLSS yield coefficient. Thus,

$$\frac{ds}{dt} = q \times X = \frac{\mu}{Y} X = \frac{\mu_m\, sx}{Y(K_s + s)} \tag{3.92}$$

In Equation 3.92, μ_m/Y is the maximum value of quantity of VSS removed per unit quantity of MLSS increased (lb VSS/lb MLSS or lb/VSS/lbBOD). Since it is an MLSS increase associated nitrogenous biofertilizers/biomanures (product P) formation system (measured in terms of sludge protein content [SPC], g/L), we may write

$$\frac{dp}{dt} = \alpha \frac{dx}{dt} \tag{3.93}$$

$$\text{or } \frac{dp}{dt} = \alpha \frac{\mu_m s}{K_s + s} \tag{3.94}$$

Taking double reciprocal, we get

$$\frac{1}{\frac{dp}{dt}} = \frac{K_s}{\alpha\mu_m}\frac{1}{s} + \frac{1}{\alpha\mu_m} \tag{3.95}$$

It shows that a double reciprocal plot of Equation 3.95 with pH as a control parameter would result in a linear profile, as given in Figure 3.73. Typical values of the parameters used in this plot are given in Table 3.20 as a function of pH.

Oxic thermophilic digestion product

Oxic digestion of waste food for a period of eight days in thermophilic conditions (60°C) in a batch autothermal digester in accordance with the above described kinetics yields

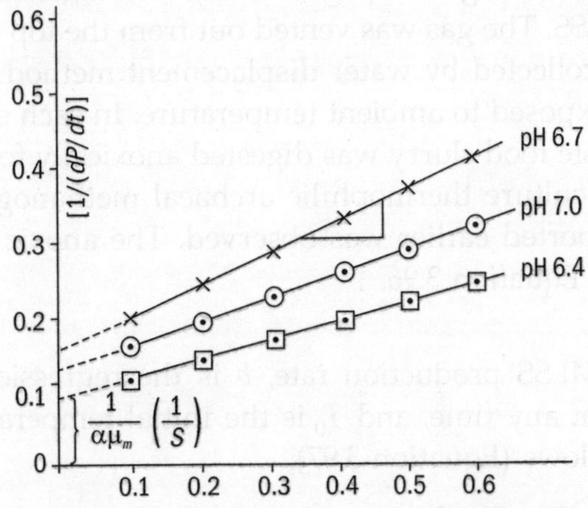

Figure 3.73 Double reciprocal plot of $1/\left(\frac{dp}{dt}\right)$ versus $1/s$

Table 3.20 Values of kinetics constants/coefficients as a function of pH of oxic thermophilic digestion

Constant/coefficient	Unit	Values at pH		
		6.40	6.70	7.00
μ_m	Per day	0.21	0.42	0.31
K_S	mg/L	68.00	120.00	83.00
Q	g/BOD/gVSS/day	12.00	18.00	14.00
Y	gVSS.gMLSSorgVSS/gBOD	0.12	0.56	0.23
SRT	Days	6.00	6.00	6.00
SVI = V_{30}/V_0X	ml/gMLSS	40.00	54.00	38.00
α'	g/L	24.00	15.00	36.00

a solid digested biomass, which was dewatered and sun-dried. A yield of 120 kg wet weight digested solid/m^3 digestion volume was measured. Absence of pathogens in the digested solid was tested using a pathogen indicator organism, which led to the assumption that the solid was pasteurized under thermic digestion conditions (60°C). Next, the N:P:K content of the solid mass was analysed and its strength was compared to different field manures. The results of the suitability of the solid as a biomanure are shown in Table 3.21.

Although this digested solid biomanure is being sold in Delhi for horticulture/nursery purposes at a price of ₹5 per kg, its thorough economic and cost-benefit estimates have to be worked out. This is primarily to examine the usefulness of the biomanure, which is found suitable as a value-added product to enrich soil with agronutrients and to improve crop yield.

d. Anoxic thermophilic digestion analysis

Thermophilic biokinetics

In ATD of waste food in a digester (Figure 3.72), the digestion product was biogas methane and anoxic MLSS. The gas was vented out from the top of the digester through a connecting pipe and collected by water displacement method. The gas collector was not insulated and was exposed to ambient temperature. In each experiment, at different temperatures, 2 L of waste food slurry was digested anoxically for about 12 days. In this mixed-substrate, mixed-culture thermophilic archaeal methanogenic digestion, similar biokinetic pattern as reported earlier was observed. The anoxic MLSS production rate can be calculated using Equation 3.96.

$$\sqrt{r} = b(T - T_0) \tag{3.96}$$

where r is the anoxic MLSS production rate, b is the regression coefficient, T is the temperature measured at any time, and T_0 is the initial temperature. This relation can also be expressed as follows (Equation 3.97).

$$\sqrt{r} = b(T - T_{min})\,[1 - \exp(T - T_{min})] \tag{3.97}$$

where T_{min} is the minimum temperature.

Table 3.21 N:P:K comparison of different field manures

Manure type	Nitrogen (%)	Phosphorus (as P_2O_5)(%)	Potassium (as K_2O)(%)
Farmyard manure	0.400	0.200	0.400
Rural/urban compost	0.600	0.500	0.600
Green manure	0.600	0.100	1.250
Biomanure (50% dry)	1.802	1.001	0.600

Sources Mukhopadhyay and Saha (2000); Mukhopadhyay (2007); Klass and Ghosh (1980); Mukhopadhyay (2006)

The biomass in MLSS is observed at T_{min}, and T_{ms} is the most suitable temperature where maximum biomass growth in MLSS of the thermophilic archeal microorganisms occurs. At temperatures above T_{ms}, Equation 3.97 ceases to apply to anoxic MLSS formation kinetics. This is probably due to the inactivation/denaturation of certain enzymic proteins or some other factors present *in vivo* in the MLSS flora. This necessitated the extension of Equation 3.97 in the form of a non-linear regression law capable of describing the growth of the thermophilic biomass throughout the entire biokinetic temperature range of this biogas. This extension equation could express its growth behaviour. Here, T_{min} and T_{max} refer to the values of minimum and maximum temperatures tolerable by this biomass, at which the rate of growth is almost zero. The parameter C enables the model extension equation to fit the data for temperatures above the most suitable temperature (T_{ms}). The temperature T_{min} may be a conceptual temperature of no metabolic significance. If T is much lower than T_{max}, the contribution of the term in brackets, $[1 - \exp(T - T_{min})]$, becomes negligible. Equation 3.97 reduces to a simpler form, which then indicates that as T increases and approaches T_{max}, the term becomes increasingly more significant until it eventually dominates. The growth rates fall as T exceeds T_{ms} and reaches zero when $T = T_{max}$. Taking logarithm of both sides of Equation 3.97, we get

$$\log [1 - \sqrt{r}/\{b\ (T - T_{min})\}] = C(T - T_{max}) \qquad (3.98)$$

The left hand side of Equation 3.98 can be numerically evaluated by any data point. From two data points in the higher temperature ranges, C and T_{max} can be evaluated. The value of C was estimated to be 0.0097. Using these values, the predicted T_{min} and T_{max} values of the mixed thermophilic archaeal flora (Table 3.22) were shown in comparison to *Methanobacterium* sp., as reported earlier (Mukhopadhyay and Saha 2000).

Anoxic thermophilic digestion products

ATD produces biogas/methane and digested solid residue. The amount of digested solid MLSS produced in ATD (74 kg wet weight/m^3 digestion volume) is much less as compared to that obtained in OTD strategy. However, the biomanural quality of this pathogen-free biomass was of equal value as that obtained in OTD in terms of N:P:K content. The biogas production in ATD of waste food solids was 200 m^3/tonne

Table 3.22 Predicted T_{min} and T_{max} values in the growth of anoxic MLSS flora in comparison to a pure culture of methane producer

Organism	Character	Predicted temperature range (*K*)		
		T_{min}	T_{ms}	T_{max}
Methanobacterium sp. (reported earlier)	Under identification	304	323	348
Thermophilic Methanogens	Anoxic MLSS flora	308	335	343

MLSS – mixed liquor suspended solids

of biodegradable matter. Biogas has a calorific value of 5200 kCal/m^3 and serves as a clean fuel. Biogas and biomanure retrieved from ATD process biotechnology are considered value-added products. Biogas is used in lighting of gas mantles, cooking in stoves as clean fuel, operating agricultural and farm machinery like flour grinders, operating internal combustion engines in place of petrol or diesel, generating electricity, and operating irrigation pumps. It is also used as a fuel in laboratory burners, heaters, incubators, and fans. Like biogas, the co-product biomanure enriches soil with agronutrients, which improve crop yields. Similar to biomethane, biomanure products are being produced through ATD in recent years, which are eco-friendly products.

OTD and ATD process biotechnology are proper methods of environmental pollution abatement, along with cooperative value-added product delivery.

3.9.6.1.5 Gas/air-lift bioreactor

One of the more recent bioreactors are air-lift bioreactors, which use air to agitate the fluid by means of draft tubes or an external recycling system. There are three types of gas-lift bioreactors (Figure 3.74). The gas–liquid contacting devices characterized by fluid circulation in a clear and defined cyclic pattern. The directionality of liquid flow is one of the most important characteristics of air-lift bioreactors. These bioreactors create unique and complex hydrodynamic conditions allowing for gas hold up, liquid velocity, gas/liquid circulation, and mass transfer of mixing. The homogeneous field of shear stress is due to mean density difference in upflow, riser, and downflow section and downcomer. In a CSTBR, turbulence dissipates with distance from impeller-sparger.

One of the important applications of air-lift bioreactors is the production of single-cell protein with bacterial cultures that are not too viscous. The most significant advantage is its high oxygen transfer. Moreover, air-lift bioreactors do not have any moving parts, resulting in high volume production. However, such bioreactors are energy intensive and have non-uniform mass transfer, leading to its limited application.

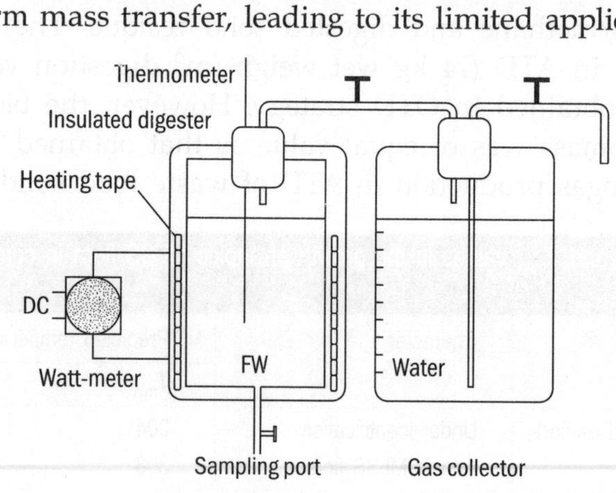

Figure 3.74 A typical gas-lift bioreactor operation principle

3.9.6.1.6 Fluidized bed bioreactor

In addition to the above-discussed bioreactors, liquid fluidized bed bioreactors are also used bioprocess engineering (Figure 3.75). The oxygen mass transfer rates of liquid fluidized bed bioreactors, which fluidize "particulately", or smoothly, have been related to those of packed beds bioreactors. This relation is given by Equation 3.99.

$$Sh = K_L d/D = 0.81/\epsilon \, N_{Re}^{1/2} \, Sc^{1/3} \tag{3.99}$$

where Sh is the Sherwood number, K_L is the mass transfer coefficient, d is the particle dimension, D is the diffusion coefficient, ϵ is the void space in packed bed/fluidized bed, N_{Re} is the Reynolds number, and Sc is the Schmidt number.

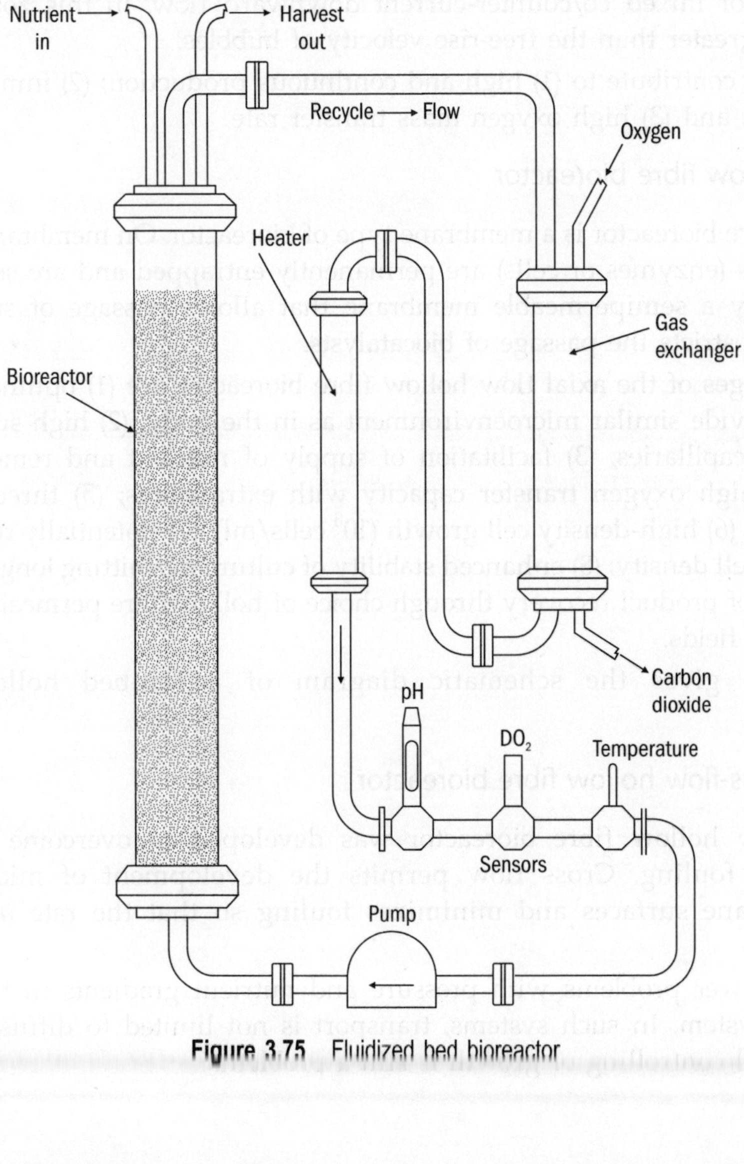

Figure 3.75 Fluidized bed bioreactor

In Equation 3.99, K_L is a function of N_{Re} and Sc.

There are three zones in a fluidized bed bioreactor (Figure 3.75).

1. **Riser**
   In this zone, the gas injection creates fluid density difference. This zone exhibits concurrent upward flow of both gas and liquid.
2. **Turbulence zone (at top of the reactor)**
   In the turbulence zone, there is a gas–liquid separator. Liquid mixing and gas recirculation take place in this zone.
3. **Downcomer zone**
   In the downcomer zone, the degassed liquid recirculates to the riser, exhibiting concurrent or mixed co/counter-current downward flow. In this zone, the liquid velocity is greater than the free-rise velocity of bubbles.

These zones contribute to (1) high and continuous production; (2) immobilization of microbial cells; and (3) high oxygen mass transfer rate.

3.9.6.1.7 Hollow fibre bioreactor

The hollow fibre bioreactor is a membrane type of bioreactor. On membrane bioreactors, the biocatalysts (enzymes or cells) are permanently entrapped and are separated from the reactant by a semipermeable membrane that allows passage of substrates and products but restricts the passage of biocatalysts.

The advantages of the axial flow hollow fibre bioreactor are (1) optimal cell-culture systems to provide similar microenvironment as in the body; (2) high surface/volume ratio-artificial capillaries; (3) facilitation of supply of nutrient and removal of waste products; (4) high oxygen transfer capacity with extra fibres; (5) three-dimensional lattice support; (6) high-density cell growth (10^8 cells/ml); (7) potentially reduced serum needs at high cell density; (8) enhanced stability of culture permitting long-term growth; (9) facilitation of product recovery through choice of hollow-fibre permeability; and (10) minimal shear fields.

Figure 3.76 gives the schematic diagram of a flat-bed hollow-fibre cell culture reactor.

3.9.6.1.8 Cross-flow hollow fibre bioreactor

The cross-flow hollow fibre bioreactor was developed to overcome the problem of membrane fouling. Cross flow permits the development of microturbulence across membrane surfaces and minimizes fouling so that the rate of permeation is enhanced.

There are fewer problems with pressure and nutrient gradients in the cross-flow hollow fibre system. In such systems, transport is not limited to diffusion, although monitoring and controlling of growth is still a problem.

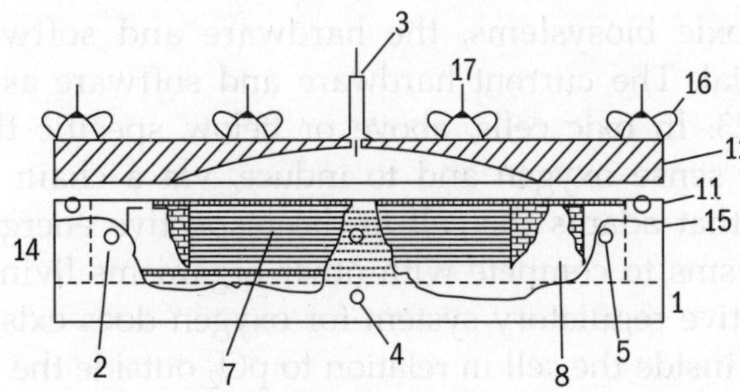

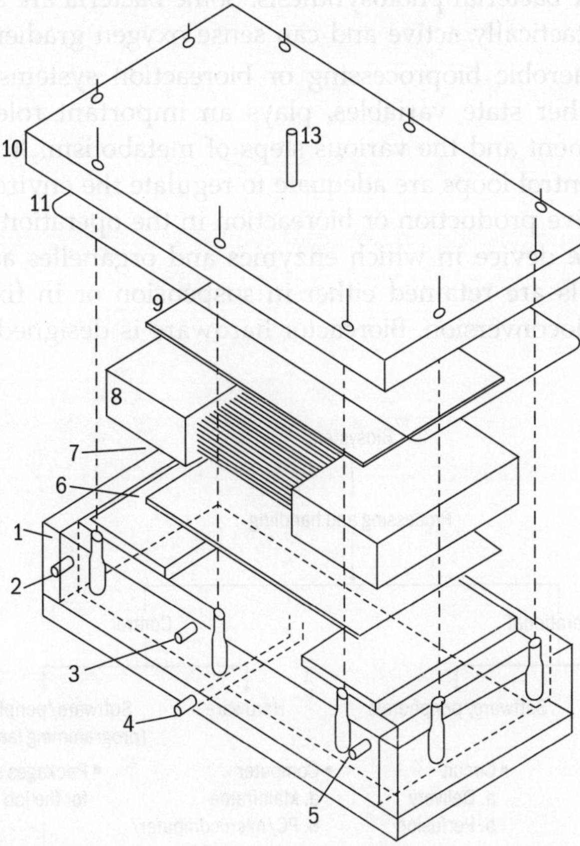

1. Lower reactor housing, 2. Air/CO_2 inlet, 3. Cell inoculation port, 4. Fresh medium inlet, 5. Air/CO_2 outlet, 6. Microporous filter, 7. Hollow fibre bed, 8. Polypropylene hollow fibre potting, 9. Open ends of hollow fibres, 10. Upper stainless steel microporous filter, 11. Gasket, 12. Upper reactor housing, 13. Spent medium oulet, 14. Pre-reactor chamber, 15. Post-reactor chamber, 16. Bolt, 17. Winged nut

Figure 3.76 Schematic diagram of a flat-bed hollow-fibre cell culture reactor

3.9.6.1.9 Biosystemic hardware and software classification

In both oxic and anoxic biosystems, the hardware and software as classified in Figure 3.77 are essential. The current hardware and software as used in biosystems are listed in Table 3.23. In oxic cells, above or below specific threshold values, the organisms are able to sense oxygen and to induce, via a chain of events, a process of cell differentiation that adapts the cell to the respective energy-producing system. This allows the organisms to compete with other organisms living in the same place. Thus, an *in vivo* inductive regulatory system for oxygen does exist. A schematic of O_2 regulation and control inside the cell in relation to pO_2 outside the cell surface has been shown in literature for bacterial photosynthesis. Some bacteria are attracted by oxygen. They are thus chemotactically active and can sense oxygen gradients.

In aerobic and anaerobic bioprocessing or bioreaction systems, the state variable K_La, among many other state variables, plays an important role in the interaction between the environment and the various steps of metabolism. It requires knowledge of whether isolated control loops are adequate to regulate the environmental conditions for optimal fermentative production or bioreaction in the operation of an individual. A bioreactor is a suitable device in which enzymes and organelles as well as microbial, animal, and plant cells are retained either in suspension or in fixed form, either for positive or negative bioconversion. Bioreactor hardware is designed for system-specific requirements.

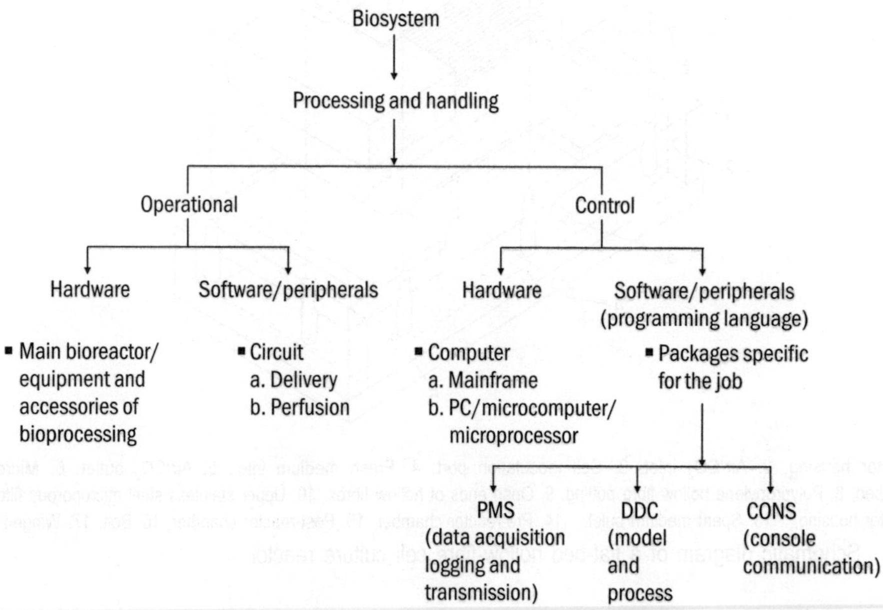

Figure 3.77 Purpose-based classification of hardware and software for biosystem

Table 3.23 Examples of current hardware and software used in a few biosystems

Biosystem	Hardware scale/mode of production	Regulator type/mode	Function	Language
Penicillin fermentation	Industrial/batch	Computer ARCII 102	DDC	Machine
Glutamic acid fermentation	Industrial/batch	Computer YODIC 500	DDC	—
Gluconic acid fermentation	Pilot	Computer PDP11/20	DDC/DO control	DAN (FORTRAN)
Blood oxygenation	Laboratory	Membrane oxygenator (Bentley-Univox)	Regulation of pO_2 in blood for clinical evaluation	
Blood oxygenation	Laboratory	Hollow fibre membrane oxygenator (IVOX)	Vena caval regulation of pO_2/pCO_2 in surgical practices	
Recombinant plasmid of *Escherichia coli*	Laboratory	Plasmid-encoded genes	Regulation of oxygen demand of cells	
Recombinant plasmid of *E. coli*	Laboratory	Plasmid-encoded genes	Regulation of metabolites/oxygen demand cells	

pCO_2 – partial pressure of carbon dioxide; pO_2 – partial pressure of oxygen; DO – dissolved oxygen

Modern microcomputer hardware chips are taking on more sophisticated software (that is, programming functions, which in the past had to be dedicated to larger hardware machines). This advancement has enabled a computer system take on extensive data acquisitions, analyses, and control tasks at a much lower cost, as compared to much higher costs for larger machines. With small personal computers (PCs), it is now possible to perform sophisticated control and acquisition functions in small laboratories and large-scale plant operations.

The first task in computer hardware selection is to define the problem and feed in instructions (data acquisition and/or process control steps) into the computer. In modern advanced bioprocessing systems such as the commercial batch fermentation system, the process control capability of a computer is an integral and important requirement.

For data acquisition tasks, it is necessary to know the hardware instruments to acquire data and the parameters that need to be recorded. It is also necessary to determine whether the instruments have either analogue or digital outputs with which to interface to a computer system. An analogue appears either as a current or a voltage signal, which changes proportionately in response to the value of the parameter being measured. A digital output appears on one of the two forms, either serial or parallel. When a serial interface is used, data are transferred to the computer in one binary bit after another. For parallel interface, all bits are presented for acquisition simultaneously. The central scheme may require only a few wires. On the other hand, a parallel scheme requires at least one wire for each bit in addition to signal ground and handshaking.

Table 3.24 lists hardware, measuring, and software instruments used in bioprocessing of cell cultivation.

The output type will have an impact on the distance of computer hardware from the instrumentation of the bioreactor. The use of a computer as a control hardware has a distinct advantage in providing the opportunity to combine data analysis capabilities and flexibility of manipulating process input variables to achieve control. Such capabilities are illustrated in Figure 3.78, which shows that measurements of the dissolved oxygen level and effluent gas oxygen concentration allow online estimation of oxygenation efficiency, K_La. The dissolved oxygen measurement information has been used to manipulate agitation rate and/or air feed rate to control dissolved oxygen at the desired level. As shown in Figure 3.78, the AD converter accepts current signals and converts instrument output voltage into current, using a V/I converter. As the computer hardware distance from instrumentation is important, it is preferable to transmit electrical analogue signals in the current form. The reason for this is to avoid line losses.

3.9.6.2 Software

Software systems are sets of programmed instructions developed by system programmer(s), which do not provide direct solutions to the problem, but perform tasks in relation to the hardware for their solution. The software consists of three subsystems: (1) data acquisition, logging, and transmission subsystem (PMS); (2) the model and process control subsystem of direct digital control (DDC); and (3) the console communication subsystem (CONS).

The use of digitally encoded data does not necessarily obviate the ability to form a computer hook-up. It does mean that the software package selected will have to support the type of interface that the instrument possesses. Usually, lower cost data acquisition packages that run on small personal computers to accept analogue input are suitable. For equipment involving both analogue and digital output, many software packages allow for so-called "hooks" to ancillary programs. These may be custom designed to interface specific equipment. After selecting the type of input, it is necessary to decide on the speed of the response and the required bandwidth of the computer data acquisition system. For digital data, it is necessary to consider and determine if the target instrument is sending its data at a rate consistent with the requirements, and if the host can retrieve the data within the constraints. With analogue data, the rate at which one can receive the data will be determined by the acquisition rate of the analogue-to-digital converter card, which will plug into the computer.

Table 3.24 Select hardware, measuring, and software instruments used in bioprocessing of cell cultivation

Operational instruments				Control instruments			
Hardware	Source	Measuring/peripheral	Source	Hardware	Source	Software/peripheral	Source
Fermenter (14 L)	Fermenter Design Inc., Allentown, PA USA (A)	Temperature: • probe • pH probe • DO probe • Agitation Servo • Sparging Servo • O_2 analyser Turbidity (Spectronic 20) • V/I converter (0.1% accuracy) • Flow meter (0–20) 1/min, model 5820 pH, air flow rpm, temperature, DO, O_2 analyser for exit air foam, feed flow, and so on	(A) (A) (A) (A) (A) (A) Bausch and Lomb Special Brokes Instrument Company, USA	Computer PDP-II	Digital Equipment Corporation, USA Multiplexer/converter Kw-11L Real	RSX-11A Real Time Executive DDC-11 DR-11A ContactI/O AD-ID AD Time clock DIDOX PMS	Special I/O Routine Special interface equipment
Fermenter (10 L)	Schweiler Keil Germany			Digital computer system 16/440 Bits processor PCX1/PCX2	Gist Brocades	DDC CONS	

DO – dissolved oxygen; rpm – revolutions per minute; I/O – input/output

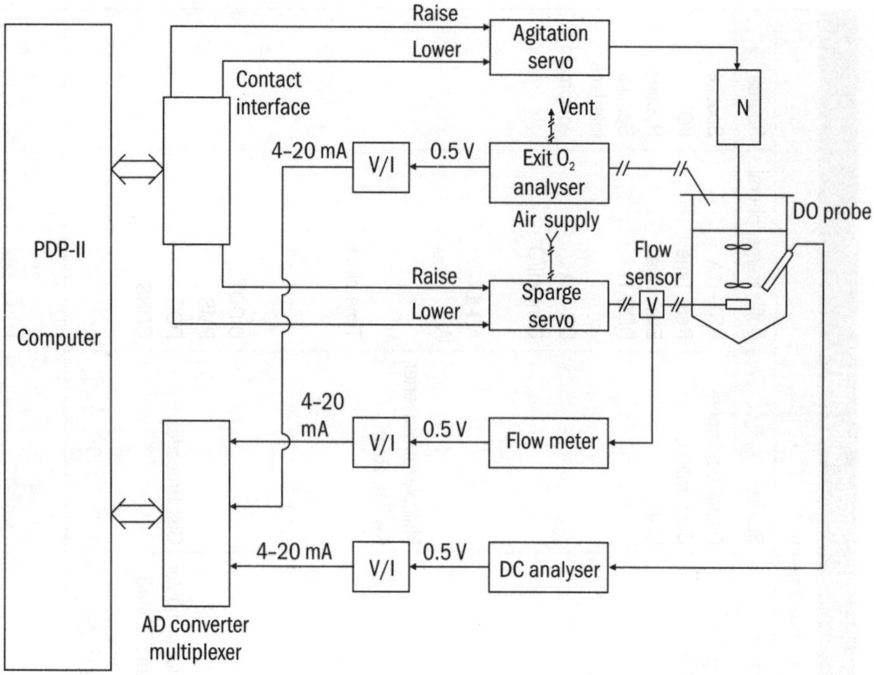

Figure 3.78 Computer-instrumented fermenter block diagram configuration

For fermentative production of pharmaceutically/medically important gluconic acid, a highly instrumented bioreactor hardware configuration and its computer control system are shown in Figure 3.78. Principles of measurements of dissolved oxygen concentration, inlet and exit gas flow rates, and compositions are known. These may be used to determine the average volumetric oxygen transfer coefficient (K_La) in a bioreactor. An online dissolved oxygen control data analysis computer software (DAN) program has been developed and used (Nyiri, Jefferis, and Humphrey 1974). However, any software engineer (as distinct from the system programming for oxygen regulation and control) will be concerned with (1) establishing user requirements; (2) verifying the user requirements; (3) assessing and choosing software design methods for the oxygenation project; (4) structuring and specifying tests; (5) planning and implementing design reviews; (6) choosing software tools (like validators, compilers, and test beds); and (7) criticizing methods at the end of the life cycle.

An example of block diagram configuration of the combination of fermenter and digital computer (PDP11) hardware used in bioprocessing is shown in Figure 3.78. The list of hardware and software instruments used in this system is given in Table 3.24. The capability of process control computers to measure and control simultaneously hundreds of plant variables is now well known, as is the equally important use of the computer for real-time data analysis, including estimation of process variables not directly measured or controlled. An example of this is the indirect estimation of active cell mass and cell growth rate in a fermenter mash. In addition, because of the accuracy and time stability of digital computers, it is possible to achieve the precise oxygen balance and mathematical integration accuracy required to implement this indirect measurement. In making this balance, it is usually assumed that all uncounted for material is utilized by the cell for growth and maintenance.

A typical digital computer instructional flow sheet used in simulating theoretical and experimental results of gluconic acid fermentation kinetics of *Pseudomonas ovalis* in HRTFB is shown in Figure 3.79.

3.9.6.2.1 Fully filled bioreactor

1. Bioreactor

A highly instrumented bioreactor, developed at the ETH, operates almost completely filled with culture medium. The low gas hold-up of 5% prevents segregation of biomass through cell wall growth and overlaying foam formation. An axial propeller circulates the liquid through a draught tube in a well-defined flow pattern. The reactor is equipped with a mechanical foam breaker. Analogue control loops are available for pH, temperature, and dissolved oxygen. Instrumentation allows the measurement of pressure stirrer speed, electrical power input, redox potential, weight of the reactor contents, gas mass flow rate, and gas analysis (O_2 and CO_2). Other measurements, possible with available equipment but not performed on a routine basis, are biomass concentration (spectrophotometric and cell counting methods) and substrate and product concentration using automated enzymatic methods show the bioreactor design and instrumentation (Figures 3.80 and 3.81).

2. Computer

The computer is a PDP11/45 process computer equipped with 64K words (16 bit) core memory, magnetic and paper tape readers, and two disc storage units. Real-time FORTRAN programming is made possible with a time-sharing software system (RSX-11 version 6), which allows several processes to be controlled in a real-time online mode.

The computer is linked to the process, which is located 1 km from the computer via an internal telephone line. Signals are transmitted from the terminal interface unit (borer) in digital form to the computer. A controller unit at the computer serves to compare the signals for error and relieves the computer from the task of data administration. The interface performs all multiplexing, analogue–digital conversion, and channel identification tasks. A keyboard provides operator communication with the computer. It is divided into function keys and decimal input keys. The function keys are redefinable according to work area (example, dissolved oxygen control) by numerical

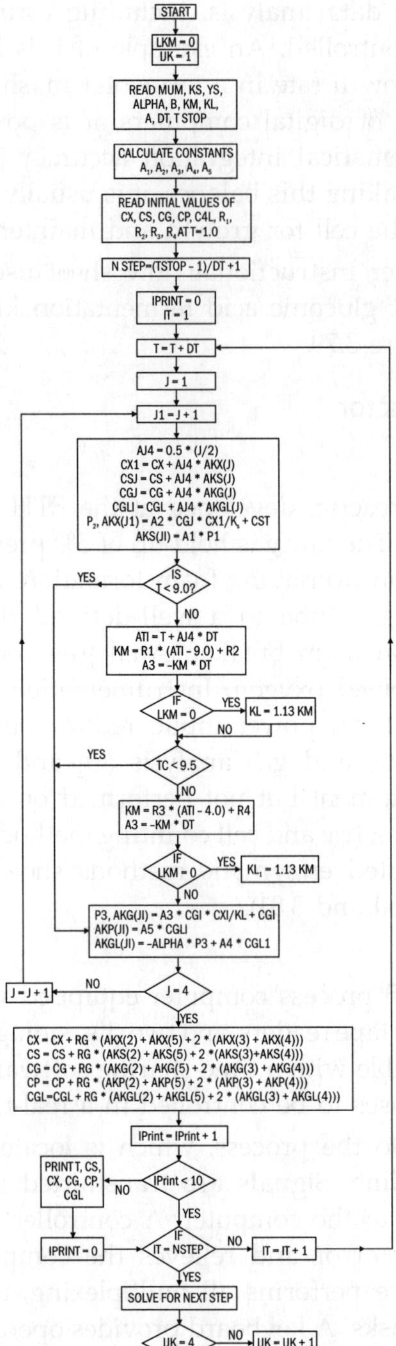

Figure 3.79 Instructional flow chart of ICL1909 digital computer processing program used in simulation to check the appropriateness of gluconic acid fermentation kinetics of *Pseudomonas ovalis* in HRTFB

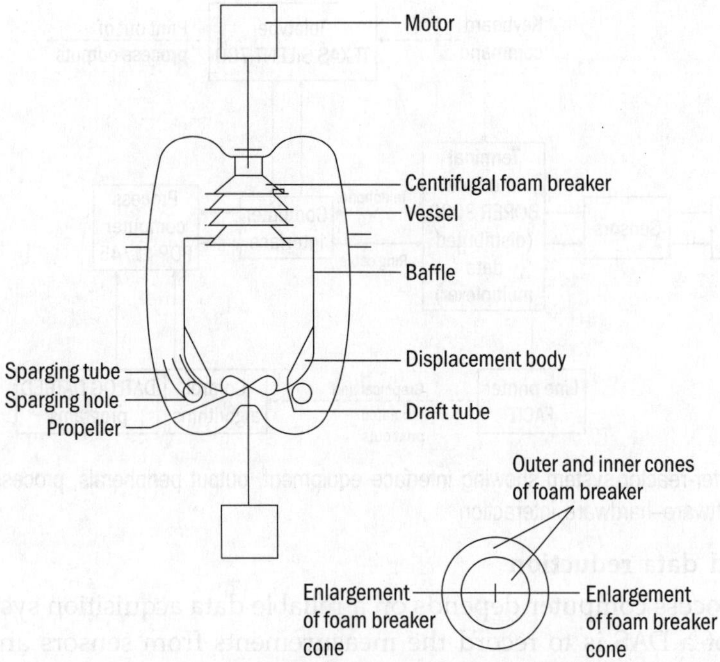

Figure 3.80 Fully filled bioreactor

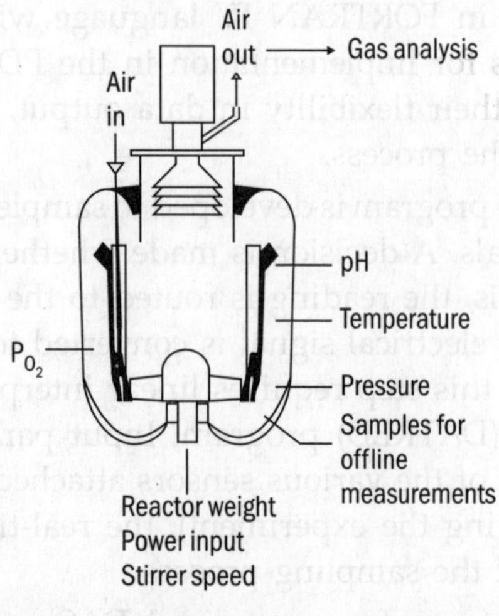

Figure 3.81 Fully filled bioreactor design hardware with instrumentation

codes. A decimal display on the interface terminal provides single-value output data. Peripheral equipment at the process are a terminal (Texas Instruments Silent 700) and a line printer (Facit 4553). The terminal serves to develop and start programs and is connected via a separate telephone line. The line printer is used to display data in tabulated and graphical form. Figure 3.82 shows how the computer and the equipment are connected to the process.

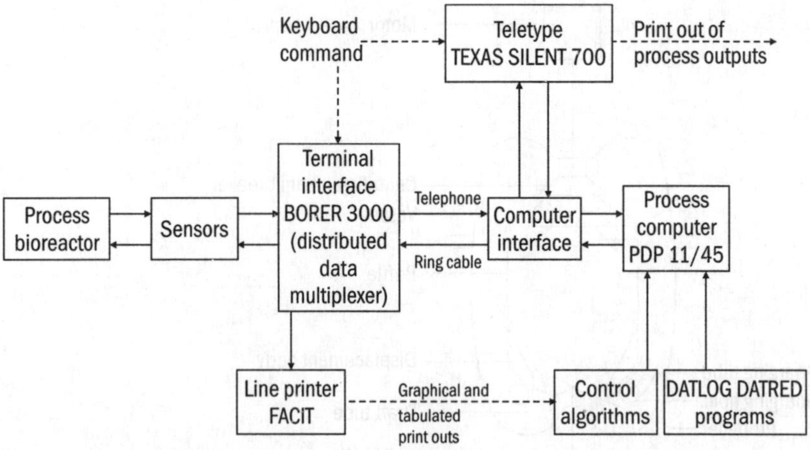

Figure 3.82 Computer-reactor system showing interface equipment, output peripherals, process linkage, and software–hardware interaction

3. Data logging and data reduction

The function of a process computer depends on a reliable data acquisition system (DAS). The primary duty of a DAS is to record the measurements from sensors and transmit them to the computer for storage and processing. The DAS is the first step towards development of DDC, online modelling, and parameter estimation. Two programs presented here are written in FORTRAN IV language with ordinary process control extensions and sub-routines for implementation in the PDP11/45 process computer. A feature of the programs is their flexibility in data output, which ensures close contact between the operator and the process.

A data logging (DATLOG) program is developed to sample atomically the measurement signals at prescribed intervals. A decision is made whether the reading is required for control purposes only. If it is, the reading is routed to the control program for further instructions. If it is not, the electrical signal is converted to the proper unit by the use of stored conversion tables; this step requires linear interpolation. The data are finally sent to the data reduction (DATRED) program. Input parameters to the program are the different sampling rates of the various sensors attached to the reactor (these can be altered by the operator during the experiment); the real-time clock inside the process computer is used to control the sampling process.

DATRED is the second stage of computerized DAS, and its functions are to store incoming data from DATLOG, process them, and on completion of the experiment present the data bank and sets of calculated variables in tabulated and graphical forms. Input parameters to the program are the experimental conditions used, the data obtained online by the DATLOG routine, and the data obtained offline by later laboratory analysis practices.

As data are entered by the DATRED program, they are stored in a matrix for later use. On request from the keyboard, the data can be made available to an operator. Thus the instantaneous value of any variable or, if desired, a graphical display of the raw data can be obtained. These data are used to evaluate sets of biological parameters of interests to the biochemical engineer and microbiologist.

The formula used in determining the biological parameters for a batch aerobic culture is as follows.

$$\mu = \frac{1}{X}\frac{dx}{dt} \tag{3.100}$$

where μ is the specific growth rate and X is the cell biomass.

The derivative term is evaluated from a quadratic equation fitted to every three points on the X versus t curve by a least square method. The kinetic parameter (μ_m) is obtained from a Michaelis–Menten-type graphical analysis.

The outlet gas molar flow rate (F_1) is calculated from the measurement of the inlet flow using a nitrogen balance.

$$F_1 = F_0/(1 - y_{O_2} - y_{CO_2}) \tag{3.101}$$

where F_0 is the initial value of the outlet gas molar flow rate. F_1 is expressed in mol/h.

$$F_0 = (G_0)(\mathrm{STPH})(0.79)/22.4 \tag{3.102}$$

$$STPH = [273/(T_0 + 273)][P_0/1][1/(1 + h)] \tag{3.103}$$

This relation converts the flow into dry conditions, standard temperature, and standard pressure. The oxygen transfer rate (OTR) is obtained from an oxygen balance on the reactor.

$$OTR = 0.21F_0 - F_1y_1 \text{ (mol } O_2\text{/h)} \tag{3.104}$$

The volumetric oxygen transfer coefficient (K_La) can be calculated as given below.

$$K_La = \mathrm{OTR}/(C_L^* - C_L)V \text{ (per hour)} \tag{3.105}$$

$$C_L^* = P_Fy_{1O_2}/H \text{ (mol } O_2\text{/L)} \tag{3.106}$$

where H is the Henry coefficient

The specific oxygen uptake rate (Q_{O_2}) is calculated as given below.

$$Q_{O_2} = OTR/(XV) \text{ (mol } O_2\text{/g biomass h)} \tag{3.107}$$

The specific CO_2 production rate (Q_{CO_2}) is calculated as given below.

$$Q_{CO_2} = (y_{1CO_2}F_1 - 0.03\ F_0)/(XV) \text{ (mol } CO_2\text{/g cell h)} \tag{3.108}$$

The respiratory quotient (RQ) is calculated as given in Equation 3.109.

$$RQ = Q_{CO_2}/Q_{O_2} \qquad (3.109)$$

The cumulative carbon losses (CLC) can be evaluated from the following equation.

$$CLC = \int F_0 y_{0CO_2} dt - \int F_1 y_{1CO_2} dt - [(S_1 - S_1)\, VS_c + (X_1 - X_1)\, VX_c] \text{ (g atom of C)} \qquad (3.110)$$

where the initial culture conditions are given by S_1 and X_1.

The carbon contents of the substrate and biomass are given by the constants S_c and X_c. The instantaneous rate of carbon losses (CLR) is given by the following equation.

$$CLR = F_0 y_{0CO_2} - F_1 y_{1CO_2} - \frac{\Delta(S_1 VS_c + X_1 VX_c)}{\Delta t \text{ (g atom C per hour)}} \qquad (3.111)$$

An important feature of the DATRED program is that under keyboard instructions from the interface, the raw data or the calculated biological parameters may be requested at any time during the experimental period. In case offline analyses must be made later because of the absence of an online sensor, the experimental data are fed into the program via the keyboard when they are available. Output at the process can be in the form of tabulated or graphical printout. At the end of the experiment, however, all the raw data and reduced data are available in tabulated and graphical forms.

Digital control

Commonly, fermentations are run batch-wise. Consequently, the conditions and the dynamic responses of the system will change with time as the reaction progresses. Conventional controller constant setting must generally be altered to compensate for this. It was desired to develop a digital control algorithm that would adapt automatically to the changing reactor conditions and would be applicable for the control of any variable. The algorithm chosen is the iterative gradient procedure described by the following equation.

$$dv = \left(\frac{\partial v}{\partial m}\right)_t dm + \left(\frac{\partial v}{\partial t}\right)_m dt \qquad (3.112)$$

The change in the control variable (dv) is expressed as the sum of two effects: a change caused by the manipulated variable (m) and the time change contributed by the batch-wise biological growth process. In practice, values of the two gradients are required to calculate the required control action to eliminate a given offset.

The computer can be programmed to initiate a control action according to the above equations as follows.

- The control variable v is allowed to change within a given dead zone without control.

- During this pause, the rate of change $(\Delta v/\Delta t)_m$ of the offset is measured when m is constant.
- On reaching the tolerance limit, the manipulated variable m is altered by an amount Δm and its effect $(\Delta v/\Delta m)_t$ is measured.
- Using Equation 3.112, the procedure is repeated until the set point is reached.

The above iterative gradient algorithm has proved itself to be a stable method in pH control. It is expected that it will prove useful in many fermenter applications (Table 3.25).

pH control application

The iterative gradient control algorithm has been applied to pH control. Fermenter pH control lends itself well to this control method because the buffer capacity of the culture as measured by ΔpH/(Δ titrated volume) is not known but must be measured directly by the computer. The rate of change of the controlled variable ($\Delta pH/\Delta t$) must also be measured. The pH digital control procedure can be summarized as follows. During the pause phase, the pH is allowed to deviate from the set point within certain dead zone limits without control action. Greater offset initiates control action, which iteratively returns the system to the set point. The volumetric addition of basic neutralization medium during the iteration step according to the iterative gradient procedure is given in the equation below.

$$\Delta v_{i+1} = \frac{(pH_{set} - pH_i) - (\Delta pH/\Delta t)(t_{i+1} + t_m)}{\Delta pH/\Delta v_i} \tag{3.113}$$

where $\Delta pH/\Delta v_i = \dfrac{(pH_i - pH_{i-1}) - (\Delta pH/\Delta t)(t_{i+1} + t_m)}{\Delta v_i}$

The term $\Delta pH/\Delta v_i$ is the buffering capacity. The transient growth acid production effects are accounted for by the term $\Delta pH/\Delta t$, which is determined during the pause

Table 3.25 Applications for the iterative gradient control algo

Control variable	Measurement method	Manipulated variable
Dissolved oxygen	Electrode (polarographic) Gas analysis	rpm, aeration rate or pressure
Respiratory quotient	CO_2 and O_2 (infrared adsorption and paramagnetism)	Critical substrate feed rate
K_La	Electrode and O_2 gas analysis Optical	rpm, aeration rate, power input
Biomass concentration	density, light scattering Resistance	Substrate feed rate
Temperature	O_2 gas balance, optical density	Water and steam flows
Specific growth rate	Power input	Substrate flow
Viscosity		Dilution flow

K_La – average volumetric oxygen transfer coefficient; rpm – revolutions per minute

period without control action. The value t_i is the time required to introduce a quantity Δv_i. The mixing time to achieve homogeneity is given by t_m. For a given dead zone, control action always begins at the same offset. The buffer capacity in a batch system of basic neutralization medium to reach the set point will also increase with time. This knowledge of the control history is used to reduce the number of iterations required by initially adding the total amount of medium used in the previous control phase. Foolproof control requires that a number of logical safety features be built into the algorithm to account for changes in set point, dead band width, noisy measurement, and operator errors.

In advancing biochemical engineering and biotechnology, online process computation and control for the involved biochemical engineers, and besides being a valuable biosystemic research tool, the strategy will provide an interface for idea exchanges between bioscientists and engineers. Therefore, the future scope of such analysis will be enormous.

3.9.7 Cell Cultivation Modes

3.9.7.1 *Batch*

In the batch mode of cell cultivation for product formation, flow neither enters nor leaves the used bioreactor. The liquid contents are well mixed. Batch fermentation of ethanol has been carried through ages in this mode in different countries.

One of the most important questions raised in alcoholic batch fermentation is that of oxygen transfer. CO_2 ventilation could be important. However, definitive studies to show the importance of the latter have not been undertaken.

The data collected by Fietchter and Von Meyenburg (1968) for the batch growth (Figure 3.83) of *Saccharomyces cerevisiae* on glucose medium are reasonably typical of the events that can occur in fermentation. In this particular case, the maximum CO_2 ventilation rate (29 mmol/L-h) is much larger than the maximum O_2 uptake rate (4.2 mmol/L-h) and is not coincidental with it. Also, neither transfer rate is very high as a well-designed fermenter should have the capability of transferring up to 250 mmol O_2/L-h.

3.9.7.2 *Fed-batch culture (without withdrawing culture medium)*

Fed-batch culture has been developed in order to achieve a high productivity of microbial cells and secondary products in fermentation industries. Baker's yeast production, for example, has been adopting this method to avoid the "glucose effect", that is, catabolite repression. By means of this device, nutrients or growth-limiting substrate is continuously fed in the culture medium.

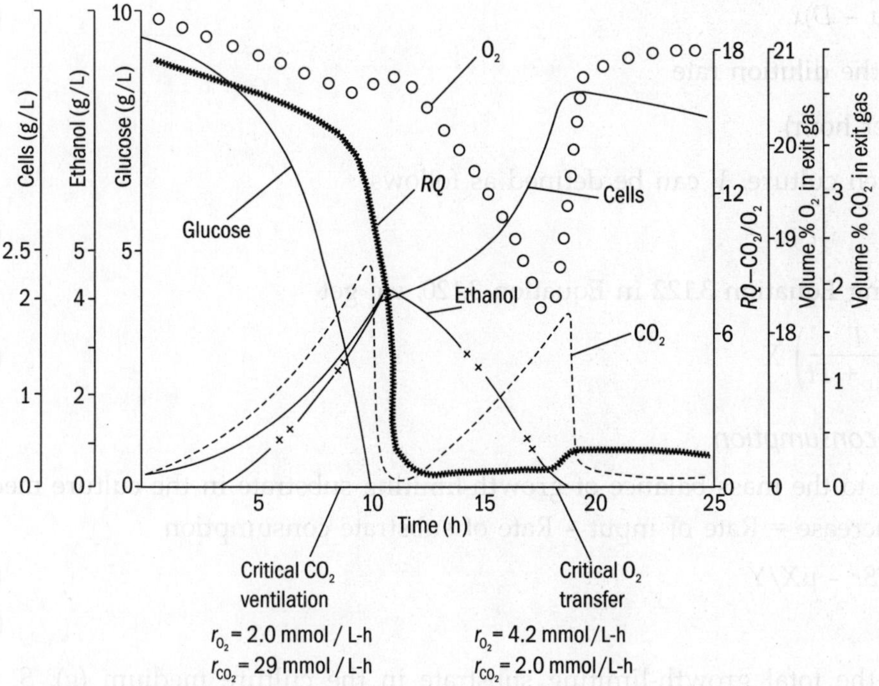

Figure 3.83 A typical batch growth of *Saccharomyces cerevisiae* on glucose in ethanol fermentation

3.9.7.2.1 Basic theory of fed batch

1. Growth

The growth of fed-batch culture can be calculated as follows.

$$dX/dt = \mu X \tag{3.114}$$

$$X = Vx \tag{3.115}$$

where X is the total biomass in culture vessel (g cell), V is the culture volume at time (L), V_0 is the initial culture volume (L), F is the flow rate of fresh medium, t is the culture time (h), x is the biomass concentration (g cell/L), μ is the specific growth rate (L/h).

$$F = dv/dt \text{ (L/h)} \tag{3.116}$$

$$V = V_0 + Ft \tag{3.117}$$

Substituting Equation 3.117 in Equation 3.116, we get

$$d(Vx)/dt = \mu Vx \tag{3.118}$$

Equation 3.118 then changes as given below.

$$X(dv/dt) + V(dx/dt) = \mu Vx \tag{3.119}$$

or $dx/dt = (\mu - D)x$ (3.120)

where D is the dilution rate

$D = F/V$ (per hour) (3.121)

In fed-batch culture, V can be defined as follows.

$V = V_0 + Ft$ (3.122)

Substituting Equation 3.122 in Equation 3.120, we get

$$\frac{dx}{dt} = \left(\mu - \frac{F}{V_0 + Ft}\right) X \quad (3.123)$$

2. Substrate consumption

With regard to the mass balance of growth-limiting substrate in the culture medium,

Rate of increase = Rate of input – Rate of substrate consumption

or $dS/dt = FSr - \mu X/Y$ (3.124)

and $S = sV$ (3.125)

where S is the total growth-limiting substrate in the culture medium (g), S_r is the concentration of growth-limiting substrate in a fresh medium (g substrate/L), Y is the growth yield for substrate

$Y = -dX/dS$ (g cell/g substrate) (3.126)

Substituting Equations 3.115 and 3.125 in Equation 3.124, we get

$d(sV)/dt = FSr - \mu xV/Y$ (3.127)

Equation (3.127) changes as shown below.

$$\frac{ds}{dt} = \frac{F}{V}(S_r - S) - \frac{\mu x}{Y} \quad (3.128)$$

Substituting Equation 3.124 in Equation 3.128, we get

$$\frac{ds}{dt} = \frac{F}{V_0 + Ft}(S_r - S) - \frac{\mu x}{Y} \quad (3.129)$$

If the feed rate of fresh medium is changed step-wise at an arbitrary time interval during the cultivation, Equations 3.123 and 3.127 can be rewritten as Equations 3.130 and 3.131, respectively. A schematic diagram of the step-change feeding in a fed-batch culture is shown in Figure 3.84.

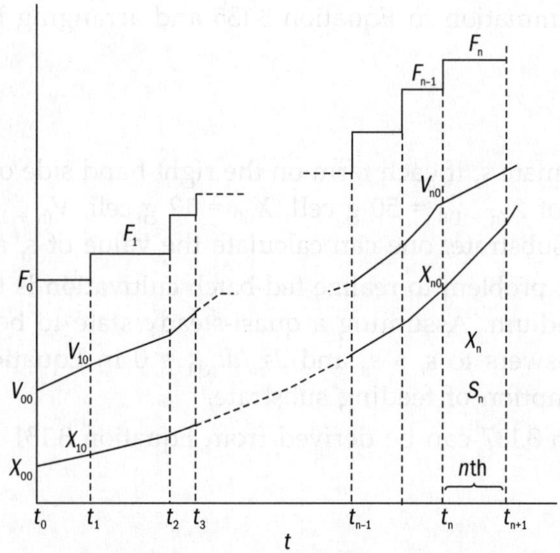

F – flow rate of a fresh medium (L/h); V – culture volume (L); X – biomass in the culture vessel (g cell); t – culture time (h)

Figure 3.84 Schematic diagram of step-change feeding in fed-batch cultivation

$$\frac{dx_n}{dt_{n+1}} = \left(\mu - \frac{F_n}{V_{n0} + F_n(t_{n+1} - t_n)}\right) X_n \tag{3.130}$$

$$\text{or } \frac{ds_n}{dt_{n+1}} = \frac{F_n}{V_{n0} + F_n(t_{n+1} - t_n)} (S_r - S_n) - \frac{\mu}{Y} X_n \tag{3.131}$$

where x_n is the cell mass concentration in the culture medium in the nth time interval, s_n is the residual substrate concentration in the culture medium in the nth time interval (g substrate/L), F_n is the flow rate of a fresh medium in the nth time interval (L/h), V_{n0} is the culture volume at $t = t_n$ (L), V_{00} (= V_0) is the culture volume at $t = t_0$ (L).

$$x_n = X_n/V_n \text{ (g cell/L)} \tag{3.132}$$

$$s_n = S_n/V_n \tag{3.133}$$

$$V_{n0} = V_{00} + \sum_{n=1}^{n} F_{n-1}(t_n - t_{n-1}) \tag{3.134}$$

Assuming growth yield (Y) to be constant and residual growth-limiting substrate concentration (s or s_n) (Equations 3.129 or 3.131) to be almost nil compared to s_r, the following mass balance will be established in fed-batch cultivation.

$$\left.\begin{aligned} F_0(t_1 - t_0)s_r &= (V_{10} - V_{00})s_r = (1/Y)(X_{10} - X_{00}) \\ F_1(t_2 - t_1)s_r &= (V_{20} - V_{10})s_r = (1/Y)(X_{20} - X_{10}) \\ F_n(t_{n+1} - t_n)s_r &= (V_{(n+1)0} - V_{n0})s_r = (1/Y)(X_{(n+1)0} - X_{n0}) \end{aligned}\right\} \tag{3.135}$$

Considering the summation in Equation 3.135 and arranging by s_r, we get

$$s_r = \frac{X_{(n+1)0} - X_{00}}{Y(V_{(n+1)0} - V_{00})} \tag{3.136}$$

It is possible to estimate s_r if each term on the right-hand side of Equation 3.136 is set. For example, in case of $X_{(n+1)0}$ = 50 g cell, X_{00} = 12 g cell, $V_{(n+1)0}$ = 6.5 L, V_{00} = 3.5 L, and Y = 0.5 g cell/g substrate, one can calculate the value of s_r as 25 g substrate/L.

Another significant problem to realize fed-batch cultivation is to choose an optimum flow rate of fresh medium. Assuming a quasi-steady state to be established in a fed-batch culture, this answers to $s_r \gg s_n$ and $ds_n/dt_{n+1} \approx 0$ in Equation 3.131 as a result of an immediate consumption of feeding substrate.

Therefore, Equation 3.137 can be derived from Equation 3.131 in terms of F_n.

$$F_n = \frac{\mu X_{(n+1)0}}{Ys_r} \tag{3.137}$$

If μ and Y are constant values and s_r is estimated beforehand (Equation 3.136), one can predict F_n as a function of total biomass in the culture vessel at $t = t_{n+1}$, $X_{(n+1)0}$.

For the estimation of $X_{(n+1)0}$, integration of Equation 3.114 gives the following.

$$\left.\begin{aligned} &2.3 \log X_{(n+1)0} = \mu(t_{n+1} - t_n) + 2.3 \log X_{n0} \\ &\text{or } X_{(n+1)0} = X_{n0} \exp\{\mu(t_{n+1} - t_n)\} \end{aligned}\right\} \tag{3.138}$$

In aerobic fermentation during yeast propagation, the ethanol yield for glucose ($Y_{p/s}$) is defined as follows.

$$\left.\begin{aligned} &(Y_{p/s})_n = \frac{Q_p}{Y_n} = \frac{k_1(Q_{CO_2} - k_2 Q_{O_2})}{Y_n} \\ &\text{or } (Y_{p/s})_n = \frac{k_1(RQ-1)(I_{O_2})_n}{\dfrac{\mu X_n}{Y}} \end{aligned}\right\} \tag{3.139}$$

Cancelling out the term $\mu X_{(n+1)0}/Y$ from Equations 3.137 and 3.139, we get

$$F_n = \frac{180\, k_1(RQ-1)}{s_r\,(Y_{p/s})_n}(I_{O_2})_n \tag{3.140}$$

where k_1 is the stoichiometric coefficient $\left(\frac{\text{mole ethanol}}{\text{mole } CO_2}\right)$, k_2 is the stoichiometric coefficient $\left(\frac{\text{mole } CO_2}{\text{mole } O_2}\right)$.

$$I_{O_2} = Q_{O_2} X \tag{3.141}$$

3. Case example

Fed-batch culture controlling the feed rate of fresh medium by the ad hoc measurement of *RQ* was conducted to achieve a high productivity in baker's yeast cultivation to ward off the "glucose effect". The feed rate throughout cultivation was increased step wise when *RQ* fell to about 1.0 and decreased step wise when *RQ* rose to about 1.2. Thus far, a typical aerobic fermentation of yeast was succeeded in retaining moderate specific growth rate 0.24 L/h and growth yield 0.55 g cell/g glucose. Since the total oxygen demand I_{O_2} could be measured easily and accurately, a perspective of minicomputer to determine the feed rate by means of I_{O_2} signal instead of cell mass concentration (*X*) was discussed with *RQ* control in the aerobic yeast production.

Theoretical considerations

A schematic diagram of step-change feeding in fed-batch culture is shown in Figure 3.85. The mass balance equations are shown below.

$$\frac{\Delta X_i}{\Delta t_i} = \left(\mu_i - \frac{F_i}{V_{0i} + F_i\,\Delta t_i}\right) X_i \tag{3.142}$$

$$\frac{\Delta S_i}{\Delta t_i} = \frac{F_i}{V_{0i} + F_i \Delta t_i}(S_R - S_i) - \frac{\mu_i}{Y_i} X_i \tag{3.143}$$

S_R and F_i can be determined from Equation 3.143, provided that $\Delta S_i/\Delta t_i \simeq 0$, $S_R \gg S_i$, and μ_i, Y_i = constant (= μ, Y).

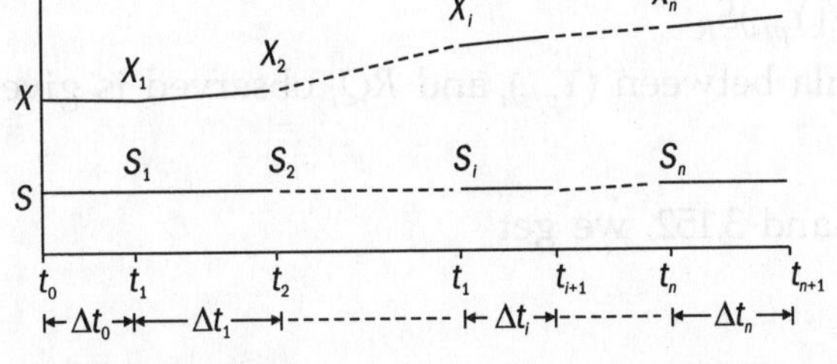

X – cell mass concentration in culture (g cell/L); *x* – total cell mass in fermentor (g cell) (*x* = *XV*); *V* – broth volume (L); *F* – feed rate of fresh medium (L/h); *S* – glucose concentration in culture (g/L); S_R – glucose concentration in fresh feed (g/L); *t* – time (h); *Y* – growth yield (g cell/g glucose); μ – specific growth rate [μ = (1/*x*) (*dx*/*dt*) (L/h)]

Figure 3.85 Schematic diagram of fed-batch culture

$$F_i S_R = \frac{F_i X_i}{Y} + \frac{1}{Y}(V_{0i} + F_i\,\Delta t_i)\,\frac{\Delta X_i}{\Delta t_i} \tag{3.144}$$

Then,

$$\sum_{i=0}^{n} F_i\,\Delta t_i\,S_R = \frac{1}{Y}(X_{n+1} - X_0) \tag{3.145}$$

$$S_R = \frac{X_{n+1} - X_0}{Y\sum_{i=0}^{n} F_i \Delta t_i} = \frac{X_F - X_0}{Y(V_F - V_0)} \tag{3.146}$$

where F is the final step in fed batch.

$$F_i = \mu X_i / Y S_R \tag{3.147}$$

Results and discussion

An example of fed-batch culture is shown in Figure 3.86 The figure shows a 10 L bench scale fermenter with 600 rpm at 30°C, pH = 4.5 and 2.5 to 3.27 L air/min. Two peristalistic pumps are present (pump A for main feeding and pump B for fine control, on when $RQ < 1.0$ and off when $RQ > 1.1$).

Procedures to control the feed rate by I_{O_2} signal instead of x_i in Equation 3.147 are discussed below.

$$(Y_{p/s})_i = Q_{pi}/\gamma_i = k_1(Q_{CO_2} - k_2 Q_{O_2})_i/\gamma_i \tag{3.148}$$

or $(Y_{p/s})_i = (I_{O_2})_i\,(RQ_i - 1)/(\mu_i x_i/180Y_i)$ (3.149)

where $Y_{p/s}$ is the ethanol yield from glucose (mol EtOH/mol glucose), Q_P is the specific rate of EtOH formation (mol/g cell/h), γ_i is the specific rate of glucose uptake (mol/g cell/h), $k_1 = 1$ (mol EtOH/mol CO_2), and $k_2 = 1$ (mol CO_2/mol O_2).

$I_{O_2} = Q_{O_2} \times$ total oxygen uptake rate (mol O_2/h) (3.150)

Substituting $F_i S_R = \mu_i X_i / Y_i$ of Equation 3.147 in Equation 3.149, we get

$$F_i = 180\,(I_{O_2})_i\,(RQ_i - 1)/(Y_{p/s})S_R \tag{3.151}$$

An empirical formula between $(Y_{p/s})_i$ and RQ_i observed is given below.

$$(Y_{p/s})_i = K(RQ_i - 1) \tag{3.152}$$

From Equations 3.151 and 3.152, we get

$$F_i = 180\,(I_{O_2})_i/KS_R \tag{3.153}$$

The control of feed rate based on I_{O_2} should be coupled with a sophisticated feed pump keeping the RQ value used. The possibility and advisability to control the fed-batch culture using either I_{O_2} and/or RQ coupled with a minicomputer are now evident.

The empirical constant K in Equation 3.153 was 3.80 for glucose and 2.65 for molasses medium.

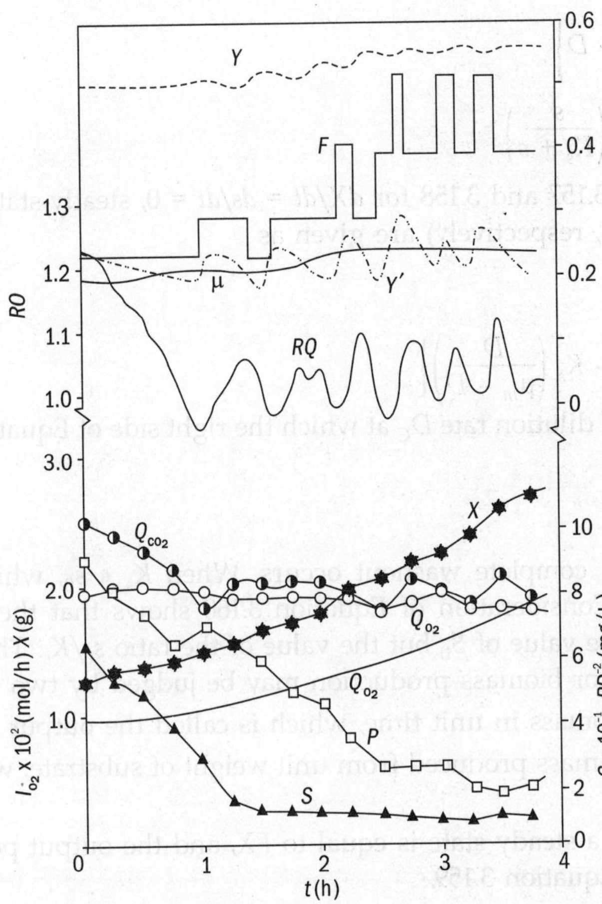

Figure 3.86 Growth patterns of baker's yeast in fed-batch culture of glucose medium

3.9.7.3 Continuous culture

In continuous culture, the generalized mass balance expression for a completely mixed system is

$$\frac{dx}{dt} = \pm \gamma X + D(X_0 - X) \tag{3.154}$$

Control of growth by the concentration of growth-limiting substrate can be either external or internal in continuous cultures. In turbidostat, growth is internally controlled as the flow rate (*F*) (or dilution rate $D \approx F$/work volume [*V*]) is controlled by means of a photodetector-controller to maintain constant turbidity of the culture. Cells grow at or near μm in the provided nutrient broth. In chemostat, growth is controlled externally, limiting the supply of growth-limiting substrate. In this case, growth may vary widely although μm value cannot be attained. Equation 3.154, the mass balance equation, is valid for both turbidostat and chemostat operations with complete mixing. Specifying Equation 3.155 and substituting Monod's well-known equation ($X_0 = 0$), we get

$$\frac{dX}{dt} = X\mu_m \left\{\left(\frac{s}{K_s + s}\right) - D\right\} \tag{3.155}$$

$$\frac{ds}{dt} = D(s_0 - s) - \frac{\mu_m X}{Y}\left(\frac{s}{K_s + s}\right) \tag{3.156}$$

Solving Equations 3.157 and 3.158 for $dX/dt = ds/dt = 0$, steady-state values of X and s (designated $\overline{X}$ and $\bar{s}$, respectively) are given as

$$\bar{s} = K_s\left(\frac{D}{\mu_m - D}\right) \tag{3.157}$$

$$\overline{X} = Y(s_0 - \bar{s}) = Y\left\{s_0 - K_s\left(\frac{D}{\mu_m - D}\right)\right\} \tag{3.158}$$

The critical value of dilution rate D_C at which the right side of Equation 3.158 vanishes is given as

$$D_C = \mu_m \frac{s_0}{K_s + s_0} \tag{3.159}$$

Over the D_C value, complete washout occurs. When $K_s \ll s_0$, which is usually the case, then $D_C \approx \mu_m$. Consideration of Equation 3.160 shows that the important factor here is not the absolute value of S_0 but the value of the ratio s_0/K_s. The performance of a continuous culture for biomass production may be judged by two criteria.

1. The quantity of biomass in unit time, which is called the output rate.
2. The quantity of biomass produced from unit weight of substrate, which is called the effective yield.

The total output in a steady state is equal to $F\overline{X}$, and the output per unit volume of culture is $D\overline{X}$. From Equation 3.159,

$$D\overline{X} = DY\left\{s_0 - K_s\left(\frac{D}{\mu_m - D}\right)\right\} \tag{3.160}$$

There is a value of D for which the $D\overline{X}$ value is maximum (D_M). This particular dilution rate is obtained by differentiating Equation 3.160 with respect to D and equating it to zero.

$$D_M = \mu_m\left(1 - \sqrt{\frac{K_s}{K_s + s_0}}\right) \tag{3.161}$$

The steady-state cell concentration at D_M is obtained by substituting the value of D_M for D in Equation 3.36.

$$X_M = Y\left\{(S_0 + K_S) - \sqrt{K_S\,(S_0 + K_S)}\right\} \tag{3.162}$$

The effective yield (Y_E) is the ratio of output of biomass to input of substrate, which is less than the Y value owing to substrate wasted in the outflow, which decreases with the dilution rate (Figure 3.87). The steady-state values of substrate concentration, bacterial concentration, and output at different dilution rates are calculated from Equations 3.162 and 3.163, for an organism with the following growth constants: μ_m = 1.0 per hour, Y = 0.5, and K_s = 0.2 g/L, and a substrate concentration in the inflowing medium of S_0 = 10 g/L.

$$Y_E = \frac{\overline{X}}{S_0} = \frac{Y(S_0 - \overline{S})}{S_0} \tag{3.163}$$

3.9.7.4 Synchronous cultures

Effects to produce the desired synchronization of cell division in microbial culture have been made. Synchronous cultures can be divided into two broad classes: induction synchrony and selection synchrony. In induction synchrony, cells of a culture are induced to divide synchronously by the same treatment. In selection synchrony, a fraction of the cells in a growing culture of particular stage of the cycle is separated off and grown separately as a synchronous culture.

3.9.7.4.1 Induction types

An inhibitor that acts at one stage of the cycle is added to a culture growing asynchronously. The cells accumulate at the stage blocked by the inhibitor and, after release from the inhibitor, divide synchronously. The block at the time of DNA synthesis can be used to produce in the fission yeast *Schizosacchatomyces pombe.*

The method of synchronizing, which is widely used with microorganisms, is to starve a culture or let it run into the stationary phase and then into the fresh medium. The culture usually grows synchronously for one or more division. One of the best examples of starvation synchrony is that developed for bud yeasts by Williamson and Scopes. A culture in the stationary phase is treated with alternating cycles of cold growth

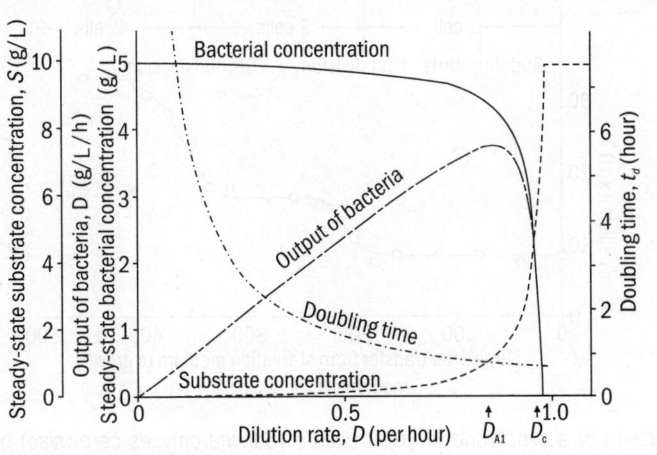

Figure 3.87 Steady-state relationships in a continuous culture (theoretical)

medium and warm starvation medium accompanied by the decanting-off of small cells (Figure 3.88). The first cell cycle is abnormally long because of the lag phase of about 75 min, which is the start of building.

In many cases, bacteria will grow for a time as a synchronous culture in stationary phase cells are transferred into a fresh medium. The exact stage the cells may be critical for successful synchrony. Periodic changes of medium can be used to induce synchrony in cultures grown in chemostats. Cultures of *E. coli* grown in this way with phosphate as the limiting nutrient divide with partial synchrony when phosphate pulses are added to the medium with a periodicity equal to the doubling time of the cultures.

The cultures of photosynthetic microbes such as *Chlorella* can be synchronized in a light–dark cycle. Temperature cycles can be applied for the induction of synchrony in bacteria. A single cold shock will synchronize bacteria but not effectively. Multiple shocks are much more efficient. Temperature cultures of 8 min at 37°C and 28 min at 25°C produce a synchronous culture in *Salmonella typhimurium*. By this procedure, biomass and DNA increases are periodic and only occur in the warm periods, whereas they are continuous in the normal cycle in closely related bacterial rods.

3.9.7.4.2 Selection types

Density gradient centrifugation for the separation of cells of the same size from a random growing culture has been developed. This method has been tested on microorganisms, *E. coli*, and budding yeast.

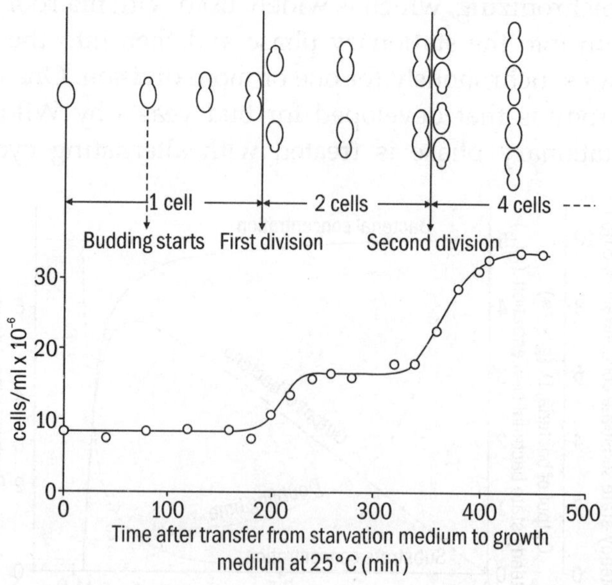

Figure 3.88 Growth of a synchronized yeast culture (*Saccharomyces cerevisiae*) on inoculation into synthetic medium

An ingenious method of synchronizing *E. coli* has been developed by Helmstetter and Cummins. A random growing culture is collected on a membrane filter. The filter is then inverted, and a fresh medium is run through slowly. After washing off the excess cells, newborn cells released from the filter surface are collected and grown up separately as a synchronous culture (Figure 3.89). The figure shows the outline of the procedure for determining the rate of incorporation of a labelled molecule into cells of different ages in an exponential unsynchronized culture.

The ideal distribution of cell ages in a random culture should be as shown in Figure 3.90(a), and the relative concentration of cells in the eluent after loading and eluting should follow the fluctuating curve in Figure 3.90(b). The rate of synthesis of macromolecules and the elution pattern of radioactivity per cell from the membrane culture, which was pulse labelled with a radioactive precursor of the macromolecules, are shown in Figure 3.91.

In practice, the assumption that there is no dispersion of individual cycle time is not completely valid, and the experimental results are somewhat different from the ideal case. However, this method gives a fairly clear idea of the synthetic pattern of DNA in different mediums.

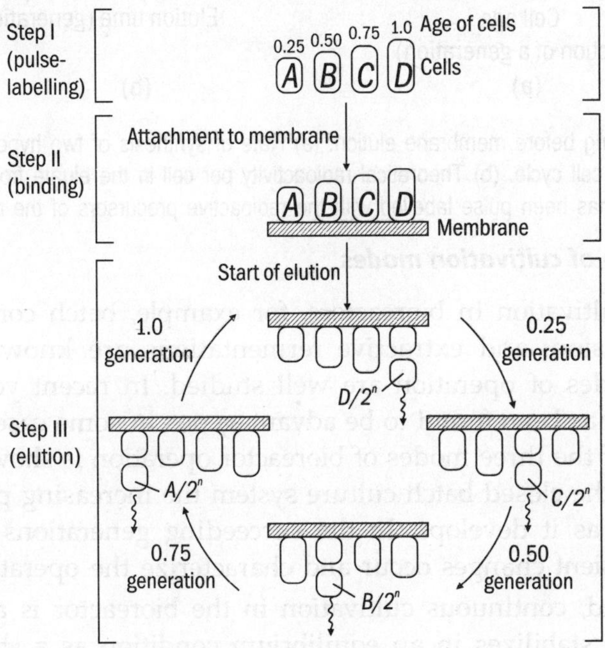

Figure 3.89 Outine of the procedure for pulse abelling before membrane elution with *Escherichia coli*.

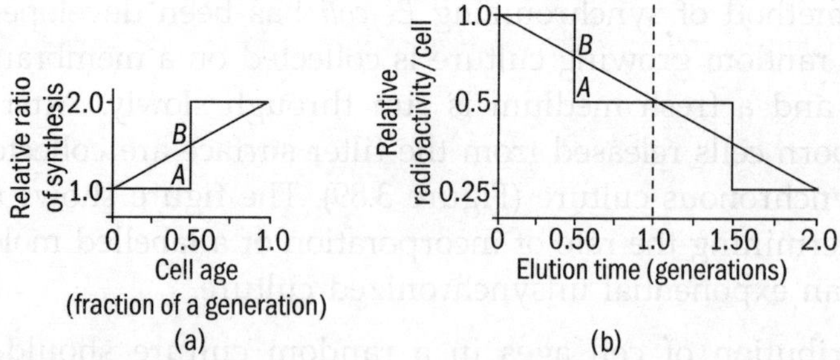

Figure 3.90 Pulse labelling before membrane elution with *Escherichia coli.* (a) Idealized age distribution in an exponential phase culture containing no dispersion of generation times of individual cells. (b) Theoretical concentration of cells in the eluate from a membrane-bound culture with an age distribution as shown in (a).

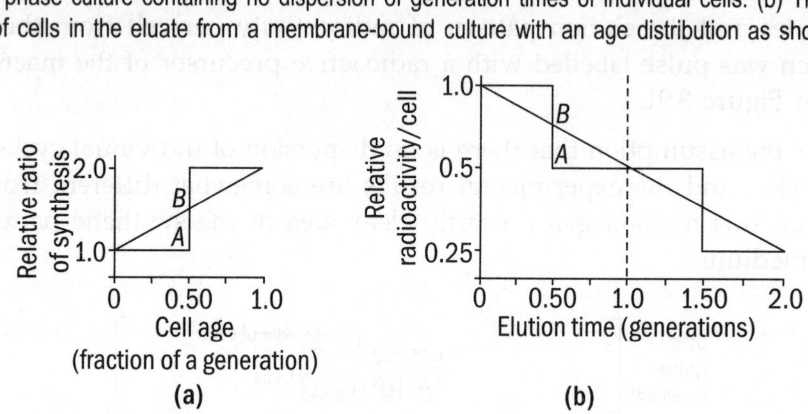

Figure 3.91 Pulse labelling before membrane elution. (a) Rate of synthesis of two hypothetical macromolecules through the cell cycle. (b) Theoretical radioactivity per cell in the eluate from a membrane-bound culture if it has been pulse labelled with the radioactive precursors of the macromolecules in (a).

3.9.7.5 Comparison of cultivation modes

A comparison of cultivation in bioreactors, for example, batch continuous, fed-batch, synchronous, perfusion, and extractive fermentations are known. Of them, batch and continuous modes of operation are well studied. In recent years, processing by synchronous mode has been found to be advantageous in some cases. A comparison of the characteristics of the three modes of bioreactor operation is shown in Figure 3.92. It can be seen that in the closed batch culture system the increasing population modifies the nutrient milieu as it develops. In the succeeding generations of cells produced, sequential and transient changes occur and characterize the operation mode.

On the other hand, continuous cultivation in the bioreactor is an open system. In this system, growth stabilizes in an equilibrium condition as a steady state. Here, a continuous addition of nutrient at a constant feed rate supplied to a growing population maintains it constant by diluting the growth increase produced by the medium flow; the rate of growth equals the dilution rate.

The resemblance of batch and continuous modes of bioreactor operation relates to the fact that the cells in the populations are randomized at all stages of growth in the cell cycle of their development, as shown in Figure 3.93.

Based on the features of the modes, their critical relative advantages and disadvantages are shown in Table 3.26.

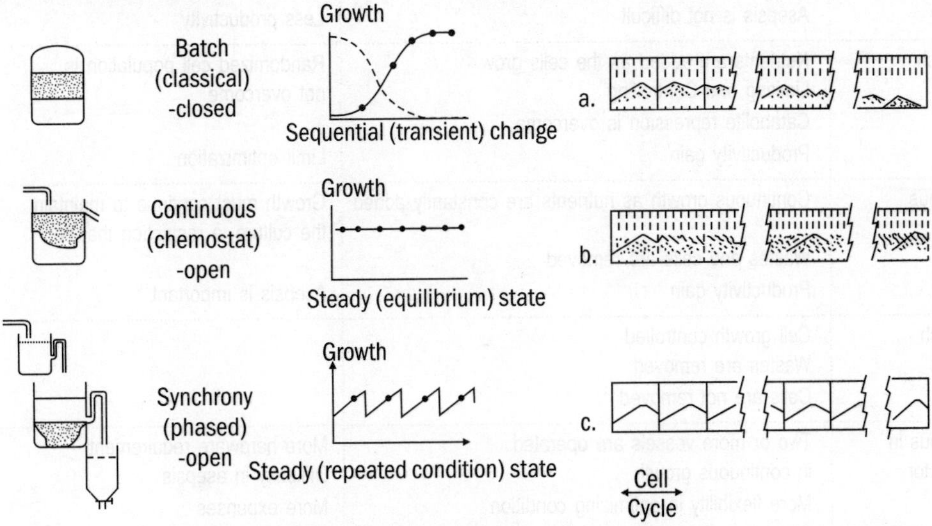

Figure 3.92 Comparison of the modes

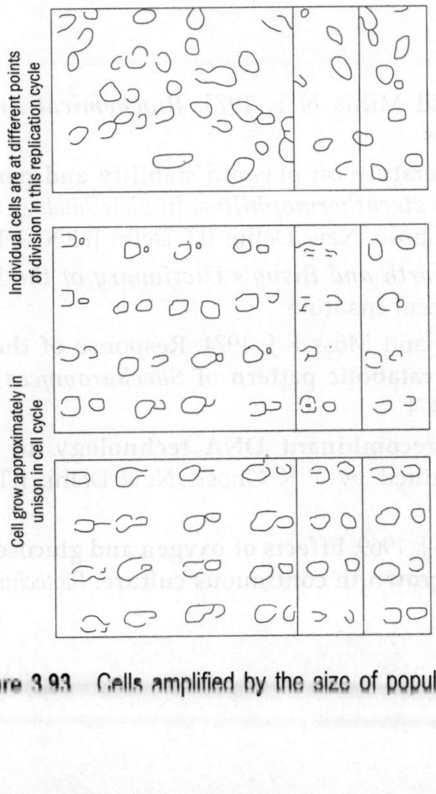

Figure 3.93 Cells amplified by the size of population

Table 3.26 Modes and features of fermentation process

Mode	Advantage	Features Disadvantage
Batch	All nutrients are loaded into the fermenter at the outset Growth occurs until nutrient is consumed Asepsis is not difficult	Randomized cell population Limit optimization Less productivity
Fed batch	Nutrients are added as the cells grow Limiting and Controlling Catabolite repression is overcome Productivity gain	Randomized cell population is not overcome Limit optimization
Continuous	Continuous growth as nutrients are constantly added Wastes and cells are removed Productivity gain	Growth must continue to maintain the culture, a restriction that can Asepsis is important
Fed-batch recycling	Cell growth controlled Wastes are removed Cells are not removed	
Continuous in multireactor	Two or more vessels are operated in continuous growth More flexibility in optimizing condition	More hardware requirement Difficulty in asepsis More expenses
Synchronous	Advantageous in cell growth and product regulation Phased growth	Developmental expertise

REFERENCES

Aiba S, Hamphrey A E, and Millis N F. 1973. ***Biochemical Engineering***, 2nd edn. Tokyo: University of Tokyo Press

Aiba S. 1984. **Effect of temperature on plasmid stability and penicillinase productivity in a transformant of *Bacillus sterothermophillus*.** In *Biotechnology and Bioprocess Engineering*, pp. 471–484, edited by T K Ghose. New Delhi: IIT Delhi [NOC VIIth IBS, IIT Delhi, India]

Ainsworth G C. 1971. ***Ainsworth and Bisby's Dictionary of the Fungi***. Kew, Surrey, England: Commonwealth Mycological Institute

Akbar M D, Rickard P A D, and Moss F J. 1974. **Response of the adenosine phosphate pool level to changes in the catabolic pattern of *Saccharomyces cerevisiae*.** *Biotechnology and Bioengineering* **16**(4): 455–474

Arber W. 1984. **Future of recombinant DNA technology**. In *Biotechnology and Bioprocess Engineering*, pp. 45–53, edited by T K Ghose. New Delhi: IIT Delhi [NOC VIIth IBS, IIT Delhi, India]

Babij T, Moss F J, and Ralph B J. 1969. **Effects of oxygen and glucose levels on lipid composition of yeast *Candida utilis* grown in continuous culture**. *Biotechnology and Bioengineering* **11**(4): 593–603

Bailey J E and Ollis D F. 1986. ***Biochemical Engineering Fundamentals,*** 2nd edn. New York: McGraw Hill Book Co. 918 pp.

Bennet G F and Kempe L L. 1964. **Oxygen transfer mechanisms in the gluconic acid fermentation by *Pseudomonas ovalis*.** *Biotechnology and Bioengineering* **6**(3): 347-360

Brown D E. 1970. **Aeration in submerged culture of microorgonisms.** In *Methods in Microbiology,* vol. 2, pp. 125-174, edited by J R Norris and D W Ribbons. London: Academic Press.

Buchman R E and Gibbons N E (eds). 1974. ***Bergey's Manual of Determinative Bacteriology.*** Baltimore, Maryland: Williams and Wilkins Co.

Button D K. 1964. PhD thesis submitted to the University of Wisconsin

Chen W L and Charalampous F C. 1969. **Mechanism of induction of cytochrome oxidase in yeast: I. Kinetics of induction and evidence for accumulation of cytoplasmic and mitochondrial precursors.** *Journal of Biological Chemistry* **244**: 2767-2776

Cole J A. 1967. **Metabolic pathways for nitrate reduction in *Escherichia coli*.** PhD thesis submitted to the University of Oxford

Cole J A and Wimpenny J W T. 1966. **The inter-relationships of low redox potential cytochrome C_{552} and hydrogenase in facultative bacteria.** *Biochimica et Biophysica Acta* **128:** 419-425

Collins F M and Lascelles J. 1962. **The effect of growth conditions on oxidative and dehydrogenase activity in *Staphylococcus aureus*.** *Microbiology* **29**: 531-535

Cruickshank R. 1970. ***Medical Microbiology.*** London: ELBS Longman

Daniels W F, Garcia L H, and Rosensteel J F. 1970. **Oxidation-reduction potential and concomitant growth patterns of cultures of Earle's "L" cells in centrifuge bottle spinners.** *Biotechnology and Bioengineering* **12**(3): 419-428

Davis B D, Dulbbecco R, Ginsberg H S, Eisen H N, Barrywood W Jr. 1968. ***Principles of Microbiology and Immunology.*** New York and London: Harper and Row

Dolin M L. 1961. ***The Bacteria,*** vol. 2. New York: Academic Press. 319 pp.

Einsele A, Fiechter A, and Knöpfel H P. 1972. **Respiratory activity of *Candida tropicalis* during growth on hexadecane and on glucose.** *Archives of Microbiology* **82**(3): 247-253

Fietchter A and Von Meyenburg K. 1968. **Automatic analysis of gas exchange in microbial systems.** *Biotechnology and Bioengineering* **10**(4): 535-549

Fox G E, Stackebrandt R B, Hespell J, Gibson I, Manetoff T A, Pyer R S. 1980. **The phylogeny of prokaryotes.** *Science* **209**: 457-463

Fredrickson A G, Megee R D III, and Tsuchiya H M. 1970. **Mathematical models for fermentation processes.** *Advances in Applied Microbiology* **13**: 419-465

Fuller R, Kaguni J, and Kornberg A. 1981. **Enzymatic replication of the origin of the *Escherichia coli* chromosome.** *Proceedings of the National Academy of Sciences* **78**(12): 7370-7374

Hambleton P. 1989. **Microbial swords into plough shares?** *Biotechnology News* **18.** UK: The University of Birmingham

Harrison D E F and Topiwala H H. 1974. **Transient and oscillatory states of continuous culture.** *Advances in Biochemical Engineering* **3**: 167-219

Harrison D E F, MacLennan D G, and Pritt S J. 1969. **Responses of bacteria to dissolved oxygen tension.** In *Fermentation Advances,* edited D Perlman. New York: Academic Press

Harrison D E F and Pirt S J. 1967. **The influence of dissolved oxygen concentration on the respiration and glucose metabolism of *Klebsiella aerogenes* during growth**. *Microbiology* **46**: 193–211

Hawker L E, Linton A H, Folks B F, Curticle M J. 1980. ***An Introduction to the Biology of Microorganisms***. London: Edward Arnold Publishers Ltd

Higgins I J and Quayle J R. 1970. **Oxygenation of methane by methane-grown *Pseudomonas methanica* and *Methanomonas methanooxidans***. *Biochemical Journal* **118**(2): 201–208

Hirose Y, Sonoda H, Kinoshita K, Okada H. 1968. **Studies on oxygen transfer in submerged fermentations. Part IX. Oxygen demand in glutamic acid fermentation**. *Agricultural and Biological Chemistry* **32**(7): 855–859

Hirsch C A, Michael R, Bernard D D, Edmund C C L. 1963. A fumarate reductase in ***Escherichia coli*** distinct from succinate dehydrogenase. *Journal of Biological Chemistry* 238(11): 3770–3774

Humphrey A E. 1964. **Biotechnology: where do we go from here?** In *Biotechnology and Bioprocess Engineering*, pp. 55–68, edited by T K Ghose. New Delhi: IIT Delhi [NOC VIIth IBS, IIT Delhi, India]

Imanaka T and Aiba S. 1981. **A perspective on the application of genetic engineering: stability of recombinant plasmid**. *Annals of the New York Academy of Sciences* **369**: 1–14

Jacobs M B and Gerstein M J. 1960. ***Handbook of Microbiology***. Princeton, New Jersey: Van Nostrand Co. Inc.

Jones W J, Leigh J A, Mayer F, Woese C F, Holfe R S. 1983. ***Methanococcus jannaschii* sp. nov., an extremely thermophilic methanogen from a submarine hydrothermal vent.** *Archives of Microbiology* **136**(4): 254–261

Khan A H. 1974. PhD thesis submitted to IIT Delhi

Khan A H and Ghose T K. 1973. **Kinetics of citric acid production by *Aspergillus niger*.** *Journal of Fermentation Technology* 51: 734–741

Klass D L and Ghosh S. 1980. **Fuels from biomass**. American Chemical Society National Meeting, 27–29 August 1980, San Francisco, California [Division of Fuel Chemistry, American Chemical Society]

Lapage S P, Clark W A, Lessel E F, Seeligkr H P R, Sneath P H A. 1973. **Proposed revision of the international code of nomenclature of bacteria**. *International Journal of Systematic Bacteriology* **23**: 83–108

Lascelles J. 1964. **Oxygen and the evolution of biochemical pathways**. In *Oxygen in Animal Organisms*, edited by F Dickens and E Neil. Oxford: Pergamon Press

Lengyel Z L and Nyiri L. 1965. **An automatic aeration control system for biosynthetic processes.** *Biotechnology and Bioengineering* 7(1): 91–100

Lodder J (ed). 1970. ***The yeasts: a taxonomic study***. Amsterdam: North Holland Publishing Co.

Maclennan D G and Pirt S J. 1970. **The dynamics of decane and glucose utilization by a *Pseudomonas* sp. in batch and chemostat cultures under controlled dissolved oxygen tensions.** *Journal of Applied Microbiology* **33**(2): 390–360

McPhedran P, Sommer B, and Edmund C C L. 1961. **Control of ethanol dehydrogenase levels in *Aerobacter aerogenes*.** *Journal of Bacteriology* **81**(6): 852–857

Mor J and Fiechter A. 1968. **Continuous cultivation of *Saccharomyces cerevisiae*. I. Growth on ethanol under steady-state conditions**. *Biotechnology and Bioengineering* **10**(2): 159–176

Mukhopadhyay S N. 1993. **Structural engineering of DNA in advance of biochemical engineering and process biotechnology**. [Auditorium of India Institute of Chemical Engineering, Jadavpur University, Calcutta, 15–16 January 1993]

Mukhopadhyay S N. 2005. ***Advanced Process Biotechnology***. New Delhi: Viva Books Pvt. Ltd.

Mukhopadhyay S N. 2007. ***Experimental Process Biotechnology Protocols***. New Delhi: Viva Books Pvt. Ltd.

Mukhopadhyay S N. 2006. **Eco-friendly products through process biotechnology in the provision of biotechnology economy recent advances.** *Technorama* **55**(T): 10–20

Mukhopadhyay S N and Ghose T K. 1993. **Oxygen response ambivalence in BamHI production.** *International Journal of Toxicology, Occupational, and Environmental Health*. New Delhi: Wiley Eastern Ltd. [Abstract no. Op. 19 in Abstracts oxygen Radicals and Antioxidants in Biotechnology and Medicine]

Mukhopadhyay S N, Maulik N, Lu D, Das D K. 1993. **Newer bioproducts by stressing of biocells.** [CEP Course, New Frontiers in Bioprocess Engineering Lecture Notes, Department of Biochemical Engineering and Biotechnology, Indian Institute of Technology Delhi, sponsored by Department of Biotechnology, Government of India, November-December 1993]

Mukhopadhyay S N and Saha N. 2000. **Process in thermophilic biotropic *Methanobacterium* sp.**, pp. 74–76. [Proceedings of Book of Abstracts, vol. 4, Dechema e.v. Germany]

Nosrati M. 2004. PhD thesis submitted to IIT Delhi

Oginsky E L and Umbreit W W. 1959. ***An Introduction to Bacterial Physiology***, 2nd edn. San Francisco: W H Freeman and Co.

Ryder D N and Sinclair C G. 1972. **Model for the growth of aerobic microorganisms under oxygen limiting conditions**. *Biotechnology and Bioengineering* **14**(5): 787–798

Salle A J. 1961. ***Fundamental Principles of Bacteriology***, 5th edn. New York: McGraw Hill Book Co. Inc.

Shibai H, Ishizaki A, and Hirose Y. 1973. **Conversions of microbial products in relation to oxygen supply in inosine fermentation.** *Agricultural and Biological Chemistry* **37**(9): 2083–2088

Somlo M and Fukuhara H. 1965. **On the necessity of molecular oxygen for the synthesis of respiratory enzymes in yeast**. *Biochemical and Biophysical Research Communications* **19**(5): 587–591

Soni V K and Ghose T K. 1974. **Studies on dynamics of continuous microbial-growth process.** *Journal of Fermentation Technology* **52**(8): 551–563

Strohm J, Dale R F, and Peppler H J. 1959. **Polarographic measurement of dissolved oxygen in yeast fermentations**. *Applied Microbiology* **7**(4): 235–238

Tengerdy R P. 1961. **Redox potential changes in the 2-keto-L-gulonic acid fermentation—I. Correlation between redox potential and dissolved-oxygen concentration**. *Journal of Biochemical and Microbiological Technology and Engineering* **3**(3): 241–253

Wiegel J and Ljungdahl L G. 1986. **The importance of thermophilic bacteria in biotechnology**. *Critical Reviews in Microbiology* **3**: 39

Zeikus J G. 1979. **Thermophilic bacteria: ecology, physiology, and technology**. *Enzyme* and *Microbial Technology*. **I**: 243–251

Zeikus J G and Bowen V G. 1975. **Comparative ultrastructure of methanogenic bacteria**. *Canadian Journal of Microbiology* **21**: 121–129

BIBLIOGRAPHY

Axe D D and Bailey J E. 1987. **Application of ^{31}P nuclear magnetic resonance spectroscopy to investigate plasmid effects on *Escherichia coli* metabolism**. *Biotechnology Letters* **9**: 83

Baker H A. 1956. ***Bacterial Fermentation***. New York: John Wiley and Sons Inc.

Balch W E, Fox G E, Magrum L J, Woese C R, Wolfe R S. 1979. **Methanogens: reevaluation of a unique biological group**. *Microbiology Reviews* **43**(2): 260–296

Cox C S Jr, Zwischenberger J B, and Kurnsz M. 1991. **Development and current status of a new intercorporeal membrane oxygenator (IVOX)**. *Perfusion* **6**: 291

Fried D W and Thompson J B. 1991. **Oxygen transfer efficiency of three microporous polypropylene membrane oxygenators**. *Perfusion* **6**(2): 105–114

Gunsalus I C and Sterier R Y. 1960. ***The Bacteria, vol. 1: Structure***. New York: Academic Press

Hersbach G J M. 1984. **Instrumentation and automation of pilot plant fermenters**. In *Innovations in Biotechnology: progress in industrial biotechnology*, vol. 20. Amsterdam: Elsevier. 305 pp.

Jones W J, Nagle D P Jr, and Whitman W B. 1987. **Methanogens and the diversity of archaebacteria.** *Microbiological Reviews* **51**(1): 135–177

Kilbum D G and Webb F C. 1968. **The cultivation of animal cells at controlled dissolved oxygen partial**. *Biotechnology and Bioengineering* **10**(6): 801–814

Kushner D J, Bayley S T, Borcing J, Kates M, Gibbons N E. 1974. **name of article??** *Canadian Journal of Microbiology* **10**: 483–497

Larsen H. 1984. **Family V. Halobacteriaceae**. In *Bergey's Manual of Systematic Bacteriology*, edited by N R Kreig and J G Holt. Baltimore: Williams and Wilkins

Lenski R E and Nauyen T T. 1988. **Stability of recombinant DNA and its effects on fitness.** *Trends in Biotechnology* **6**: 51

Merki R. 1984. **Semi-continuous Production of Monoclonal Antibodies in a Fermenter.** [B. Braun Biotechnology Report No. 4]

Mukhopadhyay S N. 1994. **Current status of biomolecular design engineering science in advancing biochemical engineering**, pp. 511–513. [Third Proceedings of the AP Biochemical EC, Singapore 1994]

Mukhopadhyay S N, Saha N, Saha L, Saha P, Kolomaznik K. 2004. **Retrieval of biodegradable polymer and value-added products from leather industry waste through process biotechnology: a progress review.** *Journal of the American Leather Chemists Association* **99**(11): 449–456

Mukhopadhyay S N and Mukhopadhyay P. 2000. **Water properties, pollution, and management.** [Proceedings of the ICEE *"Water Resource Management in the Developing World, Ecological Engineering Approach"*, University of Kalyani and IEE Society, Switzerland Pub.]

Niyri L K. 1972. **Application of computers in biochemical engineering.** In *Advance in Biochemical Engineering,* vol. 2, edited by A Fiechter and N Blakebrough. Berlin: Springer-Verlag

Nyiri L K, Jefferis R P, and Humphrey A E. 1974. **Application of computers to the analysis and control of microbiological processes.** In *Advances in Microbial Engineering,* edited by B Sikyta, A Prokop, and M Novak. New York: Wiley-InterScience

Nowak J, Pols G, Brands W, Rosscel P. 1992. **Clinical evaluation of the Bentley Univox membrane oxygenator.** *Perfusion* **7**: 59

Nosrati M, Shojaosadati S A, Sreekrishnan T R, Mukhopadhyay S N. 2004. **Inhibition of thermophilic anaerobic digestion of waste food by long chain fatty acids and propionate.** *Iranian Journal of Biotechnology* **2**(4): 261–268

Nosrati M, Sreekrishnan T R and Mukhopadhyay S N. 2007. **Energy audit, solids reduction, and pathogen inactivation in secondary sludges during batch thermophilic aerobic digestion process.** *Journal of Environmental Engineering* **133**(5): 477–484

Rouviere P E and Wolfe R S. 1987. **Use of subunits of the methylreductase protein for taxonomy of methanogenic bacteria.** *Archives of Microbiology* **148**(4): 253–259

Schumpe A, Quicker G, and Deckwer W D. 1982. **Gas solubilities in microbial culture.** *Advances in Biochemical Engineering* **24**: 1–38

Sehgal S N, Kates M, and Gibbons N E. 1962. **Lipids of *Halobacterium cutirubrum*.** *Canadian Journal of Biochemistry and Physiology* **40**: 69–81

Whitman W B. 1985. **Methanogenic bacteria.** In *The Bacteria,* vol. VIII, edited by C R Woese and R S Wolfe. New York: Academic Press

Yamashita S, Hoshi H, and Inagki T. 1969. **Automatic control and optimization of fermentation processes: glutamic acid.** In *Fermentation Advances,* edited by D Perlman. New York: Academic Press

4

Chemical and Biochemical Engineering Principles

4.1 UNITS AND DIMENSIONS

The International System of Units is the universal system of units and is abbreviated as SI units. However, older systems, particularly the CGS and FPS systems, are still in use. Thus, chemical or biochemical engineers and researchers express many physico-chemical data in CGS or FPS units. To make use of these data and convert them from one system of units to another, it is necessary to have an excellent understanding of the three systems.

4.1.1 Concept of Units and Dimensions

A measured quantity has a numerical value and unit. It is engineering practice to write both the value and the unit of each quantity appearing in an equation (for example, 10 h, 5 km, and 5 m/s).

A dimension is a property that can be measured, such as length, time, mass, and temperature, or calculated by multiplying or dividing other dimensions, such as velocity (length/time), volume (length^3) or density (mass/length^3).

Units can be treated like algebraic variables when quantities are added, subtracted, multiplied or divided. The numerical values of two quantities may be added or subtracted only if the units are the same. But numerical values and their corresponding units may always be combined by multiplication and division.

4.1.2 System of Units

Base units: These units are dimensionally independent. They are used to designate only one dimension such as mass, length, time, temperature, electrical current, and light intensity.

Multiple units: These units are defined as multiples or fractions of base units such as minutes, hours, and milliseconds. Multiple units are used for convenience (for example, it is more convenient to refer to 5 years than to 157 680 000 seconds).

Derived units: These units are a combination of various dimensions obtained by multiplying and dividing base or multiple units (for example, cm^2, gm-cm/s, and kg/m^3). Derived units of this type are referred to as compound units, while others are called defined units. Examples of defined units are 1 erg (defined as 1 g-cm/s) and 1 lb_f (defined as 32.174 lb_m-ft/s^2). Derived units are also used for convenience (for example, it is more convenient to write 1000 g-cm/s as 1000 ergs).

The base units in the SI system are metre (m) for length, kilogram (kg) for mass, second (s) for time, and Kelvin (K) for temperature. However, base units for electrical current, ampere (A), and light intensity, candela (cd), are not of interest to process engineers. The principal SI, CGS, and FPS units and prefixes are given in Tables 4.1, 4.2, and 4.3. An SI convection is to use blanks rather than commas when writing large numbers. For example, 41 417 instead of 41,417.

4.1.3 Force and Weight

According to Newton's second law of motion, force is proportional to the product of mass and acceleration ($length/time^2$). Therefore, the natural units of force are (1) kg-m/s^2 (SI), (2) g-cm/s^2 (CGS), and (3) lb_m-ft/s^2 (FPS).

Table 4.1 SI, CGS, and FPS base units and their prefixes

Quantity	Base unit	Symbol
Length	Metre (SI)	m
	Centimetre (CGS)	cm
	Foot (FPS)	ft
Mass	Kilogram (SI)	kg
	Gram (CGS)	g
	Pound mass (FPS)	lb_m
Time	Second	s
Temperature	Kelvin (SI), Kelvin (SI, CGS)	K (or °K) °R
	Rankine (FPS)	

Table 4.2 Multiple unit prefixes

Unit	Symbol	Equivalence
Mega	M	10^0
Kilo	k	10^3
Centi	c	10^{-2}
Milli	m	10^{-3}
Micro	μ	10^{-6}
Nano	n	10^{-9}

Table 4.3 Derived units and their prefixes

Quantity	Unit	Symbol	Equivalent in terms of base units
Volume	Litre	L (or lit)	0.001 m^3
Force	Newton (SI)	—	1 kg-m/s^2
	Dyne (CGS)	—	1 g-cm/s^2
Pressure	Pascal (SI)	Pa	1 N/m^2
Energy, work	Joule (SI)	J	1 N-m = 1 kg-m^2/s^2
	Erg (CGS)	—	1 dyn-cm = 1 g-cm^2/s^2
	Gram calorie	cal	4.184 J = 41.84 kg-m^2/s^2
Power	Watt	W	1 J/s = 1 kg-m^2/s^2

Since it is awkward to carry these units in various calculations, derived force units have been defined for each system and are (1) 1 Newton (N) = kg-m/s^2 (SI), (2) 1 dyn = 1 g-cm/s^2 (CGS). In FPS, the derived unit of force is pound-force (lb_f), which is the product of unit mass (1 lb_m) and acceleration of gravity at sea level and 45° latitude (32.174 ft/s^2).

To convert a force from a defined unit (such as newton) to a natural unit (kg-m/s^2), a conversion factor, usually denoted by the symbol g_c, is used.

$$g_c = \frac{1 \text{ kg-m/s}^2}{\text{N}}$$

$$\text{or } g_c = \frac{1 \text{ g-cm/s}^2}{\text{dyn}}$$

$$\text{or } g_c = 32.174 \frac{\text{lb}_\text{m}\text{-ft/s}^2}{\text{lb}_\text{f}}$$

Equation 4.1 relates force in defined units to mass and acceleration.

$$F = ma/g_c \tag{4.1}$$

The weight of an object is the force exerted on the object by the gravitational attraction of earth. If m is the mass of the object, power can be expressed as shown in Equation 4.2.

$$W = ma/g_c \tag{4.2}$$

The value of g varies proportionally with the mass of an attracting body and inversely with the square of the distance between the centres of masses of the two bodies. The value given in Equations 4.3–4.5 may be used for most conversions of mass and weight.

$$g = 9.8066 \text{ m/s}^2 \tag{4.3}$$

$$g = 980.66 \text{ cm/s}^2 \tag{4.4}$$

$$g = 32.174 \text{ ft/s}^2 \tag{4.5}$$

$$g/g_c = 9.8066 \text{ N/kg} \tag{4.6}$$

$$g/g_c = 980.66 \text{ dyn/g} \tag{4.7}$$

$$g/g_c = 1 \text{ lb}_f/\text{lb}_m \tag{4.8}$$

4.2 DIMENSIONAL ANALYSIS

Dimensional analysis is a simple mathematical tool that does not require deep knowledge of complicated mathematics. Knowledge of its use will help a user save considerable time in planning experiments, correlating results, and using correlations developed by others.

This is a method intermediate between mathematical development and a completely empirical study. It does not yield an equation, and the experiment is required to complete the solution to the problem. The method is based on the concept of dimension and the use of dimensional formulae. In this procedure, known variables are arranged in a number of dimensionless combinations or groups. The fundamental equation may be expressed in terms of these groups rather than in terms of individual variables. The resulting equation will be simpler as the number of dimensionless groups will be less than the number of variables, generally given by the number of dimensions used.

4.2.1 Components in Dimensional Analysis

4.2.1.1 Physical quantities

Several physical quantities are required in dimensional analysis. These quantities may be divided into two groups.

Primary quantities: These are small number of independent quantities chosen arbitrarily, both in number and type, as per convenience in forming groups. The minimum list of primary quantities is length (L), mass (M), time (t), temperature (T), heat (H), force (F), and so on. The SI system does not require H and F in the test.

Secondary quantities: These quantities are expressed in terms of primary quantities. The units in the set defining a primary quantity must meet the requirement of existence of a constant conversion factor between any two units contained in the set.

4.2.1.2 Dimensions

Every primary quantity is represented by a letter to symbolize the dimension of that quantity. It designates the set of all units that have been and/or may be used to measure that quantity. For example, if L is the dimension of length, then L can be expressed in foot, inch, metre, mile, yard, cubic, and so on. Similar definitions are made for the dimensions of mass (M), time (t), temperature (T), force (F), and heat (H).

The dimension of a secondary quantity shows how that quantity is formed from primary quantities. For example, the velocity of any object, regardless of unit or magnitude, is found by dividing the length (distance) by time (Equation 4.9).

$$\text{Velocity} = \frac{\text{Length}}{\text{Time}} \tag{4.9}$$

Therefore, the dimension of velocity is $\frac{L}{t}$ or Lt^{-1}.

Similarly, the dimension of acceleration is $\frac{L}{t^2}$ or Lt^{-2} and that of pressure is $\frac{F}{L^2}$ or FL^{-2}.

4.2.1.3 *Dimensional formulae*

The dimensions of any quantity may be shown by the use of a bar on top of the symbol, thus $L = \overline{L}$ and $P = \overline{F}\,\overline{L}^{-2}$. This indicates that the dimensions of pressure are force times length to the power minus 2. After choosing the primary quantity, the dimensional formula for any secondary quantity can be found from its definitions. The general dimensional formula is given by Equation 4.10.

$$G = \overline{L}^a\,\overline{M}^b\,\overline{t}^c\,\overline{F}^d\,\overline{T}^e\,\overline{H}^f \tag{4.10}$$

where *a, b, c, d, e,* and *f* are exponents. These are always positive or negative integers, small positive or negative integral fractions or zero.

4.2.1.4 *Principles of dimensional analysis*

Three principles are involved in developing dimensionless equations for analysing problems relating variables of the process. These are as follows.

(a) Each of the physical quantities or measurements may be expressed as a product of the powers of the fundamental dimensions (Tables 4.1, 4.2, and 4.3).

(b) Equations of physical quantities are homogeneous in the net dimension of all additive terms. All terms added together must have same dimensions when expressed in terms of the fundamental dimensions selected.

(c) Any general relation between physical quantities can be expressed in a generalized manner independent of any particular unit to evolve only the dimensionless product of all physical variables and the necessary conversion constants.

4.2.1.5 *Dimensionless groups*

Dimensional analysis is based on the fact that if a theoretical equation does not exist among the variables affecting a physical process, the equation must be dimensionally homogenous. Because of this requirement, it is possible to group many factors into a smaller number of groups called dimensionless groups. Dimensionless groups may be formed by analysing the dimensions of the term appearing in the differential equations governing various operations. Such a process emphasizes the physical significance of the dimensionless group and guards against inadvertent omission of important variables.

The numerical value of a dimensionless group for a given case is independent of the units chosen for the primary quantities, provided consistent units are used within the group. The units chosen for one group need not be consistent with those of another.

4.2.1.6 Buckingham Pi method of dimensional analysis

The Buckingham *Pi* method is applicable even when the relevant differential equation is not known. In this method,

$$P = f(L, u, \rho, \mu, g, \sigma, g_c) \tag{4.11}$$

where P is a function of all variables and dimensional constants.

Finding the form of this function by pure empiricism would involve a number of experiments. By dimensional analysis, the number of experiments may be reduced. The *Pi* method is based on the principle that any equation representing the behaviour of a system must be dimensionally homogeneous, so that it would be possible to write it in the form of dimensionless groups by a procedure of functional relationship. Using π_i as a symbol for a dimensionless group, we may generalize (Equation 4.12).

$$\pi_i = f(\pi_2, \pi_3, \pi_4, \ldots, \pi_i) \tag{4.12}$$

Equation 4.11 can be written in the form of Equation 4.12, involving the dimensionless groups $\pi_1, \pi_2, \pi_3, \pi_4, \ldots, \pi_i$. In order to do so, it is necessary to know how many groups of Equation 4.11 will yield the value of i, which can be determined by Equation 4.13.

$$i = n - r \tag{4.13}$$

where n is the number of variables and r is number of variables that will not form a dimensionless group. The quantity r is often equal to the number of fundamental dimensions; thus it is equal to the number of repeating variables.

4.2.1.7 Illustration of the Pi theorem

Microbiological engineering reactions are carried out in a bioreactor. To carry out pure culture cultivation reactions, the medium of cultivation, along with the bioreactor, has to be sterilized to avoid contamination. Let us consider a bioreactor fitted with mixing impellers and baffles of a given size, around the wall of which heating/cooling pipe coils are arranged. To control the temperature during a microbial reaction, hot/chilled water is circulated in the pipe. Thus, heat transfer from or to the liquid contained in the bioreactor takes place. The correlations of the heat transfer coefficients for fluids in turbulent conditions in the reactor, degree of turbulence, properties of the material of the pipe, and properties of the fluid may be developed by dimensionless analysis and further extended by incorporating dimensionless factors and viscosity factors.

In the temperature control phenomena in bioreactors, the heat transfer coefficient (h) under turbulent flow conditions depends on the diameter of the pipe (D), velocity of flow of fluid (u), density of the fluid (ρ), viscosity of the fluid (μ), specific heat of the fluid (C_P), and thermal conductivity of the fluid (K). The dimensions of these factors are listed in Table 4.4.

Thus, the dimensional relation is

$$HtL^{-2}T^{-1} = \alpha L^{a}(Lt^{-1})^{b}\,(ML^{-3})^{c}\,(ML^{-1}t^{-1})^{d}\,(Ht^{-1}L^{-1}T^{-1})^{e}\,(HM^{-1}T^{-1})^{f}$$

Here,

Total number of variables = 7

Number of repeating variables = Number of basic dimensions = 4

According to the *Pi* theorem, the number of dimensionless groups is 7 – 4 = 3.

Let the dimensionless groups be π_1, π_2, π_3 and the repeating variables be D, K, ρ, and μ.

We may write

$\pi_1 = f\,(h, D, K, \mu)$

$\pi_2 = f\,(u, D, K, \mu)$

$\pi_3 = f\,(C_P, D, K, \mu)$

$\pi = h\,D^{a}, K^{b_1}, \rho^{c_1}, \mu^{d_1}$

Transforming in dimensions, we have

$$H^0M^0L^0T^0t^0 = (Ht^{-1}L^{-2}T^{-1})\,L^{a_1}(Ht^{-1}L^{-2}T^{-1})^{b_1}\,(ML^{-3})^{c_1}\,(ML^{-1}t^{-1})^{d_1}$$

Table 4.4 Dimensions of select factors

Factor	Significant heat	Dimension
H	$\frac{\text{Energy}}{\text{Temp-Time-Length}^2}$	$\frac{H}{T \cdot t \cdot L^2} = hT^{-1}t^{-1}L^{-2}$
D	Length	L
u	$\frac{\text{Length}}{\text{Time}}$	$\frac{L}{t} = Lt^{-1}$
ρ	$\frac{\text{Mass}}{\text{Length}^3}$	$\frac{M}{L^3} = ML^{-3}$
μ	$\frac{\text{Mass}}{\text{Length-Time}}$	$\frac{M}{Lt} = ML^{-1}t^{-1}$
K	$\frac{\text{Heat}}{\text{Time-Length-Temp}}$	$\frac{H}{t \cdot L \cdot T} = Ht^{-1}L^{-1}T^{-1}$
C_P	$\frac{\text{Heat}}{\text{Mass-Temp}}$	$\frac{H}{M \cdot T} = HM^{-1}T^{-1}$

Equating the coefficients on both sides, we get

$H, 0 = 1 + b_1 \ldots b_1 = -1$ (4.14)

$M, 0 = c_1 + d_1 \ldots c_1 = -d_1$ (4.15)

$L, 0 = -2 + a_1 - b_1 - 3c_1 - d_1$

or $L, 0 = -2 + a_1 + 1 - 3c_1 + c_1$

or $L, 0 = -1 + a_1 - 2c_1$ (4.16)

$T, 0 = -1 - b_1 \ldots b_1 = -1$

$t, 0 = -1 - b_1 - d_1 \ldots d_1 = -1 - b_1 = -1 + 1 = 0$ (4.17)

Thus, from Equation 4.15, we get

$c_1 = -d_1 = 0$

From Equation 4.16, we get

$a_1 = 1 + 2c_1 = 1$

Thus,

$\pi_1 = hD^1K^{-1}\rho^0\mu^0$

or $\pi_1 = \dfrac{hD}{K}$

Calculating similarly for π_2 and π_3, we get

$\pi_2 = \dfrac{Du\rho}{\mu}$

$\pi_3 = C_P\,\mu/K$

Now,

$\pi_1 = \varnothing(\pi_2, \pi_3)$

or $\dfrac{h_iD}{K} = \left(\dfrac{Du\rho}{\mu}\right)^a \left(\dfrac{C_p\mu}{K}\right)^b$

or $N_{Nu} = \alpha(N_{Re})^a\,(N_{Pr})^b$

The values of α, *a*, and *b* can be determined experimentally. Taking into consideration the viscosity change of the fluid, the above correlation can be further expressed as

$N_{Nu} = 0.023\, N_{Re}^{0.8}\, N_{Pr}^{1/3}\, (\mu/\mu_w^{0.14})$ (4.18)

Determined experimentally, it was found that $\alpha = 0.023$, $a = 0.8$, and $b = 1/3$.

In Equation 4.18, μ is the viscosity of fluid at the average temperature of fluid and μ_w is the viscosity of fluid at the wall temperature.

Dividing both sides of Equation 4.18 by $1/N_{Re}$ and $1/N_{Pr}$, we get

$$\frac{h_i D}{K} \cdot \frac{\mu}{DG} \cdot \frac{K}{C_p \mu} = 0.023 \frac{DG^{0.8}}{\mu} \frac{(C_p \mu)^{1/3}}{K} \left(\frac{\mu}{\mu_w}\right)^{0.14} \frac{\mu}{DG} \frac{k}{C_p \mu} \tag{4.19}$$

The left-hand side of Equation 4.19 is called the J_H factor. Thus,

$$J_H = \frac{h_i}{C_p G} \frac{(C_p \mu)^{2/3}}{K} \left(\frac{\mu_w}{\mu}\right)^{0.14}$$

4.2.1.8 Tutorial

Question 1 Cell sedimentation is carried out in the preliminary steps of downstream processing of many fermented broth for isolation and purification of product. In this settling operation, the velocity of settling of spherical cells in the fermented broth depends on cell diameter, density, and viscosity of the liquid, and acceleration due to gravity. By dimensional analysis, derive the equation for the velocity of cell settling in terms of these variables.

Answer The symbols and dimensions of the variables required to solve this problem are listed in Table 4.5.

Thus,

$$V \propto f(D, \rho_1, \rho_2, \mu, g) \tag{4.20}$$

Therefore,

Total number of variables, $n = 6$.

Number of repeating variables, $r = 3$ (M, L, t)

Thus, the number of dimensionless groups is

$i = n - r = 6 - 3 = 3$

We may write Equation 4.20 as

$V = K, D^a, \rho_1^b, \rho_2^c, \mu^d, g^e$

Let the dimensionless groups be π_1, π_2, and π_3 and the number of repeating variables be D, ρ_1, and g.

Table 4.5 Symbols and dimensions of variables for solution

Variables	Notation	Dimensions
Settling velocity	V	Length/time (L/t)
Cell diameter	D	Length (L)
Cell density	ρ_1	Mass/length^3 (M/L^3)
Liquid density	ρ_2	Mass/length^3 (M/L^3)
Liquid viscosity	μ	Mass/length × time (M/LT)
Acceleration due to gravity	g	Length/time^2 (L/t^2)

Thus,

$\pi_1 = \varnothing_1(V, D, \rho_1, g)$

$\pi_2 = \varnothing_2(\rho_2, D, \rho_1, g)$

$\pi_3 = \varnothing_3(\mu, D, \rho_1, g)$

Therefore,

$\pi_1 = VD^a\rho_1^b g^c$

or $M^0L^0T^0 = (Lt^{-1}L^a)(ML^{-3})^b(Lt^{-2})^c$

or $M^0L^0T^0 = L^{1+a-3b+c}\, M^b\, t^{-1-2c}$

Therefore,

$M: 0 = b$

$L: 0 = 1 + a - 3b + c$

$t: 0 = -1 - 2c$

Therefore,

$C = -1/2$

$\pi_1 = VD^{-1/2}\rho_1{}^0 g^{-1/2} = V\sqrt{Dg}$

Similarly,

$\pi_2 = \rho_2 D^{a_1}\rho_1{}^{b_1}g^{c_1}$

or $M^0L^0T^0 = ML^{-3}L^{a_1}(ML^{-3})^{b_1}(Lt^{-2})^{c_1}$

Therefore,

$M: 0 = 1 + b_1$

Thus,

$b_1 = -1$

$L: 0 = -3 + a_1 - 3b_1 + c_1$ or $a_1 + c_1 = 0$

Thus,

$a_1 = 0$

$t: 0 = -2c_1$

Therefore,

$c_1 = 0$

Now,

$\pi_2 = \rho_2 D_1^0 \rho_1^{-1}\, g_1^0$

Therefore,

$\pi_2 = \rho_2/\rho_1$

$\pi_3 = \mu\, D^{a_2}\, \rho_1^{b_2}\, g^{c_2}$

Thus,

$M^0L^0T^0 = ML^{-1}t^{-1}L^{a_2}(ML^{-3})^{b_2}(Lt^{-2})^{c_2}$

Therefore,

M: $0 = 1 + b_2$ or $b_2 = -1$

L: $0 = -1 + a_2 - 3b_2 + c_2$ or $a_2 + c_2 = -2$

t: $0 = -1 - 2c_2$ or $c_2 = -1/2$

Therefore,

$$a_2 = -2 + \frac{1}{2} = \frac{-4 + 1}{2} = \frac{-3}{2}$$

Therefore,

$$\pi_3 = \mu D^{-3/2} \rho^{-1} g^{-1/2}$$

$$\text{or } \pi_3 = \mu / D^{3/2} \rho g^{1/2} = \mu / \sqrt{a^3} g \rho_1$$

Now,

$$\pi_1 = \varnothing\ (\pi_2, \pi_3)$$

$$\text{or } \frac{V}{\sqrt{Dg}} = \varnothing \left(\frac{\rho_2}{\rho_1} \frac{\mu}{D^3 \rho_1 g} \right)$$

$$\text{or } \frac{V}{\sqrt{Dg}} = K \left(\frac{\rho_2}{\rho_1} \right)^a (\mu \sqrt{D^3} g \rho_1)^b$$

Question 2 For long, smooth pipes carrying an incompressible fluid, the skin factor loss per unit length of the pipe (H_{fs}/L) depends on the viscosity of the fluid (μ), density of the fluid (ρ), velocity of the fluid ($\overline{V}$), diameter of the pipe (D), and Newton's conversion factor (g_c).

By dimensional analysis, find out the representative equation between H_{fs}/L and other variables.

Answer We have

$$\frac{H_{fs}}{L} = \varnothing\ (\mu, \rho, \overline{V}, D, g_c)$$

The variables and their notations and dimensions are listed in Table 4.6.

Total number of variables = 6

Number of repeating variables = 4

Thus, number of dimensionless groups = 6 – 4 = 2

Thus,

$$\lambda_1 = \varnothing(\lambda_2)$$

$$\lambda_1 = \psi \left(\frac{H_{fs}}{2}, \rho, \overline{V}, D, g_c \right)$$

$$\lambda_2 = \psi\ (\mu, \rho, V, D, g_c)$$

Table 4.6 Variables: notations and dimensions

Variable	Notation	Dimension
Friction loss per unit length	$\frac{H_{fs}}{L}$	$\frac{\text{ft}(\text{lb}_f/\text{lb})}{\text{Length}} = \frac{L\bar{F}}{ML} = \bar{F}M^{-1}$
Viscosity	μ	$M/L\theta$
Density	ρ	M/L^3
Velocity	$\bar{V}$	L/θ
Diameter	D	L
Newton's law conversion	g_c	$\frac{\text{ft-lb}}{\text{lb}_f\ \text{s}^2} = \frac{LM}{\bar{F}\theta^2} = LM\bar{F}^{-1}\theta^{-2}$

Thus,

$$\lambda_1 = \frac{H_{fs}}{2}\rho^a\bar{V}^bD^cg_c^d$$

Thus,

$$\bar{F}^0M^0L^0\theta^0 = \bar{F}M^{-1}(ML^{-3})^a(L\theta^{-1})^bL^cL^dM^d\bar{F}^{-d}\theta^{-2d}$$

or $\bar{F}^0M^0L^0\theta^0 = \bar{F}^{1-d}M^{-1+a+d}L^{-3a+b+1+c}\theta^{-b-2d}$

Thus,

$\bar{F}$: $0 = 1 - d$

Thus,

$d = 1$

M: $0 = -1 + a + d$

Thus,

$a = 1 - d = 1 - 1 = 0$

L: $0 = -3a + b + 1 - c$

Thus,

$b - c = -1$

or $c = b + 1 = -2 + 1$

or $c = -1$

θ: $0 = -b - 2d$

Thus,

$b = -2d = -2$

$$\lambda_1 = \frac{H_{fs}}{L}\rho^0\bar{V}^2D^1g_c^1$$

$$\text{or } \lambda_1 = \frac{H_{fs}Dg_c}{\bar{V}^2L}$$

$$\lambda_2 = \mu\rho^{a_1}\bar{V}^{b_1}D^{c_1}g_c^{d_1}$$

Thus,

$\overline{F}^0 M^0 L^0 \theta^0 = ML^{-1}\theta^{-1} (ML^{-3})^{a_1} (L\theta^{-1})^{b_1} L^{c_1} (LM\overline{F}^{-1}\theta^{-1})^{d_1}$

or $\overline{F}^0 M^0 L^0 \theta^0 = M^{1+a_1+d_1} \overline{F}^{-1} L^{-1-3a_1+b_1+c_1+d_1} \theta^{-1-b_1-2d_1}$

Thus,

$\overline{F}$: $0 = -d_1$

or $d_1 = 0$

M: $0 = -1 + a_1 + d_1$

or $1 + a_1 = 0$

Thus,

$a_1 = -1$

L: $0 = -3a_1 + b_1 + c_1 + d$

Thus,

$-1 + 3 + b_1 + c_1 = 0$

or $b_1 + c_1 = -2$

or $c_1 = -2 - b_1$

or $c_1 = -2 + 1$

Thus,

$c_1 = -1$

θ: $0 = -1 - b_1 - 2d_1$

Thus,

$b_1 = -1$

Thus,

$$\lambda_2 = \mu\rho^{-1}\overline{V}^{-1}D^{-1}g_c^{\,0} = \frac{\mu}{D\overline{V}_\rho}$$

Thus,

$\lambda_1 = \varnothing\ (\lambda_2)$

$$\frac{H_{fs}g_cD}{\overline{V}^2K} = K\left(\frac{\mu}{D\overline{V}_\rho}\right)^a = K\left(\frac{D\overline{V}_\rho}{\mu}\right)^{a_1} = K(N_{Re})^{a_1}$$

Thus,

$$\frac{H_{fs}g_cD}{\overline{V}^2L} = \varnothing(N_{Re})$$

4.3 FUNDAMENTALS OF MATERIAL BALANCE

Material balance of a process is the exact accounting of all the materials that enter, leave, accumulate or are depleted in the course of a given time interval of the operation. In accounting terms, it is law of conservation of mass.

Chemical/biochemical processes may be classified as batch, continuous or semi-batch processes and as steady state or transient (unsteady state). In a steady-state process, the variables do not change with time. The general balance equation of such a state may be given by Equation 4.21.

$$\text{Input} + \text{Generation} - \text{Output} - \text{Consumption} = \text{Accumulation} \tag{4.21}$$

For a continuous steady-state process, the general balance equation simplifies as given by Equation 4.22.

$$\text{Input} + \text{Generation} = \text{Output} + \text{Consumption} \tag{4.22}$$

The general procedure for material balance calculation is summarized below.

- Draw flow chart.
- Choose basis of calculation.
- Label known and unknown variables on the chart.
- Convert known stream volumes or volumetric flow rates to mass or molar quantities. Convert all quantities to one unit.
- Write material balance equations. Solve the equations for unknown quantities.

4.3.1 Recycle

Recycling is common practice to increase yields, enrich a product or improve operations. A labelled flow chart of a process involving reaction, product separation, and recycle is shown in Figure 4.1.

In a recycling operation, the combined feed is made up of a mixture of fresh feed and recycle stream.

$$\text{Overall conversion} = \frac{\text{Reactant input to process} - \text{Reactant output from process}}{\text{Reactant input to process}}$$

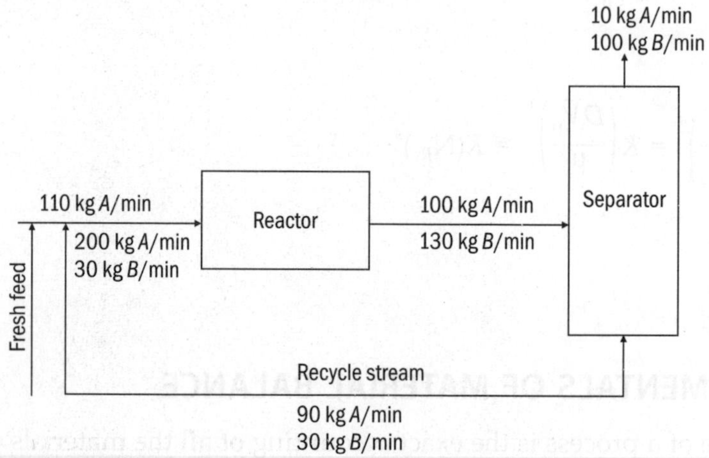

Figure 4.1 Process involving reaction, product separation, and recycle

$$\text{Single pass conversion} = \frac{\text{Reactant input to reactor} - \text{Reactant output from reactor}}{\text{Reactant input to reactor}}$$

$$\text{Recycle ratio} = \frac{\text{Quantity of a reactant recycled}}{\text{Quantity of same reactant entering a recycling operation as fresh feed}}$$

$$\text{Combined feed ratio} = \frac{\text{Reactor feed of a recycling operation}}{\text{Quantity of same reactant entering the operation as fresh feed}}$$

4.3.2 Limiting and Excess Reactants

Two reactants *A* and *B* are said to be in stoichiometric proportion if the ratio (moles *A* present/moles *B* present) equals the stoichiometric ratio obtained from the balanced reaction equation. The reactant that gets utilized first if the reaction proceeds to completion is termed limiting reactant, while the other reactants are termed excess reactants. If all reactants are present in stoichiometric proportion, then no reactant is limiting. In many chemical/biochemical systems, undesirable side reactions occur besides the desired reaction. These side reactions are an economic loss (as less desired product is obtained for a given quantity of raw material). The terms yield and selectivity are defined as follows.

$$\text{Yield (based on feed)} = \frac{\text{Moles of desired product formed}}{\text{Moles of limiting reactant feed}}$$

$$\text{Yield (based on reactant consumption)} = \frac{\text{Moles of desired product formed}}{\text{Moles of limiting reactant consumed}}$$

$$\text{Selectivity} = \frac{\text{Moles of desired product formed}}{\text{Moles of undesired product formed}}$$

4.3.3 Simple Equation of State

An equation that relates quantity (mass or moles) and volume of a gas to the temperature and pressure of the gas is called the equation of state. The ideal gas law (Equation 4.23) is the simplest and most widely used equation of state.

$$PV = nRT \tag{4.23}$$

where *P* is the absolute pressure of the gas, *V* is the volume of the gas, *N* is the number of moles of the gas, *R* is the gas constant (value depends on the units of *P*, *V*, *n*, and *T*), and *T* is the absolute temperature of the gas.

The value of the gas constant R depends on the units of P, V, n, and T.

R = 8.314 m^3 Pa/mol-K = 0.082 L-atm/mol-K = 62.36 L-mm-Hg/mol-K

1 g mol of an ideal gas at 0°C and 1 atm occupies 22.4 L.

At low temperature and high pressure, the ideal gas law provides a poor description of gas behaviour. For such real gases, more complex equations of states are used to represent the P–V–T behaviour.

4.3.4 Material Balance

In the design of a separation unit, certain physical properties of the system should be known. For example, the velocity of each feed component should be known while designing a distillation column. This requires knowledge of the vapour pressure of each substance as a function of temperature. Similarly, for the design of a crystallizer, the solubilities of the salts as a function of temperature should be known.

A phase diagram of a pure substance is a plot of pressure versus temperature, which shows the condition of the substance when it exists as a solid, liquid, and gas. The phase diagram of water is shown in Figure 4.2.

Several terms may be defined with reference to the phase diagram.

- If T and P correspond to a point on the vapour–liquid equilibrium curve for a substance, P is the vapour pressure of the substance at T and T is its boiling point at pressure P. At P = 1 atm, the boiling point of a substance is its normal boiling point.

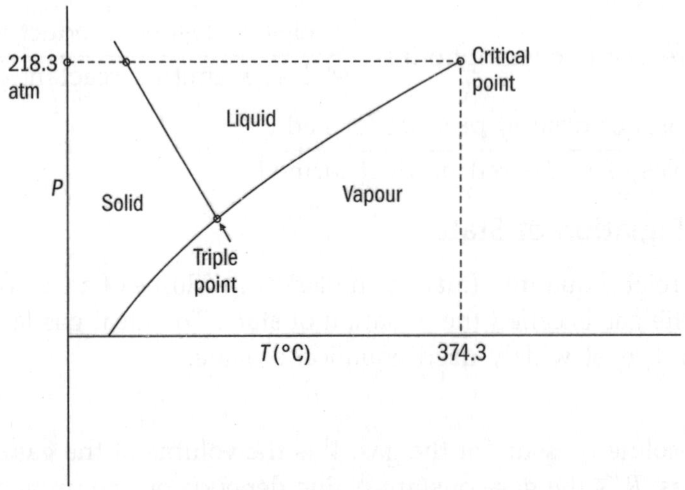

Figure 4.2 Phase diagram of water

- If T and P fall on the solid–vapour equilibrium curve, P is the vapour pressure of the solid and T is its sublimation point at pressure P.
- If T and P fall on the solid–liquid equilibrium curve, T is the melting point or freezing point of the solid at pressure P.

The Clausius–Clapeyron relation (Equation 4.24) gives an approximate relation between vapour pressure of a pure substance (p^*) and the absolute temperature (T).

$$\ln p^* = -(\Delta H_v/RT) + B \tag{4.24}$$

where ΔH_v is the latent heat of vaporization, R is the gas constant, and B is a constant that varies from one substance to another.

In industrial processes, we encounter systems containing several components, of which only one is capable of existing as a liquid under process conditions. Such systems include evaporation, drying, humidification (transfer of liquid into gas phase), condensation, and humidification (transfer of condensable species from gas to liquid phase).

If a gas at temperature T and pressure P contains a saturated vapour whose mole fraction is Y_v (mole vapour/mole total gas) and if the vapour is the only species that would condense if the temperature is lowered, the partial pressure of the vapour in the gas (p_v) equals the pure component vapour pressure $p_v{}^*(K)$ at the system temperature.

$$p_v = Y_v P = p_v^* \ (T)$$

If $p_v = p_v^*$, then the vapour is saturated. A vapour present in a gas in less than its saturation amount is referred to as superheated vapour. For such a vapour,

$$p_v = Y_v P \ \rho < p_v^* \ (T)$$

If a gas containing a single superheated vapour is cooled at constant pressure, the temperature at which the vapour becomes saturated is referred to as dew point of the gas (Equation 4.25).

$$p_v = Y_v P = p_v^* \ (T_{dp}) \tag{4.25}$$

The difference between the temperature of gas and its dew point is called the degree of superheat.

If a substance is contained in a gas–liquid system in equilibrium at temperature T and pressure P, then according to Raoult's law

$$p'_A = Y_A P = X_A \ p_A^* \ (T)$$

where p'_A is the partial pressure of A in the gas phase, Y_A is the mole fraction of A in the gas phase, X_A is the mole fraction of A in the liquid phase, and $p_A{}^*$ is the vapour pressure of pure liquid at T.

4.3.4.1 *Example of material balance in a microbial reaction*

Material balance approach is entirely based on laws of conservation of mass of each element taking part in a bioreaction. It is essential to balance the elemental requirements for growth and product formation. For example, in hetero-fermentative anaerobic bioreaction for the production of propionic acid by *Propionibacter acidipropionici* ATCC 25562 (at 30°C, pH 6.8 for 34 h), the minimum carbon requirement in the medium may be determined stoichiometrically. In this bioprocessing, the carbon content of the substrate (glucose) was found to be distributed mainly in the products (propionic acid, acetic acid, CO_2, and cell mass). On elemental analysis, the carbon content in the cell mass was found to be 44%. Propionic acid is rarely formed as a sole product in an acidogenic bioprocessing. It is usually accompanied by the formation of acetic acid. This occurs for stoichiometric reasons and to maintain hydrogen and redox balance.

About 95% of the initial substrate carbon present could be accounted in various products. Trace amounts of succinic and isovaleric acids, which require sensitive methods of estimation, probably account for the rest of the 5% carbon. In anoxic bioprocessing, every oxidation is accompanied by a reduction that yields products formed from the substrate. Thus, the balance between the oxidized and reduced products provides an important criterion for evaluating the accuracy of the analysis. An oxidation–reduction (O/R) balance in propionic acid production by *P. acidipropioninici* revealed the values given in Tables 4.7 and 4.8.

The balance of hydrogen and oxygen was obtained by comparing their ratios in the substrate and product. When $2H = 0$ in a substrate, its oxidation value is zero. Every additional pair of hydrogen atom constitutes O/R number of –1 and every additional oxygen atom constitutes O/R number of +1. An O/R index of 0.94 is observed from Table 4.8.

An O/R index close to 1 indicates that all products were accounted for in right proportions. Deviation from unity accounted for experimental error.

Cell composition and product composition data can enable one to assess the elemental requirements in medium design considering requirements and constraints.

Thus, mass and energy balances associated with growing cultures play a key role in media design. The heat of reaction per electron transferred to oxygen for a variety of organic molecules, number of available electrons per carbon atom in biomass, and the weight fraction of carbon in biomass have been found to be relatively constant.

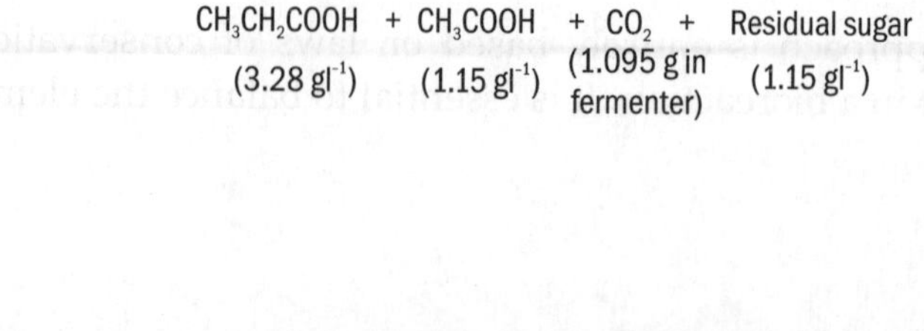

Table 4.7 Substrate carbon distribution in propionic acid fermentation by *P. acidipropioninici*

Product	g/L	g in fermenter	g°C
CO_2	—	1.095	0.298
Propionic acid	3.28	4.52	2.198
Acetic acid	1.15	1.587	0.634
Cell mass	2.24	3.09	1.359
Residual sugar	0.1	0.138	0.055
Total			4.544

Note Substrate carbon distribution at the end of 34 h/g C in fermenter at 0 h is 4.75.

Table 4.8 Oxidation–reduction balance in propionic acid fermentation based on 100 mmol of glucose (30°C, initial pH 6.8)

Product	mmol of product per 100 mmol of glucose	O/R number	Milliequivalent	
			Oxidized	Reduced
Acetic acid	32.95	0	—	—
Propionic acid	92.0	–1	—	92.00
CO_2	43.26	2	86.52	—
		—	86.52	92.00
O/R index = 0.94				

O/R – oxidation-reduction

4.3.5 Basics of Energy Balance

We can no longer afford to waste energy, and its inefficient utilization can have a strong bearing on the product cost. Besides material balance, energy balance is also required for a complete design.

The total energy of a system has three components: kinetic energy, potential energy, and internal energy. For a closed system where no mass is transferred across its boundaries while the process is taking place, energy may be transferred in two ways: (1) as heat (heat is said to be positive when it is transferred to the system from the surroundings), and (2) as work (when work is done by the surroundings on the system, it is said to be positive work).

The energy balance equation can be written as follows.

Final system energy - initial system energy = Net energy transferred to the system

Initial system energy = $U_i + E_{Ki} + E_{pi}$

Final system energy = $U_f + E_{Kf} + E_{pf}$

Energy transferred = $Q + W$

Thus,

$$U + \Delta E_K + \Delta E_P = Q + W \tag{4.26}$$

The internal energy of the system (U) depends on the chemical composition, state of aggregation, and temperature of the system and is independent of pressure for liquids and solids. If a system and its surroundings are at the same temperature ($Q = 0$), then the system is termed adiabatic.

Specific enthalpy (H) is a property that occurs in the energy balance equation for open systems (Equation 4.27).

$$H = U + PV \tag{4.27}$$

For steady-state open systems,

$$\Delta H + \Delta E_K + \Delta E_P = Q + W$$

The term sensible heat denotes the heat that must be transferred to raise or lower the temperature of a substance or mixture of substances.

$Q = \Delta U$ (closed system)

$Q = \Delta H$ (open system)

The specific internal energy of a substance depends strongly on temperature. Heat at constant volume of a substance is defined as given by Equation 4.28.

$$C_V(T) = \frac{d\overline{U}}{dT} \text{ or } \Delta\overline{U} = \int_{T_1}^{T_2} C_v\,(T)\,dT \tag{4.28}$$

The heat capacity at constant pressure (Cp) is defined as given by Equation 4.29.

$$C_p(T) = d\overline{H}/dT \tag{4.29}$$

$$\Delta\overline{H} = \int_{T_1}^{T_2} Cp(T)\,DT, \text{ for an ideal gas}$$

$$\Delta\overline{H} = \overline{V}\Delta P + \int_{T_1}^{T_2} Cp(T)\,dT, \text{ for solid or liquid}$$

For liquid and solid, $C_p \approx C_V + R$; for an ideal gas, $C_p = C_V + R$.

Phase changes such as fusion and vaporization are usually accompanied by large changes in initial energy and enthalpy, and heat duties in phase change operations tend to be substantial.

Trouton's rule is a simple formula for estimating standard heat of vaporization (Equations 4.30 and 4.31).

$$\Delta H_V \text{ (kJ/mol)} = 0.088\ T\sigma \text{ (K), for non-polar liquids} \tag{4.30}$$

$$\Delta H_V \text{ (kJ/mol)} = 0.109\ T\sigma \text{ (K), for water, low molecular weight alcohols} \tag{4.31}$$

Large enthalpy changes are commonly associated with chemical reactions. The heat of a reaction is the difference in the enthalpy of products and reactants ($H_{products} - H_{reactants}$) for a reaction that takes place under the following conditions.

- Stoichiometric quantities of reactants are fed and the reaction proceeds to completion.
- Reactants and products emerge at same temperature and pressure.

If ΔH_r is negative, the reaction is said to be exothermic; if it is positive, the reaction is said to be endothermic.

If the reaction takes place in a closed reactor at constant volume, the heat released or absorbed is determined by the change in the internal energy (ΔU) between reactants and products.

The internal energy of the reaction is related to the heat of the reaction as given by Equation 4.32.

$$\Delta \overline{U}_r(T) = \Delta \overline{H}_r(T) - RT(\underbrace{\Sigma \upsilon_i}_{\text{Gaseous products}} - \underbrace{\Sigma \upsilon_i}_{\text{Gaseous reactants}}) \tag{4.32}$$

The energy balance equation for reactive processes can be written as shown in Equation 4.33.

$$\Delta H = \underbrace{\Sigma n_i \overline{H}_i}_{\text{Outlet}} - \underbrace{\Sigma n_i \overline{H}_i}_{\text{Inlet}} \tag{4.33}$$

The energy changes associated with the formation of a solution from the solutes and solvent at 25°C is called the standard heat of formation of the solution (Equation 4.34).

$$(\Delta \overline{H}^0{}_f)_{\text{solution}} = (\Delta \overline{H}^0{}_f)_{\text{solute}} + \Delta \overline{H}^0_S\,(n) \tag{4.34}$$

where n is the moles of solvent per mole of solute and $\Delta \overline{H}^0_S$ is the heat of solution at 25°C.

4.4 CONCEPT OF MODELLING OF BIOLOGICAL REACTION SYSTEMS

A model of a biological reaction system is a set of relationships between the variables of interest in the system being studied.

The relationships include equations (algebraic and differential), graphs, and tables. The variables of interest include operation, economic evaluation, and safety. The objectives of studying these systems are to produce more cells, make chemicals, and consume unwanted materials. The importance of these studies relates to the exploration of ideas, aid to thought, biological information, quantification, parameter determination, sensitivity studies, and global system studies.

It is important to choose the simplest adequate model firmly rooted in known fundamental principles of physical, chemical, and biological interactions. Model refinement is also necessary, depending on the system. A simplistic approach of refinement is depicted in Figure 4.3.

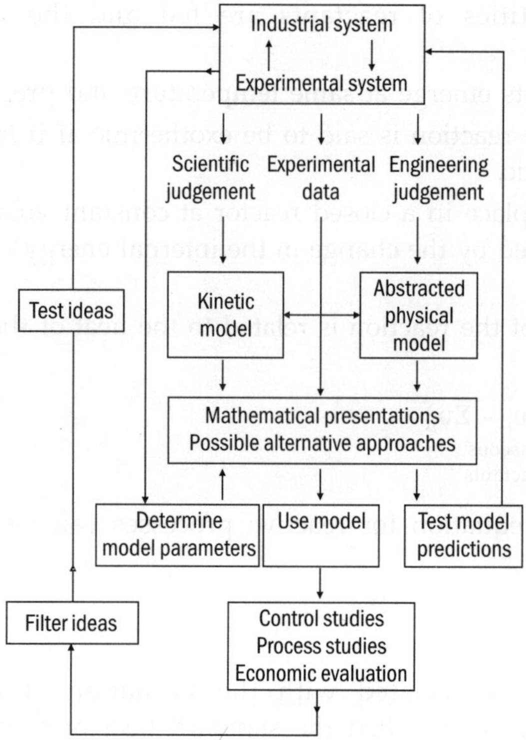

Figure 4.3 Block diagram of model refinement steps

4.4.1 Boundaries of the Control Region

The following phase boundaries may arise in the control system.

- Phase boundaries that define the region but across which no exchange takes place (for example, liquid-containing vessel or bioreactor walls).
- Phase boundaries across which exchange of mass or energy takes place (for example, gas bubble–liquid interface).
- Geometrically defined boundaries within one phase across which exchange takes place by bulk flow or molecular diffusion (for example, nutrients inlet and outlet pipes).

4.4.2 Mathematical Model Base

In order to construct a conventional mathematical model, we have to write a set of equations for each control region. This set consists of conservation equations, rate equations, and thermodynamic equations.

4.4.2.1 Conservation equations

Conservation equations are meant for each conserved quantity, such as mass energy, momentum or individual species. For any conserved quantity in any control region, rate of accumulation = rate of input – rate of output

The input terms are (1) bulk flow across geometrical boundaries, (2) diffusion across geometrical boundaries, (3) transfer across phase boundaries, and (4) generation within the control region.

The output terms are (1) bulk flow across geometrical boundaries, (2) diffusion across geometrical boundaries, (3) transfer across phase boundaries, and (4) consumption within the control region.

Diffusion across geometrical boundaries is not important with finite control regions.

4.4.2.1.1 Individual terms in conservation equations

The individual terms in conservation equations are all in mass units but could also be in energy units; all per unit volume of control region.

Rate of accumulation term

The rate of accumulation in $kg/m^3/h$ or in $g/L/h$ is as follows.

$$\text{Rate of accumulation} = \frac{\text{Change in kg/m}^3}{\text{Time elapsed}} = \frac{\text{kg/m}^3}{\text{t}}$$

$$\text{or rate of accumulation} = \frac{d}{dt}\,\text{kg/mg}^3$$

$$\text{or rate of accumulation} = \frac{d}{dt}\,(\text{amount of conserved quantity in unit volume of control region})$$

If the control region volume is changing (fed batch), then a more general approach can be used as follows.

$$\text{Rate of accumulation} = \frac{\frac{d}{dt}\,(\text{amount of conserved quantity in whole of control region})}{\text{Volume of control region}}$$

For cells,

$$\frac{1}{v}\frac{d(VX)}{dt} = \frac{1}{v}\left(V\frac{dx}{dt} + X\frac{dv}{dt}\right)$$

$$\text{or } \frac{d(VX)}{dt} = \frac{dx}{dt} + \frac{X}{V}\frac{dv}{dt}$$

4.4.2.2 Input and output terms

All the input and output terms are expressed per unit volume of the control region.

1. **Bulk flow:** The bulk flow (in $kg/m^3/h$ or $g/L/h$) is expressed as follows.

$$\text{Bulk flow} = \frac{\text{Flow rate (m}^3\text{/h) concentration (kg/m}^3)}{\text{Control region volume (m}^3)}$$

For example,

$$\frac{FX}{V} = DX$$

2. **Transfer across phase boundary:** The transfer across phase boundary

$$(\text{kg/m}^3\text{/h}) = \text{Transfer coefficient } \frac{\text{kg}}{\text{h}\cdot\text{m}^2\ (\text{D}\cdot\text{F}\cdot\text{units})}\ \frac{\text{m}^2}{\text{m}^3} \times \text{Area per unit volume } \frac{\text{m}^2}{\text{m}^3}$$

$$\times \text{ Driving force (intensity gradient) (kg O}_2\text{/m}^3)$$

For example, $N = k_L\, a\, (C_g^* - C_L)$

3. **Generation or consumption:** Generation or consumption (kg/m^3/h) = Specific rate on cell basis (kg/kg-cell-h) × Cell concentration (kg-cell/m^3)

   For example,

   $Y_{0_2} = Q_{0_2}x$, for oxygen

   or $Y_x = \mu x$, for cells

   Generation or consumption = Stoichiometric constant $\left(\frac{Kg\ A}{Kg\ B}\right)$ × Rate of generation or consumption $\left(\frac{Kg\ B}{\text{m}^3\text{h}}\right)$

   For example,

   $\gamma_p = 1/Y_{X/P}\, \gamma_x$

   or $\gamma_s = a_{p/x} \cdot r_x$

   Generation or consumption = sum of above terms

   For example,

$\gamma_s = 1/Y_{X/S}\, r_X + m_X X$

4.4.2.3 Rate equations

The rate equations focus on the following two factors.

1. Rates of transfer of mass, energy, momentum or individual species across the boundaries of the region.
2. Rates of generation or consumption of individual species within the control region.

4.4.2.4 Thermodynamic equations

Thermodynamic equations relate to thermodynamic properties, such as P, T, V, concentration, and viscosity, either (1) within the control region (for example, gas laws) or (2) on either side of a phase boundary (for example, Henry's Law).

Thermodynamic equations are used to related non-measurable thermodynamic state variables to measurable ones using known physical laws.

For example,

1. $C^* = P_{O_2}/H$ (Henry's Law)
   where P_{O_2} is the partial pressure of oxygen in gas phase.
   or C^* = O_2 concentration at gas–liquid interface
2. $\rho_g = P/RT$
   where ρ_g is the density of gas, P is the pressure of gas, and T is the temperature of the gas.

Let us take the following two examples.

1. **Ethanol production**

For cell growth rate,

$$\frac{dx}{dt} = r_x - D_x$$

where dx/dt is the accumulation of cells, r_x is the generation of cells, and D_x is the bulk flow out of cells.

If $r_x = \left(1 - \frac{p}{p_m}\right)$

where P/P_m is the inhibition of growth.

Likewise, for substrate uptake rate

$$\frac{ds}{dt} = D(S_0 - S) - \frac{r_p}{Yp/s}$$

where dS/dt is the accumulation of substrate, $D(S_0 - S)$ is the difference between the bulk flow in and the bulk flow out, and $r_p/Y_{p/s}$ is the consumption of substrate for production.

Similarly, for product formation rate

$dP/dt = r_P - DP$

where dP/dt is the accumulation of alcohol, r_P is the production of alcohol, and DP is the bulk flow out of product.

Here,

$r_P = \mu'(1 - P/P_m')X \approx Kr_x$

2. **Waste water treatment**

$dS_a/dt = Q_i/V(S_i - S_a) + Q_r/V(S_R - S_a) + 1/R(\mu x_a)$

where dS_a/dt is the accumulation of substrate, $Q_i/V(S_i - S_a)$ is the bulk flow in minus the bulk flow out, $Q_r/V(S_R - S_a)$ is the bulk flow in minus the bulk flow out (recycle), $1/R(\mu x_a)$ is the rate of consumption of substrate, and R is the recycle ratio.

$dx_a/dt = Q_r/Vx_r + (Q_i - Q_r)/Vx_a + \mu x_a - k_d x_a$

where dx_a/dt is the accumulation of cells, Q_r/Vx_r is the bulk flow in of cells with recycle, $(Q_i - Q_r)/Vx_a$ is the bulk flow out of cells, μx_a is the generation of cells by growth, and $k_d x_a$ is the consumption of cells by death.

4.4.3 Complete Model Requirements

The requirements for a complete model are (1) initial conditions, which are related to values of variables at time zero (for example, after inoculation); (2) constraints, that is, ranges of the variables within which they make physical sense.

To check model, the difference in the number of variables and equations should be equal to the number of variables that must be independently fixed ($\neq 0$).

That is,

Number of variables – Number of equations = Number of variables that must be independently fixed

$\neq 0$

4.4.4 Future Development Needs

Current models rarely take into account what is happening inside the cell. These models are based on the black box concept and are, therefore, not perfect. With the availability of adequate information on the molecular basis of cells and the sequences of biological synthesis and biochemical reactions, such knowledge can be incorporated into the models. Much more cooperation and active interaction between microbiologists, biochemists, and biochemical engineers is required.

4.4.5 Kinetics: Batch Growth

Cell growth in batch occurs in geometrical progression (Equation 4.35).

$$1P \rightarrow 2P \rightarrow 4P \rightarrow 8P \rightarrow 16P \quad (4.35)$$

or $$2^0P \rightarrow 2^1P \rightarrow 2^2P \rightarrow 2^3P \rightarrow 2^4P \ldots 2^nP \quad (4.36)$$

where n is the doublings after an interval of time t and P is the population of cells.

Hence, $$n = t/t_d \quad (4.37)$$

The final population after time t (h) is related to the initial population.

$$P_t = P_0 2^n = P_0 2^{t/t_d} \quad (4.38)$$

$$P_t/P_0 = 2^{t/t_d} \quad (4.39)$$

or $$\ln (P_t/P_0) = t/t_d (\ln 2) \quad (4.40)$$

or $$\frac{\ln(p_t/p_0)}{t} = \frac{0.693}{t_d} \quad (4.41)$$

Thus, the slope of a plot of (ln P) versus t equals to $0.693/t_d$ and $0.301/t_d$ in a plot of log P versus t. From Equation 4.41, we get

$$\frac{d \ln P}{dt} = \frac{0.693}{t_d} \tag{4.42}$$

$$\text{or } d\ln P/dt = (d\ln P/dt)\ (dP/dt) \tag{4.43}$$

Since the differential coefficient of (ln P) is $1/P$, we get

$$(1/P)(dP/dt) = 0.693/t_d \tag{4.44}$$

Assuming dry cell density is equal to population density, Equation 4.44 can be written in terms of specific growth rate (Equation 4.45).

$$(1/x)(dx/dt) = \mu = 0.693/t_d \tag{4.45}$$

and based on (log P) versus t

$$(1/x)(dx/dt) = \mu = 0.301/t_d \tag{4.46}$$

Let us consider the following problem.

One litre of nutrient medium inoculated with 10 μg of cells (≡ 50 × 10^6 cells) having a doubling time (t_d) of 20 min is cultivated for 24 h. What is the size of the final population?

$$\mu = \frac{0.693}{0.333} = 2.08 \text{ per hour}$$

$$\text{or } \mu = (\ln x - \ln x_0)/t \tag{4.47}$$

Thus,

$$\ln (x/x_0) = 2.08\ t = 49.92$$

$$\text{or } x/x_0 = e^{49.92} = 4.8 \times 10^{21}$$

$$\text{and } x = 4.8 \times 10^{21}\ x_0 = 4.8 \times 10^{22}\ \mu g \equiv 48 \text{ billion tonnes}$$

This high cell population is unrealistic and reflects that cell growth is bound by certain limitations.

The logic of growth dynamics is "as exponential growth proceeds in a batch culture, the depletion of the essential nutrients follows the same finger prints".

The material balance of cell synthesis is expressed in Equation 4.48.

$$\mu = (1/x)(dx/dt) - (1/x)\ (ds/dt) \tag{4.48}$$

$$\text{or } \mu = Yq \tag{4.49}$$

where Y is the yield factor and q is the specific rate of substrate uptake.

The molar yield factor for *Escherichia coli* with respect to glucose in salt medium is 90 g of cells per mole of glucose consumed.

$\ln t_d = 0.693$

μ *E. coli* = 1.0 per hour

q = 11.1 mmol/g equivalent of dry cell of *E. coli* per hour.

With all nutrients in excess and glucose 10 g/L (55.6 mmol) and inoculum ≡ 0.01 mg cell per ml (50×10^6 cell/ml), we have

cell density for glucose exhaustion = 90 (55.6/1000) = 5.0 g/L

From Equation 4.47, we have

$\ln (x/x_o) = \mu t$

or $\ln (5.0/0.01) = t \ (\mu = 1.0)$

or $\ln (5.0/0.01) = 6.21 \text{ h} = t$

4.4.6 Dynamics of Growth: Continuous Culture

4.4.6.1 Continuous culture devices

Continuous culture devices are based on monitoring of turbidity caused due to growth by control of dilution rate of incoming nutrient solution in transient state. A batch culture system may be transformed into a continuous culture either by turbidostat or by chemostat.

Batch culture is a closed system, while continuous culture is an open system in which all the elements controlling the cell growth remain constant at any time. Thus, input of fresh medium need not be made at an exponentially changing rate but at a constant rate specified by a total culture mass concentration.

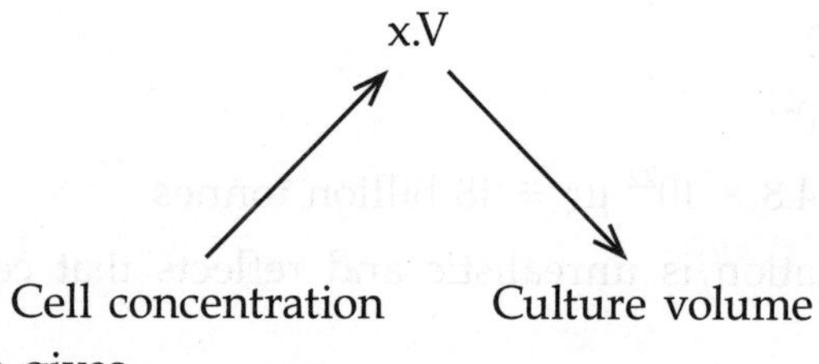

The material balance gives

rate of change = input – output – consumption

or $dS/dt = FS_0 - (FS + q \cdot x \cdot v)$ (4.50)

where S_0 is the substrate concentration in the reactor input (g/L), S is the substrate concentration in the reactor output (g/L), F is the substrate flow rate (L/h), x is the cell concentration (mg dry cell/L), v is the culture volume (L), q is the specific rate of substrate uptake $\left(-\frac{1}{x}\frac{dS}{dt}\right)$.

At steady state,

$dS/dt = 0$

$$q \cdot x \cdot V = F(S_0 - S) \tag{4.51}$$

or $q = D(S_0 - S)/x$ (4.52)

Since $\dfrac{S_0 - S}{x}$ is the reciprocal of yield factor Y, we have

$$q = D/Y \tag{4.53}$$

The culture is growing exponentially while fresh medium is added at a constant rate (linear) (ml/h). Although it appears anomalous, it is not because the culture is effectively diluted at an exponential rate by virtue of the presence of the overflow device, which removes excess culture at the same rate at which fresh nutrients enter the vessel. An addition of a drop of medium will reduce the cell concentration by a factor $V/(V + dV)$. Similarly, after the removal of one drop of culture and the addition of a second drop of the medium, the dilution factor changes to $\left(\dfrac{V}{V + dV}\right)^2$ and so on, until it becomes equal to $\left(\dfrac{V}{V + dV}\right)^1$.

Here, $n = \dfrac{V}{dV}$. That is, n = one culture volume.

To be able to determine the extent to which the culture has been diluted, it is necessary to compute the numerical equivalent of the expression $\left(\dfrac{V}{V + dV}\right)^{V/dV}$.

The extent to which the culture has been diluted can also be determined by solving Equation 4.54.

$$\left(\frac{V}{V + dV}\right)^{V/dV} = \left(\frac{V/dV}{V/dV + 1}\right)^{V/dV} = \left(\frac{n}{n + 1}\right)^n + \left(\frac{n}{1 + 1}\right)^n = (1 + 1/n)^{-n} \tag{4.54}$$

Since e is defined as $\lim\limits_{n \to \infty} \left(1 + \dfrac{1}{n}\right)^n$ or $e = 2.718$ by binomial expansion, we can write

$$\lim_{n \to \infty} \left(1 + \frac{1}{n}\right)^{-n} = e^{-1} = 0.368 \tag{4.55}$$

Therefore, after one culture volume is passed through the vessel, the concentration (population) of non-viable (non-growing) cells will decrease from x' to $0.368\ x_0'$ due to dilution effect.

That is,

$$x' \rightarrow 0.368\ x_0' \tag{4.56}$$

where x' is the density of non-viable population.

The number of medium volumes into the culture vessel during a given period of time t will be

$$t \cdot F/V \text{ or } t \cdot D \tag{4.57}$$

The concentration of cells present at any time will be quantitatively related to the initial cell concentration.

$$x' = x_0' e^{-tD} \tag{5.58}$$

Equation 4.58 applies to only non-growing cell population when cells grow such that their steady-state concentration does not change with time ($ds/dt = 0$), which is the characteristic maximum population phase. It must be understood that μ becomes equal to the wash out rate (rate at which the cells leave the vessel instead of growing).

$$\mu = D \tag{4.59}$$

$$dx/dt = \mu_x - D_x \tag{4.60}$$

In other words, when $\mu > D$, cell density in the vessel increases, and when $\mu < D$, cell density in the vessel decreases.

The logic of expressing in terms of natural logarithm should now be clear from what has so far been discussed.

The cell growth process is directed by a set of complex biological phenomena at the molecular level. The growth process cannot be optimized unless there is a complete understanding of the pathways through which carbon, energy, phosphorus, and other growth parameters flow and perform their functions (Figures 4.4–4.9).

$$\frac{d}{dt}\left[\begin{pmatrix}\text{Culture}\\ \text{volume}\end{pmatrix}\begin{pmatrix}\text{Molar concentration}\\ \text{of component p}\end{pmatrix}\right] = \left[\begin{pmatrix}\text{Culture}\\ \text{volume}\end{pmatrix}\left(\frac{\text{Moles of } P \text{ formed}}{\text{Unit culture volume} \times \text{unit time}}\right)\right]$$

$$\frac{d}{dt}(V_R \cdot Cp) = V_R \cdot r_{fp}$$

where V_R is culture volume, Cp is moles of P per unit culture volume, and

$$r_{fp} = \frac{\text{Moles of } p \text{ formed by conversion}}{\text{Unit culture volume} \times \text{unit time}}$$

The culture is contained in a flask (F) to which medium (M) is pumped at a rate determined by the photocell (L) and the controller (C). The culture is vigorously stirred by means of a magnetic stirrer (S) and aerated by means of injecting air (A), along with the medium. Excess culture is removed by means of an overflow device (O). Two pumps (P) are required: one to deliver the medium to the culture, and the second to circulate culture through the photometer cell.

4.4.6.2 Application of continuous culture in selection of mutants

Many yeasts are able to utilize D-xylose, but recently some have been identified to directly ferment D-xylose to ethanol. *Pachysolen tannophilus* is one of the most efficient among these, but development of this organism for industrial use may require alteration of several characteristics that make its immediate use unfeasible. More recent results

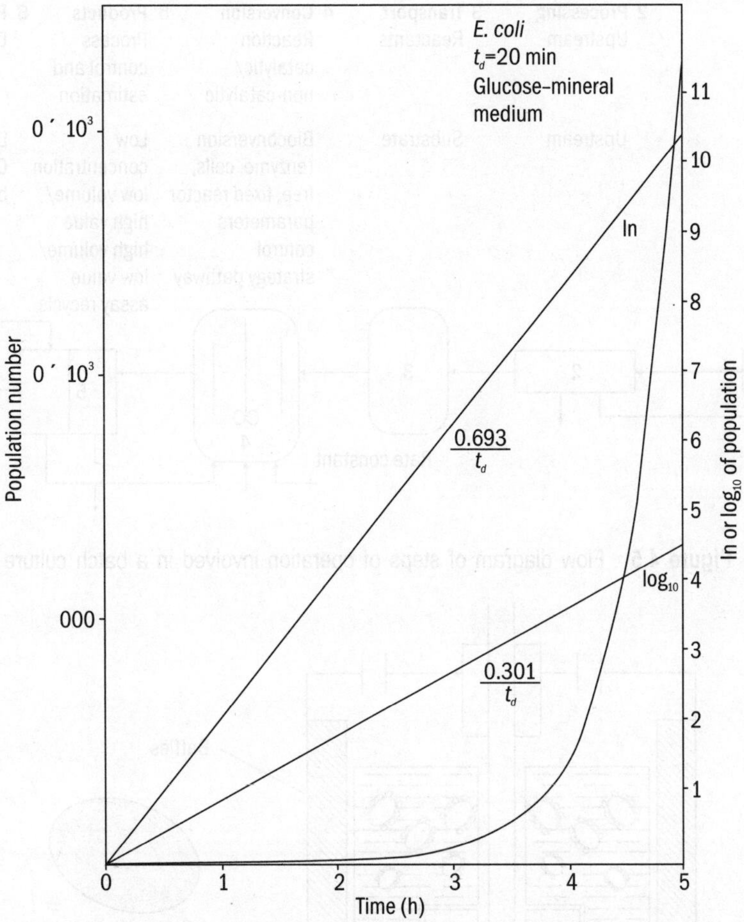

Figure 4.4 Logarithmic and linear plot of a bacterial growth

have shown that ethanol production by *P. tannophilus* is adversely affected when D-xylose concentrations exceed 50 g/L and that D-xylose uptake is inhibited when ethanol concentrations become greater than 20 g/L.

In addition to inhibition by relatively low concentrations of substrate or product, *P. tannophilus* has a slow growth rate, less than half that attained by *Saccharomyces cerevisiae* on D-glucose. For industrial purposes, a slow growth rate is not desirable because it increases the production time and, consequently, the cost of production. Furthermore, a slower growth rate increases the possibility of significant contamination, resulting in production loss. However, successful isolation of mutants capable of rapid aerobic growth on D-xylose is discussed in subsequent paragraphs.

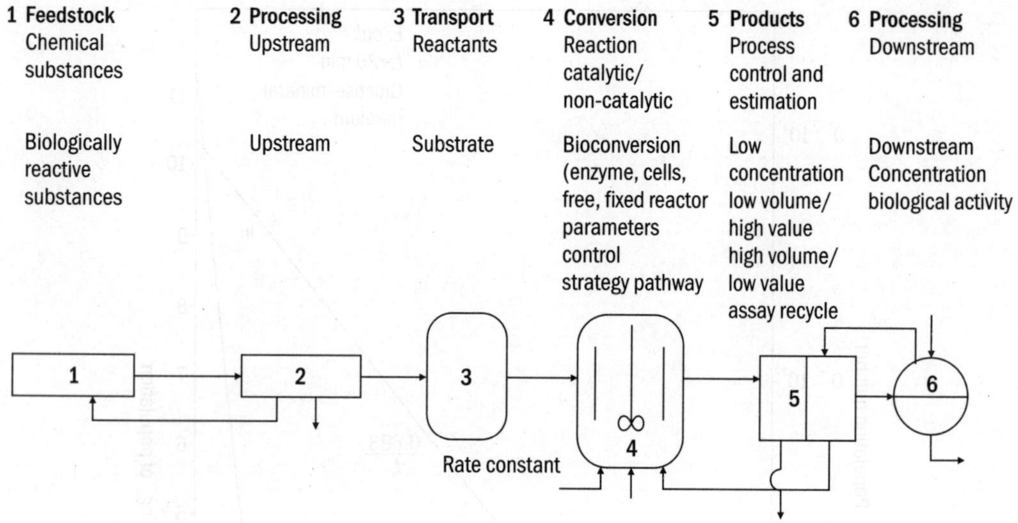

Figure 4.5 Flow diagram of steps of operation involved in a batch culture

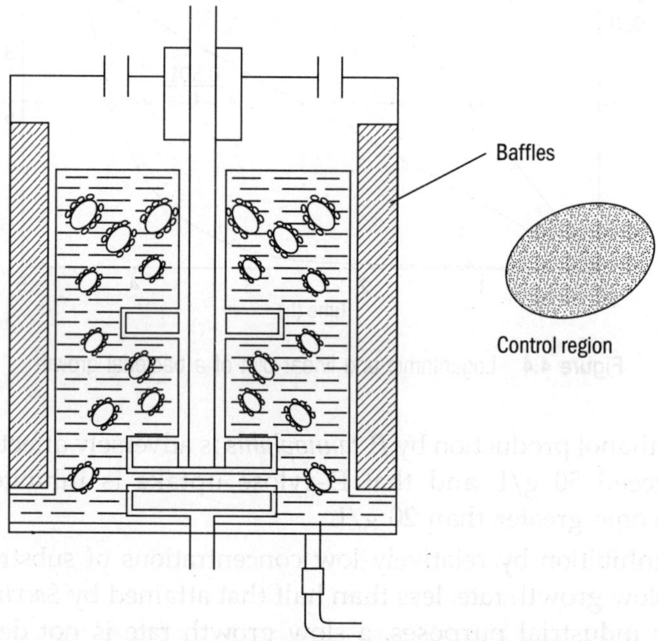

Figure 4.6 Schematic diagram of a physical hardware system

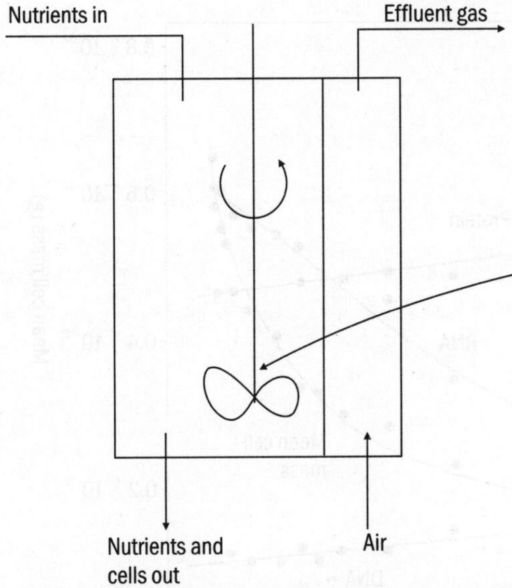

Figure 4.7 Schematic diagram of a simple hardware showing inputs and outputs (batch system)

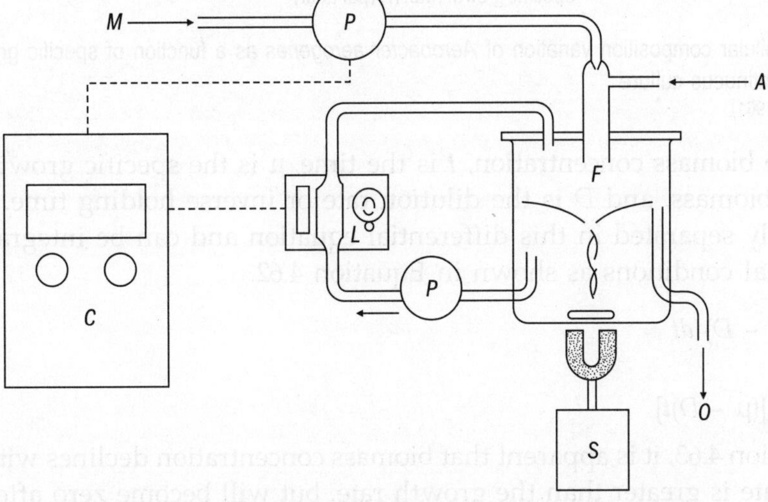

Figure 4.8 Continuous culture operation schematic in a turbidostat

Selection of growth rate mutants of *P. tannophilus* was accomplished by subjecting a mutagenized cell population in continuous culture to high rates of substrate dilution over an extended period of time. The effect of dilution rate on biomass concentration in continuous cell culture can be determined from Equation 4.61.

$$\frac{db}{dt} = \mu b - Db \qquad (4.61)$$

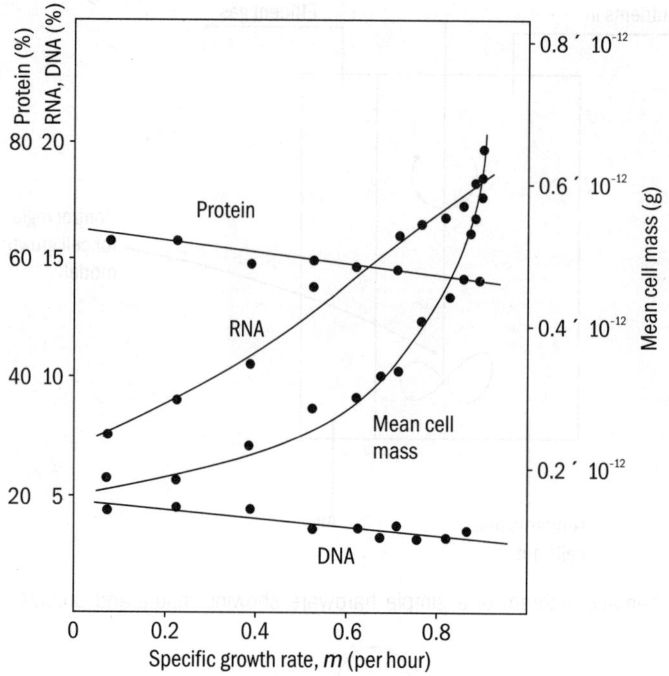

Figure 4.9 Cellular composition variation of *Aerobacter aerogenes* as a function of specific growth rate in a continuous culture

Source Herbert (1961)

where b is the biomass concentration, t is the time, μ is the specific growth rate or the rate per unit biomass, and D is the dilution rate or inverse holding time. The terms b and t are easily separated in this differential equation and can be integrated between initial and final conditions as shown in Equation 4.62.

$$\int_{b_0}^{b} d \ln b = \int_{t_0}^{t} (\mu - D)\, dt \tag{4.62}$$

$$\text{or } b = b_0 \exp [(\mu - D)t] \tag{4.63}$$

From Equation 4.63, it is apparent that biomass concentration declines with time when the dilution rate is greater than the growth rate, but will become zero after an infinite amount of time. The relative rate at which biomass concentration declines is dependent on the magnitude of ($\mu - D$). This is so because fast-growing mutants could be selected by maintaining the continuous culture at a high dilution rate over a sufficient period of time. This allowed the individual growth rate to affect the potential of an individual cell to be retained within the culture. Mutants with an improved growth rate characteristic, selected in this manner, were further evaluated for their ability to retain an improved growth rate following a common storage procedure, that is, lyophilization, and in their ability to ferment D-xylose to ethanol.

4.4.6.2.1 Experimental reports

1. Organisms and growth conditions

The Agricultural Research Service Culture Collection at the Northern Regional Research Centre, Peoria, Illinois, supplied wild type *P. tannophilus* (NRRL Boidin et Adzet Y-2460), *Saccharomyces uvarum* (NRRL Y-1347), and *S. cerevisiae* (NRRL Y-2235). A commercial wine-yeast preparation of *S. cerevisiae*, trade name Fermivin, was obtained from Gist-Brocades Fermentation Industries Inc., Delft, Holland. Stock cultures of these wild-type yeasts and derived *P. tannophilus* mutants were maintained at 32°C on agar slants containing 6.7 g/L yeast-nitrogen base (YNB) (Difco Laboratories, Detroit, Michigan) 50 g/L carbohydrate, and 20 g/L agar. The pH of the medium was 4.5. D-glucose (Difco Laboratories) served as the carbohydrate source for all *Saccharomyces* strains, and D-xylose (Sigma Chemical Co., St Louis, Missouri) served as the carbohydrate source for wild-type and derived mutants of *P. tannophilus*. Liquid cultures were carried out at 32°C in the same medium minus the agar.

2. Mutagenesis

Cells of *P. tannophilus* from a YNB of 5% D-xylose agar slant were used to inoculate 20 ml YNB of 5% D-xylose liquid medium contained in a 50 ml Erlenmeyer flask fitted with a silicone-sponge closure (Bellco Glass, Inc., Vincland, New Jersey). Following 72 h of incubation at 32°C on a gyratory water bath operating at 200 rpm, 0.5 ml was used to inoculate a second 50-ml flask containing 20 ml YNB of 5% D-xylose liquid medium. Following overnight incubation under the same conditions, cell counts were made by haemacytometer. Two 10 ml volumes of a 6×10^7 cells/ml suspension were transferred to sterile Petri dishes for mutagenesis. The Petri dishes were placed on a mechanical rotary shaker and agitated at 120 rpm to provide uniform exposure of the cells to irradiation during treatment. Each 10 ml volume was exposed to 33 s of 1900 erg/cm^2/s ultraviolet (UV) light generated by a general electric bulb (G-1578) positioned 16 in above the surface of the mechanical shaker. Pre- and post-irradiation viability platings on YNB of 5% D-xylose agar plates were used to calculate survival rate, which in this case was 57%. Thus, a total population of approximately 6.84×10^8 cells survived mutagenesis.

Following mutagenesis, the total population, that is, 20 ml by total volume, was added to 65 ml YNB of 5% D-xylose liquid medium in a 250 ml Erlenmeyer flask fitted with a Bellco silicone-sponge closer and incubated at 32°C for 16 h in the dark on a mechanical shaker operating at 200 rpm. The following day, the entire culture was transferred aseptically to a spinner flask for continuous culture.

3. Continuous culture selection

Continuous culture of mutagenized *P. tannophilus* was carried out at 32°C in a jacketed Bellco spinner flask modified with additional entrance ports. The culture was stirred at 800 rpm by a magnetic stirrer and aerated by pumping air at 120 ml/min through a sterile cotton filter into a 400 ml headspace above the stirred culture. Sterile disposable syringes were used to draw samples through a 4 inch stainless steel hypodermic needle permanently mounted on a rubber septum closing one of the access ports. When samples were not being taken, a sterile syringe was left attached to the needle to prevent contamination. Peristaltic pumps were used to deliver sterile liquid growth medium to the spinner flask at specific rates and to maintain the holding volume at 85 ml.

An 85 ml volume containing the irradiated mutant-cell population was transferred aseptically to the spinner flask, and the continuous culture was operated over a 42-day period according to the following dilution rate schedule: 0.071 ml/h on day 1, 0.32 ml/h on days 2–11, 0.53 ml/h on days 12–21, and 1.0–1.3 ml/h on days 22–42. Dilution rates were calculated daily from the ratio of the overflow collection volume to the holding volume. Samples were withdrawn daily for microscopic examination and inoculation of slants and/or liquid cultures for future testing.

Progress in the selection process was monitored by comparing growth rates of samples taken on a weekly basis. Continuous culture selection was terminated when samples of the population failed to show improvement in specific growth rate from one week to the next. Isolates of the long-term, continuous culture were presumed mutant by virtue of their display of an altered phenotypic trait, that is, growth rate and their recovery from a mutagenized population.

4. Preparation of single-cell isolates

The slant prepared from the continuous culture sample taken on day 42 served as the source of mutant isolates to be tested and compared to similar wild-type isolates. Presumed single-cell-derived isolates of wild-type and 42-day selected mutant *P. tanophilus* were obtained by dilution plating class of colonies (that is, large, small, and so on), 10 colonies of each type were picked from one area of a plate; therefore, the sample should be representative of the existing population.

5. Growth rate measurement

All growth cultures were carried out in 50 ml Erlenmeyer flasks containing 20 ml of liquid medium. Flasks were plugged with Bellco silicone-sponge closures and incubated at 32°C on a mechanical rotary shaker operating at 200 rpm. Each isolate was grown 48 h in a 20 ml batch culture from a dilute inoculum, and its optical density (OD) at 620 nm was determined. A sample was then used to incubate a second 20 ml growth culture to a final optical density of 0.1 at 620 nm and incubated as before. Over the first 12–14 h of the second culturing, absorbance at 620 nm was monitored and the data were plotted on semi-log paper. Linear regression analysis of the data obtained during the logarithmic growth phase yields a best-fit straight line whose slope represents the initial specific growth rate.

6. Fermentation rate and yield measurement

For fermentation assays, all batch experiments were carried in 500 ml Erlenmeyer flasks initially containing 400 ml of the medium. Incubations were placed on a rotary shaker operating at 150 rpm at 32°C. To prepare the inocula, isolates lf-w (wild type) and lc-m (mutant) were grown from dilute inocula in 500 ml Erlenmeyer flasks capped with Bellco silicone-sponge closures. Following 48 h of incubation, absorbance at 620 nm was determined and used to calculate an inoculum volume required to obtain a final 620 nm absorbance of 0.1 in 400 ml of medium. This volume was then used to inoculate six 500 ml Erlenmeyer flasks capped with Bellco silicone-sponge closures and incubated as described above. Following 48 h incubation, Bellco closures from three of the six cultures were replaced with Alwood valves filled with concentrated sulphuric acid to produce an anaerobic environment. Samples were taken every 24 h and analysed for biomass, D-xylose, and ethanol content. Since each flask was opened daily, cultures described as anaerobic were actually microaerophilic.

Specific ethanol production and D-xylose uptake rates were calculated over each sampling interval as $\Delta E/(\bar{b} \cdot \Delta t)$ or $\Delta X/(\bar{b} \cdot \Delta t)$, respectively, where Δt is the time between samplings, ΔE and ΔX are the corresponding concentration changes for ethanol and D-xylose, and $\bar{b}$ is the average biomass concentration over the interval. Average rates over the first 192 h of each fermentation were computed as the time-weighted average of rates over individual intervals. The yield was computed as g ethanol/g D-xylose at the point of highest ethanol concentration.

7. Lyophilization

Prior to lyophilization, yeast cultures were grown 48 h at 32°C on YNB of 5% D-xylose agar slants. The growth from two slants was resuspended in 0.9 ml sterile bovine (ox) serum (CS-1337, Colorado Serum Co. Laboratories, Denver, Colorado) and lyophilized according to the procedure of Wickerham and Andreasen (1942). Lyophilized cultures were held at room temperature approximately three days prior to resuspension in 0.5 ml YM broth containing 0.3% Difco yeast extract, 0.3% Difco malt extract, 0.5% Difco Bactopeptone, and 1% Difco Bactodextrose and replating on YM agar slants to recover colonies. Recovered cultures were regrown on YNB of 5% D-xylose agar prior to retesting.

8. Optical density measurements

Sample absorbance was measured at 620 nm on a Bausch and Lomb Spectronic 2000 spectrophotometer. When necessary, samples were diluted to values between 0.05 and 0.4 units where optical density is a linear function of the biomass concentration.

9. Ethanol and D-xylose measurements

Samples were prepared for ethanol and D-xylose analyses by centrifugation for 10 min in room temperature in a clinical centrifuge (International Equipment Co., Needham, Massachusetts) operating at maximum speed (1640 × g). Supernatants were then filtered through a 0.45 μm filter (Millex HV, Millipore Corp., Bedford, Massachusetts) and frozen at –20°C until analysis.

Ethanol concentration was determined on a Packard model 428 Gas Chromatograph equipped with a 6 ft Porapak Q column operating at 150°C.

D-xylose concentration was determined on a Waters High Performance Liquid Chromatographic system equipped with a Biorad HPX-87H cation exchange column and a refractive index detector. The column was operated at 28°C, and water was used as the mobile phase.

10. Reported results

a. Growth rates of wild-type and mutant isolates in aerobic batch culture

To evaluate the effect of long-term, high-dilution-rate, continuous culture selection on growth rate, initial specific growth rates were determined for each of 10 wild-type and nine derived mutant isolates (pre-lyophilization columns, Table 4.9). Both the average initial specific growth rate and variation (estimated by variance) of the mutant isolates were significantly greater than those of the wild-type isolates. Presumably, significant variation in individual initial specific growth rate was induced in the wild-type population by UV treatment, and subsequent selection through continuous culture led to a derived population with a much higher average initial specific growth rate and a lower degree of individual growth-rate variability compared to the initial, mutagenized population. The fact that individual variation remained high compared to wild-type indicates that further improvement might be expected with continued selection.

To assess stability of growth rates in wild-type and mutant isolates following a common storage procedure, each was frozen, lyophilized, and subsequently regrown and tested (post-lyophilization columns, Table 4.9). As before, the group of mutant isolates attained both a higher average initial specific growth rate and demonstrated a higher degree of individual variation compared to that attained by the group of wild-type isolates. Initial specific growth rates of individual wild-type isolates were very similar to those attained prior to lyophilization, whereas mutant isolates exhibited much greater individual variability in both pre- and post-lyophilization rates. However, paired *t* tests of pre- versus post-lyophilization rates from either the wild-type or mutant group of isolates failed to demonstrate significant differences. Thus, lyophilization did not strongly affect the ability of an isolate to retain and express a particular growth rate.

Table 4.9 Initial specific growth rates of wild-type and mutant isolates measured before and after lyophilization

Wild-type isolate	Initial specific growth rates (per hour)		Mutant isolate	Initial specific growth rates (per hour)	
	Pre-lyophilization	Post-lyophilization		Pre-lyophilization	Post-lyophilization
1a-w	0.24	0.26	1a-m	0.46	0.34
1b-w	0.24	0.26	1b-m	0.43	0.39
1c-w	0.25	0.25	1c-m	0.40	0.39
1d-w	0.26	0.26	1d-m	0.32	0.38
1e-w	0.24	0.24	1e-m[a]	–	–
1f-w	0.24	0.26	1f-m	0.30	0.37
1g-w	0.25	0.25	1g-m	0.27	0.31
1h-w	0.26	0.24	1h-m	0.36	0.35
2a-w	0.25	0.27	2a-m	0.33	0.28
2b-w	0.25	0.25	2b-m	0.39	0.39
Mean	0.248	0.254	Mean	0.362	0.354
variance = S^2	0.000062	0.000093	S^2 =	0.00394	0.00155

S – standard deviation

[a] Did not survive lyophilization

A t test of the pooled pre- and post-lyophilization rates of mutant versus wild-type isolates yielded a value of 9.06 with 17 degrees of freedom, a value significant at the 1% level. Thus, we conclude that long-term, high-dilution-rate, continuous culture of a mutagenized population of *P. tannophilus* allows selection for isolates with an improved oxidative growth rate.

For comparative purpose, initial specific growth rates of three recognized "fast-growing" strains of *Saccharomyces* were determined. With D-glucose as substrate, strain NRRL Y-1347 had an initial specific growth rate of only 0.27 per hour, whereas those of NRRL Y-2235 and Fermavin were 0.41 per hour and 0.42 per hour, respectively. The average rates attained by both pre-lyophilized (0.362 per hour) and post-lyophilized (0.354 per hour) mutant isolates growing in D-xylose approach those of the faster growing strains of *Saccharomyces* growing on D-glucose.

b. Fermentation rates and yields of wild-type and mutant isolates in batch culture

P. tannophilus, like other yeasts and fungi, converts D-xylose to D-xylulose via a two-step reduction/oxidation process (Smiley and Bolen 1982). D-xylose is first reduced to D-xylitol by NADPH-linked aldose reductase and then oxidized to D-xylulose by an NDA-linked D-xylitol dehydrogenase. As in other yeasts, further metabolism is believed to proceed by phosphorylation of D-xylulose to D-xylulose-5-phosphate with subsequent conversion by well-characterized reactions of the pentose-phosphate shunt and Embden–Meyerhoff glycolytic pathways (Bolen and Slininger 1984). Although *P. tannophilus* converts D-xylose to ethanol under both anaerobic and aerobic conditions, oxygen is required for growth on D-xylose. Thus, the metabolic pathway for oxidative utilization of D-xylose for growth must differ in part from the pathway taken for ethanol production. Since it is likely that only a part of the pathway of D-xylose catabolism is common for both aerobic growth and ethanol production, selection based on an improved ability to grow on D-xylose under aerobic conditions may result in the simultaneous selection of isolates with an altered ability, either improved or impaired, to produce ethanol. In preliminary experiments, each of the mutant isolates was found to retain the ability to produce ethanol from D-xylose under anaerobic conditions (Table 4.10). In a more extensive analysis, aerobic and anaerobic ethanol-production rates and yields of mutant isolate 1e-m and wild-type isolate 1f-w were determined to ascertain whether these capabilities were affected in the selected mutant isolate (Figure 4.10 and Table 4.10). In the first 192 h of culture, both wild-type and mutant isolates per unit biomass took up D-xylose more quickly and produced ethanol at higher rates under anaerobic conditions. This was also reflected in higher average yields, which were attained in anaerobic cultures at 288 h and aerobic cultures at 240 h. Under aerobic conditions, the average ethanol-production rate and yield of mutant cultures were slightly better than those of wild-type cultures. Under anaerobic conditions, however, the converse was true. Not only did wild-type cultures have a higher average ethanol-production rate and yield compared to those attained by mutant cultures, but these differences were significant at the 5% level. Clearly, the mutant is impaired, although only slightly, in its ability to produce ethanol under anaerobic conditions. Only slight differences between mutant and wild-type D-xylose uptake rates were observed under both anaerobic and aerobic conditions. Similar evaluations of other mutant isolates are in progress to determine whether any are improved in their ability to anaerobically produce ethanol from D-xylose.

4.4.6.3 Fed-batch culture (without withdrawing culture medium)

Fed-batch culture has been developed in order to achieve a high productivity of microbial cells and secondary products in fermentation industries. Baker's yeast production, for example, has been adopting this method to avoid a so-called glucose-effect, that is,

Table 4.10 Average specific ethanol production rates (0–192 h), specific D-xylose uptake rates (0–192 h), and yields of wild-type isolate 1f-w and mutant isolate 1c-w cultures grown under aerobic and anaerobic conditions

Isolate	Culture	0–192 h average specific ethanol-production rate (g ethanol/g biomass/hour)	0–192 h average specific D-xylose uptake rate (g D-xylose/g biomass/hour)	Average yield
1f-w	Aerobic	0.0640	0.306	0.218
1f-w	Anaerobic	0.0941	0.342	0.297
1e-m	Aerobic	0.0704	0.286	0.229
1c-w	Anaerobic	0.0799	0.374	0.263

Note Each value represents the average of determinations on the three cultures.

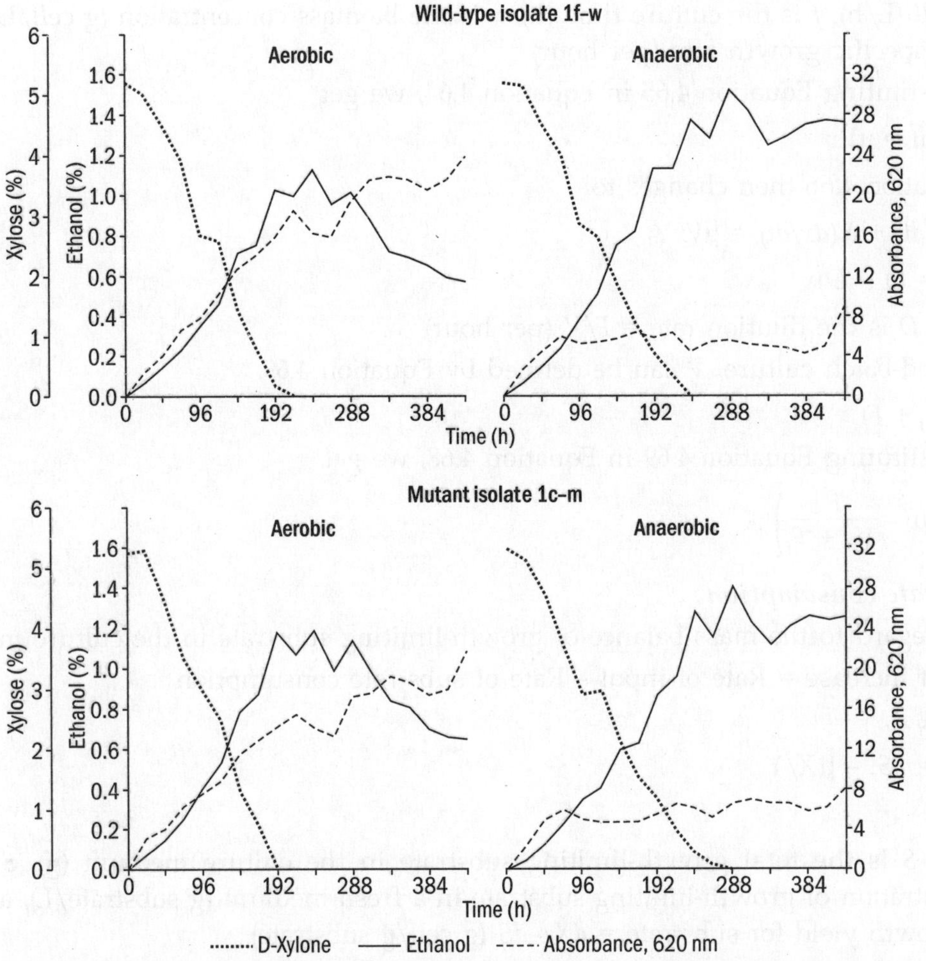

Figure 4.10 Aerobic and anaerobic production of ethanol from *P. tannophilus* wild-type isolate 1f-w and select mutant isolate 1c-m

Note Each point represents the average value of the three cultures.

catabolite repression. By means of this device, nutrients or a growth-limiting substrate is continuously fed in the culture medium.

4.4.6.3.1 Basic theory

Growth of microbes in this cultivation mode may be given as below.

$$dX/dt = \mu X \tag{4.64}$$

$$X = Vx \tag{4.65}$$

where X is the total biomass in the culture vessel (g cell), V is the culture volume at time $t = V_0 + Ft$ (L), V_0 is the initial culture volume (L), F is the flow rate of fresh medium

= dV/dt (L/h), t is the culture time (h), x is the biomass concentration (g cell/L), and μ is the specific growth rate (per hour).

Substituting Equation 4.65 in Equation 4.64, we get

$$d(Vx)/dt = \mu Vx \tag{4.66}$$

Equation 4.66 then changes to

$$X\,(dV/dt) + V(dx/dt) = \mu Vx \tag{4.67}$$

$$dx/dt = (\mu - D)x \tag{4.68}$$

where D is the dilution rate = F/V (per hour)

In fed-batch culture, V can be defined by Equation 4.69.

$$V = V_0 + Ft \tag{4.69}$$

Substituting Equation 4.69 in Equation 4.68, we get

$$\frac{dx}{dt} = \left(\mu - \frac{F}{V_0 + F_t}\right) X \tag{4.70}$$

Substrate consumption

With regard to the mass balance of growth-limiting substrate in the culture medium, Rate of increase = Rate of input – Rate of substrate consumption

That is,

$$dS/dt = FS_r - \mu X/Y \tag{4.71}$$

$$S = sV \tag{4.72}$$

where S is the total growth-limiting substrate in the culture medium (g), S_r is the concentration of growth-limiting substrate in a fresh medium (g substrate/L), and Y is the growth yield for substrate = $dX/-dS$ (g cell/g substrate).

Substituting Equations 4.65 and 4.72 in Equation 4.71, we get

$$d(sV)/dt = FS_r - \mu xV/Y \tag{4.73}$$

This changes to

$$\frac{ds}{dt} = \frac{F}{V}(S_r - S) - \frac{\mu x}{Y} \tag{4.74}$$

Substituting Equation 4.69 in Equation 4.74, we get

$$\frac{ds}{dt} = \frac{F}{V_0 + F_t}(S_r - S) - \frac{\mu x}{Y} \tag{4.75}$$

If the feed rate of fresh medium is changed step wise at an arbitrary time interval during the cultivation, Equations 4.70 and 4.75 can be rewritten as shown in Equations 4.76 and 4.77, respectively. A schematic diagram of the step-change feeding in a fed-batch culture is shown in Figure 4.11.

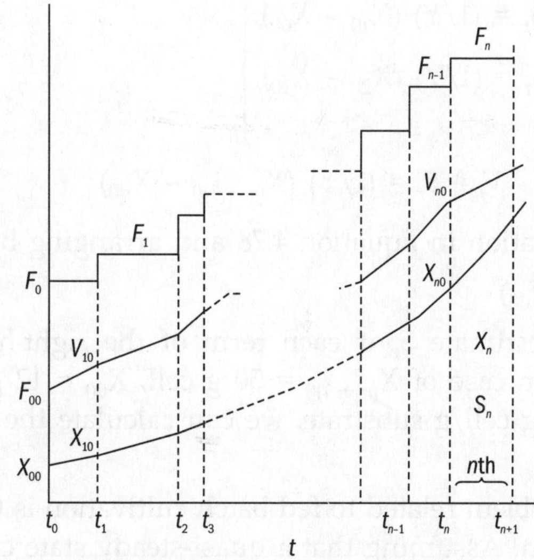

F – flow rate of a fresh medium (L/h); V – culture volume (L); X – biomass in the culture vessel (g cell); t – culture time (h)

Figure 4.11 Schematic diagram of step-change-feeding in fed-batch cultivation

$$\frac{dx_n}{dt_{n+1}} = \left(\mu - \frac{F_n}{V_{n0} + F_n\,(t_{n+1} - t_n)}\right)X_n \tag{4.76}$$

$$\frac{ds_n}{dt_{n+1}} = \left(\frac{F_n}{V_{n0} + F_n(t_{n+1} - t_n)}\right)(S_r - S_n) - \frac{\mu x}{Y}X_n \tag{4.77}$$

where

X_n = cell mass concentration in the culture medium in nth time interval

= X_n/V_n (g cell/L),

S_n = residual substrate concentration in the culture medium in nth time interval

= S_n/V_n (g substrate/L),

F_n = flow rate of a fresh medium in nth time interval (L/h),

V_{n0} = culture volume at $t = t_n = V_{00} + \sum_{n=1}^{n} F_{n-1}\,(t_n - t_{n-1})$ (L),

V_{00} (= V_0) = culture volume at $t = t_0$ (L)

Assuming growth yield Y to be constant and the residual growth-limiting substrate concentration, S or S_n (Equation 4.75 or Equation 4.77), to be almost nil compared to S_r, the following mass balance will be established in fed-batch cultivation (Figure 4.11).

$$\left.\begin{aligned} F_0(t_1 - t_0)S_r &= (V_{10} - V_{00})S_r = (1/Y)(X_{10} - X_{00}) \\ F_1(t_2 - t_1)S_r &= (V_{20} - V_{10})S_r = (1/Y)(X_{20} - X_{10}) \\ &\vdots \\ F_n(t_{n+1} - t_n)S_r &= (V_{(n+1)0} - V_{n0})S_r = (1/Y)(X_{(n+1)0} - X_{n0}) \end{aligned}\right\} \tag{4.78}$$

Considering the summation in Equation 4.78 and arranging by S_r, we get

$$S_r = (X_{n+1} - X_{00})/Y(V_{n+1} - V_{00}) \tag{4.79}$$

It can be possible to estimate S_r if each term of the right-hand side of Equation 4.79 is set. For example, in case of $X_{(n+1)0} = 50$ g cell, $X_{00} = 12$ g cell, $V_{(n+1)0} = 6.5$ L, $V_{00} = 3.5$ L, and $Y = 0.5$ g cell/g substrate, we can calculate the value of S_r ($S_r = 25$ g substrate/L).

Another significant problem related to fed-batch cultivation is to choose an optimum flow rate of fresh medium. Assuming that a quasi-steady state can be established in a fed-batch culture, $S_r \gg S_n$ and $dS_n/dt_{n+1} \approx 0$ as calculated from Equation 4.77. This is due to an immediate consumption of feeding substrate. Therefore, Equation 4.80 can be derived from Equation 4.77 in terms of F_n.

$$F_n = \frac{\mu X_{(n+1)0}}{YS_r} \tag{4.80}$$

In case μ and Y are assumed to be constant values and can be estimated beforehand (Equation 4.79), we can predict F_n as a function of the total biomass in the culture vessel at $t = t_{n+1}$, $X_{(n+1)0}$.

For the estimation of $X_{(n+1)0}$, integration of Equation 4.64 gives

$$\left.\begin{aligned} &2.3 \log X_{(n+1)0} = \mu(t_{n+1} - t_n) + 2.3 \log X_{n0} \\ &\text{or } X_{(n+1)0} = X_{n0} \exp\{\mu(t_{n+1} - t_n)\} \end{aligned}\right\} \tag{4.81}$$

In aerobic fermentation during yeast propagation, the ethanol yield for glucose, $Y_{p/s}$, can be expressed by Equation 4.82.

$$(Y_{p/s})_n = \frac{QP}{Y_n} = \frac{(k_1 Q_{co_2} - k_2 Q_{o_2})}{Y_n}$$

$$\text{or } (Y_{p/s})_n = \frac{k_1(RQ - 1)(Io_2)n}{\dfrac{\mu Xn}{Y}} \tag{4.82}$$

Cancelling the term $\mu X_{(n+1)0}/Y$ from Equations 4.80 and 4.82, we get

$$F_n = 180\,\frac{k_1(RQ - 1)}{S_r(Y_{p/s})_n}(Io_2)_n \tag{4.83}$$

where k_1 and k_2 are stoichiometric coefficients.

$$k_1 = \left(\frac{\text{Mole ethanol}}{\text{Mole } CO_2}\right)$$

$$k_2 = \left(\frac{\text{Mole } CO_2}{\text{Mole } O_2}\right)$$

$$Io_2 = Qo_2 X$$

4.4.6.3.2 Tutorials

Question 1 Compare the rate of metabolism in a man whose surface area is about 260 cm^2/kg of body weight with that of a cylinder-shaped bacteria whose size is 1μ × 2μ (assume metabolism to be proportional to the surface area, specific gravity of bacteria = 1, and μ = 10^{-4} cm).

Answer Volume of bacteria $= \pi r^2 h = \pi\left(\frac{\mu}{2}\right)^2 \times 2\mu$

or volume of bacteria $= \dfrac{\pi \times 10^{-12}}{2}$ cm^3

Thus,

Weight of the bacteria $= \dfrac{\pi \times 10^{-12}}{2}$ g [since specific gravity = 1]

Surface of bacteria $= 2\pi r(r + h) = 2\pi\left(\frac{\mu}{2}\right)\left(\frac{\mu}{2} + 2\mu\right)$

or surface of bacteria $= \dfrac{5\mu \times 10^{-8}}{2}$ cm^2

Thus,

surface per gram of bacteria $= \dfrac{5\pi \times 10^{-8}}{\pi \times 10^{-12}}$ $cm^2 = 5 \times 10^4$ cm^2

Surface per kilogram of bacteria $= 5 \times 10^4 \times 10^3$ $cm^2 = 5 \times 10^7$ cm^2

Thus,

$$\frac{\text{Surface per kilogram of bacteria}}{\text{Surface per kilogram of man}} = \frac{5 \times 107}{260} \approx 2 \times 10^4 \text{ or } 200000$$

Comment It is seen that metabolism in bacteria is about 200000 times faster than in a human body. Due to this enormous rate of metabolism and absorption of food, there is rapid growth in bacteria and bacterial division takes place at the rate of about once in half an hour.

Question 2 An organism has a generation time of 15 min. A single cell (weighing about 10^{-12} g) was transferred to a fresh medium, where it immediately started growing exponentially. What will be the weight of the crop harvesting after 36 h, assuming it has been growing continuously (McCabe, Smith, and Harriott 1985).

Answer We have

$N = N_0 2^n$

where

$$n = \frac{\text{Elapsed time}}{\text{Generation time}} = \frac{36 \times 60}{15} = 144$$

Thus,

$N = N_0\, 2^{144} = 1 \times 2^{144}$ cells

Weight of 1 cell = 10^{-12} g

Therefore,

weight of 2^{144} cells = $2^{144} \times 10^{-12}$ g = 2.911×10^{19} g = 2.911×10^{16} kg

The answer deduced (implying weight) is a huge number, which shows that this cannot be achieved practically by growing a single cell continually for 36 h. This means that after the declining phase, there was a death phase of cells in the cell growth phase and the cells could not multiply to such an amount in weight.

Question 3 A two-stage system with fermenter volumes V_1 = 100 L and V_2 = 150 L is compared with a single-stage system with fermenter volume V_1 = 250 L. The volumetric flow rate, F = 40 L/h, is the same for both the systems. Which system will have higher productivity of cells? Comment on your answer. Calculate the cell concentration in the effluent stream for each stage. The growth curve data for batch process is given in Table 4.11. (Aiba, Humphrey, and Millis 1973)

Comment In the present system, maximum cell productivity is desired. It is evident from the solution that the productivity of the two-stage system ($N_1D_1 + N_2D_2$ = 0.2 + 2.309 = 2.523 UOD/ml-h), which is much higher than the productivity of the single-stage system (ND = 1.136 UOD/ml-h), is desirable.

Answer From the batch data (t versus N), each value of dn/dt is calculated and plotted against N as shown in the graph (Table 4.12 and Figure 4.12).

We have,

$$N_n - N_{n-1} = \frac{[dn_n/dt]}{F/V}$$

Thus, the slopes of N versus dn/dt curve will give the value of F/V. From the graphical analysis,

$$\text{Slope } \frac{F}{V_1} = \left[\frac{2.09}{5.35}\right] = 0.40 \text{ per hour} = D_1$$

From the graph (Figure 4.12), the cell concentration in the first stage of the two-stage system is N_1.

Table 4.11 Growth curve data for batch process

t (h)	*N* (UOD/ml)
1.00	0.12
1.50	0.14
2.00	0.17
2.50	0.22
3.00	0.28
3.50	0.35
4.00	0.45
4.50	0.57
5.00	0.72
5.50	0.91
6.00	1.15
6.50	1.39
7.00	1.76
7.50	2.21
8.00	2.79
8.50	3.46
9.00	4.31
9.50	5.35
10.00	6.50
10.50	7.00
11.00	7.60
11.50	8.14
12.00	8.51
12.50	8.84
13.00	9.13
13.50	9.40

$$N_1 = 5.35 \frac{\text{UOD}}{\text{ml}}$$

The productivity of cell concentration in the first stage is as follows.

$$N_1 \times D_1 = 5.35 \times 0.40 = 0.214 \frac{\text{UOD}}{\text{ml–h}}$$

For the second stage,

F = 40 L/h and V_2 = 150 L

Therefore,

$$D_2 = \frac{F}{V_2} = \frac{40}{150} = 0.267 \text{ per hour}$$

From point N_1 (Figure 4.12), a line with a slope of 0.40 per hour is drawn to touch the curve dN/dt versus N. A vertical line is drawn from the point of intersection to the x-axis, which gives the value of N_2 for the second stage. From the graph, we have

Table 4.12 Calculated values of dn/dt

t (hours)	*N* (UOD/ml)	*dn/dt* (UOD/ml)
1.00	0.12	0.04
1.50	0.14	0.05
2.00	0.17	0.06
2.50	0.22	0.09
3.00	0.28	0.13
3.50	0.35	0.16
4.00	0.45	0.21
4.50	0.57	0.27
5.00	0.72	0.35
5.50	0.91	0.44
6.00	1.15	0.55
6.50	1.39	0.67
7.00	1.76	0.85
7.50	2.21	1.06
8.00	2.79	1.33
8.50	3.46	1.63
9.00	5.35	2.09
10.00	6.30	1.70
10.50	7.00	1.30
11.00	7.60	1.06
11.50	8.14	0.89
12.00	8.51	0.74
12.50	8.84	0.62
13.00	9.13	0.54
13.50	9.40	0.44

$$N_2 = 8.65\,\frac{\text{UOD}}{\text{ml}}$$

Thus, productivity in the second stage is N_2D_2.

$$N_2D_2 = 8.65 \times 0.267 = 2.309\,\frac{\text{UOD}}{\text{ml–h}}$$

For the single-stage system,

$F = 40$ L/h and $V = 250$ L.

Thus,

$$D = \frac{F}{V} = \frac{40}{250} = 0.16 \text{ per hour}$$

From the graph, the intersection of the curve is at point p and the vertical from this point on the x-axis gives the value of effluent cell concentration, N.

$$N = 7.1\,\frac{\text{UOD}}{\text{ml}}$$

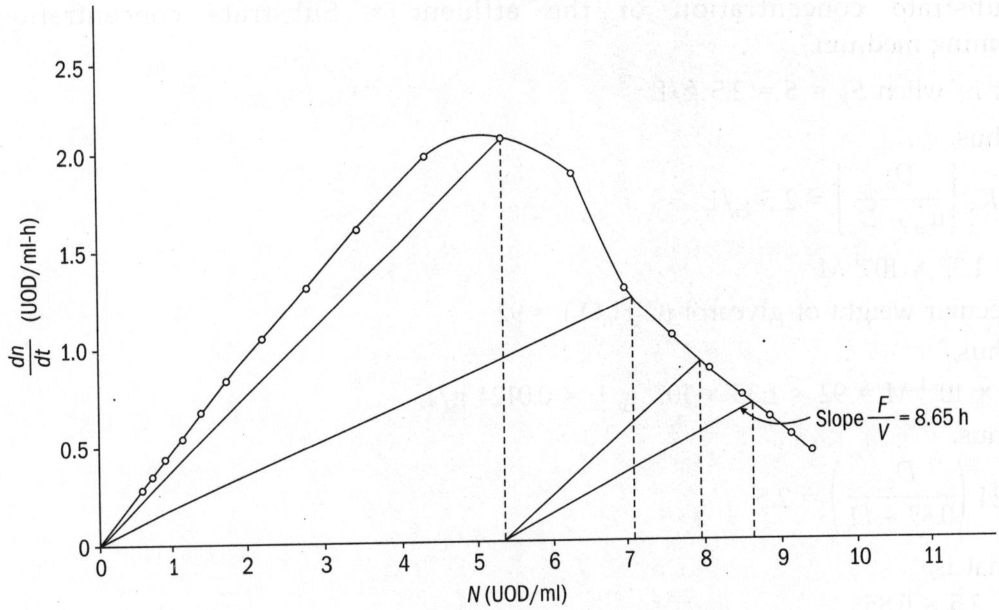

Figure 4.12 Plot of *N* versus *dn*/*dt* from the calculated values

Thus, productivity of the single-stage system is *ND*.

$$ND = 7.1 \times 0.16 = 1.136 \frac{\text{UOD}}{\text{ml–h}}$$

Question 4 *Aerobacter aerogenes* was grown in a continuous culture in a single pilot plant having a culture vessel of 20 L capacity. The carbon source in an ammonium salt medium was glycerol at a concentration of 2.5 g/L. What bacterial densities would be obtained by adjusting the flow rate of the incoming medium to values of 5, 10, and 20 L/h and what would be the corresponding outputs of the system?

Data In a number of batch culture experiments, it was ascertained that 1 L of the medium yielded 1.325 g dry weight of bacterium and the maximum specific growth rate was 0.85 per hour. $K_S = 1.35 \times 10^{-4}$ *M*.

Answer Dilution rates corresponding to flow rates 5, 10, and 20 L/h are as follows.

$$D = \frac{F}{V} \; [V = 20 \text{ L}]$$

where *D* is the dilution rate.

1. $D = 5/20 = 0.25$ per hour
2. $D = 10/20 = 0.05$ per hour
3. $D = 20/20 = 1.0$ per hour

Critical dilution rates (at which wash out occurs) will be reached when the following condition is satisfied.

Substrate concentration of the effluent = Substrate concentration of incoming medium.

That is, when $S_R = \tilde{S} = 2.5$ g/L

Thus,

$$\tilde{S} = K_S\left[\frac{D_c}{\mu_m - D_c}\right] = 2.5 \text{ g/L}$$

$K_s = 1.35 \times 10^{-4}$ M

Molecular weight of glycerol ($C_3H_6O_3$) = 92

Thus,

1.35×10^{-4} M = $92 \times 1.35 \times 10^{-4}$ g/L = 0.0124 g/L

Thus,

$$0.0124\left(\frac{D_c}{0.85 - D_c}\right) = 2.5$$

That is,

$$D_c = \frac{2.5 \times 0.85}{2.512} = 0.846 \text{ per hour}$$

Thus, washout will occur with dilution rates ≥0.846 per hour (as with the highest flow rates; 20 L/h ≈ D of 1.0 per hour). For other two flow rates, the steady-state bacterial concentration $\tilde{X}$ is obtained from Equation 4.84.

$$\tilde{X} = Y(S_R - S) = Y\left[S_R - K_S\left(\frac{D}{\mu_m - D}\right)\right] \tag{4.84}$$

where Y is the gram dry weight of bacteria produced per gram of glycerol consumed.

$$Y = \frac{1.325}{2.5} = 0.53 \text{ (since } S_r = 2.5 \text{ g/L)}$$

The steady-state bacterial concentration for D = 0.2 is

$$\tilde{X} = 0.53\left[2.5 - 0.0124\left(\frac{0.25}{0.85 - 0.25}\right)\right]$$

or $\tilde{X}$ = 0.53 [2.5 – 0.005] = 1.323 g/L

For D = 0.50 per hour,

$$\tilde{X} = 0.53\left[2.5 - \left(\frac{0.0124 \times 0.50}{0.35}\right)\right]$$

or $\tilde{X}$ = 0.53 [2.5 – 0.018] = 1.316 g/L

The outputs of bacteria at each flow rate will be $D.\tilde{X}$ g dry wt/L-h.

Thus, for D = 0.25 per hour,

output = 0.25 × 1.323 = 0.331 g/L-h

For D = 0.50 per hour,

output = 0.50 × 1.316 = 0.658 g/L-h

Since volume of the culture vessel is 20 L, the total outputs of the apparatus per hour are 6.62 g and 13.16 g dry weight of cells.

4.5 FLUID STATICS AND APPLICATIONS

Fluid statics is concerned with the static properties and behaviour of fluids. In case of liquids, it is known as hydrostatics and in case of gases, it is known as pneumatics.

4.5.1 Fluid

Fluid does not permanently resist distortion and easily takes the shape of the container. Shear stresses exist when a fluid attempts to change in shape, but when a final shape is attained, the shear stresses disappear. A fluid in equilibrium is free from all shear stresses.

4.5.1.1 Incompressible and compressible fluids

At a given temperature and pressure, a fluid possesses a definite density, which is usually measured in kilograms per cubic metre (kg/m^3) or grams per cubic centimetre (g/cm^3) or pounds per cubic foot (lb/ft^3). Although the density of a fluid depends on temperature and pressure, variation in density with changes in these variables may be large or small. If the density is not affected appreciably by moderate changes in temperature and pressure, the fluid is said to be incompressible. However, if the density change is sensitive to changes in these variables, the fluid is said to be compressible. Normally, liquids are considered incompressible and gases compressible.

4.5.2 Pressure

The basic property of a static fluid is pressure. Pressure is the surface force exerted by a fluid against the walls of its container. Pressure also exists at every point within the volume of the fluid. A static fluid pressure is independent of the orientation of any internal surface on which the pressure is assumed to act. It is one way of stating Pascal's law.

Pressure is a scalar quantity and changes into a vector quantity, called force, when multiplied by the surface area (with the direction perpendicular and towards the fluid surface).

4.5.2.1 Gas pressure

Gas pressure is the difference between atmospheric pressure and a given fluid pressure. The readings of pressure gauges are commonly positive. A vacuum gauge may be used to show negative gauge pressures, that is, fluid pressures less than that of the atmosphere.

4.5.2.2 *Absolute pressure*

Absolute pressure is the true total pressure. It is equal to gauge pressure (taken with proper sign) plus atmospheric pressure.

4.5.3 Hydrostatic Equilibrium

In a stationary mass of a single static fluid, the pressure is constant in any cross section parallel to the earth's surface but varies with height. Suppose the pressures at heights h_1 and h_2 above the earth surface are P_1 and P_2 (Figure 4.13), respectively, then

$$P_1 + h_1\rho g/g_c = P_2 + h_2\rho\, g/g_c \tag{4.85}$$

where ρ is the density of fluid (assumed constant) and g is the gravitational force.

Equation 4.85 expresses mathematically the condition of hydrostatic equilibrium.

4.5.3.1 *Hydrostatic equilibrium in a centrifuged field*

In a rotating centrifuge, a layer of liquid is thrown outwards from the axis of rotation and is held against the wall of the bowl by centrifugal force. The free surface of the

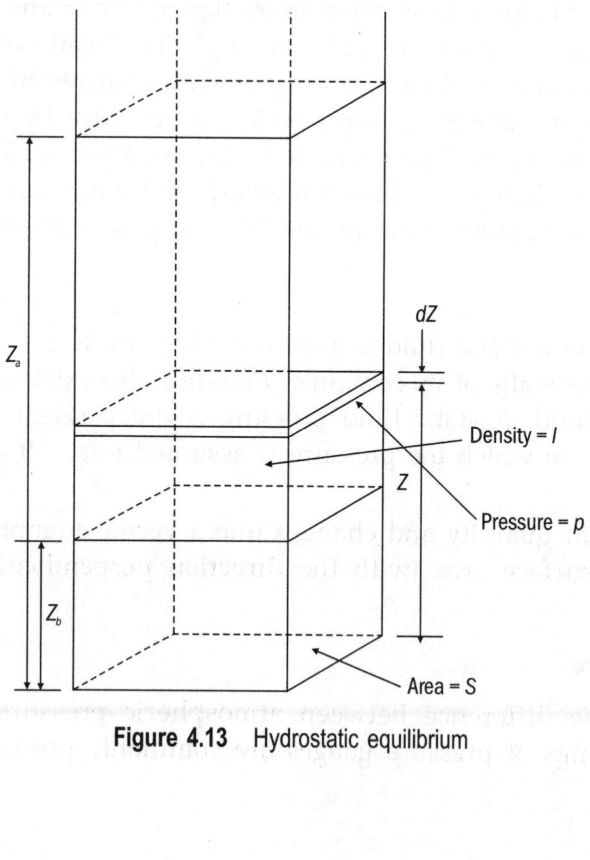

Figure 4.13 Hydrostatic equilibrium

liquid takes the shape of the paraboloid of revolution. In industrial centrifuges, the rotational speed is so high and the centrifugal force so much greater than the force of gravity that the liquid surface is virtually cylindrical and coaxial with the axis of rotation. This is illustrated in Figure 4.14, where r_1 is the radial distance from the axis of rotation to the free liquid surface and r_2 is the radius of the centrifuge bowl.

At equilibrium, the entire mass of liquids indicated in Figure 4.14 is assumed to be rotating like a rigid body, with the liquid layers not sliding over each other. Under these conditions, the pressure distribution in the liquid may be found from the principles of fluid statics and is given by Equation 4.86.

$$P_2 - P_1 = \frac{2w^2\rho\,(r_2^2 - r_1^2)}{2\,g_c} \tag{4.86}$$

where P_1 is the pressure at r_1, P_2 is the pressure at r_2, ρ is the density of the fluid (assumed constant), and ω is the angular speed in radians/second.

4.5.4 Applications of Fluid Statics

4.5.4.1 Manometer

A manometer is an important device for measuring pressure differences. Figure 4.15 shows the simplest form of a manometer. The shaded portion of the U tube is filled with liquid A (say) having density ρ_A, and the arms of the U tube above the liquid are filled with fluid *B* having density ρ_B. Fluid *B* is immiscible with liquid *A* and less dense; it is often a gas such as air or nitrogen.

A pressure P_a is exerted in one arm of the U tube and pressure P_b in the other. As a result of the difference in pressure (P_a - P_b), the meniscus in one branch of the U tube is higher than the other. The vertical distance between the two menisci, R_m, may

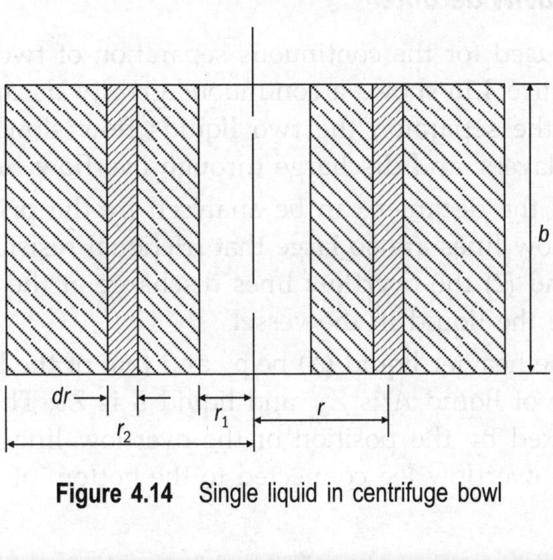

Figure 4.14 Single liquid in centrifuge bowl

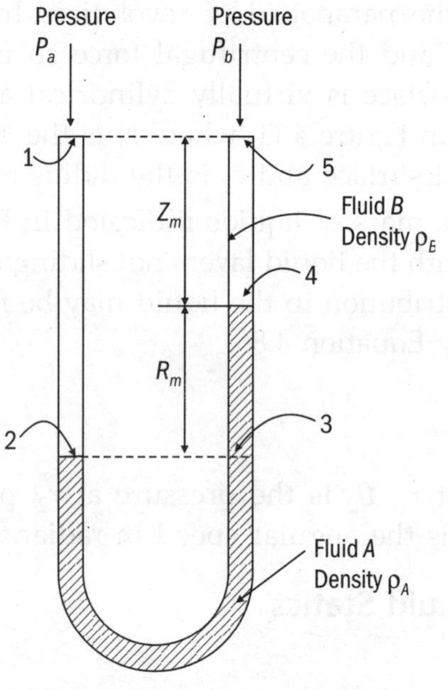

Figure 4.15 Simple manometer

be used to measure the difference in pressure. The relationship between $P_a - P_b$ and R_m is derived from the fluid statics as given by Equation 4.87.

$$P_a - P_b = (g/g_c)\, R_m(\rho_A - \rho_B) \tag{4.87}$$

If P_b is the atmospheric pressure, then $P_a - P_b$ is the gauge pressure.

For measuring small differences in pressure, an inclined manometer may be used.

4.5.4.2 *Continuous gravity decanter*

A gravity decanter is used for the continuous separation of two immiscible liquids of differing densities. Figure 4.16 shows a continuous gravity decanter. The feed mixture enters at one end of the separator, the two liquids flow slowly through the vessel, separate into the two layers, and discharge through overflow lines at the other end.

The performance of the decanter can be analysed by the principles of fluid statics provided (1) the overflow lines are so large that frictional resistance to the flow of the liquids is negligible and (2) the overflow lines discharge at the same pressure as that in the gas space above the liquid in the vessel.

Let the density of the heavier liquid (*A*) be ρ_A and that of the lighter liquid (*B*) be ρ_B. The depth of the layer of liquid *A* is Z_{A_1} and liquid *B* is Z_B. The total depth of liquid in the vessel, Z_T, is fixed by the position of the overflow line for liquid *B*. Liquid *A* discharges through an overflow leg connected to the bottom of the vessel and rises to

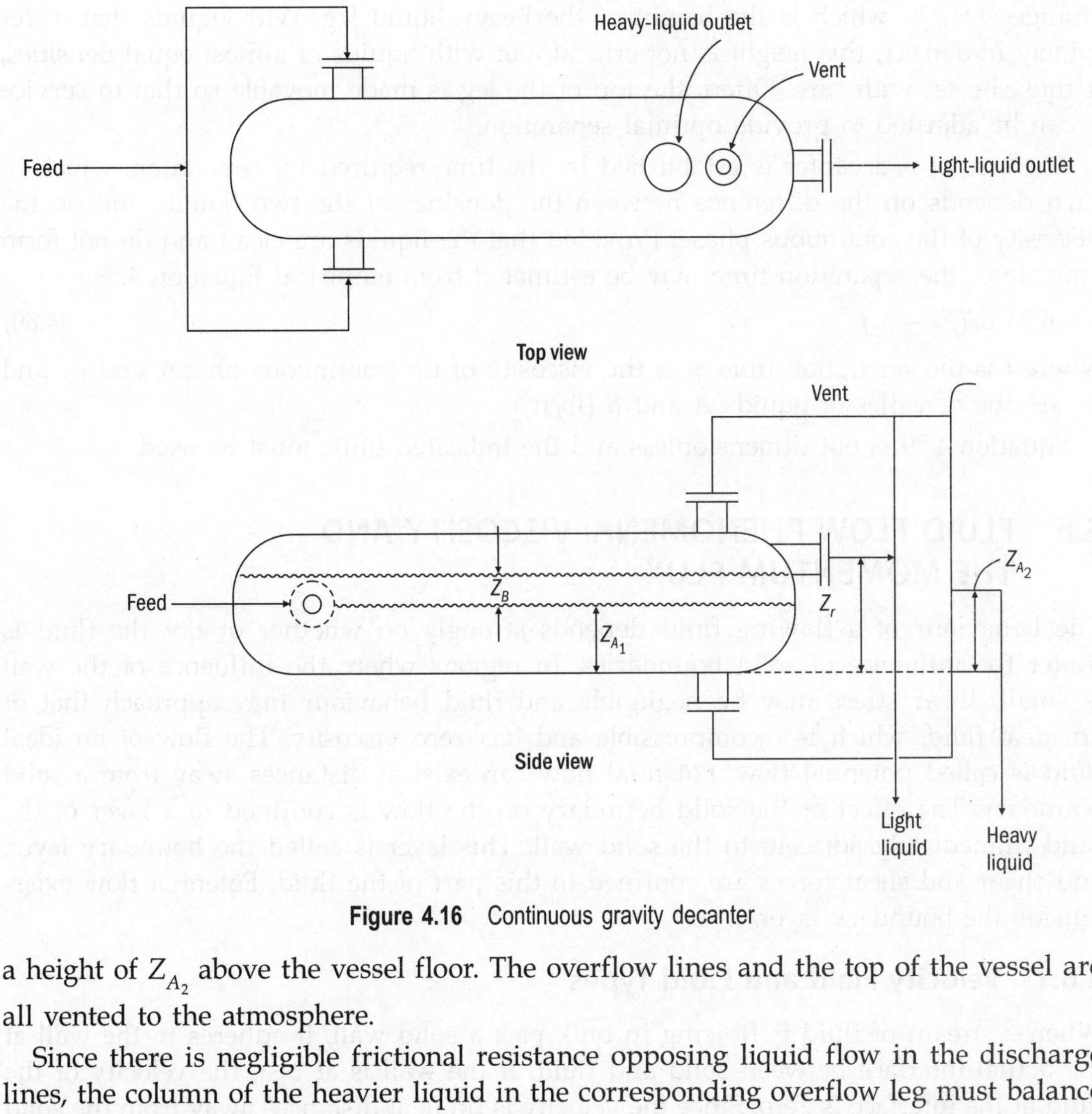

Figure 4.16 Continuous gravity decanter

a height of Z_{A_2} above the vessel floor. The overflow lines and the top of the vessel are all vented to the atmosphere.

Since there is negligible frictional resistance opposing liquid flow in the discharge lines, the column of the heavier liquid in the corresponding overflow leg must balance the somewhat greater depth of the two liquids in the vessel. A hydrostatic balance leads to Equation 4.88.

$$A_{A_1} = \frac{Z_{A_2} - Z_T(\rho_B/\rho_A)}{1 - \rho_B/\rho_A} \tag{4.88}$$

Equation 4.88 shows that the position of the liquid–liquid interface in the separator depends on the ratio of the densities of the two liquids and on the elevations of the overflow lines. It is independent of the rates of flow of the liquids. Equation 4.88 also shows that as ρ_A approaches ρ_B, the position of the interface becomes sensitive to the changes in Z_{A_2}, which is the height of the heavy liquid leg. With liquids that differ widely in density, this height is not critical, but with liquids of almost equal densities, it must be set with care. Often, the top of the leg is made movable so that in service it can be adjusted to provide optimal separation.

The size of a decanter is established by the time required for separation, which in turn depends on the difference between the densities of the two liquids and on the viscosity of the continuous phase. Provided that the liquids are clean and do not form emulsions, the separation time may be estimated from empirical Equation 4.89.

$$t = 6.24\ \mu/(\rho_A - \rho_B) \tag{4.89}$$

where t is the separation time, μ is the viscosity of Cp (continuous phase), and ρ_A and ρ_B are the densities of liquids A and B (lb/ft^3).

Equation 4.89 is not dimensionless and the indicated units must be used.

4.6 FLUID FLOW PHENOMENA: VISCOSITY AND THE MOMENTUM FLUX

The behaviour of a flowing fluid depends strongly on whether or not the fluid is under the influence of solid boundaries. In regions where the influence of the wall is small, shear stress may be negligible and fluid behaviour may approach that of an ideal fluid, which is incompressible and has zero viscosity. The flow of an ideal fluid is called potential flow. Potential flow can exist at distances away from a solid boundary. The effect of the solid boundary on the flow is confined to a layer of the fluid immediately adjacent to the solid wall. This layer is called the boundary layer, and shear and shear forces are confined to this part of the fluid. Potential flow exists outside the boundary layer.

4.6.1 Velocity Field and Fluid Types

When a stream of fluid is flowing in bulk past a solid wall, it adheres to the wall at the actual interface between solid and fluid. If the wall is at rest, the velocity of the fluid at the interface is zero. Since the velocity is finite at distances away from the solid wall, there must be variations in velocity at different points in the flowing stream. In such situations, the velocity field exists in the space occupied by the fluid. The velocity in the space occupied by the fluid and at a given location may also vary with time. When the velocity at each location is constant, the field is invariant with time and the flow is said to be steady; otherwise, it is called unsteady flow.

4.6.1.1 Laminar flow

At low velocities, fluids tend to flow without lateral mixing and adjacent layers slide past one another like playing cards. There are neither cross-current nor eddies. This regime is called laminar flow.

4.6.1.2 *One-dimensional flow*

Velocity is a vector and in general, the velocity at a point has three components—one for each space coordinate. In most simple situations, like laminar flow, all velocity vectors in the field are parallel and only one velocity component is required. This situation, which obviously is much simpler than the general vector field, is called one-dimensional flow (for example, fluid flow through straight pipes).

The following discussion is based on the assumptions of steady-state one-dimensional flow.

4.6.1.2.1 Velocity gradient and rate of shear

Consider the steady-state, one-dimensional laminar flow of an incompressible fluid along a solid plane surface. Figure 4.17 shows the velocity profile for such a stream. The abscissa U is the velocity, and ordinate Y is the distance measured perpendicularly from the wall and, therefore, at right angles to the direction of velocity.

At $Y = 0$, $U = 0$ and U increases with distance from the wall but at a decreasing rate. Let us consider two nearby planes A and B (ΔY apart), where the velocities along the planes are U_A and U_B, respectively. Assuming that $U_B > U_A$, $\Delta U = U_B - U_A$. The velocity gradient at Y_A, dU/dY, is defined as given by Equation 4.90.

$$\frac{dU}{dY} = \lim_{Y \to 0} \left(\frac{\Delta U}{\Delta Y}\right) \tag{4.90}$$

The local velocity gradient is also called the shear rate or time rate of shear.

4.6.2 Shear Stress Field

Since an actual fluid resists shear, a shear force must exist wherever there is a time rate of shear. In one-dimensional flow, the shear force acts parallel to the plane of the shear. For example, at plane C and at a distance Y_C from the wall, the shear force F_S acts in the direction parallel to the plane as shown in Figure 4.17(a). This force is exerted by the fluid outside of plane C on the fluid between plane C and the wall. By Newton's

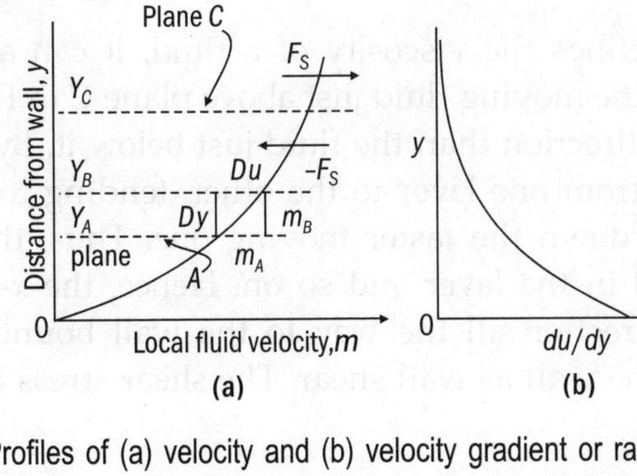

Figure 4.17 Profiles of (a) velocity and (b) velocity gradient or rate of shear in layer flow

third law, an equal and opposite force, $-F_S$, acts on the fluid outside of plane C, from the fluid inside plane C. It is convenient not to use the total force F_S. The force per unit area of shear plane is called shear stress and denoted by τ (Equation 4.91).

$$\tau = F_S/A_S \tag{4.91}$$

where A_S is the area of the plane.

Shear forces are generated in both laminar and turbulent flow.

4.6.3 Newton's Law of Viscosity

A fluid is said to follow Newton's law of viscosity when the corresponding shear stress is proportional to the velocity gradient or shear rate. Such fluids are called Newtonian fluids. Newton's law of viscosity is mathematically expressed by Equation 4.92.

$$\tau = -\mu\left(\frac{dU}{dY}\right) \tag{4.92}$$

where μ is the constant of proportionality and is called the viscosity of fluid. The negative sign on the right-hand side ensures that μ remains positive.

All gases and most simple liquids are mathematically described by Equation 4.92. Fluids that do not obey this simple law are called non-Newtonian fluids.

In SI units, τ is measured in newtons per square metre (N/m^2) and in kilograms per metre second (kg/m-s) or pascal-second (P-s). In the CGS system, viscosity is expressed in grams per centimetre-second (g/cm-s); this unit is called poise (P). Viscosity is generally reported in centipoise, or cP (1 cP = 0.01 poise), since most fluids have viscosities much less than 1 P.

In some situations, it is useful to have a symbol to represent the viscosity divided by the mass density (mass per unit volume) of the fluid. The new symbol so defined is called kinematic viscosity and is represented by *ν*. In SI units, *ν* is measured in square metres per second. In the CGS system, kinematic viscosity is called stoke (*St*), which is equal to 1 cm^2/s.

4.6.4 Viscosity

Although Equation 4.92 defines the viscosity of a fluid, it can also be interpreted in terms of momentum flux. The moving fluid just above plane C in Figure 4.17 has slightly more momentum in the *x*-direction than the fluid just below it. By molecular collisions, momentum is transferred from one layer to the other, tending to speed up the slower moving layer and to slow down the faster moving one. Thus, the momentum passes across plane C to the fluid in the layer and so on. Hence, the *x*-direction momentum is transferred in the $-y$ direction all the way to the wall bounding the fluid, where $U = 0$ and is delivered to the wall as wall shear. The shear stress at the wall is denoted by τ_w.

The unit of momentum flux is kilograms per metre square second (kg/m-s^2). It is equivalent to the unit for shear stress (τ), since N/m^2 equals 1 kg/m-s^2. Thus, Equation 4.92 states that the momentum flux normal to the direction of flow of the fluid is proportional to the velocity gradient, with the viscosity as the proportionality factor. Momentum transfer is analogous to conductive heat transfer resulting from a temperature gradient, where the proportionality factor between the heat flux and temperature gradient is called thermal conductivity. In laminar flow, momentum is transferred as a result of the velocity gradient and viscosity may be regarded as the conductivity of momentum transferred by this mechanism.

4.6.5 Viscosity of Gases and Liquids

The viscosity of a Newtonian fluid depends primarily on temperature and, to a lesser degree, pressure. The viscosity of a gas increases with temperature in accordance with Equation 4.93.

$$\mu/\mu_0 = (T/273)^n \tag{4.93}$$

where μ is the viscosity at absolute temperature T K, μ_0 is the viscosity at 0°C or 273 K, and n is a constant.

The constant n is a complicated parameter and ranges in magnitude from about 0.65 to 1.0. The viscosity of a gas is almost independent of pressure in regions where the gas laws apply. In such regions, the viscosities of gases are generally between 0.01 cP and 0.1 cP. At higher pressures, gas viscosity increases with pressure, especially in the neighbourhood of the critical point.

The viscosities of liquids are generally much greater than those of gases and cover several orders of magnitude. Liquid viscosities decrease significantly with increase in temperature. The viscosity of a liquid increases with pressure, but the effect is generally insignificant at pressures less than 40 atm.

4.6.6 Non-Newtonian Fluids

When the sample relation between shear rate and shear stress, as given by Equation 4.92, is not followed, the fluid is said to be a non-Newtonian fluid. For such fluids, various relationships between shear rate and shear stress other than that given by Equation 4.92 are followed.

Figure 4.18 shows the various types of non-Newtonian fluids in which curve A (a straight line passing through the origin) depicts Newtonian fluid behaviour. Curve B is an example of Bingham plastics fluid, where threshold shear stress (τ_0) is required before the fluid flows, after which the velocity gradient is linearly related to the shear stress. Curve C represents a pseudoplastic fluid; the curve passes through the origin, is concave downwards at low shears, and becomes linear at high shears. Rubber

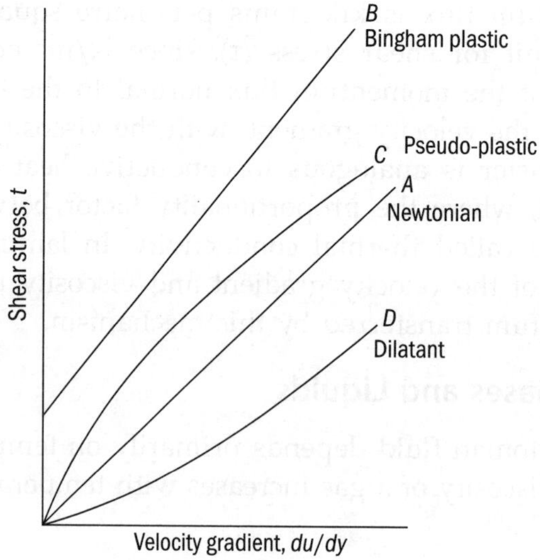

Figure 4.18 Shear stress versus velocity gradient for Newtonian and non-Newtonian fluids

latex is an example of such a fluid. Curve *D* represents a dilatant fluid; the curve is concave upwards at low shears and becomes linear at high shears. Quicksand and some sand-filled emulsion show this behaviour. Pseudoplastics are said to be shear-rate-thinning and dilatant fluids shear-rate-thickening.

None of the curves given in Figure 4.18 depend on the history of the fluid, and a given sample of material shows the same behaviour regardless of the amount of time the shearing stress has been applied. However, such is not the case for some non-Newtonian liquids whose shear stress versus shear rate curve depends on how long the shear has been active. Thixotropix liquids break down under continued shear and on mixing give lower shear stresses for a given shear rate. Rheopectic substances behave in the reverse manner, and the shear stress at constant rate increases with time.

4.6.7 Tutorials

Question 1 A liquid with density 40.0 lb/h is flowing through a pipe of diameter 2 in. A 1 in orifice is located in the line of the orifice and is indicated by an Hg monometer in which fluid from the pipe and the Hg are in contact. The reading from the manometer is 4 in Hg. Assuming the orifice weight to be 0.61, calculate the flow rate per hour.

Answer From the given data, we have

$$\sqrt{u_B^2 - u_A^2} = C_0\sqrt{2g\Delta H}$$

Now,

$$\frac{D_1}{D_2} = \frac{2}{1}$$

$u_A D_1^2 = u_B D_2^2$

or $u_B 4 = u_A 4^2$

Thus,

$u_B = 4u_A$

$\Delta P = \frac{4}{12}[13.6 \times 62.4 - 40]$

Thus,

$\Delta H = \frac{\Delta\rho}{\rho} = \frac{13.6 \times 62.4 - 40}{3 \times 40}$

or $\Delta H = \frac{1}{3}(21.21 - 1) = \frac{1}{3} \times 20.21 = 6.74$

Thus,

$\sqrt{16\,u_A^2 - u_A^2} = C\sqrt{2g\Delta H}$

or $\sqrt{16\,u_A^2 - u_A^2} = 0.61\,\sqrt{64.4 \times 6.74}$

Thus,

$15\,u_A^2 = 0.61^2 \times 64.4 \times 6.74$

Thus,

$u_A = \sqrt{\frac{(0.61)^2 \times 64.4 \times 6.74}{15}} = 3.281 \text{ ft/s}$

Thus, flow in pound per hour = velocity $\frac{\text{ft}}{\text{hr}}$ × cross-sectional area × ft^2 × density (lb/ft^3)

or flow in pound per hour = $3.281 \times 60 \times 60 \times \frac{3.14 \times 2 \times 2}{4 \times 12 \times 12} \times 40 = 10\,300$ lb/h

Question 2 A pump takes water at 50°F from a large reservoir and delivers it to the bottom of an open elevated tank. The level in the tank averages 160 ft above the surface of the reservoir. The line is 3 in of pipe and consists of 550 ft of pipe. There are six elbows, two T joints, and two gate valves. The pump delivers 150 gpm. What is the HP consumed if the pump has an efficiency of 55%?

Answer According to Bernoulli's theorem,

$$X_A + \frac{u_A^2}{2g} + \frac{P_A}{\rho_A} + W - \Sigma F = X_B + \frac{u_B^2}{2g} + \frac{P_B}{\rho_B} \tag{4.94}$$

Since A is on the ground level, $X_A = 0$

As the reservoir is large,

$u_A = 0$, $P_A = P_B$, and $\rho_A = \rho_B$

$X_B = 160$ ft

Thus, Equation 4.94 becomes

$$W - \Sigma F = 160 + \frac{u_B^2}{2g} \text{ in/s}$$

Cross-sectional area of the pipe $= \frac{(3.05)^2\pi}{4} = 7.39 \text{ in}^2 = 0.0513 \text{ ft}^2$

Thus,

$$u_B = 150 \text{ gpm} = \frac{150 \times 231}{60} \times \frac{1}{7.39} \times \frac{1}{12} \text{ ft/s}$$

where the cross-sectional area of the pipe is 7.39 in^2

$$N_{Re} = \frac{\Delta u \rho}{\mu}$$

where μ of water at 50°F is 0.00088

Thus,

$$N_{Re} = \frac{0.2556 \times 6.5 \times 62.4}{0.00088} = 118\ 000$$

Friction loss in 3 in pipe

$$\frac{\Delta H_f D}{L u^2} = 0.00033$$

Now,

$$L = [500 + 6 \times 5.72 + 2 \times 2.81 + 2 \times 2.13] = 544.2 \text{ ft}$$

Thus,

$$\Delta H_f = \frac{0.00033 \times 544.2 \times (6.5)^2}{0.2556}$$

or $\Delta H_f = 30.67$ ft of water

Contraction losses

$$\Delta H_f = X\frac{u^2}{2g} = \frac{0.5 \times (6.5)^2}{2 \times 32.2} = 0.33 \text{ ft of water}$$

Thus,

$$\Sigma F = \Delta H_f + \Delta H_c$$

or $\Sigma H_f = 30.67 + 0.33 = 31$ ft of water

Thus,

$$W - \Sigma F = 160 + \frac{u_B^2}{2g}$$

or $W - \Sigma F = 160 + \frac{(6.5)^2}{64.4}$

Thus,

$W = 160 + 0.658 + 31 = 191.658$ ft of water $= 191.658 \times 62.42$ lb/ft^2 of water

$$\frac{\text{lb}}{\text{ft}^2} \times \frac{\text{ft}^3}{\text{s}} = \frac{\text{ft} \times \text{lb}}{\text{s}}\frac{\text{ft}^3}{\text{s}} = \text{ft}^2 \times 6.5\frac{\text{ft}}{\text{s}} = 0.513 \times 6.\ 5 \times 0.333$$

Thus, work done by the pump = 191.658 × 6.5 × 0.333

or work done by the pump is

$$\frac{191.655 \times 62.42 \times 0.333}{550} = 7.25 \text{ HP}$$

Thus, actual HP = $7.25 \times \frac{100}{55} = 13.8$

$$\frac{\text{ft}^3}{\text{s}} = \text{ft}^2 \times \frac{\text{ft}}{\text{s}} = 0.0513 \times 6.5 = 0.3335 \frac{\text{in–ft}}{\text{s}}$$

Work done = $190.62 \times 62.42 \times 0.3335 \frac{\text{ft–lb}}{\text{s}}$

Thus,

$$\text{HP consumed by the pump} = \frac{190.62 \times 62.42 \times 0.3335}{550} \times \frac{100}{55} = 12.7 \text{ HP}$$

Question 3 Water at 40°F flows out of an elevated tank through 1000 ft of a standard 4 in pipe. The level of the tank is 20 ft above the discharge end of the pipe. What is the rate of flow in gpm?

Answer We have,

$u_B = 4$ ft/s

Thus,

$$\frac{u_B^2}{y} = \frac{\text{lb}}{64.4} = 0.25 \text{ ft}$$

$$\Delta u_C = \frac{0.5 \times 4^2}{64.4} = 0.12 \text{ ft}$$

$$\frac{Du\rho}{\mu} = \frac{0.3355 \times 4 \times 62.43}{0.000104} = 80\,560$$

Thus,

$$\frac{\Delta H_f D}{u^2 L} = 80\,560$$

$$\text{or } \Delta H_f = \frac{0.00035 \times 42 \times 1000}{0.3355} = 16.69 \text{ ft}$$

$F = 16.69 + 12 = 16.80$ ft

Thus,

$X_A = 16.18 + 25 = 17.05$

Since X_A should be 20 ft, u must be corrected. Say, $u = 4.5$ ft/s

Thus,

$$\frac{u_B^2}{2g} = \frac{4.5^2}{64.4} = 0.35$$

Thus,

$$\Delta H_c = \frac{0.5 \times 4.5^2}{64.4} = 0.16 \text{ in}$$

To calculate ΔH_f

$$\frac{Du\rho}{\mu} = \frac{0.3355 \times 4.5 \times 62.43}{0.000104} = 93\,000 \tag{4.95}$$

$$\frac{\Delta H_f D}{u^2 L} = 0.00034$$

or $\Delta H_f = 20.6$ ft (4.96)

$X_A = 20.6 + 0.16 + 0.32 = 21.08$ ft

Thus, velocity must be 4.4 ft/s

$$\frac{Du\rho}{\mu} = 91\,500$$

$$\frac{\Delta H_f D}{u^2 L} = 0.00034$$

$\Delta H_f = 19.6$

$X_A = 19.6 + 0.4 = 20$ ft

Thus,

$$\text{velocity} = \frac{4.4 \times 60 \times 12 \times 12.73}{231} \text{ ft/s}$$

Question 4 Salt brine at 60°F (specific gravity is 1.21 μ = 1.7 cP) is pumped at the rate of 160 gal/min through a 3 in line whose equivalent is 350 ft. There is a vertical leg 80 ft between the level of the feed and discharge level. Steam at 100 lb gauge pressure is available. The following is a list of simple pumps with 3 in discharge.

5 × 6 × 8	6 × 7 × 8	7 × 8 × 8	1st figure = diameter of standard cylinder
5 × 7 × 8	6 × 8 × 8	8 × 6½ × 8	2nd figure = diameter of standard cylinder
5 × 8 × 8	7 × 6 × 8	8 × 7 × 8	
6 × 6 × 8	7 × 7 × 8	8 × 8 × 8	

If water end efficiency of 75% and steam end efficiency of 60% is assumed, which pump should be specified?

Answer

Velocity through 3 in line

The inside section through 3 in line = 7.393 in^2

Thus,

$$\text{flow rate} = \frac{160 \times 231}{60} \times \frac{1}{7.393} \times \frac{1}{12} = 6.913 \text{ ft/s}$$

Friction in 3 in pipe

$D = 3.068 \text{ in} = \dfrac{3.068}{12} \text{ ft}$

$u = 6.913$ ft/s

$\rho = 1.21 \times 62.37$ lb/ft

$\mu = 1.7 \text{ cP} = 1.70 \times 0.000672\ \Sigma u$

$L = 350$ ft

Thus,

$$N_{Re} = \frac{D\mu\rho}{\mu} = 116\,600$$

$$\frac{\Delta H_f D}{u^2 L} = 0.00031$$

$$\Delta H_f = \frac{0.00031 \times 350 \times 12 \times 6.913^2}{3.068} = 19.99 \text{ ft} \approx 20 \text{ ft}$$

Now,

$$X_A + \frac{u_A^2}{2g} + P_A N_A + w - F = X_B^2 + \frac{u_B^2}{2g} + P_B V_B$$

Thus,

$$w - F = \frac{u_B^2}{2g} + X_B$$

$$\text{or } w - 20 = 80 + \frac{6.913^2}{64.4}$$

or $w = 80 + 20 + 0.8 = 100.8$ ft of solution of specific gravity 1.2

Question 5 Write down the Bernoulli's equation for incompressible flow and state what each term represents. How is the loss of head due to friction taken into account while using this equation?

Answer Bernoulli's equation for incompressible fluid flow is given by Equation 4.97.

$$X_A + \frac{u_A^2}{2\alpha g_c} + \frac{P_A}{\rho_A} + w - \Sigma F = X_B + \frac{u_B^2}{2\alpha g_c} + \frac{P_B}{\rho_B} \tag{4.97}$$

$$\text{or } X_A + \frac{u_A^2}{2\alpha g_c} + P_A V_A + w - F = X_B + \frac{u_B^2}{2\alpha g_c} + P_B V_B$$

where the various terms are often called heads.

- X_A and X_B terms represent the elevation of points A and B from an arbitrary datum plane. These terms are called potential heads and are measured in feet.
- The $u^2/2\alpha g_c$ term is called velocity head. If X is measured in feet, then these are also measured in feet. In this term, α is called the kinetic energy correction factor. Its value

varies with N_{Re} and ranges from 0.88 to 0.97. It can be taken as 1 without appreciable error unless the kinetic energy term is important in comparison with other terms in the Bernoulli's equation. g_c is the Newton's law conversion factor, which is equal to 32.17 ft/s^2.

- The P/ρ term represents the pressure needs. It is expressed in lb/ft^2 × ft^3/lb (that is, in feet). This term represents the mechanical work done by forces external to the stream tube on the fluid in pushing it into the tube or the work recovered from the fluid when it leaves the tube.
- The term ΣF is called the friction head and represents the total friction that occurs in the fluid between the two stations, A and B.

4.7 WEIRS

Weirs are used to measure the flow of liquids in open channels and not in closed pipelines. The weir consists of a partition or obstruction with a notch cut in it to carry the liquid. The amount of flow is determined by observing the level of the liquid above the edge of the opening in the weir at a point far enough upstream to be free from the disturbances that occur in the neighbourhood of the notch. Weirs may be of various types, for example, rectangular, *N*-notch, and curved side notches.

The equations of flow for various weirs, or notches, are as follows.

4.7.1 Rectangular Weirs

4.7.1.1 Francis formula

$$\frac{V}{\theta} = 3.33\,[L - 0.2\,H]\,H^{3/2}$$

where V is the volume flow rate (ft^3), θ is the time (s), L is the width of the weir (ft), and H is the height of liquid above the obstruction (ft).

$L - 0.2\,H$ is due to contraction at η_{app}. The flow for rectangular weirs is mathematically given by Equation 4.98.

$$V = A\sqrt{2g_c}\,LH^{3/2} \tag{4.98}$$

where A is a constant.

4.7.2 V-notch Weirs

$$V = 2.505\left[\tan\frac{\alpha}{2}\right]^{0.996} \times H^{2.47}$$

where α is the angle of notch and H is the height from the vertex of the notch.

The flow for rectangular weirs is given by Equation 4.99.

$$V = B\sqrt{2g_c}\left[\tan\frac{\alpha}{2}\right]H^{2.5} \tag{4.99}$$

where B is a constant.

4.7.3 Notch Weirs

In notch weirs, the energy of the fluid due to its hydrostatic pressure is converted into kinetic energy.

$\mu = 1.31$ cP $= 0.00088\ \Sigma u$

The total pipe length is 500 + 6 × 5.72 + 2 × 2.81 + 2 × 2.13 ft = 544.20 ft

$\rho_{50°F} = 62.42$

$$D = \frac{3.068}{12} = 0.256$$

$$\text{Cross-sectional area} = \frac{\pi}{9} \times (0.256)^2 = 0.0513 \text{ ft}^2 = 7.393 \text{ in}^2$$

$$\mu = \frac{150 \times 231}{60} \times \frac{1}{7.393} \times \frac{1}{12} = 6.5 \text{ ft/s}$$

$$NR_e = \frac{Du\rho}{\mu} = \frac{0.256 \times 6.5 \times 62.42}{0.00088} = 105400$$

The average velocity of liquid in the pipe is 6.831 ft/s.

In turbulent motion,

average velocity = 0.8 × maximum velocity

Thus, maximum velocity is

$$\frac{6.831}{0.8} = \frac{68.31}{8} = 8.539 \text{ ft/s}$$

$$\Delta H = \frac{m^2}{2g_c} = \frac{8.54^2}{2 \times 32.2} = \frac{72.932}{64.4} = 1.132 \text{ ft}$$

$P_a - P_b$ = pressure drop $= \Delta H \times \rho = 1.132 \times 62.42$ lb/ft^2

If R is the reading of the manometer, then

$$R(2.97 - 1) \times 62.42 = 1.132 \times 62.42$$

$$\text{or } R = \frac{1.132}{1.97} \text{ ft} = \frac{1.132}{1.97} \times 12 = 6.88 \text{ in}$$

4.8 LAMINAR AND TURBULENT FLOWS

4.8.1 Nature of Fluid Flow Over a Surface

When a fluid is flowing through a tube or over a surface, the pattern of flow will vary with the velocity, physical properties of the fluid, and geometry of the surface. This was first examined by Reynolds in 1883 using the apparatus shown in Figure 4.19. A glass tube with a flared entrance was immersed in a glass tank fed with water. By means of the valve, the rate of flow from the tank through the glass tube could be controlled. By introducing a fine filament of coloured water from a small reservoir centrally into the

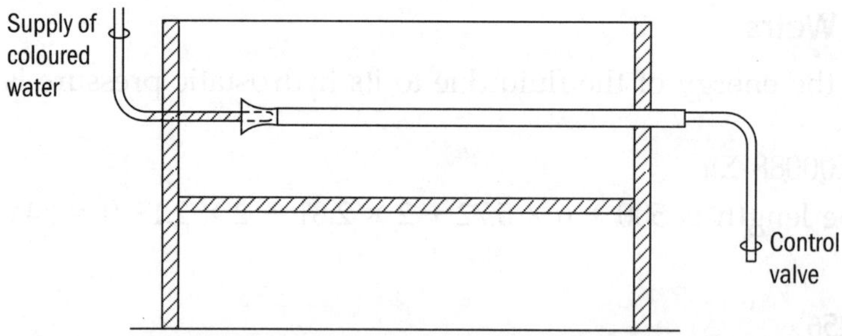

Figure 4.19 Reynolds apparatus for tracing flow patterns

flared entrance of the glass tube, the nature of the flow could be observed. At low flow rates, the coloured filament remained at the axis of the tube, indicating that the flow was in the form of parallel streams that did not interfere with each other. Such flow is called laminar or streamline and is characterized by the absence of bulk movement at right angles to the main stream direction, though a small amount of radial dispersion may occur as a result of diffusion. As the flow rate was increased, a condition was reached in which oscillations appeared in the coloured filament, which broke up into eddies causing dispersion across the tube section. This type of flow is known as turbulent flow and is characterized by the rapid movement of fluid as eddies, causing dispersion in random directions across the tube (Figure 4.20). These experiments clearly show the nature of the transition from streamline to turbulent flow. Below critical velocity, oscillations in the flow were unstable and any disturbance would quickly die out. At higher velocities, however, the oscillations were stable and increased in amplitude, causing a high degree of radial mixing.

In another set of experiments by Reynolds and others, the drop in pressure for flow through a pipe is plotted against the average velocity through the tube (Figure 4.21[a]). At low velocities, a straight line is obtained, indicating that pressure is directly proportional to velocity. At higher velocities, however, the pressure drop increases more rapidly. If a logarithmic plot is used, as in Figure 4.21(b), the results will fall into three sections. Over the region of low velocity, the line *PB* has a slope of unity. However, beyond this

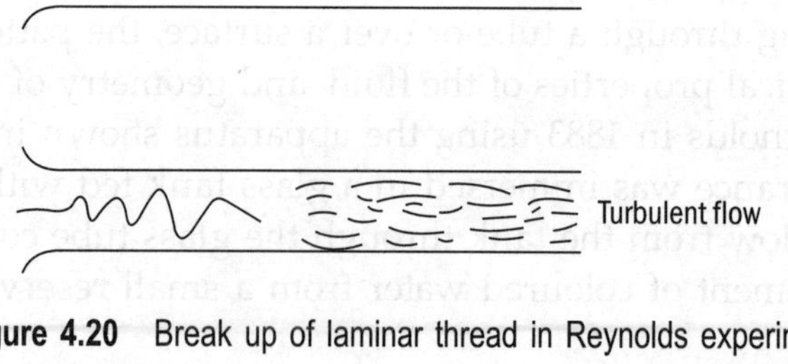

Figure 4.20 Break up of laminar thread in Reynolds experiment

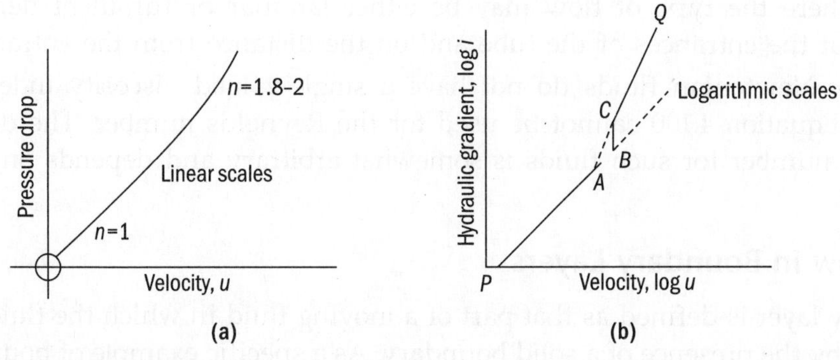

Figure 4.21 Pressure drop versus velocity: (a) simple plot and (b) logarithmic plot

region, over the section *BC*, there is instability with poorly defined points. At higher velocities, the line *QC* has a slope of about 1.8. If *QC* is produced backwards, it cuts at the point *A*, corresponding to the change from laminar to turbulent flow in Reynolds earlier experiments. The point *A* represents critical velocity. Thus, for streamline flow, the pressure gradient is directly proportional to velocity and, for turbulent flow, it is proportional to velocity raised to the power 1.8.

4.8.2 Reynolds Number and Transition from Laminar to Turbulent Flow

In Figure 4.21(b), the velocity corresponding to points *A* and *B* are taken as the lower and higher critical velocities, respectively. Reynolds studied the conditions under which one type of flow changes into the other. He found that the critical velocity at which laminar flow changes into turbulent flow depends on four quantities—diameter of the tube, and viscosity, density, and average linear velocity of the fluid. Furthermore, he found that these four factors can be combined into one group and that the change in type of flow occurs at a definite value of the group. The grouping of variables so found is expressed in Equation 4.100.

$$N_{Re} = \frac{du\rho}{\mu} \tag{4.100}$$

where *d* is the diameter of tube, *u* is the average velocity of fluid, μ is the viscosity of fluid, and ρ is the density of the fluid.

The dimensionless group of variables defined by Equation 4.100 is called the Reynolds number, N_{Re}. Its magnitude is independent of the units used, provided the units are consistent. As evident from Figure 4.21(b), the transition from laminar to turbulent flow (that is, critical point) may occur over a wide range of Reynolds numbers. Laminar flow is always encountered at Reynolds numbers below 2100, but it can persist up to Reynolds numbers of several thousands under special conditions of well-rounded tube entrances and fluid flow in the tank. Under ordinary conditions, the flow is turbulent at Reynolds numbers above about 4000. Between 2100 and 4000, a transition region is found where the type of flow may be either laminar or turbulent depending on conditions at the entrances of the tube and on the distance from the entrance.

Since non-Newtonian fluids do not have a single-valued viscosity independent of shear rate, Equation 4.100 cannot be used for the Reynolds number. The definition of a Reynolds number for such fluids is somewhat arbitrary and depends on the nature of the fluid.

4.8.3 Flow in Boundary Layers

A boundary layer is defined as that part of a moving fluid in which the fluid motion is influenced by the presence of a solid boundary. As a specific example of boundary layer formation, consider the flow of fluid parallel to a thin plate, as shown in Figure 4.22. The velocity of the fluid upstream from the leading edge of the plate is uniform across the entire fluid stream.

In Figure 4.22, the dashed line *OL* is so drawn that the velocity changes are confined between this line and the trace of the wall. Because the velocity lines are asymptotic with respect to distance from the plate, it is assumed that to locate the dashed line definitely, the line passes through all points where the velocity is 99% of the bulk velocity U_∞.

Line *OL* represents an imaginary surface that separates the fluid stream into two parts: one in which the fluid velocity is constant and the other in which the velocity varies from zero at the wall to that of an undisturbed fluid. The imaginary surface separates the fluid that is directly affected by the plate from that in which the local velocity is constant and equal to the initial velocity of the approaching fluid. The zone, or layer, between the dashed line and the plates constitutes the boundary layer. The formation and behaviour of the boundary layer are important not only to the flow of the fluid but also to the transfer of heat and mass.

4.8.4 Laminar and Turbulent Flows in Boundary Layers

The fluid velocity at the solid–fluid interface is zero, and the velocities close to the solid surface are, of necessity, small. Flow in this part of the boundary layer (near the surface),

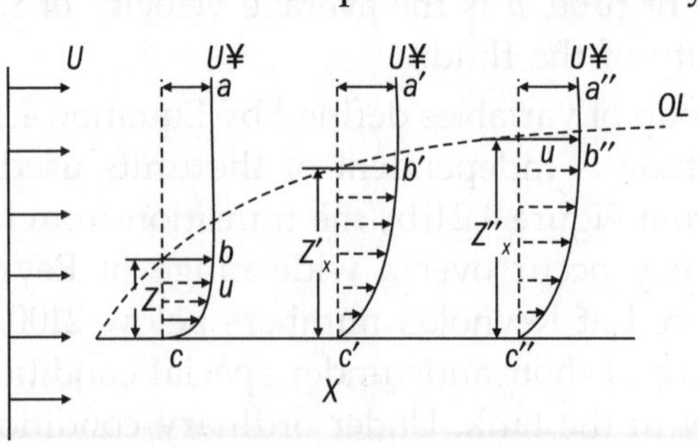

Figure 4.22 Boundary layer

therefore, is laminar. Further away from the surface, though less than the velocity of undisturbed fluid, the fluid velocity may be fairly large and the flow in this part of the boundary layer may become turbulent. There is a transition, or buffer, layer of intermediate character between the zone of fully developed turbulence and the region of laminar flow. Thus, a turbulent boundary layer is considered to consist of three zones: the viscous sublayer, the buffer layer, and the turbulent zone. In some cases, the boundary layer may be entirely laminar, but in most cases of engineering interest, it is part laminar, part turbulent.

The boundary layer near the leading edge of a flat plate immersed in a fluid of uniform velocity is thin, and the flow in the boundary layer is entirely laminar. However, as the layer thickens at distances farther from the leading edge, a point is reached where turbulence appears. The onset of turbulence is characterized by a sudden rapid increase in the thickness of the boundary layer (Figure 4.23).

When flow in the boundary layer is completely laminar, the thickness Z_x of the layer increases with $x^{0.5}$, where x is the distance from the leading edge of the plate. For a short time after turbulence appears, Z_x increases with $x^{1.5}$, and after the turbulence is fully developed, it increases with $x^{0.8}$.

The initial, fully laminar part of the boundary layer may grow to a moderate thickness of perhaps 2 mm, with air or waste moving at moderate velocities. Once turbulence begins, however, the thickness of the laminar part of the boundary layer diminishes considerably, typically to about 0.2 mm.

4.8.4.1 Transition from laminar to turbulent flow in boundary layers

The factors that determine the points at which turbulence appears in a laminar boundary layer are coordinated by the dimensionless Reynolds number, which is defined by Equation 4.101.

$$N_{Re,\,x} = \frac{x u_\infty \rho}{\mu} \tag{4.101}$$

where x is the distance from the leading edge of plate, u_∞ is the bulk fluid velocity, ρ is the density of fluid, and μ is the viscosity of fluid.

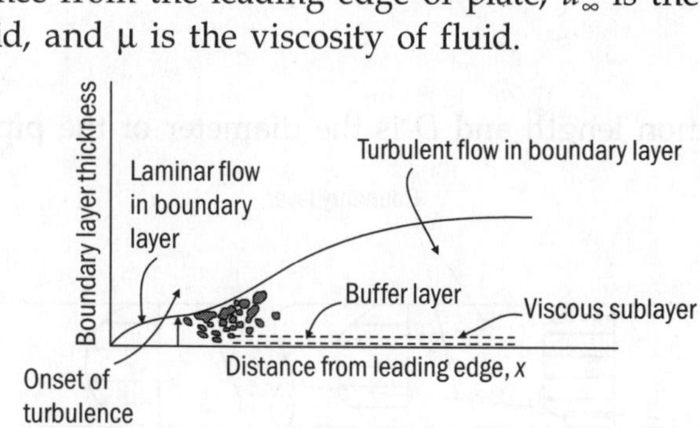

Figure 4.23 Development of turbulent boundary layer on a flat plate

With parallel flow along a plate, turbulent flow first appears at a critical Reynolds number, which is between 10^5 and 3×10^6. Transition occurs at lower Reynolds numbers when the plate is rough and the intensity of turbulence in the approaching stream is high. It occurs at higher values when the plate is smooth and the intensity of turbulence in the approaching stream is low.

4.8.5 Boundary Layer Formation in Straight Tubes

Consider a straight, thin-walled tube with fluid entering it at a uniform velocity as shown in Figure 4.24. A boundary layer begins to form at the entrance to the tube, and the layer thickens as the fluid moves through the first part of the channel. During this stage, the boundary layer occupies only part of the cross section of the tube. The total stream consists of fluid flowing in a road-like manner at constant velocity and an annular boundary layer between the wall and core. In the boundary layer, the velocity increases from zero at the wall to a constant velocity existing in the core. As the stream moves farther down the tube, the boundary layer occupies an increasing portion of the cross section. Finally, at a point well downstream from the entrance, the boundary layer reaches the centre of the tube, the road-like core disappears, and the boundary layer occupies the entire cross section of the stream. At this point, the velocity distribution in the tube reaches its final form (as shown by the last curve in Figure 4.24) and remains unchanged for the remaining length of the tube. Such flow with an unchanging velocity distribution is called fully developed flow.

4.8.5.1 *Transition length for laminar and turbulent flows*

The length of the entrance necessary for the boundary layer to reach the centre of the tube and for fully developed flow to be established is called the transition length. Since the velocity varies not only with the length of the tube but also with the radial distance from its centre, the flow in the entrance region is two dimensional.

For laminar flow, the approximate length of a straight pipe required for completion of the final velocity distribution is given by Equation 4.102.

$$\frac{x_1}{D} = 0.05\, N_{Re} \tag{4.102}$$

where x_1 is the transition length and D is the diameter of the pipe.

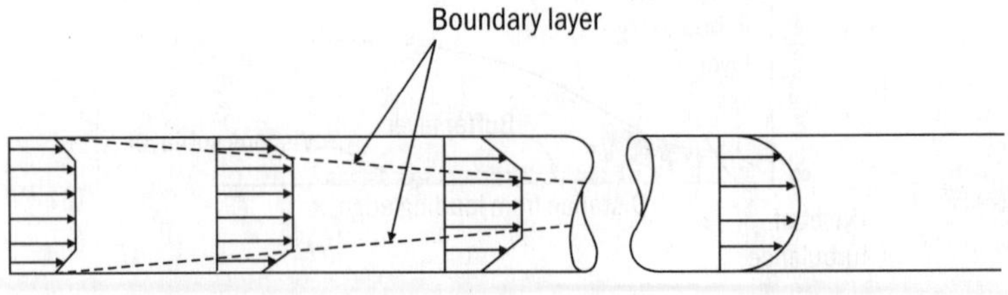

Figure 4.24 Development of boundary layer flow in pipe

For a pipe of internal diameter (ID) 50 mm (2 in) and a Reynolds number of 1500, the transition length is 3.75 m (12.3 ft). If the fluid entering the pipe is turbulent and the velocity is above critical velocity, the transition length is almost independent of the Reynolds number and is about 40 to 50 pipe diameters, with little difference between the distribution at 25 diameters and at greater distances from the entrance. For a 50 mm ID pipe, 2–3 m of straight pipe is sufficient. If the flow of fluid entering the tube is laminar and becomes turbulent on entering it, a longer transition length as large as 100 pipe diameters is required. The transition length for turbulent flow is less than that for laminar flow unless the Reynolds number is small.

4.8.6 Boundary Layer Separation and Wake Formation

The growth of boundary layers has been discussed in preceding sections. We will now consider what happens at the far side of a submerged object, where the fluid leaves the solid surfaces.

At the trailing edge of the flat plate parallel to the direction of flow, the boundary layers on the two sides of the plate have grown to a maximum thickness. For a time after the fluid leaves the plate, the layers and velocity gradients persist. Soon, however, the gradients fade out; the boundary layers intermingle and disappear, and the fluid once more moves with uniform velocity. This is shown in Figure 4.25(a).

Let us now consider a situation where the plate is turned at right angles to the direction of flow as shown in Figure 4.25(b). A boundary layer forms as before in the fluid flowing over the upstream face. When the fluid reaches the edge of the plate, however, its momentum prevents it from making the sharp turn around the edge. It

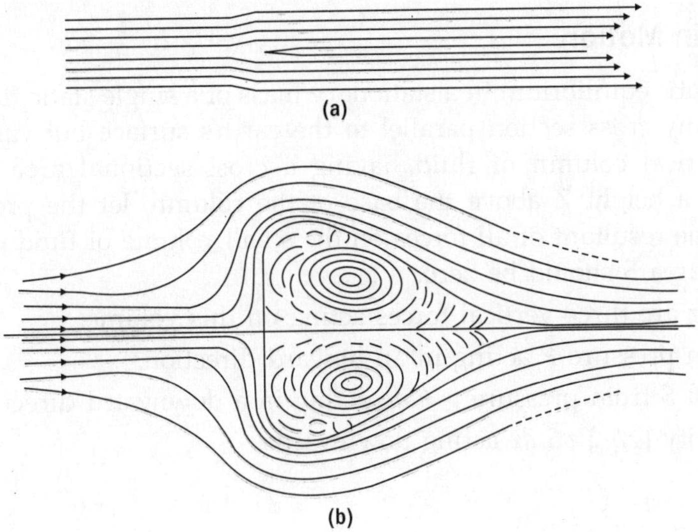

Figure 4.25 Flow past flat plate: (a) flow parallel with plate and (b) flow perpendicular to plate

separates from the plate and proceeds outwards into the bulk of the fluid. Behind the plate is the backwater zone of strongly decelerated fluid, in which large eddies called vortices are formed. This zone is known as the wake. Eddies in the wake are kept in motion by the shear stresses between the wake and the separated current. They consume considerable amounts of mechanical energy and may lead to a large pressure loss in the fluid.

Boundary layer separation occurs whenever the change in velocity of the fluid, either in magnitude or direction, is too large for the fluid to adhere to the solid surface. It is most frequently encountered when there is an abrupt change in the flow channel, like a sudden expansion or contraction, a sharp bend, or an obstruction around which the fluid must flow. Separation may also occur from decrease in velocity in a smoothly diverging channel. Because of the large energy loss resulting from the formation of a wake, it is often desirable to minimize or prevent boundary layer separation. In some cases, this can be done by suction, that is, by drawing part of the fluid into the solid surface at the area of potential separation. Most often, however, separation is minimized by avoiding sharp changes in the cross-sectional area of the flow channel and by streamlining any objects over which the fluid must flow. For some purposes, such as the promotion of heat transfer or mixing in a fluid, boundary layer separation may be desirable.

4.9 BASIC EQUATIONS OF FLUID FLOW

The basic equations of fluid mechanics may be classified under two different conditions—static and dynamic, or fluid, in motion.

4.9.1 Static in Motion

During hydrostatic equilibrium in a stationary mass of a single static fluid, the pressure is constant in any cross section parallel to the earth's surface but varies with height. Consider a vertical column of fluid having a cross-sectional area S, as shown in Figure 4.26. At a height Z above the base of the column, let the pressure be P and density be ρ. The resultant of all forces on the small volume of fluid of height dZ and cross-sectional area S should be zero.

The following are three vertical forces acting on this volume.

1. Force pS from pressure P acting in an upward direction.
2. Force $[p + dp]\ S$ from pressure $p + dp$ acting in a downward direction.
3. Force of gravity $[g/g_c]\ pS\ dz$ acting downwards.

Then,

$$+ p^S - (P + dp)S - \frac{q}{g_c} P^S dZ = 0 \tag{4.103}$$

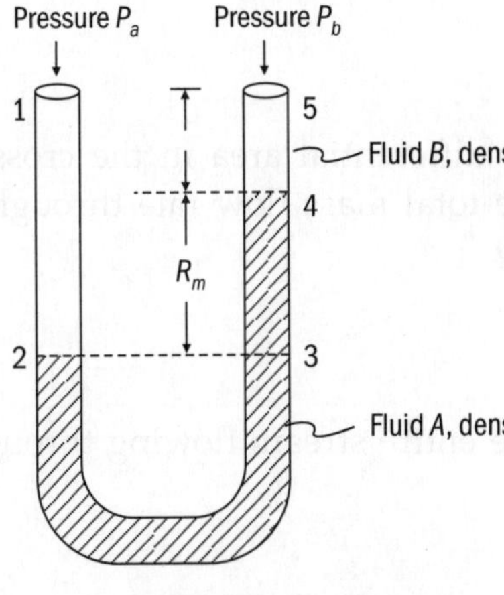

Figure 4.26 Manometer

In Equation 4.103, the forces acting upwards are taken as positive and those acting downwards are taken as negative. After simplification and division by S, Equation 4.103 changes to

$$dp + \frac{g}{g_c} p\, dZ = 0 \tag{4.104}$$

Between the two definite heights Z_a and Z_b,

$$\frac{P_b}{P} - \frac{P_a}{P} = \frac{g}{g_c} = (Z_a - Z_b) \tag{4.105}$$

Equation 4.105 mathematically expresses the condition of hydrostatic equilibrium.

4.9.1.1 Manometers

A manometer is a device for measuring pressure difference. Figure 4.26 shows the simplest form of a manometer. A pressure P_a is exerted on one arm of the U tube and pressure P_b on the other. As a result of the difference in pressure $(P_a - P_b)$, the meniscus in one branch of the U tube is higher than in the other. The vertical distance between the two menisci, R_m, is used to measure the difference in pressure.

At point 1, the pressure is P_a; then, as shown by Equation 4.105, the pressure at point 2 is $P_a + (g/g_c)(Z_m + R_m)P_b$. By the principles of hydrostatics, this is also the pressure at point 3. The pressure at point 4 is less than that at point 3 by an amount $(g/g_c)R_mP_a$, and the pressure at point 5, which is P_b, is still less by an amount $(g/g_c)Z_mP_b$. These statements can be summarized by Equation 4.106.

$$P_a + \frac{g}{g_c}[(Z_m + R_m)\,\rho_B - R_m\rho_a - Z_m\rho_b] = P_b \tag{4.106}$$

Simplification of Equation 4.106 gives Equation 4.107.

$$P_a - P_b = \frac{g}{g_c} R_m(\rho_a - \rho_b) \tag{4.107}$$

4.9.2 Dynamic in Motion

4.9.2.1 Average velocity

The mass flow rate through a differential area in the cross section of a stream tube is given by Equation 4.108. The total mass flow rate through the entire cross-sectional area is given by Equation 4.109.

$$d\dot{m} = \rho u \; dS \tag{4.108}$$

$$\dot{m} = \rho \int S \; udS \tag{4.109}$$

The average velocity $\overline{V}$ of the entire stream flowing through the cross-sectional area S is defined by Equation 4.110.

$$\overline{V} = \mathrm{m}/\rho \cdot S = \frac{1}{S} \int S \; udS \tag{4.110}$$

It is equal to the total volumetric flow rate divided by the cross-sectional area through which it is flowing (Equation 4.111).

$$\overline{V} = \frac{q}{S} \tag{4.111}$$

The continuity equation for flow through a finite tube is given by Equation 4.112.

$$\dot{m} = \rho_a \overline{V}_a S_a = \rho_b \overline{V}_b S_b = \rho \overline{V} S \tag{4.112}$$

For circular cross section,

$$\dot{m} = \frac{1}{a} \pi \; D_a^2 \; \rho_a \; \overline{V}_a = \frac{1}{4} \pi \; D_b^2 \; \rho_b \; \overline{V}_b \tag{4.113}$$

$$\text{or } \frac{\rho_a \overline{V}_a}{\rho_b \overline{V}_b} = \left(\frac{D_b}{D_a}\right)^2 \tag{4.114}$$

The friction forces acting on the volume element of stream tube are shown in Figure 4.27.

The rate of momentum flow at the tube entrance, M_a, is mu and at the tube exit, M_b, is $m(u)$. The momentum balance between the two points can be written as given by Equation 4.115.

$$\overline{Z} \; Fg_c = M_a - M_b \tag{4.115}$$

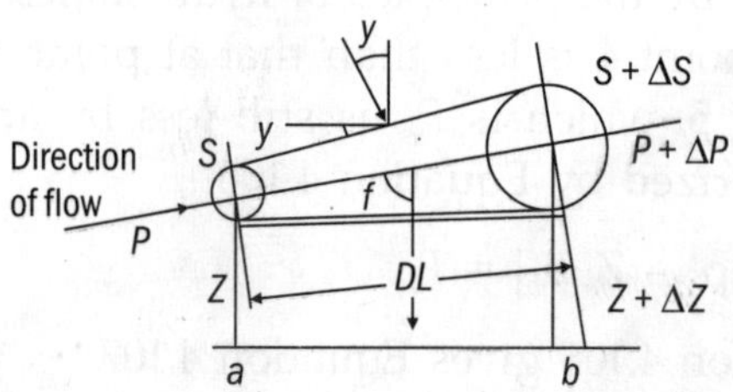

Figure 4.27 Forces acting on the volume element of stream tube

$$\text{or } F = \frac{\dot{m}u}{g_c} \tag{4.116}$$

The pressure forces normal to the cross section of the tube at the inlet and outlet are related as given by Equation 4.117.

$$P_aS_a = P^S \; P_bS_b = (P + \Delta P)\,(S + \Delta S) \tag{4.117}$$

Since the total projected area from the side of the tube is ΔS, the total side force can be calculated as given by Equation 4.118.

$$F_w = \int_0^{\Delta S} P'dS = \overline{P}'\,\Delta S \tag{4.118}$$

where $\overline{P}'$ is the average value of the pressure surrounding the tube. $\overline{P}'$ has a value between P and $(P + \Delta P)$.

The only other force acting on the flowing fluid is the component of gravity acting along the axis. The volume of the tube may be written as $\overline{S}L$, where $\overline{S}$ is the average cross section, which has a value between S and $(S + \Delta S)$. Then the mass of fluid in the tube is $\overline{S}_pL$. The component of gravitational force opposite to the direction of flow is given by Equation 4.119.

$$F_g = \frac{g}{g_c}\,\overline{S}\,\rho\Delta L\,\cos\theta \tag{4.119}$$

Since $\cos\theta = \Delta Z/\Delta L$

$$F_g = \frac{g}{g_c}\,\overline{S}\rho\Delta L\,\frac{Z}{L} = \frac{g}{g_c}\,\overline{S}\rho\Delta Z \tag{4.120}$$

Substituting from Equations 4.113–4.115 in Equation 4.120 gives Equation 4.121.

$$\frac{\dot{m}}{g_c}\,\Delta u = \Delta S\,(\overline{P} - P) - S\Delta P - \Delta P\Delta S - \frac{g}{g_c}\,\overline{S}_p\,\rho\Delta Z \tag{4.121}$$

Dividing Equation 4.121 by $\rho^S\Delta L$ gives Equation 4.122.

$$\frac{\dot{m}}{g_c\rho^S L}\,\frac{\Delta u}{\Delta L} = \frac{\overline{P} - P\Delta S}{\rho^s\Delta L} - \frac{1\Delta P}{p\Delta L} + \frac{\Delta S\Delta P}{\rho^s\Delta L} - \frac{g\overline{S}\Delta Z}{g_c\,S\Delta L} \tag{4.122}$$

Now, finding the limits of all terms in Equation 4.122 as $\Delta L \rightarrow 0$, we have $\Delta S \rightarrow 0$, $\overline{S} \rightarrow S$, and $(\overline{P} - P) \rightarrow 0$. The ratios of increments become corresponding differential coefficients, so that in the limit,

$$\frac{\dot{m}}{g_cP^S}\,\frac{du}{dL} = \frac{1}{P}\,\frac{dp}{dL} - \frac{g}{g_c}\,\frac{dZ}{dL} \tag{4.123}$$

The mass flow rate is given by Equation 4.124.

$$\dot{m} = upS \tag{4.124}$$

Substituting Equation 4.124 in Equation 4.123, we get

$$\frac{u\rho S}{g_c\rho S}\frac{du}{dL} = \frac{1}{\rho}\frac{dp}{dL} - \frac{g\,dZ}{g_c dL} = 0 \quad (4.125)$$

and

$$\frac{1}{P}\frac{dp}{dL} + \frac{g\,dZ}{g_c dL} + \frac{d\,(u^2/2)}{g_c dL} = 0 \quad (4.126)$$

Equation 4.126 is the point form of the Bernoulli equation without friction.

The differential form of Equation 4.126 is given by Equation 4.127.

$$\frac{dp}{\rho} + \frac{g}{g_c}dZ + \frac{1}{g_c}d\left(\frac{u^2}{2}\right) = 0 \quad (4.127)$$

Between two definite points in the tube, say stations a and b, Equation 4.127 can be integrated, since P is a constant, to give Equation 4.128.

$$\frac{P_a}{\rho} + \frac{gz_a}{g_c} + \frac{u_a^2}{2g_c} = \frac{P_b}{\rho} + \frac{gz_b}{g_c} + \frac{u_b^2}{2g_c} \quad (4.128)$$

Equation 4.128 is known as the Bernoulli equation without friction.

4.9.2.1.1 Correction of Bernoulli equation for fluid friction

Friction manifests itself by the disappearance of mechanical energy. In frictional flow, the quantity $\frac{p}{\rho} + \frac{u^2}{2g_c} + \frac{g}{g_c}Z$ is not constant along a streamline, as indicated by Equation 4.128 but always decreases in the direction of flow.

For incompressible fluids, the Bernoulli equation is corrected for friction by adding a term (kinetic energy correction factor) on the right-hand side of Equation 4.128, which then changes to

$$\frac{P_a}{\rho} + \frac{gz_a}{g_c} + \frac{\alpha_a\overline{V}a^2}{2g_c} = \frac{P_b}{\rho} + \frac{gz_b}{g_c} + \frac{\alpha_b\overline{V}b^2}{2g_c} + h_f \quad (4.129)$$

The units of h_f and all other terms in Equation 4.129 are energy per unit mass. The term h_f represents the overall friction generated per unit mass of fluid (and, therefore, all the conversion of mechanical energy into heat) that occurs in the fluid between stations a and b.

4.9.2.1.2 Pump work in Bernoulli equation

In a flow system, a pump is used to increase the mechanical energy of the flowing fluid. The increase in energy is used to maintain flow. If W_p is the work done by the pump per unit mass of fluid and h_{fp} is the total friction in the pump per unit mass of fluid, then the net work done to the fluid is $W_p - h_{fp}$. In practice, in place of h_{fp}, a pump efficiency denoted by η is used, defined by Equation 4.130.

$$W_P - h_{fp} = \eta W_P$$

$$\text{or } \eta = (W_P - h_{fp})/W_p \tag{4.130}$$

The mechanical energy delivered to the fluid is then ηW_p, when η is 1. Equation 4.129 corrected for pump work is given by Equation 4.131.

$$\frac{P_a}{\rho} + \frac{gz_a}{g_c} + \frac{\alpha_a \overline{V}a^2}{2g_c} + \eta W_P = \frac{P_b}{\rho} + \frac{gz_b}{g_c} + \frac{\alpha_b \overline{V}b^2}{2g_c} + h_f \tag{4.131}$$

Equation 4.130 is the final working equation for problems on the flow of incompressible fluids.

4.10 MASS TRANSFER PRINCIPLES

When a system contains two or more components whose concentrations vary from point to point, there is a natural tendency for mass to be transferred, minimizing the concentration differences within the system. The transport of one component from a region of higher concentration to that of lower concentration is called mass transfer. In most mass transfer operations used for separating the components of a solution, the result is achieved by bringing the solution to be separated into contact with another insoluble phase and allowing the material to be transferred from one homogeneous phase to another. The driving force for mass transfer is a concentration gradient similar to temperature gradient, which provides the driving force for heat transfer. Mass transfer stops when equilibrium is reached, that is, the chemical potentials in the two phases become the same.

Mass transfer also occurs when water flows in a pipe. However, the mass transfer referred here is the migration of a component in a mixture, either within the phase or from one phase to another, because of a difference in physical equilibrium.

The occurrence of mass transfer processes throughout the biological, chemical, physical, and engineering fields is extremely widespread. Biological involvements include respiratory mechanisms and oxygenation of blood, kidney functions, and food and drug assimilation. In chemical engineering, mass transfer finds widespread applications. For example, the process is used in obtaining desired constituents from naturally occurring materials like natural gas, crude oil, and coal; for easy handling and recovery of components; and in recovering the desired components from a chemical reaction when many by-products are also formed. Some examples are as follows.

In distillation, intimate contact between saturated vapour and boiling liquid phases is facilitated for the transfer of the more volatile component from the liquid to gas phase and the less volatile component from vapour to liquid phase. Gas absorption refers to the transfer of a soluble component from the gas phase into a non-volatile liquid solvent.

The reverse process is called stripping. In liquid–liquid extraction, transfer of a common component takes place from one liquid phase to an insoluble liquid phase. In leaching, a solvent is used to dissolve a soluble component from its mixture with another inert solid component. In adsorption, advantage is taken of the preferential adsorption of a component from a gaseous or liquid mixture onto a solid absorbent. In humidification, air conditioning, and evaporative cooling of water, there is a simultaneous transfer of heat and mass. Drying operations involve transport of both liquid and vapour within the solid and of vapour in the drying gas. There are many other applications of mass transfer such as ion exchange, sublimation chromatography, and reverse osmosis.

Mass transfer can be treated in much the same way as heat or momentum transfer as driving forces, resistance, fluxes, boundary layer, and similar factors are involved. The key difference is that a transfer of mass always involves the motion of the medium as otherwise mass cannot be transferred.

4.10.1 Diffusion

4.10.1.1 Laws

The flow of mass on a molecular level occurs because of diffusion. In the absence of turbulence, the rate of diffusion of component A, expressed as flux J_A, is defined as given by Equation 4.132.

$$J_A = -D_{AB}\frac{\alpha}{\partial} \tag{4.132}$$

Equation 4.132 is Ficks' first law. Here, D_{AB} is the diffusion coefficient of A in B,

and $\partial C_A/\partial$ is the concentration gradient in the z-direction. This definition is applicable when there is no net flow, that is, in case of equimolar counterdiffusion of A and B. In many diffusional operations, there is a net flow of the entire phase in the direction of diffusion. For example, a fish swimming along a stream appears to be swimming faster to a stationary observer than to an observer swimming along the same path. This leads to two kinds of fluxes: one relative to a stationary set of coordinates (N) and the other relative to the average molar velocity of overall constituents (J).

Consider a section through a fluid phase of two components, A and B, diffusing from the interface in a fluid (Figure 4.28).

The volume of the thin section is AdZ, where A is the area perpendicular to the direction of diffusion, $C_{A.}\ AdZ$ is the number of moles in it, and C_A is the concentration of A in moles per unit volume. If the flux of A is N_A relative to the interface, the time of passage of A through the slice is $(C_AAdZ)/N_AA = C_AdZ/N_A$.

The velocity u_A of A relative to the interface is, therefore, given by Equation 4.133.

$$u_A = \frac{dZN_A}{C_AdZ}\frac{N_A}{C_A} \tag{4.133}$$

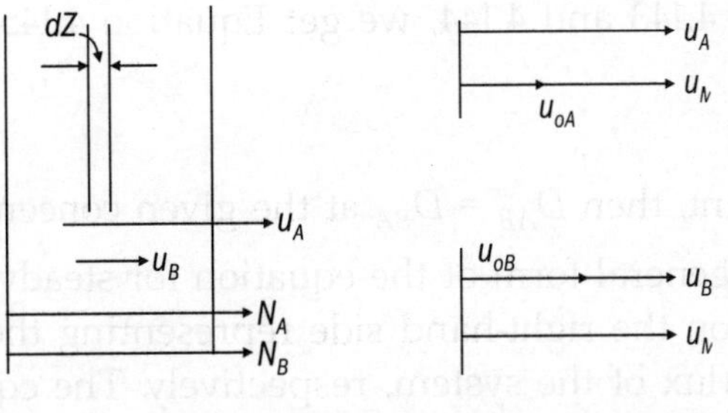

Figure 4.28 Fluxes and velocities through a single phase

This leads to the relationship given by Equation 4.134.

$$N_A = u_A C_A \tag{4.134}$$

Similarly, for the component B,

$$N_B = u_B C_B \tag{4.135}$$

The molar average velocity u_M is given by

$$u_M = \frac{u_A C_A + u_B C_B}{C} \tag{4.136}$$

$$\text{or } u_M = \frac{N_A + N_B}{C} \tag{4.137}$$

where C is the total molar concentration.

The velocities U_{0A} and U_{0B} of A and B, respectively, with respect to the system as a whole are given by Equations 4.138 and 4.139.

$$U_{0A} = J_A/C_A \tag{4.138}$$

$$U_{0B} = J_B/C_B \tag{4.139}$$

$$\text{Since } U_{0A} = U_A/U_A \tag{4.140}$$

$$\frac{J_A}{C_A} = \frac{N_A}{C_A} - \frac{N_A + N_B}{C} \tag{4.141}$$

$$\text{or } J_A = N_A - (N_A + N_B)\frac{C_A}{C_B} \tag{4.142}$$

This leads to a useful relationship given by Equation 4.143.

$$N_A = (N_A + N_B)\frac{C_A^{-D_{AB}}}{C_B}\frac{dC_A}{dZ} \tag{4.143}$$

Similarly, for B

$$N_B = (N_A + N_B)\frac{C_B}{C} - D_{BA}\frac{dC_B}{dZ} \tag{4.144}$$

By adding Equations 4.143 and 4.144, we get Equation 4.145.

$$-D_{AB}\frac{dC_B}{dZ} = -D_{BA}\frac{dC_B}{dZ} \tag{4.145}$$

If $C_A + C_B$ is a constant, then $D_{AB} = D_{BA}$ at the given concentration and temperature.

Equation 4.143 is the general form of the equation for steady-state diffusion, with the first and second terms on the right-hand side representing the contribution from bulk flow and the diffusion flux of the system, respectively. The equation can be integrated if the relationship between N_A and N_B are known.

4.10.1.2 Steady-state diffusion of A through stagnant B

The steady-state diffusion of *A* through stagnant *B* arises in many binary systems. For example, absorption of a gas from an inert gas in a solvent, leaching and extraction where one component (*B*) is taken as stationary and only one component (*A*) diffuses. For gaseous systems, the equation can be written in the following form, with concentrations expressed in partial pressures.

$$N_A = (N_A + N_B)\frac{P_A}{P} - \frac{D_A B^P}{RT}\frac{dP_A}{dZ} \tag{4.146}$$

where *P* is the total pressure, P_A is the partial pressure of *A*, *T* is the temperature on absolute scale, and *R* is the gas constant.

Since $N_B = O$,

$$N_A\frac{(1 - P_A)}{P} = \frac{-D_{AB}}{RT}\frac{dP_A}{dZ} \tag{4.147}$$

Integrating Equation 4.147 for the diffusion of *A* through fixed boundaries, that is, $P_A = P_{A_1}$ when $Z = Z_1$ and $P_A = P_{A_2}$ when $Z = Z_2$, we get

$$N_A = \frac{D_A B^P}{RT\,(Z_2 - Z_1)}\ln\frac{P - P_{A_2}}{P - P_{A_1}} \tag{4.148}$$

We have $P - P_{A_2} = P_{B_2}$ and $P - P_{A_1} = P_{B_1}$. Multiplying and dividing the right-hand side by $P_{A_1} - P_{A_2}$, we get

$$N_A = \frac{D_A B^P}{RT\,(Z_2 - Z_1)}\frac{(P_{A_1} - P_{A_2})}{P_{B_2} - P_{B_1}}\ln\frac{P - P_{A2}}{P - P_{A1}} \tag{4.149}$$

Now, $P_{BM} = \dfrac{P_{B_2} - P_{B_1}}{\ln(P_{B_2}/P_{B_1})}$ and $(Z_2 - Z_1)\ \Delta Z_f$ is the film thickness through which diffusion takes place.

$$N_A = \frac{D_A B^P}{RT\,ZP_{BM}}(P_{A_1} - P_{A_2}) \tag{4.150}$$

The partial pressures are shown graphically in Figure 4.29.

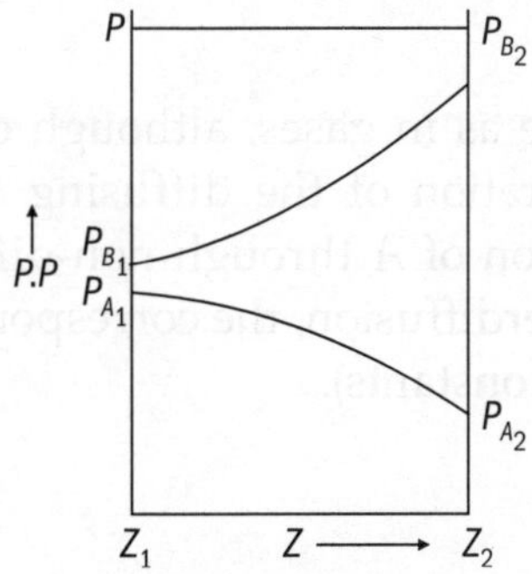

Figure 4.29 Diffusion of A through stagnant B

4.10.1.3 Steady-state equimolar counterdiffusion

Steady-state equimolar counterdiffusion is often encountered in a fractionating column as a result of approximately equal values of molal latent heats of components involved, that is, for each mole of the more volatile component vaporizing, approximately a mole of the less volatile component condenses.

Since $N_A = -N_B$, Equation 4.143 reduces to

$$N_A = \frac{-D_{AB}}{RT}\frac{dP_A}{dZ}$$

On integration with boundary conditions $P_A = P_{A_1}$ at $Z = Z_1$, $P_A = P_{A_2}$ at $Z = Z_2$, and setting $Z_2 - Z_1 = Z_f$, we get

$$N_A = \frac{D_{AB}}{RTZ}(P_{A_1} - P_{A_2}) \tag{4.151}$$

This is graphically represented in Figure 4.30.

The diffusion equation can be solved for other mass transfer interfaces and for different relationships between N_A and N_B in a reacting system, depending on the stoichiometry of the reaction between A and B.

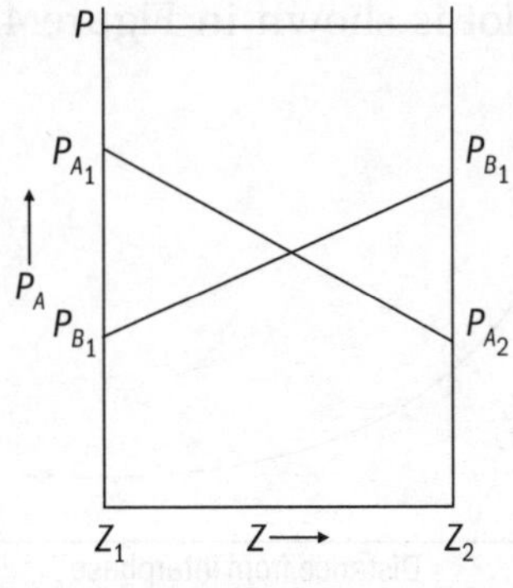

Figure 4.30 Equimolal counterdiffusion

4.10.2 Diffusion in Liquids

The treatment is much the same as in gases, although diffusivities in liquids are often strongly dependent on concentration of the diffusing component. The corresponding equation for steady-state diffusion of A through non-diffusing B is Equation 4.152, and for steady-state equimolal counterdiffusion, the corresponding equation is Equation 4.153 (assuming that D_{AB} and C are constants).

$$N_A = \frac{D_{AB}}{ZX_{BM}} \frac{(\rho)}{M} (x_{A_1} + x_{A_2}) \tag{4.152}$$

$$N_A = \frac{D_{AB}}{Z} \frac{(\rho)}{M} (x_{A1} + x_{A2}) \tag{4.153}$$

where Z is the thickness of the film, x_{A_1} and x_{A_2} are the the mole fractions of A at the ends of the liquid film, ρ is the density of the solution, and M is the molecular weight of the solution.

The average of ρ/M values represents the average at $Z = Z_1$. D_{AB} is the diffusivity applicable at $\frac{\rho}{M_{av}}$.

4.11 MASS TRANSFER PRINCIPLES ADVANCES

When mass transfer takes place across an interface between a stationary phase and a fluid phase and the flow is in turbulent conditions, there is a thin film of fluid in laminar flow adjacent to the interface. This is followed by transition and turbulent regions. In the laminar region, the mechanism of mass transfer can be assumed to take place by molecular diffusion as described. However, in the turbulent region, the flow is not orderly and eddies move from one point to another. The mass carried by the moving eddies contributes significantly to the mass transport in the direction of the mass transfer. The mass transfer is greater in the turbulent region than in the laminar region because of rapid eddy motion. Thus, the concentration gradients existing in the former are smaller. A typical plot is shown in Figure 4.31.

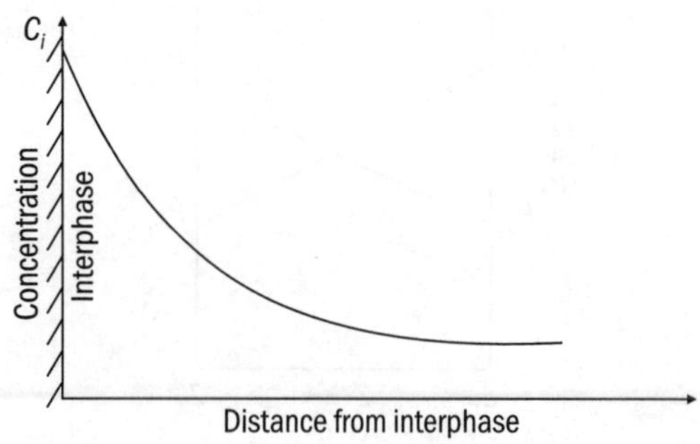

Figure 4.31 Concentration gradient in a moving fluid

In the turbulent region, the diffusion equation takes the form of Equation 4.154.

$$N_A = -(D_{AB} + \epsilon_D)\frac{dC_A}{dZ} \tag{4.154}$$

where ϵ_D is the eddy diffusion coefficient.

In the turbulent region, $\epsilon_D \gg D_{AB}$, but in the laminar region, $D_{AB} \gg \epsilon_D$. The value of ϵ_D depends on the extent of turbulence.

4.11.1 Film Theory

If we consider a liquid flowing in turbulent flow conditions past a stationary solid surface (the solid is soluble in the liquid), a typical concentration gradient is established under steady-state conditions (Figure 4.32).

A majority of the concentration gradient will be in the laminar film of thickness Z_F. The film theory postulates that the entire concentration difference ($C_{A_i} - C_{A_b}$) lies in a film of thickness Z_D, whose entire resistance to mass transfer is equivalent to the combined resistance offered by the viscous, buffer, and turbulent layers. Since the film Z_D is thin, it is assumed to retain negligible quantity of solute compared to the quantity passing through it, and the concentration gradients can be easily established. It is, therefore, a steady-state model. It is further assumed that transfer is unidirectional and the concentration levels are low such that essentially equimolal counterdiffusion conditions apply. Transfer is entirely by molecular diffusion in the film, that is, ϵ_D=0 in the film $0 - Z_D$ and $\epsilon_D \to \infty$ beyond Z_D.

If we consider a thin slice dZ within the liquid film and equate the rate of input of solute into the film at Z = Z to the rate of output solute at Z = Z + ΔZ (accumulation is zero and A is neither generated nor consumed by a chemical reaction), we get

$$-D_{AB}A\frac{dC_A}{dZ} - \left[-D_{AB}A\frac{dC_A}{dZ}\left(-D_{AB}A\frac{dC_A}{dZ}\right)\Delta Z\right] = 0 \tag{4.155}$$

or $$\frac{d^2\,C_A}{dZ^2} = 0 \tag{4.156}$$

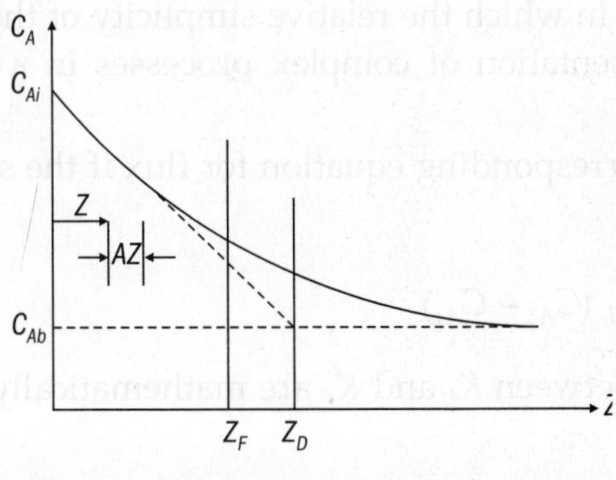

Figure 4.32 Concentration versus distance (film model)

Integrating Equation 4.156 twice and applying boundary conditions, we get

$$\left.\begin{aligned} C_A &= C_{A_i} \text{ at } Z = 0 \\ C_A &= C_{A_b} \text{ at } Z = Z_D \end{aligned}\right\} \tag{4.157}$$

$$C_A = C_{A_i} - \left(\frac{C_{A_i} - C_{A_b}}{Z_D}\right) Z \tag{4.158}$$

Due to low levels of mass transfer, the flux N_{A_i} at the interface is given by Equation 4.159.

$$N_{A_i} = -D_{AB}(dC_A/dZ) \tag{4.159}$$

From Equations 4.158 and 4.159, we get

$$N_{A_i} = \frac{D_{AB}}{ZD}(C_{A_i} - C_{A_b}) \tag{4.160}$$

Since D_{AB} and Z_D are constants under a given set of conditions, we get

$$N_{A_i} = K_L\,(C_{A_i} - C_{A_b}) \tag{4.161}$$

Equation 4.161 gives the definition of mass transfer coefficient K_L applied to a liquid film; the concentrations are expressed in moles/volume. The individual (liquid film) mass transfer coefficient is, therefore, the flux divided by driving force. From Equations 4.160 and 4.161 and according to the film theory, it can be seen that

$$K_L = D_{AB}/Z_D$$

That is, $K_L \propto D_{AB}$ (4.162)

Equation 4.162 is found to be experimentally incorrect. The value of K_L is found to be proportional to $D_{AB}^{0.5}$. A more developed penetration theory with different models for surface renewal predicts the exponent on D_{AB} as 0.5, which is more difficult to apply to practical situations because of difficulty in measuring the contact time of eddies at the interface and their rates of renewal.

The film theory, however, incorporates an essential feature of mass transfer, that is, the solute has to get into the system by dissolution and subsequent molecular diffusion. There are many instances in which the relative simplicity of the film theory contributed to the quantitative representation of complex processes in a way that is consistently useful in design.

Equation 4.163 is the corresponding equation for flux if the system is one of diffusion of A through stagnant B.

$$N_{A_i} = \frac{D_{AB}\,(C_{A_i} - C_{A_b})}{Z_D\,(C_{BM}/C)} = K_L\,(C_{A_i} - C_{A_b}) \tag{4.163}$$

Thus, the relationship between K_L and K_L' are mathematically expressed by Equations 4.160–4.162.

$$K_L' = K_L\,(C_{BM}/C) = K_L X_{BM} \tag{4.164}$$

The corresponding equation for mass transfer coefficients in gas phase for the equimolal counterdiffusion of A and B is Equation 4.165 and for the diffusion of A through stagnant B is Equation 4.166.

$$N_{A_i} = \frac{D_{AB}}{RTZ}(P_{A_i} - P_{A_G}) = K_G'(P_{A_i} - P_{A_G}) \quad (4.165)$$

$$N_{A_i} = \frac{D_{AB}}{RTZ\, P_{EM}}(P_{A_i} - P_{A_G}) = K_G'(P_{A_i} - P_{A_G}) \quad (4.166)$$

It is clear from Equations 4.165 and 4.166 that

$$K_G' = K_G \frac{P_{BM}}{P} \quad (4.167)$$

Different mass transfer coefficients can be defined depending on the system of units chosen for expressing the concentrations and whether it is equimolal counterdiffusion or diffusion of A through stagnant B. For example,

$N_A = K_Y'(Y_{A_i} - Y_{A_b}) = K_Y(Y_{A_i} - Y_{A_b})$

or $N_A = K_X(X_{A_i} - X_{A_b}) = K_x(X_{A_i} - X_{A_b})$

where Y_A and X_A are the mole fractions of A in gas and liquid, respectively, and the subscripts i and b refer to the interface and bulk, respectively. The prime refers to equimolal counterdiffusion. It can also be shown that

$$K_Y' = K_Y \frac{P_{BM}}{P} = K_G P_{BM} \quad (4.168)$$

$$\text{and } K_x' = K_x X_{BM} = K_L C \quad (4.169)$$

Empirical correlations are available for the prediction of mass transfer coefficients for different situations. In many cases, mass transfer coefficients can be calculated based on the analogy between heat, mass, and momentum transfers via friction factor and heat transfer coefficients.

4.11.2 Interface and Overall Mass Transfer Coefficients

So far, only transfer in a single phase has been considered, and the individual mass transfer coefficients K_x and K_y have been defined. There are many situations in which resistances are present in both fluid phases; for example, absorption of a soluble gas from an inert gas into a liquid solvent. In such a case, there are resistances on the gas side as well as the liquid side. In fact, in almost all practical situations of mass transfer, resistances are present in both fluid phases.

Figure 4.33 shows a concentration profile established at a particular cross section of a gas absorber operating under steady-state conditions. The concentrations in the bulk and at the interface are given. It is assumed that the interface itself offers no resistance to mass transfer. That is, the mole fractions Y_{A_i} and X_{A_i} in the gas and

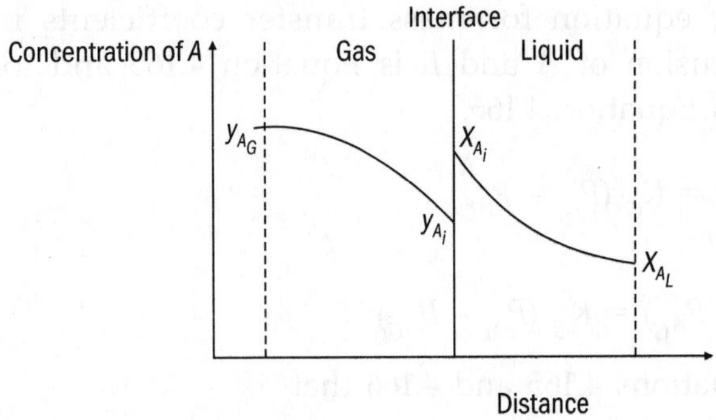

Figure 4.33 Mass transfer across an interface

liquid phases, respectively, are always in equilibrium and can be obtained from the equilibrium distribution applicable to the system. The entire resistance to mass transfer is then supposed to be vested in the gas and liquid films present on either side of the interface. That is, on the gas side the entire concentration change $(Y_{AG} - Y_{A_i})$ is present exclusively on the gas film and, similarly, $(X_{A_i} - X_{AL})$ is present exclusively on the liquid film. Beyond these films, the gas and liquid concentrations are constant at Y_{AG} and X_{AL}, respectively.

The rate at which A reaches the interface at steady state must be equal to the rate at which A is carried away from the interface to the bulk of the liquid by diffusion (there is no accumulation at the interface).

$$N_A = K_Y(Y_{AG} - Y_{A_i}) = K_X(X_{A_i} - X_{A_L}) \tag{4.170}$$

$$\text{or} \quad \frac{Y_{A_G} - Y_{A_i}}{X_{A_L} - X_{A_i}} = -\frac{K_x}{K_y} \tag{4.171}$$

In Figure 4.34, points P and M represent bulk and interfacial concentrations, respectively.

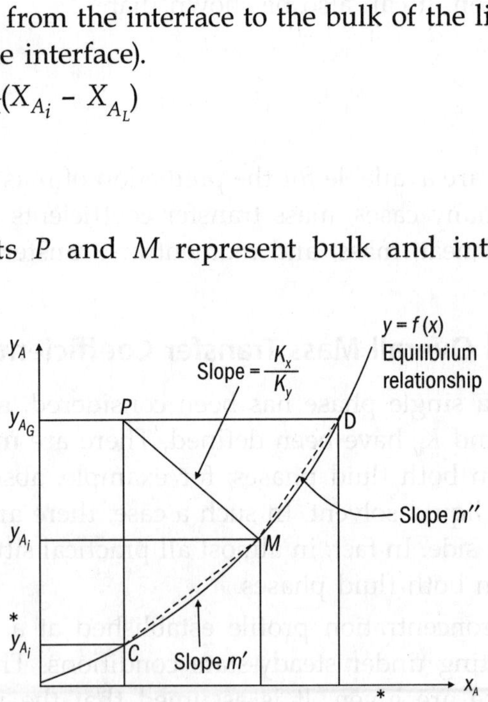

Figure 4.34 Bulk and interfacial concentrations in interface mass transfer

4.11.3 Local and Overall Mass Transfer Coefficients

Since it is not possible to measure interfacial concentrations, the overall effects are determined by defining overall mass transfer coefficients. The overall gas and liquid transfer coefficients K_Y and K_X, respectively, are defined as given by Equation 4.172.

$$N_A = K_Y(Y_{AG} - Y_A^*) = K_X(X_A^* - X_{A_L}) \tag{4.172}$$

where Y_A^* is the fictitious gas concentration in equilibrium with X_{A_L}, although they are not in contact with each other. Similarly, X_A^* is the fictitious liquid concentration in equilibrium with Y_{AG}.

$$(Y_{AG} - Y_A^*) = (Y_{A_G} - Y_{A_i}) + (Y_{A_i} - Y_A^*) \tag{4.173}$$

$$\text{or } (Y_{AG} - Y_A^*) = (Y_{AG} - Y_{A_i}) + m'(X_{A_i} - X_{A_L}) \tag{4.174}$$

Thus,

$$\frac{N_A}{K_Y} = \frac{N_A}{K_Y} + m'\frac{N_A}{K_X} \tag{4.175}$$

$$\text{or } \frac{1}{K_Y} = \frac{1}{K_Y} + \frac{m'}{K_X} \tag{4.176}$$

where M' is the slope of the chord CM.

This gives the concept of addition of resistances, that is, the total resistance $1/K_Y$ is equal to the sum of the gas film resistance $1/K_Y$ and the liquid film resistance m'/K_X. Similarly, it can be shown that

$$\frac{1}{K_X} = \frac{1}{m''K_Y} + \frac{1}{K_X} \tag{4.177}$$

From Equation 4.177 it is clear that the total resistance $1/K_X$ is equal to the sum of the gas film resistance $1/m''K_Y$ and liquid film resistance $1/K_X$. Equations 4.176 and 4.177 indicate that the relative magnitudes of the gas and liquid film resistances are dependent on the magnitudes of m' and m''.

The combined gas and liquid film resistances, $1/K_X$, for the absorption of a highly soluble gas (like NH_3) in a solvent (like water) is approximately equal to $1/K_Y$, which is the gas side resistance, since m'' is small (and m'/K_X can be neglected) and $K_X \approx K_Y$. Such systems are supposed to be gas film controlled.

Similarly, for the absorption of a sparingly soluble gas (like CO_2) in water, where the value of m'' is high, it is seen that $1/K_X \approx 1/K_X$. That is, the combined resistance is equal to the liquid film resistance. Such systems are supposed to be liquid film controlled.

For gas film controlled mass transfer processes, decrease in the gas side resistance improves the mass transfer rate, that is, by decreasing the gas film thickness. It can also be achieved by increasing the gas velocity and, therefore, its turbulence. This can be achieved by a shower of liquid through a column of gas rather than bubbling the gas through the liquid. The latter methodology is more effective in improving the rate of mass transfer in liquid film controlled situations, like absorption of a sparingly soluble gas in water, in which optimum results are obtained by decreasing the liquid film resistance by creating turbulence on the liquid side.

The controlling resistances are dependent on the relative magnitudes of K_Y and K_X and on the value of the slope of equilibrium relation m (expressed as $y = mx$). The equipment for measuring the resistance is dictated not only on the relative magnitude of the resistances as described above but also on the considerations of the maximum possible interfacial area. Volumetric mass transfer coefficients like K_La and K_ga, where a is the interfacial area per unit volume of a continuous contacting equipment (for example, packed bed), are used for such measurements.

4.12 HEAT TRANSFER BY CONDUCTION

Conduction is the transfer of heat from one part of a body to another part of the same body, or from one body to another in physical contact with it, without appreciable displacement of the particles within the body (Perry 1963).

4.12.1 Fourier's Law

The rate of heat flow is directly proportional to the temperature difference between two points of a body, directly proportional to the surface area perpendicular to the direction of flow, and inversely proportional to the length of flow (thickness of the slab). Heat flows from higher temperatures to lower temperatures at steady-state conditions. A schematic representation of Fourier's law is given in Figure 4.35.

$$\frac{Q}{t} = K\frac{A\ (T_1 - T_2)}{X} \tag{4.178}$$

where Q is the amount of heat transferred in time t and K is the proportionality constant. K is defined as the thermal conductivity of the material of the body and is dependent on temperature.

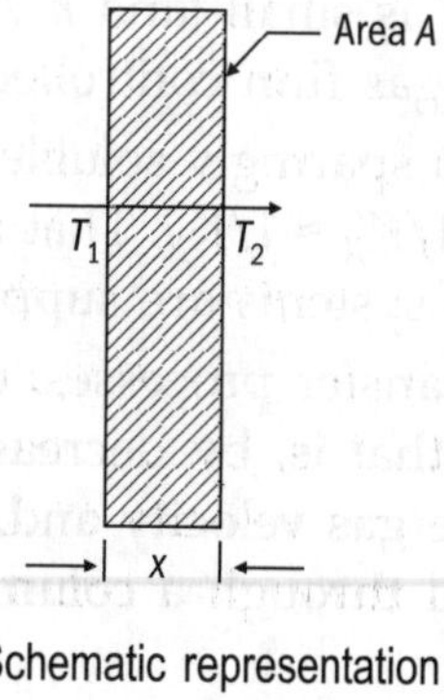

Figure 4.35 Schematic representation of Fourier's law

Fourier's law is the fundamental differential equation (Equation 4.179) derived from Equation 4.178.

$$\frac{dQ}{dt} = - K\,A\,\frac{dt}{dx} \tag{4.179}$$

In a solid body, heat energy is associated with molecular vibrations; higher temperatures characterize higher degrees of vibration. Consequently, with heat flow, higher vibration energy is transmitted to adjacent molecules with lower vibration energy and so on. In a liquid, molecules are much more loosely bound and the energy is associated with translation, rotation, and vibration. In a gas, the molecules are in random motion and the energy is transmitted through collision of molecules, the rate of which is higher in higher temperature zones.

The conduction mode of heat transfer is always related to molecular movement. In liquids and gases, bulk molecular movement is easily possible. However, heat transfer through bulk movement is defined as convection and should not be confused with conduction. For this reason, conductivity measurements in liquids and gases are difficult.

Conductive heat transfer in a highly conductive solid material, such as a metal, is largely affected by the movement of electrons as well as molecular vibration. The movement of electrons is similar to electrical conduction. Hence, good electrical conductors are also good thermal conductors, and a definite proportionality exists between electrical and thermal conductivities of highly conducting materials.

In most actual situations, heat transfer is complex as there are other modes of heat transfer involved besides conduction. In such cases, the concept of resistance is used in which temperature is the driving force of heat transfer. The conduction resistance $\frac{X}{KA}$ is given by Equation 4.180.

$$\frac{Q}{t} = \frac{(T_1 - T_2)}{\dfrac{X}{KA}} \tag{4.180}$$

Other forms of heat transfer can be expressed in similar ways, and the resistances can then be added to have combined heat transfer rates. In addition, conductive heat transfer in composite solid bodies can be expressed as additive resistance terms.

4.12.2 Three-dimensional Heat Transfer

A generalized unsteady state equation of conduction in three dimension is normally used to deal with conductive heat transfer in complex situations, where heat transfer may be unsteady (that is, temperatures vary with time), and in the presence of heat sources or sinks (Equation 4.181 and Figure 4.36).

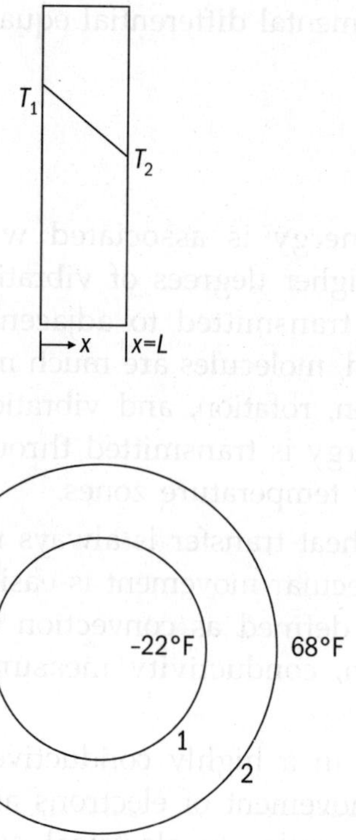

Figure 4.36 Schematic representation of three-dimensional heat transfer

$$\frac{\partial^2 T}{\partial X^2} + \frac{\partial^2 T}{\partial Y^2} \frac{\partial^2 T}{\partial Z^2} + \frac{W_i}{K} = \frac{1}{\alpha} \frac{\partial T}{\partial t} \qquad (4.181)$$

where W_i is the internal heat generation rate per unit volume

$$\alpha = \frac{K}{\rho C_P}$$

where K is the thermal conductivity of the material, ρ is the density, and C_P is the specific heat under constant pressure. Equation 4.181 can be solved under innumerable conditions with appropriate mathematical tools. Similar equations in other coordinates are as follows.

Cylindrical (r, θ, z) coordinates

$$\left[\frac{1}{r}\frac{\partial}{\partial r}\left(r\frac{\partial T}{\partial r}\right) + \frac{1}{r^2}\frac{\partial^2 T}{\partial \theta^2} + \frac{\partial^2 T}{\partial Z^2}\right] + \frac{W_i}{K} = \frac{1}{\alpha}\frac{\partial T}{\partial t} \qquad (4.182)$$

Spherical (R, θ, ϕ) coordinates

$$\frac{1}{r^2}\left[\frac{\partial}{\partial r}\left(r^2\frac{\partial T}{\partial r}\right) + \frac{1}{\sin\theta}\frac{\partial}{\partial\theta}\left(\sin\theta\frac{\partial T}{\partial\theta}\right) + \frac{1}{\sin^2\theta}\frac{\partial^2 T}{\partial\theta^2}\right] + \frac{W_i}{K} = \frac{1}{\alpha}\frac{\partial T}{\partial t} \qquad (4.183)$$

Equations 4.182 and 4.183 can be transformed into single-state systems. For example, for an infinite flat plate under steady conditions, the boundary conditions are $x = 0$, $T = T_1$, and $x = L$, $T = T_2$. The equation for such a case is

$$\frac{\partial^2 T}{\partial X^2} = 0 \tag{4.184}$$

Double integration of Equation 4.184 gives $T = C_1X + C_2$, where C_1 and C_2 are constants. At $X = 0$, $T_1 = T_2$ and at $X = L$, $T_2 = C_1L + T_1$; $C_1 = (T_2 - T_1)/L$.

The temperature distribution is

$$T = (T_2 - T_1)\frac{X}{L} + T_1$$

or
$$\frac{\partial T}{dX} = \frac{T_2 - T_1}{L}$$

From Fourier equation,

$$\frac{Q}{t} = -KA\frac{\partial T}{\partial X} = \frac{KA}{L}(T_2 - T_1)$$

Similarly, for steady-state heat transfer through the wall of an infinite hollow cylinder, under boundary conditions $r = r_1$, $T = T_1$ and $r = r_2$, $T = T_2$,

$$\frac{1}{r}\frac{\partial}{\partial r}\left(r\frac{\partial T}{\partial r}\right) = 0$$

4.13 TUTORIALS

Question 1 Problem on heat transfer in a fermenter

(a) Calculate the Btu/h that must be removed to hold constant temperature in an aerobic fermentation stirred with 0.7 HP/100 gal, with glucose consumption of 10 g/L-h, of which 50% is converted to CO_2 and water and the rest is assimilated. Assume that the 10000 gal tank is filled to 7.5 gal.

(b) Assuming 2 in 5.5 tubing as coils (ID = 2 in approximately), find the length of coil for $U = 100$, $\Delta t = 12$°F and make reasonable assumptions as desired.

(c) As a design engineer, would you recommend fabrication?

Answer Heat evolution from agitator = 0.7 HP × 2545 = 1780 Btu/h

From respiration:

$$= \frac{5\ \text{g}}{\text{L} \times \text{h}} \times \frac{1\ \text{mol}}{180\ \text{g}} \times \frac{378.5\ \text{L}}{100\ \text{gal}} = \frac{10.5\ \text{g-mol}}{100\ \text{gal-h}}\ \text{[since 50\% is assimilated]}$$

$$\text{Heat generated} = \frac{673\ \text{K-cal}}{\text{g-mol}} \times \frac{10.5\ \text{g-mol}}{100\ \text{gal-h}} \times \frac{3.96\ \text{Btu}}{\text{K-cal}} = 28\,000\ \text{Btu/h}$$

(a) Basis: 378 L
Total heat generated = 1780 + 28 000 ≈ 29 800 Btu/h
This is the amount of heat to be removed.

(b) Length of coil (in feet) required = $\dfrac{q}{U\Delta t\ (\text{ft}^2/\text{ft})}$

or $\dfrac{\text{ft}}{7500 \text{ gal}} = \dfrac{75 \times 29800}{100 \times 12 \times 0.524}$ [for 2 in tubing, $\text{ft}^2/\text{ft} = 0.524$]

Thus,

$$\text{length (in feet) required} = \frac{75 \times 29800}{100 \times 12 \times 0.524} \times 7500 \text{ ft} = 3550 \text{ ft}$$

(c) 28 387 L ≈ 1000 ft^3
Let $z/D = 2$ [height/diameter]
Assume therefore,
2.7432 m in diameter × 5.4864 m tall per coil
7.62 m ≈ 25 ft tubing
5 coils/ft × 15 ft height = 75 coils
Thus, total reasonable length is 573 m. This indicates that something should be done about U or the temperature of the water should be decreased.

4.14 HEAT TRANSFER BY CONVECTION

Convection is the transfer of heat from one point to another within a fluid (gas or liquid) by the mixing of one portion of the fluid with another. In natural convection, the motion of the fluid is entirely due to the difference in density between the two points within the fluid, resulting from temperature differences. In forced convection, the motion is produced by mechanical means.

Most useful situations of heat transfer are involved in heating and cooling of materials (in most cases, fluids). Convection is always associated with conduction and sometimes with radiation.

Fluid motion is an extremely complex phenomenon, and it is not always possible to express complex heat transfer situations in rigorous mathematical terms. When possible, such mathematical expressions are extremely difficult to solve for actual situations. Hence, most relations and equations expressing convective heat transfer are empirical in nature. Such empirical equations are derived from dimensional analysis and experimental measurements to arrive at groupings of variables (mostly in dimensionless groups) and empirical constants.

4.14.1 Forced Convection

Let us consider a situation of forced convection, where a cold fluid is forced past a hot solid surface. In the process, the cold fluid is heated by heat transfer from the solid surface. The physical situation can be approximated as follows. The solid surface of a finite area can be at constant temperature T_1. There will be a thin film of fluid, nearly stagnant, close to the solid surface. The bulk of the fluid across the hot area may be well mixed and assigned a fixed temperature T_2. It can be assumed that bulk of the resistance to heat flow will be within this film. Equation 4.185 represents the conductance/resistance form of heat transfer.

$$Q/t = hA(T_1 - T_2) = (T_1 - T_2)/(1/hA) \quad (4.185)$$

where h is the heat transfer coefficient and $1/hA$ is the inverse of heat transfer resistance. This is the film theory of convective heat transfer.

The heat transfer coefficient poses difficulty in estimation, as the film thickness cannot be accurately measured or estimated. Also, the physical state of the entire thickness of the film may not be constant or predictable. Furthermore, the constancy of temperatures cannot be assumed except in the differential area of surface and the differential volume of the liquid. In many situations, averaging of temperatures is adequate.

4.14.2 Application of Dimensional Analysis in Heat Transfer

By dimensional analysis, the relationship of the heat transfer coefficient with other variables in case of forced convection is given in Equation 4.186.

$$hD/K = C(DG/\mu)^a \, (Cp\,\mu/K)^b \quad (4.186)$$

where the dimensionless groups are hD/K, known as Nusselt number; DG/μ, known as Reynolds number; and $Cp\,\mu/KP$, known as Prandtl number; D is the characteristic length of fluid across the heat transfer surface (diameter in case of a fluid flowing through a pipe); K is the thermal conductivity; G is the mass velocity (u); μ is the viscosity; and C_p is the heat capacity.

The constants C, a, and b are obtained from actual experiments, and these values are listed for many industrial situations for steady state.

4.14.3 Average and Overall Heat Transfer Coefficients

The temperature difference between a solid surface and a fluid, with uniformly variable temperatures in the direction of flow under steady state, is averaged on the basis of logarithmic mean (Equation 4.187).

$$T_{av} = \frac{\Delta T_1 - \Delta T_2}{\ln (T_1/T_2)} \quad (4.187)$$

where ΔT_1 and ΔT_2 are the temperature differences between the solid surface and the fluid at two points across the direction of flow.

In situations where phase transformation occurs, such as evaporation or condensation, the correlations become more complex. However, similar correlations are employed for individual heat transfer coefficients. In case of natural convection, the correlation takes the form as given by Equation 4.188.

$$hD/K = C\,(D^3\,\rho^2 g\,\beta\Delta T/\mu^2)^a\,(C_p\,\mu/K)^b \qquad (4.188)$$

where the Reynolds number is replaced by Grashoff number, ρ is the volumetric expansion of the fluid, ΔT is the temperature (β is coefficient) difference between solid and fluid. Depending on the situation, the constants are different.

In an actual situation, the simplest problem could be heat transfer from a heating fluid to a coolant through an intervening conducting metal wall. In such cases, the concept of overall heat transfer coefficient (U) is used (Equation 4.189).

$$U = \frac{1}{h_1\,A_1} + \frac{L}{KA_{av}} + \frac{1}{h_2\,A_2} \qquad (4.189)$$

and $Q/t = UA\,(T_1 - T_2)$

where A is an average or area of one side, with applicable correction used.

In complex industrial situations, many more factors are introduced, which are mostly empirical (but well defined), depending on the complexity of the problem.

4.15 HEAT TRANSFER BY RADIATION

Heat or energy is carried from the sun to the surface of the earth by electromagnetic waves that are transferred freely through the vacuum of the intervening space. This process of heat or thermal energy transfer is known as radiation. Heat transfer by thermal radiation involves the transport of electromagnetic energy from a source to a sink. Thus, radiation is heat transfer via the emission of electromagnetic energy. All objects radiate/emit energy; the amount of energy emitted increases with an increase in temperature. Objects that absorb all the incident energy are called black bodies. However, no real surface behaves as a black body. Heat transfer by radiation is governed by Kirchhoff's law, which states that at thermal equilibrium, the emissivity of a body equals the absorptivity.

4.15.1 Influential Factors

4.15.1.1 Environmental factors

Since radiation consists of wavelengths predominantly in the infrared region of the spectrum, it travels in straight lines until it is obstructed by other materials. Accordingly, environmental conditions play a significant role on the quantum of radiation energy received by an object. The object that receives radiation reflects, transmits or absorbs it. The amount of the radiation absorbed is transformed into thermal energy. Thus, a continuous exchange of radiant energy exists between bodies.

4.15.1.2 Geometric

In case of black bodies, the net exchange of radiant energy depends on the geometric configuration of the radiator as well as the absorber.

4.15.1.3 Surface properties and colour

Surface properties and colour relate to the following factors.

Absorbance: If the body absorbs radiation of all wavelengths with the same absorptivity, it is called a grey body. For such bodies, absorptivity is independent of the energy distribution in the incident radiation. In such cases, the emissivity (*e*) may be equal to absorptivity (*a*).

A black body is a theoretical idealization for the prefect radiator. It is a hypothetical material that absorbs 100% of the incident radiation.

Refractance: For refractory materials, emittance and absorptance depend on the grain size and impurity concentration in the body.

Body opaqueness: Depending on the opaqueness of the absorber, the emittance and absorptance varies considerably.

4.15.2 Fundamental Theoretical Aspects

According to energy balance at thermal equilibrium, Equation 4.190 shows the relationship between the radiant energy incident (*I*) on the unit surface of a body in an enclosed space and the emissive power.

$$Ia_1 = E_1 \tag{4.190}$$

where a is the absorptivity and E_1 is the emissive power. For a black body where $a = 1$ and $E_B = I$, the emissive power is given as shown in Equation 4.191.

$$e_1 = \frac{E_1}{E_B} = \frac{Ia_1}{I} = a_1 \tag{4.191}$$

Thus, emissivity, *e*, can be defined as the ratio of energy that a body emits relative to the energy emitted by a perfect radiator/black body of equal area and under similar conditions. Absorptivity, a, is the ratio of the energy absorbed by a body to the energy absorbed by a black body of equal area and under similar conditions. The emissive power, *E*, of the surface of the body is the total energy emitted per unit area per unit time. The intensity, *I*, of radiation is the amount of energy emitted per unit area per unit time per unit solid angle.

The energy emitted by a plane surface is given by Equation 4.192.

$$q = A_{\epsilon\sigma} T_A^4 \tag{4.192}$$

where q is the rate at which energy is radiated (in *Btu/h*), A is the area of the surface (in ft^2), ϵ is the emissivity, σ is the Stefan–Boltzmann constant, and T_A is the absolute temperature of the body (°*R*).

$\sigma = 1730 \times 10^{-12}$ *Btu/ft²-h-°R⁴*

Thus, $E = q/A$ (4.193)

From Equations 4.192 and 4.193, it is obvious that since the value of q is proportional to the fourth power of absolute temperature, a small increase in surface temperature significantly increases the amount of radiation emitted. Since bodies that emit radiant energy also absorb it, the net loss or gain in energy by a surface is equal to the difference between the energy emitted and absorbed. For a body (1) at temperature T_1 enclosed by its surroundings and (2) at temperature T_2, the rate of heat transfer from stage 1 to stage 2 is given by Equation 4.194.

$$q_{1,2} = A_1\sigma\left[e_1\,T_1^4 - a_{1\leftarrow 2}\,T_2^4\right] \quad (4.194)$$

where $a_{1\leftarrow 2}$ is the absorptivity.

If absorptivity is independent of the wavelength of incident radiation, then $a_{1\leftarrow 2}$ is equal to $\in_1$.

Equation 4.194 applies only to surfaces that are adjacent to each other. In actual practice, however, the equation needs modification since geometrical relationships exist between two surfaces.

$$q_{1,2} = 0.173\ \in_1\ A_1 - \left[\left(\frac{T_1}{100}\right)^4 - \left(\frac{T_2}{100}\right)^4\right] F$$

where F is the geometric factor.

The values of F for many geometric relationships are given in *Perry's Chemical Engineers' Handbook.*

4.15.2.1 Tutorials

Question 1 Two concentric spheres of diameters 8 in and 12 in, with the space between them evacuated, are used to store liquid air at –220°F in a room at 68°F. If the surfaces of the spheres are coated with aluminium and the latent heat of liquid air is 90 Btu/ *lb*, estimate the following.

(a) F_{1-2}, F_{2-1}.

(b) Number of pounds of air vaporized per hour.

Assume surface resistance to be negligible.

Answer

(a) $F_{1-2} = 1$

Since all radiation from surface 1 is absorbed by surface 2,

$$A_1 = 4\pi \times \left(\frac{4}{12}\right)^2$$

$$A_2 = 4\pi \times \left(\frac{6}{12}\right)^2$$

Now,

$$\frac{F_{1-2}}{F_{2-1}} = \frac{A_2}{A_1}$$

Thus,

$$F_{2-1} = F_{2-1} = F_{1-2} \times \frac{A_1}{A_2}$$

$$\text{or } F_{2-1} = 1 \times \frac{4\pi\,(4/12)^2}{4\pi(6/12)^2}$$

$$\text{or } F_{2-1} = \left(\frac{4}{6}\right)^2 = \frac{4}{9} = 0.444$$

(b) $\sigma = 0.173 \times 10^{-8}$ Btu/h-ft²/°R

$$q = F_{1\text{-}2}F_{\epsilon}\sigma\, A_1(T_1^4 - T_2^4)$$

$\alpha = \epsilon = 0.05$ for polished aluminium.

Thus, $\rho_1 = 1 - 0.05 = 0.95 = \rho_2$

$$A_1F_{1\text{-}2}F_{\epsilon} = \frac{1}{\dfrac{\rho_1}{A_1\,\epsilon_1} + \dfrac{1}{A_1} + \dfrac{\rho_2}{A_2\,\epsilon_2}}$$

$$\text{or } A_1F_{1-2}F_{\epsilon} = \frac{1}{\dfrac{0.95}{0.05 \times 4\pi\left(\dfrac{4}{12}\right)^2} + \dfrac{1}{4\pi\left(\dfrac{4}{12}\right)^2} + \dfrac{0.95}{0.05 \times 4\pi\left(\dfrac{6}{12}\right)^2}}$$

$$\text{or } q = \frac{144 \times 144 \times 3.14}{389}\left(T_1^4 - T_2^4\right) \text{ lb/h}$$

Thus,

$$q = 23.18 \times 10^8 \text{ lb/h}$$

Question 2 The black, flat roof of a building has an emission of 0.9 and absorptivity of 0.8. Solar radiation of intensity 300 Btu/ft²-h beats down on it. If the temperature of air and surroundings is 80°F, wind velocity negligible, and if heat cannot penetrate the roof, what is its equilibrium temperature?

For the rate of heat transfer by conduction–convection, use Equation 4.195.

$$\frac{q}{A} = 0.38\ \Delta T^{1.25} \tag{4.195}$$

where Δt is the temperature drop between the roof and air in Fahrenheit.

Answer The amount of heat absorbed by the roof is

$300 \times \alpha = 300 \times 0.8 = 240$ Btu/h-ft²

Heat rejected by conduction and radiation is $q_c + q_r$

$q_c + q_r = 0.38(\Delta t)^{1.25} + q_r$

or $q_c + q_r = 0.38\ (T - 80)^{1.25} + q_r$

where T is the temperature of the roof.

$q_r = A\sigma F_\epsilon F_{12}\ (T_1^4 - T_2^4)$

or $\dfrac{q_r}{A} = \sigma F_\epsilon F_{12}(T_1^4 - T_2^4)$

Here,

$T_1 = T + 460$

$T_2 = 460 + 80 = 540$

F_{12} is the view factor.

$F_{12} = 1$

Thus,

$$F_\epsilon = \frac{1}{\dfrac{1}{\epsilon_1} + \dfrac{1}{\epsilon_{2-1}}}$$

$$\text{or } F_\epsilon = \frac{1}{\dfrac{1}{0.9} + \dfrac{1}{1} - 1} = 0.9$$

Thus,

$\sigma = 0.173 \times 10^{-8}$

$\dfrac{r}{A} = 0.173 \times 10^{-8} \times 0.9 \times 1\ [(T + 460)^4 - (540)^4]$

Thus,

$q_c + q_r = 0.38\ (T - 80)^{1.25} + 0.173 \times 10^{-8} \times 0.9\ [(T + 460)^4 - (540)^4]$

At equilibrium, the amount of heat absorbed by the roof must be equal to the heat rejected ($q_c + q_r$) by conduction and radiation. Thus,

$$240 = 0.38\ (T - 80)^{1.25} + 0.173 \times 10^{-8} \times 0.9 \times 1[(T + 460)^4 - (540)^4] \quad (4.196)$$

T can be evaluated from Equation 4.196.

Question 3 Estimate the maximum temperature attained by a flat roof in clear weather. The roof is coated with aluminium and has an absorptivity of 0.03 for solar radiation and an emissivity of 0.07. Radiation from sources other than the sun may be neglected and a convection heat transfer coefficient of 2.0 Btu/h-ft^2-°F may be assumed. A maximum temperature of 100°F may be assumed for the surrounding air. Solar radiation may be assumed to be 450 Btu/h-ft^2.

Answer Heat absorbed by the roof = $q_{convection} + q_{radiation}$

Now,

$$q_{convection} + q_{radiation} = 450 \times 0.03 \text{ Btu} = 13.5 \text{ Btu}$$

Heat rejected by convection and radiation = $q_i + q_r$

Now,

$$\frac{q_t}{A} = 2 \times 1(T - 100)$$

$$\frac{q_r}{A} = \sigma F_{12} F_{\epsilon}\left(T_1^4 - T_2^4\right)$$

$$\text{or } \frac{q_r}{A} = 0.173 \times 10^{-8} \times 1 \times F_{\epsilon}[(T + 460)^4 - (560)^4]$$

$$F_{\epsilon} = \frac{1}{\frac{1}{\epsilon_1} + \frac{1}{\epsilon_2} - 1} = \frac{1}{\frac{1}{0.07} + \frac{1}{1} - 1} = 0.07$$

$$\text{or } F_{\epsilon} = 2T - 200 + 0.173 \times 10^{-8} \times 0.07\ [(T + 460)^4 - (560)^4]$$

Now,

$$13.5 = 2T - 200 + 0.173 \times 10^{-8} \times 0.07\ [(T + 460)^4 - (560)^4] \quad (4.197)$$

T can be evaluated from Equation 4.197.

Question 4 Show that for infinite parallel plates of emissivity ϵ_1 and ϵ_2,

$$F_{\epsilon} = \frac{1}{\frac{1}{\epsilon_1} + \frac{1}{\epsilon_2 - 1}}$$

Answer We have

$$\frac{1}{F_{12}} = \left(\frac{1}{\epsilon_1} - 1\right) + \frac{A_1}{A_2}\left(\frac{1}{\epsilon_2} - 1\right) + \frac{1}{F_{12}} \quad (4.198)$$

Now, for infinite parallel plates

$$F_{12} = 1$$

$$\frac{A_1}{A_2} = 1$$

Thus,

$$\frac{1}{F_{12}} = \frac{1}{\epsilon_1} - 1 + \frac{1}{\epsilon_2} - 1 + 1$$

$$\text{or } \frac{1}{F_{12}} = \frac{1}{\epsilon_1} + \frac{1}{\epsilon_2} - 1$$

$$\text{or } F_{12} = \frac{1}{\frac{1}{\epsilon_1} + \frac{1}{\epsilon_2} - 1}$$

4.16 EVAPORATION

The concentration of solutions of non-volatile solutes through heat-induced vaporization of the solvent (generally water) is called evaporation. Evaporation is a relatively expensive operation, and its application in waste treatment is limited primarily to the recovery of valuable by-products from waste liquids and to the treatment of wastes where no alternative methods are available.

Essentially, evaporation involves the supply of heat to a boiling solution held in a container that is vented to permit the continuous removal of vapour collecting over the liquid. Heat must be supplied in quantities sufficient to raise the temperature of the solution to its boiling point and to furnish the latent heat required by the fraction of the solvent that vaporizes. Although hot gases, solar irradiation, and direct fire are sometimes used to meet these requirements, indirect steam is employed as a rule.

Evaporation is often carried out at pressures less than atmospheric pressure. Operations under a vacuum establish a larger temperature difference between the heat source and the boiling liquid. This, in turn, increases the rate of evaporation. Furthermore, reduced boiling temperatures serve to protect any organic fraction of the evaporating solution from thermal decomposition. Evaporators may be employed in batches or in a continuous operation. Among the evaporators used extensively in food processing biotechnology, climbing and falling film type are most suitable. Fruit juices are prepared in concentrated form in this type of evaporator for the ease of transportation and marketing.

In order to produce a satisfactory tomato juice concentrate, the final specific gravity of the juice should be about 1.19. Fresh juice has a specific gravity of about 1.01. In order to retain the quality of the juice with respect to flavour, odour, and taste on subsequent reconstitution with water, evaporation must be carried out at as low a temperature as possible and in as short a time as possible. With this objective in mind, the tomato juice should not be heated beyond 130°F (54.4°C). Consequently, evaporation is carried out under vacuum. Steam temperatures should be as low as possible, consistent with the size of the heating surface. To produce concentrated juice, either a falling film-type evaporator or a climbing film-type evaporator is used. However, the following discussion refers to a climbing film-type evaporator.

In a climbing film evaporator, the liquid to be evaporated travels up the tubular surface in the form of a film, so that evaporation takes place in a single passage. The climbing film evaporator may also be arranged for evaporation in effects. The single effect is discussed in subsequent paragraphs.

Evaporation in the climbing film evaporator is similar to the long tube natural circulation evaporation, except that the tubes of the calandria are 15–20 in long, with a preferred diameter of 2 in. A centrifugal separator is invariably fitted. A detachable cover for the tube cleaning is placed above the tubes of the calandria.

The operation of the plant is carried out by feeding the liquid at such a rate that the required concentration is affected in one passage of the tubes. Steam pressure is set to provide the desired temperature drop. The main characteristics of the climbing film evaporators are (1) they have relatively high overall heat transfer coefficients, and (2) there is a short time of contact between the liquid and the evaporating surface (of the order of 10–20 s).

The evaporator is operated under vacuum because liquids boil at a much lower temperature under vacuum than in the open and water is removed with less injury by heat to the juice. Due to the operation under vacuum, caramelization of the sugar contained in the juice, loss in flavour and colour as a result of oxidation are also minimized. To obtain the best results, a vacuum of at least 28 in Hg should be employed; this will result in a sufficiently low boiling point so that the juice is not injured materially.

4.16.1 Steps in Designing

The various factors required in designing are as follows.

- Fruit juice solid concentration fed into the evaporator should be known.
- Solid material should be balanced to know the feed rate of the juice.
- The amount of evaporation (that is, feed rate of the liquid-production rate of concentrated liquid) should be estimated from the feed rate and production rate.
- The amount of raw fruit needed to produce the required amount of juice per day from the experiment should be estimated.
- The number of effects of evaporator required to achieve the desired concentration should be estimated.

4.16.2 Typical Design Calculation

About 2000 lb of concentrated tomato juice of 35% total solid content is produced per day (24 h). The vacuum is produced by means of a two-stage ejector system. Design and draw the evaporator, barometric condenser, and two-stage ejector unit.

4.16.2.1 Scope of the design

The items in design are listed as follows.

1. Simplifying assumptions in the design.
2. Material balance in the evaporator.
3. Heat balance in the evaporator.
4. Material and heat balance for the condenser and ejector.
5. Determination of economic number of effects.

6. Simplifying assumptions.
7. Heat balance calculations.
8. Heating area calculations.
9. Number of heating tubes calculations.
10. Arrangement of the tubes.
11. Shell diameter.
12. Thickness of evaporative body.
13. Outer diameter of the shell.
14. Height of the evaporative body.
15. Feed pipe diameter of the shell.
16. Concentrated liquor outlet pipe diameter.
17. Vapour outlet pipe diameter.
18. Condensate outlet pipe diameter.
19. Separator design.
20. Barometric condenser design.
    (a) Cooling water requirement.
    (b) Cooling water inlet pipe diameter.
    (c) Tail pipe size.
    (d) Air outlet pipe size.
    (e) Condenser dimensions.
    (f) Baffle plates in the condenser.
    (g) Hot well.
21. Design of two-stage steam ejector.
    (a) Steam consumption.
    (b) Diameter of the steam nozzle.
    (c) Cooling water requirement in the intercondenser.
    (d) Length of the diffuser body.
    (e) Cooling water pump.
22. Materials of construction.

4.16.2.2 Simplifying assumptions

The following simplifying assumptions have been made to use the equations involved in the calculations.

- Heat consumed in the evaporation of 1 lb of water in the evaporator from the juice is equal to the heat of evaporation of 1 lb of pure water when evaporated at superincumbent pressure.

- Feed temperature to the evaporator is assumed to be 86°F.
- Steam at 14.7 lb/in^2 absolute is used as the heating agent in the evaporator.
- During heat exchange, steam is just condensed and no subcooling takes place thereafter.
- The heat capacity of feed solution is assumed to be 0.60.
- The overall heat transfer coefficient in the evaporator is taken to be 150 Btu/ft^2-°F-h.

4.16.2.3 *Material balance for the evaporator*

Tomato juice contains 4.8%–6.8% total solids (Pearson 1970). Let us assume that the tomato juice to be concentrated contains 6% total solids. By solid material balance,

$F \times 0.06 = P \times 0.35$

where F is the total feed/day and P is the product/day.

or $F \times 0.06 = 2000 \times 0.35$

Thus,

$$F = \frac{2000 \times 0.35}{0.06} \text{ lb/day}$$

or $F = 11\,667$ lb/day

$$\text{or } F = \frac{11\,667}{24} \text{ lb/h} = 482 \text{ lb/h}$$

Amount of water evaporated = 11 667 – 2000 = 9667 lb/day

That is,

$$\text{amount of water evaporated} = \frac{9667}{24} = 402 \text{ lb/h}$$

Now, we have to find the weight of raw tomato that will produce 482 lb of tomato juice per day. From experimental data, 1220 lb of raw tomato produced 520 lb/h of tomato juice containing 6% total solids.

Therefore, the amount of raw tomatoes that will produce 482 lb/h of tomato juice is $\frac{482 \times 1220}{520}$ lb/h = 944 lb/h.

Or the amount of raw tomatoes required per day is 944 × 24 = 22 656 lb/day. By economic considerations, the optimum number of effect is found to be one. Steam enters the calandria of the evaporator at 14.7 psia or 0 psig (temperature = 212°F).

The boiling point of the solution in the evaporator is equal to the temperature corresponding to vacuum 28 in Hg (that is, 192 mm Hg pressure or 0.942 lb/in^2 absolute pressure). Thus, boiling point of solution in evaporator is 106.6°F.

Amount of evaporation = 402 lb/h

Steam consumption = 434 lb/h

Thus,

$$\text{evaporation per lb of steam} = \frac{402}{434} = 0.93 \text{ lb}$$

4.16.2.4 Heat balance in the evaporator

Amount of evaporation = 402 lb/h
Amount of steam supplied = 434 lb/h
Boiling point of solution in the evaporator = 106.6°F
Temperature of steam = 212°F
Latent heat of steam = 970 Btu/lb

Latent heat of vaporization of water corresponding to the boiling point in evaporator is 1029 Btu/lb.

Heat supplied by steam = 434 × 970 = 424 580 Btu

Heat utilized for the evaporation of 402 lb of water and to bring the temperature of feed to boiling point in the evaporator + heat loss to surroundings = 402 × 1029 + 482(106 – 86) Btu = 423 587 Btu

4.16.2.5 Material and heat balance for condenser and ejector

4.16.2.5.1 Condenser

Amount of vapour to be condensed = 402 lb/h
Temperature of vapour = 106°F
Water inlet temperature = 86°F
Latent heat at 106°F = 1029 Btu/lb
Cooling water required = 24 700 lb/h
Temperature of saturated vapour at 1.92 in of Hg (0.942 lb/in^2 absolute) = 106°F
Water outlet temperature = 101°F (assumed)
Heat supplied by vapour = 416 070 Btu
Heat utilized by cooling water = 24 700(101 – 86) Btu = 24 700 × 15 Btu = 38 000 Btu

4.16.2.5.2 Ejector

Table 4.13 shows the pressure, steam consumption, and nozzle diameter of the first two stages in the ejector.

Table 4.13 Pressure, steam consumption, and nozzle diameter of the first and the second stages

Parameter	First stage	Second stage
Suction pressure	0.942 psia	9.7 psia
Discharge pressure	9.7 psia	14.7 psia
Steam consumption	5.1 lb/h	15.0 lb/h
Nozzle diameter	0.04 in	0.05 in

Total air coming from evaporator top = 2.01 lb/h
Steam to be condensed in the intercondenser = 152 lb/h
Cooling water required = 742 lb/h

4.16.2.6 Determination of economic number of effects

Evaporation may be carried on by single operation or multiple effect. Multiple effect evaporation is the most common method of increasing the steam economy of the evaporator. The capacity does not increase proportionately with an increase in the number of effects. However, fixed capital and fixed charges increase with the increased number of effects, while the cost of steam cooling water decreases. The choice of the number of effects to be used in series in an evaporation system becomes an economic balance of cost of heat against fixed charges. Thus, a cost balance must be done to find the economic number of effects.

Feed rate = 11 667 lb/day containing 6% solid to be concentrated to 35% solid

$$\text{Steam cost} = \frac{11\,667 \times 0.06}{0.35\,n}\left[\frac{94}{6} - \frac{55}{35}\right] \times \frac{0.40}{1000}$$

where n is the number of effects.

$$\text{or steam cost} = \frac{11\,667 \times 6}{35\,n}\left[\frac{658 - 78}{42}\right] \times \frac{0.40}{1000}$$

$$\text{or steam cost} = \frac{11\,667 \times 6 \times 580 \times 0.40}{35 \times 42 \times 1000 \times n}$$

Thus,

$$\text{steam cost} = \frac{11.04}{n}$$

$$\text{Fixed charge} = \frac{0.45 \times 6400 \times n}{300} \quad \text{(assuming 300 operating days in a year)}$$

Thus, fixed charge is \$9.6n.

Table 4.14 lists the fixed charge and steam cost of the first five stages.

Thus, the total cost is least in the first effect and increases progressively from the second effect onwards. Therefore, under given conditions, the single effect evaporator is the best one.

Table 4.14 Fixed charge and steam constant of the first five stages

Parameter	First effect ($)	Second effect ($)	Third effect ($)	Fourth effect ($)	Fifth effect ($)
Fixed charge	9.60	19.20	28.80	38.40	48.00
Steam cost	11.04	5.52	3.68	2.76	2.20
Total	20.64	24.72	32.48	41.16	50.20

4.16.2.7 Heat balance calculations

Amount of water evaporated in the evaporator = 402 lb/h

Let W be the pound of steam furnished in the evaporator per hour.

The equation for heat balance is given by Equation 4.199.

$$FC_{pf}(T_f - 32) + W_s hf_g = Vh_g + PC_p(T_p - 32) \tag{4.199}$$

where F is the feed rate, V is the amount of vapour (in pound), W_s is the amount of steam (in pound), C_{pf} is the specific heat of feed, C_p is the specific weight of product, T_f is the temperature of feed, T_p is the temperature of product (that is, boiling temperature in evaporation), h_{fg} is the heat of condensation of steam, and h_g is the enthalpy of vapour leaving the evaporator (Keenan and Keyes 1961).

Here,

F = 482 lb/h

V = 402 lb/h

C_{pf} = 0.60

T_f = 86°F

h_{fg} = heat of condensation of steam at 212°F = 970 Btu/lb

$P = 2000 \text{ lb/day} = \frac{2000}{24} = 83.33 \text{ lb/h}$

T_p = 106.6°F corresponding to vacuum

h_g = 1029 Btu/lb

C_p = 0.35

Substituting these values in Equation 4.199, we get

$482 \times 0.60(86 - 32) + W_s \times 970 = 402 \times 1029 + 83.33 \times 0.35(106.6 - 32)$

or $15616 + 970W_s = 413658 + 2175$

or $970W_s = 415833 - 15616 = 400217$

Thus,

$$W_s = \frac{400217}{970} = 434 \text{ lb/h}$$

4.16.2.8 Calculation of heating area in evaporator

The rate of heat transfer to the evaporator from steam is $W_s h_{fg}$.

Now,

$$W_s h_{fg} = qUA(T_s - T_p) \tag{4.200}$$

where U is the overall heat transfer coefficient, A is the heat transfer area in evaporator, T_p is the temperature of boiling liquid in the evaporator, T_s is the temperature of steam, and W_s is the mass of steam condensed in steam chart.

Thus, from Equation 4.200, we have

$150 \times A \times (212 - 106.6) = 434 \times 970$

or $A = \dfrac{434 \times 970}{106 \times 150} = 40 \text{ ft}^2$

Assuming 20% extra for future provision,

area = 40 + 20 = 60 ft^2

4.16.2.9 *Calculation of number of heating tubes*

Tubes of 2 in normal size with a height of 18 ft are chosen in the present design (Clarke 1958).

Inner diameter of tube = 2.067 in

Outer diameter of the tube = 2.375 in

Thus, actual diameter is

$$\frac{2.375 - 2.067}{2.303 \log \dfrac{2.375}{2.067}} = 2.2 \text{ in}$$

Area of the tubes = $\pi \times \dfrac{2.2}{12} \times 18$

or area of the tubes = $\dfrac{3.14 \times 1.1 \times 8}{2}$

or area of the tubes = $\dfrac{1557}{2} = 7.73 \text{ ft}^2$

Thus, number of tubes = $\dfrac{60}{7.73} = 8$

4.16.2.10 *Arrangement of tubes*

Tubes are laid out on either square or triangular patterns. However, use of square pattern is advantageous because the pitch tubes are accessible for external cleaning and cause a lower pressure drop. The common pitch (centre-to-centre distance between adjacent tubes) for square layout is 1.25 times the outer diameter of the 2 in tubes.

Tube pitch = 1.25 × outer diameter of tubes

or tube pitch = 1.25 × 2.375 in = 2.97 in

Thus, the area required for the tube layout = $\dfrac{(\text{Tube pitch})^2 \times N}{144} \text{ ft}^2$

or area = $\dfrac{(2.97)^2 \times 14}{144} \text{ ft}^2$

or area = 0.85 ft^2

4.16.2.11 *Shell diameter*

Total area of tubes = 0.85 ft^2

Let D be the inner diameter of the shell. Therefore,

$$\frac{\pi D^2}{4} = 0.85$$

Thus,

$$D = \sqrt{\frac{0.83 \times 4}{3.14}} = \sqrt{\frac{3.4}{3.14}} = 1.08 \text{ ft}$$

4.16.2.12 Thickness of the evaporator body

The evaporator body should be thick enough to withstand the pressure in the steam chest, vapour space, and weight of the liquid, tubes, and so on. The body should be constructed of stainless steel. Considering the portion of the evaporator in direct contact with steam, we have

$$pd = 2fte$$

$$\text{or } t = \frac{pd}{2fe}$$

where f is the working stress in psi for stainless steel = 60 000 psi, e is the efficiency of the welded joints = 80%, p is the pressure to which the evaporator body is subjected, and d is the evaporator diameter.

Factor of safety = 6

Ultimate tensile stress for stainless steel = 60 000 psi

Thus,

$$f = \frac{60\,000}{6} = 10\,000 \text{ psi}$$

Now,

$$t = \frac{40 \times 1 \times 12}{10\,000 \times 0.08 \times 2}$$

$$\text{or } t = \frac{14}{800} = \frac{1.2}{40} = \frac{0.12}{4} = 0.03 \text{ in}$$

Designed on the basis that the steam chest can withstand a pressure of up to 40 lb/in^2.

4.16.2.13 Outer diameter of shell

Outer diameter of the shell = inner diameter of the shell + 2 × thickness of the shell

$$\text{or outer diameter of the shell} = 1 + \frac{0.03 \times 2}{12} \text{ ft} = 1.005 \text{ ft}$$

4.16.2.14 Height of evaporator

Height of evaporator = Height of tube + Height of upper vapour space
+ Height below tubes in which feed liquor is introduced

or height of evaporator = 18 + 1 + 19½ ft

4.16.2.15 Feed pipe diameter

Feed rate = 482 lb/h
Density of feed = 1.06 × 62.3 = lb/ft^3
Economic pipe diameter = 3 in
(Perry 1963)

4.16.2.16 Concentrated liquor outlet pipe diameter

Product rate = 83.33 lb/h
Density of product = 1.19 × 62.3 = 73 lb/ft^3
(Perry 1963)
Economic pipe diameter = 2½ in

4.16.2.17 Vapour outlet pipe diameter

Vapour flow rate = 402 lb/h
Specific volume of steam at 106.6°F = 372 ft^3/lb
Volumetric rate of flow = 402 × specific volume
Thus, volumetric rate of flow per second = $\dfrac{402 \times 372}{60 \times 60}$ ft^3/s = 35.5 ft^3/s

Let d be the diameter of the outlet pipe (in inches).

Thus,

$$\frac{\pi d^2}{4} = \frac{35.5 \times 144}{25} \text{ (assuming steam velocity 25 ft/s)}$$

$$\text{or } d = \sqrt{\frac{35.5 \times 144 \times 4}{3.14 \times 25}} = 6 \text{ in}$$

The vapour outlet pipe diameter appears to be high for the steam velocity. We assume the vapour outlet pipe diameter to be 4 in.

4.16.2.18 Condensate outlet pipe diameter

The condensate outlet pipe in the evaporator is of diameter 2 in.

4.16.2.19 Design of separator

For the present design, the diameter of the separator is taken to be 1 ft. It is now a usual practice to keep the ratio of diameter to height as 2:3.

Therefore, the height of the separator = $1 \times \dfrac{3}{2}$ = 1½ ft

An extra 1/2 ft is taken for the lower conical portion.

So the actual height is 1½ + ½ = 2 ft

4.16.2.20 Design of barometric condenser

A complete unit of evaporator consists of many accessories besides the main evaporator body itself. As evaporation is carried out under reduced pressure, a condenser is fitted to the vapour outlet pipe of the evaporator to drive out the vapours as quickly as they are formed to maintain a vacuum. The condensers are classified into two groups.

1. Surface condenser
2. Candace condenser

Surface condensers are used where there is need for condensed vapours. In the case discussed above, a barometric condenser of the jet type is used as the performance in terms of heat transfer per unit of investment cost of such a condenser is superior to others. This can be attributed to the direct contact between condensing vapour and cooling water without the resistance of an intervening wall, even if the construction is comparatively simple and inexpensive.

Table 4.15 gives the vapour pipe size for various steam flow rates.

In the design discussed, the amount of evaporation is 402 lb/h. It is seen that the evaporation rate does not fall in the range shown in Table 4.15. To obtain the vapour pipe size, the pipe size is plotted (in/s) against the steam flow rate (in lb/h). The vapour pipe size deduced is 3.6 in. Taking the next standard size, which is 4 in/s, and using the standard steel pipe of normal size 4 in.

4.16.2.20.1 Cooling water requirement

Temperature of water vapour = 106.5°F

Vacuum in evaporator = 28 in Hg

Thus, pressure in the evaporator = 29.92 – 28 = 1.92 in Hg or 0.942 psia

Latent heat of vaporization at 0.942 psia = 1029 Btu/lb

Amount of water to be condensed = 402 lb/h

Water inlet temperature = 86°F (room temperature)

Assuming cooling water to be heated within 5°F of saturated vapour temperature at 0.942 psia pressure, the water outlet temperature is 106.6 – 5 = 101.6°F

Table 4.15 Calculations of vapour pipe size for various steam flow rates

Pipe sizes (in/s)	Steam flow rates at 27 in (vacuum lb/h)
6	828
8	1 530
10	2 550
12	3 860
13	5 760
16	7 370

By energy balance,

$Q = m \times C_p \times dt$

or $402 \times 1029 = m \times 1 \times (106.6 - 86)$

or $402 \times 1029 = m \times 20.6$

or $m = \dfrac{402 \times 1020}{20.6}$ lb/h

Thus, $m = 24700$ lb/h

4.16.2.20.2 Cooling water inlet pipe size

1 gal water = 10 lb

Thus, water rate $= \dfrac{24700}{10 \times 60} = 42$ gal/min

Table 4.16 gives the cooling water inlet pipe size for different flow rates.

From Table 4.16, the inlet pipe size is found to be 2 in/s.

4.16.2.20.3 Tail pipe size

Tail pipe size should be equal to the water inlet pipe size of up to 4 in or greater, otherwise the tail pipe size will be one pipe size larger. We, therefore, assume the tail pipe size to be 3 in/s, which is approximately 34 ft-tonne.

4.16.2.20.4 Air outlet pipe size

Roughly, 5 lb of free air and 1000 lb of steam enters the condenser. Unusual outlet size ranges from 2 in on smaller condenser to 6 in on large condense, that is, with a 30% vapour pipe. In our case, we will take the pipe size as 3 in.

4.16.2.20.5 Condenser dimension

The ratio between pipe diameter and body diameter is usually 5:8.

Vapour pipe diameter assumed earlier is 4 in.

Thus, body diameter $= \dfrac{8 \times 4}{5} = 6.4$ in $= 7$ in (say)

$$\frac{\text{Height of the condenser body}}{\text{Diameter of the condenser body}} = 2 \text{ to } 5 = 3 \text{ (say)}$$

Table 4.16 Cooling water inlet pipe size for different flow rates

Pipe size (in/s)	Water rate (in gpm)
2.0	50
2.5	80
3.0	120
4.0	200

Thus, height of the condenser body is 3 × 7 = 21 in.

Thickness of the condenser body is 0.14 in. Allowing for corrosion, it may be taken as 0.20 in (ASME 1955).

4.16.2.20.6 Baffle plates and their placing

Placing of the baffle plate should be greater than one-fifth of the diameter of the condenser.

That is,

$$\frac{1}{5} \times 7 = 1\frac{2}{5} = 1.4 \text{ in}$$

Let us assume the baffle placing to be 1/3 ft.

Height of the condenser = 21 in

$$\text{Thus, number of baffles} = \frac{21 \times 3}{12} - 1 = 4$$

4.16.2.20.7 Capacity of hot well

Hot well provides means for discharging the heated condensed water and seals the bottom of the tail pipe. Its capacity should be greater than that of the tail pipe.

Inside cross-sectional area of 3 in normal pipe size = 0.0513 ft^2

Thus, capacity of the tail pipe = 0.0513 × 34 × 62.2 = 108 lb

Let the factor of safety be 5.

Thus, capacity of hot well is 5 × 108 = 540 lb

$$\text{Volume of hot well} = \frac{540}{62.2} = 8.7 \text{ ft}^3$$

If the hot well is a cube, then each side of the well will be 3 × 8.7 = 26.1.

Outlet water pipe is one pipe size greater than the tail pipe.

Hence, outlet water pipe size is 4 in.

4.16.2.21 Cooling water pump

Cooling water rate = 24 700 lb/h

$$\text{or cooling water rate} = \frac{24\,700}{60 \times 60} = \frac{247}{36} = 7 \text{ lb/s}$$

Table 4.17 gives the cooling water pipe sizes for different flow rates, along with the flow velocity, water head, and friction for 100 ft pipe.

From Table 4.17, the corresponding to pipe size is 2 in.

Velocity head = 0.40 ft

Loss in head due to friction per 100 ft of pipe = 10 ft of water

For a gate valve fitting, the equivalent resistance pipe diameter is 7.

Table 4.17 Cooling water pipe sizes for different flow rates

Pipe size (in/s)	Wager rate (gpm)	Flow velocity used per second	Water head (ft)	Friction for 100 ft pipe
2.0	50	5.1	0.40	10.0
2.5	80	5.3	0.43	8.0
3.0	120	5.5	0.45	7.0
4.0	200	5.1	0.41	4.5

For a 90° elbows fitting, the equivalent resistance pipe diameter is $2 \times 32 = 64$. The total equivalent resistance pipe diameter is $7 + 64 = 71$.

$$\text{Thus, equivalent length} = \frac{71 \times 2.5}{12} = 14$$

Total length of pipe and fitting = 34 + 14 = 48 ft

$$\text{Loss in head due to friction} = \frac{10 \times 48}{100} = 4.8 \text{ ft}$$

Pump work = Potential head + Velocity head + Loss in head due to friction

or pump work = 34 + 4.8 + 0.04 = 39.2 ft of water

Water rate = 7 lb/s

$$\text{Thus, pump work} = \frac{39.2 \times 7}{550} = \frac{27.44}{55} = 0.5 \text{ HP}$$

Considering the combined efficiency of pump and motor to be 60%,

$$\text{HP} = \frac{0.5}{0.6} = 0.84 \text{ HP} \approx 1 \text{ HP}$$

4.16.2.22 Design of two-stage steam jet ejector

Steam ejectors are widely used in chemical as well as biochemical industries for vacuum applications. They are generally more suitable for the lower absolute, that is, higher vacuum, than reciprocating vacuum pumps. However, ejectors can be used for any suction and discharge pressure and capacity. They may be single or multistage. Suitability of an ejector for a given application depends on (1) steam pressure, (2) water available, (3) highest suction pressure, and (4) required capacity.

During evaporation, water vapour enters the condenser along with free air. The water vapour condenses in the condenser, leaving aside air. To maintain vacuum, free air must be removed either by ordinary vacuum pump or by means of a steam ejector. The following discussion refers to a steam ejector because it produces constant vacuum and has no moving parts. The initial investment for a steam ejector is also lower than for a stationary vacuum pump.

4.16.2.22.1 Principle of operation

The basic principle of operation of an ejector is the transfer of a portion of the momentum from a high velocity steam jet to a stream of air or vapour entering the ejector and subsequent recompression of the mixture by the conversion of kinetic energy into potential energy. In the ejector, there is a divergent steam nozzle in which operation steam expands and converts its pressure energy into kinetic energy, or velocity. This velocity may exceed 4000 ft/s. In the converging section of the ejector, the high steam jet entrains the air drawn in through the suction branch. This section consists of a diffusion or compression tube of the ejector. In this, the kinetic energy of the mixture of steam and air is reconverted into pressure energy. The performance of an ejector depends on the accurate proportioning of the steam nozzle and the diffuser throat to suit the desired performance.

4.16.2.22.2 Number of stages and steam consumption

The vacuum is 28 in Hg. A two-stage steam ejector unit is used to maintain a vacuum of 26.5–25.3 in Hg (Kern 1950).

Vacuum = 28 in Hg

Thus, pressure = 29.92 – 28 = 1.92 in Hg, 48.8 mm Hg, or 0.942 psia

Pressure of steam used = 150 lb/in^2

Corresponding to suction pressure 48.8 mm Hg, the steam consumption is 10 lb/lb of steam, and the number of states is 2.

Hence, the ejector is of two stages with the intercondenser working in between the stages. About 5 lb of saturated air is present per 100 lb of vapour.

Evaporation rate = 402 lb/h

Thus, air entering $= \dfrac{5 \times 402}{1000} = 2.01$ lb/h

Total steam consumption = 10 × 2.01 = 20.10 lb/h

Table 4.18 shows the assumed distribution of pressure.

Steam consumption corresponding to suction pressure of 500 mm Hg is 2.5 lb/lb of water.

Thus, steam consumption for second stage = 2.5 × 2.01 = 5.025 lb/h = 5.1 lb/h (say)

That is, steam consumption for second stage = 20.10 – 5.1 = 15 lb/h

Table 4.18 Assumed distribution of pressure

Type of pressure	First stage	Second stage
Suction pressure	48.8 mm Hg	500 mm Hg
	0.942 psia	9.7 psia
Discharage pressure	500 mm Hg	760 mm Hg
	9.7 psia	14.7 psia

4.16.2.22.3 Area of cross section of the steam nozzle

1. First stage

For dry saturated steam, let P_s (= 100 psia) be the steam pressure, P_1 (= 0.942 psia) be the suction pressure, P_2 (= 9.7 psia) be the discharge pressure, V_s (= 4.44 ft^3/lb) be the specific volume of steam at 100 psia, and R (= 1.135) be the specific heat ratio of steam.

The adiabatic work is given by Equation 4.201.

$$U_s = 144\ P_s\ V_s \frac{r}{r-1}\left[1 - \left(\frac{P_1}{P_2}\right)^{\frac{r-1}{r}}\right] \text{ ft/lb/lb of steam} \tag{4.201}$$

$$\text{or } U_s = 144 \times 100 \times 4.44 \times \frac{1.135}{1.135 - 1}\left[1 - \left(\frac{0.942}{9.7}\right)^{\frac{0.135}{1.135}}\right] \text{ ft/lb/lb of steam}$$

$$U_s = 144 \times 444 \times \frac{1.135}{0.135}[1 - 0.76] \text{ ft/lb/lb of steam}$$

$$\text{or } U_s = 144 \times 444 \times 8.4 \times 0.24 \text{ ft/lb/lb of steam}$$

$$\text{or } U_s = 128\,895 \text{ ft/lb/lb of steam}$$

The terminal nozzle velocity C is given by Equation 4.202.

$$C = \sqrt{2g\ U_s} \text{ ft/s} \tag{4.202}$$

$$\text{or } C = \sqrt{2 \times 32.2 \times 128\,895} \text{ ft/s}$$

$$\text{or } C = 2881 \text{ ft/s}$$

Assuming C to be 70% theoretical,

actual C = 0.7 × 2881 = 2016 ft/s

Now,

$$m = C \times A \times P$$

where C is the actual nozzle velocity, A is the nozzle area, P is the steam density, and m is the mass rate flow of steam (lb/s).

Here,

$$m = 15 \text{ lb/h}$$

Thus,

$$C = \frac{15}{60 \times 60} \text{ lb/s}$$

$$\text{or } C = 0.0041 \text{ lb/s} = 2016 \text{ ft/s}$$

$$P = \frac{1}{\text{Specific volume}} = \frac{1}{4.44}$$

$$A = \frac{m}{C \times P}$$

$$\text{or } A = \frac{0.0041}{2016 \dfrac{1}{4.44 \times 144}}$$

$$\text{or } A = \frac{0.0041 \times 4.44 \times 144}{2016}$$

or $A = 0.0012 \text{ in}^2$

Also,

$$A = \frac{\pi D^2}{4}$$

Thus,

$$D = \sqrt{\frac{4A}{\pi}}$$

where D is the nozzle diameter.

$$D = \sqrt{\frac{4 \times 0.0012}{3.14}}$$

or $D = \sqrt{0.0015} = 0.04$ in

2. Second stage

We have,

$P_s = 100$ psia, $P_1 = 9.7$ psia, $P_2 = 14.7$ psia, and $r = 1.135$ (for dry saturated steam).

$$U_s = 144 P_s V_s \frac{r}{r-1}\left[1 - \left(\frac{P_1}{P_2}\right)^{\frac{r-1}{r}}\right] \text{ ft/lb/lb of steam}$$

$$\text{or } U_s = 144 \times 100 \times 4.44 \times \frac{1.135}{0.135}\left[1 - \left(\frac{9.7}{14.7}\right)^{\frac{0.135}{1.135}}\right] \text{ ft/lb/lb of steam}$$

or $U_s = 24864$ ft/lb/lb steam

The theoretical nozzle velocity is given by Equation 4.203.

$$C = \sqrt{2\,g\,U_s} \text{ ft/s} \tag{4.203}$$

or $C = \sqrt{2 \times 32.2 \times 24864}$ ft/s

or $C = 1266$ ft/s

Assuming 70% of the nozzle velocity to be theoretical,
actual $C = 0.70 \times 1266$ ft/s.

That is,
actual $C = 887$ ft/s

The mass flow rate is given by Equation 4.204.

$$m = 5.1 \text{ lb/h} \tag{4.204}$$

or $m = \dfrac{5.1}{60 \times 60}$ lb/s

or $m = 0.0014$ lb/s

Now, $m = C \times A \times P$

Therefore,

$$A = \frac{m}{C \times P}$$

or $A = \dfrac{0.0014}{\dfrac{1}{4.44 \times 144} \times 887}$

or $A = \dfrac{0.0041 \times 4.44 \times 144}{887}$

or $A = 0.00101$ in^2

Also,

$$A = \frac{\pi D^2}{4}$$

Thus,

$$D = \sqrt{\frac{4 \times 0.00101}{3.14}} \text{ in}$$

or $D = \sqrt{\dfrac{0.00404}{3.14}}$ in

or $D = 0.04$ in

4.16.2.22.4 Area of cross section of nozzle at the commencement of compression

4.16.2.22.4.1 First stage

$$\text{Steam ratio} = \frac{\text{Steam consumption (in lb/h)}}{\text{Mass of air drawn (in lb/h)}}$$

or steam ratio $= \dfrac{15}{2.01} = 7.5$

Velocity ratio $= \dfrac{U}{C} = 0.7$

Thus,

$U = C \times 0.7 = 2016 \times 0.7 = 1411$ ft/s

Again,

$$U = \sqrt{2\, gU_m}$$

Thus,

$$U_m = \frac{U^2}{2g}$$

$$\text{or } U_m = \frac{(1411)^2}{32.2 \times 2} = 30\,901$$

U_m can also be expressed as given by Equation 4.205.

$$U_m = 144 P_1 V_m \frac{n}{n-1}\left[\left(\frac{P_2}{P_1}\right)^{\frac{(n-1)}{n}} - 1\right] \qquad (4.205)$$

where V_m is the specific volume of air–steam mixture at the commencement of compression and n is the specific heat ratio for air–steam mixture.

Calculation of *n*

The air–steam mixture contains 2.01 lb/h of air and 15 lb/h of steam.

$$\text{Percentage of steam in the mixture} = \frac{15}{15 + 2.01} \times 100 = 88.7\%$$

$$\text{Percentage of air in the mixture} = \frac{2.01}{17.01} \times 100 = 11.3\%$$

Due to throttling, steam in the diffuser must be superheated

r for superheated steam = 1.3

n for air = 1.4

Thus,

$$n = 0.887 \times 1.3 + 0.113 \times 1.42 = 1.312$$

P_1 = 0.942 psia

P_2 = 9.7 psia

Now, from Equation 4.205,

$$30\,901 = 144 \times 0.942 \times V_m \times \frac{1.312}{0.312}\left[\left(\frac{9.7}{0.942}\right)\frac{0.312}{1.312} - 1\right]$$

$$\text{or } 30\,901 = 144 \times 0.942 \times V_m \times 4.2 \times 0.74$$

Thus

$$V_m = \frac{30\,901}{144 \times 0.942 \times 4.2 \times 0.74} = 75 \text{ ft}^3/\text{lb}$$

Now,

$$m = U \times A \times P$$

Here,

$$\frac{1}{V_m} = \frac{1}{75} = 0.0133 \text{ lb/ft}^3$$

or $\dfrac{15}{3600} = 0.0041$ lb/s

$$A = \frac{m}{U}$$

or $A = \dfrac{0.0041 \times 144}{1411 \times 0.0133}$

or $A = 0.03$ in^2

Also,

$$A = \frac{\pi d^2}{4}$$

Thus,

$$d = \sqrt{\frac{4A}{\pi}}$$

or $A = \sqrt{\dfrac{4 \times 0.030}{3.14}} = 0.18$ in

Hence, the diameter at the commencement of compression is 0.18 in.

Diameter of narrowest section = 0.8 × diameter at the beginning of compression

or diameter of narrowest section = 0.8 × 0.18 = 0.15 in

Thus, the narrowest diameter is 0.15 in.

3. Second stage

Steam ratio $= \dfrac{5.1}{2} = 2.5$

Velocity $= \dfrac{U}{C} = 0.57$

Thus,

$$U = 0.57 \times C$$

or $U = 0.57 \times 887 = 506$

The mixture entering the diffuser tube contains 5.1 lb steam per hour and 2 lb air per hour.

Thus,

percentage of steam $= \dfrac{5.1}{5.1 + 2} \times 100 = 71.8\%$

Percentage of air $= \dfrac{2}{5.1 + 2} \times 100 = 28.1\%$

Thus,

$$n = 0.718 \times 1.3 + 0.281 \times 1.4$$

or $n = 1.327$

$$U = \sqrt{2g\, U_m}$$

$$\text{or } U_m = \frac{U^2}{2g} = \frac{(506)^2}{2 \times 32.2} = 3972$$

$$\text{or } U_m = 44 \times 9.7 \times V_m \times \frac{1.327}{0.327}\left[\left(\frac{14.7}{9.7}\right)^{\frac{0.327}{1.327}} - 1\right]$$

$$\text{or } 3972 = 309.7\ V_m$$

Thus,

$$V_m = \frac{3972}{309.7} = 12.8\ \text{ft}^3/\text{lb}$$

Now,

$$m = U \times A \times P$$

$$\text{or } A = \frac{m}{U}$$

Here,

$$1/V_m = 1/12.8$$

$$m = \frac{5.1}{3600}\ \text{lb/s}$$

$$U = 506$$

Thus,

$$A = \frac{5.1 \times 144 \times 4.44}{3600 \times 506}$$

$$\text{or } A = \frac{94}{36 \times 506} = 0.0051\ \text{in}^2$$

$$\text{or } A = \frac{\pi d^2}{4}$$

Thus,

$$d = \sqrt{\frac{4A}{3.14}}$$

$$\text{or } d = \sqrt{\frac{4 \times 0.0051}{3.14}}$$

$$\text{or } d = \sqrt{0.0065} = 0.08\ \text{in}$$

The total length of the diffuser body is usually 50 times the diameter at the commencement of compression (McCabe, Smith, and Harriott 1985).

4.16.2.22.5 Length of diffuser body

First stage = 50 × 0.18 = 9 in

Second stage = 50 × 0.08 = 4 in

4.16.2.22.6 Cooling water requirement for the intercondenser

Cooling water temperature = 86°F

Assuming 20°F rise in temperature of outlet water, the exit water temperature is 86 + 20 = 106°F

Enthalpy of saturated steam at 150 psig,

H = 1198 Btu

Enthalpy of liquid water at 106°F,

h = 74 Btu/lb

Heat to be removed during condensation = 1198 – 74 = 1124 Btu/lb

Steam to be condensed in first stage = 15 lb/h

Thus,

$$m = \frac{15 \times 1124}{20}$$

or $m = 0.67 \times 1124$

or m = 742 lb/h

$$\text{or } m = \frac{742}{10 \times 60} \text{ gpm}$$

or m = 1.23 gpm

4.16.2.22.7 Cooling water pump

Neglecting frictional head and velocity head, work required to pump 742 lb/h, or 0.208 lb/s, of water through a height of 34 ft (ejector is placed 34 ft above ground level for ease of passage of water) is calculated.

$W = 0.208 \times 34$

or W = 7.072 ft-lb/s

Taking the combined efficiency of pump and motor to be 50%, we have

$$W = \frac{7.072}{550 \times 0.5} = \frac{7.072}{275}$$

or W = 0.086 HP ≈ 0.10 HP

The same pump can supply water to the ejector intercondenser and cascade-wise condenser.

The total HP of the meter = 1.0 + 0.1

or total HP = 1.1 HP

or total HP = 1.5 HP (say)

4.16.2.23 Materials of construction

Evaporator body: Evaporators made of stainless steel are used for concentration of fruit juices and other food materials. Other materials like mild steel and copper cannot be used because the metal will react with the contents of the food material. The steam tubes are also made of stainless steel.

Pumps and pipings: The pumps and pipings are constructed of steel. Product withdrawal pipings are also made of stainless steel.

Condenser: Copper is successfully used for the construction of the condenser. Nowadays, stainless steel, stainless clad steel, and other metals have been introduced but copper has not lost its significance in the construction of condensers.

Based on the calculations given in Table 4.19, the design drawings of the climbing film evaporator and its connecting barometric condenser and steam ejector are shown in Figures 4.37 and 4.38.

Table 4.19 Different materials of construction for the evaporator, pump, pipes, and condenser

Description	Material	Number	Remark
Evaporator main body	Stainless steel	1	–
Upper dome	Stainless steel	1	–
Feed inlet pipe	Stainless steel	1	–
90° elbow	Stainless steel	1	–
Valve	Stainless steel	–	–
Heating tubes	Stainless steel	–	2 in diameter
Tube sheet	Stainless steel	–	–
Vapour outlet pipe from the evaporator	Mild steel	1	3 in diameter
Pressure gauge	Bronze	1	150 lb/in^2
Vacuum gauge	Bronze	1	30 in of Hg
Condenser body	Copper 1	7 in × 21 in	
Inlet water pipe to the condenser	Condenser inlet	–	2 in/s diameter
Valve	Condenser inlet	1	Gate type
Pipe	Condenser inlet	–	34 ft
Cooling water pump (centrifugal)	Condenser inlet	–	1 HP (7 lb/s)
Condensed water outlet pipe	Condenser inlet	–	3 in diameter
Hot well	Concrete	1	–
Baffles in the condenser for cascading the water	4	–	–
Air outlet pipe	Mild steel	–	3 in diameter
Two stage ejector with an intercondenser (it has both nozzle and diffuser)	Bronze	1	–
Pump for intercondenser (centrifugal)	1	–	0.1 HP

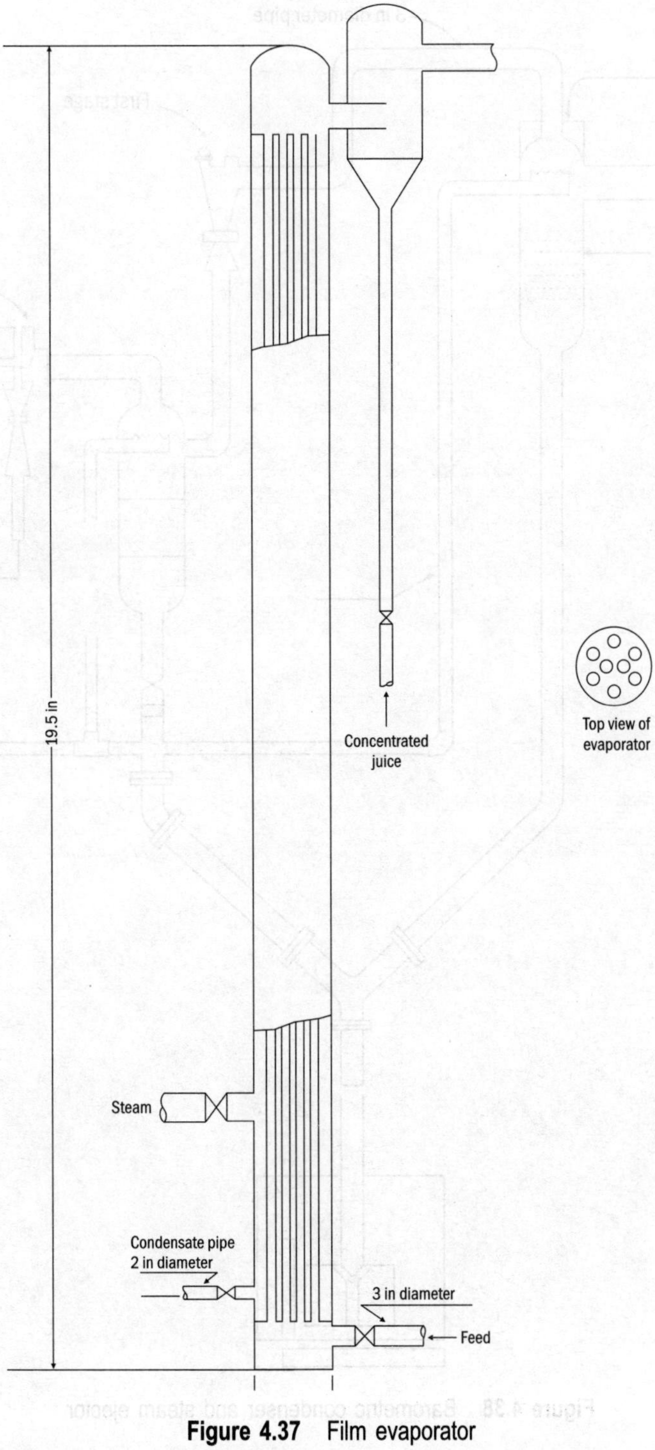

Figure 4.37 Film evaporator

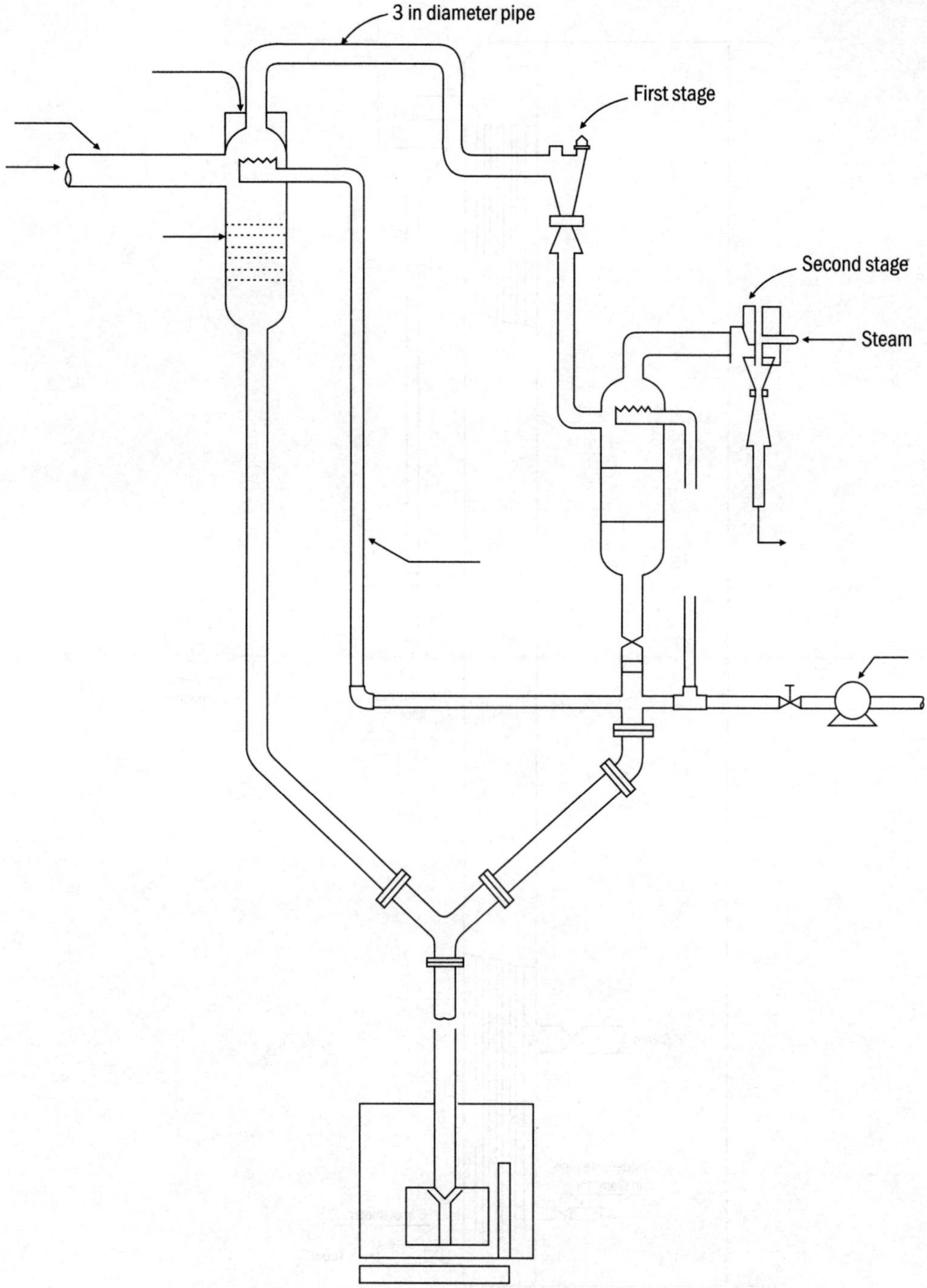

Figure 4.38 Barometric condenser and steam ejector

4.16.3 Tutorials

Question 1 About 50000 lb/h of a chemical solution is concentrated at 100°F and 10% solids to a product that contains 50% solids. Steam is available at 12 psig, and the final effect of a triple effect evaporator with equal heat transfer surfaces in each effect is assumed to operate at a vacuum of 26 in Hg referred to a 30 in barometer. Water is available at 85°F for use in the barometric condenser.

Assuming the condensate from each effect leaves at its saturation temperature, a negligible Btu/h, an average specific heat of 1.0 in all effects, and a negligible radiation loss, calculate (a) the heating surface required for each effect and (b) the condense water requirement.

The overall heat transfer coefficients for different effects will be U_1 = 600 Btu/h-ft^2-°F, U_2 = 250 Btu/h-ft^2-°F, and U_3 = 125 Btu/h-ft^2-°F.

Answer

(a) Total feed = 50000 lb/h (given)

Total solid in feed = 50000 × 0.10 = 5000 lb/h

$$\text{Total product} = \frac{5000}{0.5} = 10000 \text{ lb/h}$$

Total evaporation = $W_{1\text{-}3}$ = 50000 – 10000

or $W_{1\text{-}3}$ = 40000 Btu/h

Thus, C_F = 1.0

where C_F is the coefficient of friction.

Enthalpy balances

First effect: $W_S\lambda_S + W_F(t_F - t_1) = W_1\lambda_1$

Second effect: $W_1\lambda_1 + (W_F - W_1)(t_1 - t_2) = W_2\lambda_2$

Third effect: $W_2\lambda_2 + (W_F - W_1 - W_2)(t_2 - t_3) = W_3\lambda_3$

Material balance, $W_1 + W_2 + W_3 = W_{1\text{-}3}$

t_F = 100°F

T_s at 12 psig = 244°F

T_s at 26 in Hg (1.95 psig) = 125°F

Thus, total temperature difference = 244°F – 125°F = 119°F

$$\text{Thus, the average pressure difference} = \frac{26.70 - 1.95}{3} = 8.25 \text{ psi/effect}$$

The enthalpy balances of the first three effects are given in Table 4.20.

$$949W_s + 50000\,(100 - 224) = 961W_1 \quad (4.206)$$

Table 4.20 Enthalpy balance of the first three effects

Steam chart	Pressure (psia)	ΔP psia	Steam (°F)	λ(Btu/lb)
First effect	26.70	–	$T_s = 244$	$\lambda_s = 949$
Second effect	18.45	8.25	$t_1 = 224$	$\lambda_1 = 961$
Third effect	10.20	8.25	$t_2 = 194$	$\lambda_2 = 981$
Vapour to condenser	(20.7 in Hg) 1.95	8.25	$t_3 = 125$	$\lambda_3 = 1022$

$$961W_1 + (50000 - W_1)(224 - 194)_1 = 981W_2 \tag{4.207}$$

$$981W_2 + (50000 - W_1 - W_2)(194 - 125) = 1022W_3 \tag{4.208}$$

$$W_1 + W_2 + W_3 = 40000$$

Solving simultaneously for Equations 4.206–4.208, we get

$W_1 = 12400$

$W_2 = 13300$

$W_3 = 14300$

$W_s = 19100$

Now

$$A_1 = \frac{\lambda_s W_S}{\mu_1 (t_s - t_1)}$$

$$\text{or } A_1 = \frac{946 \times 19100}{600 \times (244 - 224)} \text{ ft}^2$$

$$\text{or } A_1 = \frac{949 \times 19100}{600 \times 20} \text{ ft}^2$$

or $A_1 = 1510 \text{ ft}^2$

Similarly,

$$A_2 = \frac{W_1 \lambda_1}{\mu_2 (t_1 - t_2)}$$

$$\text{or } A_2 = \frac{12400 \times 961}{250 \times (224 - 194)}$$

or $A_2 = 1590 \text{ ft}^2$

$$A_3 = \frac{W_2 \lambda_2}{\mu_3 (t_2 - t_3)}$$

$$\text{or } A_3 = \frac{13300 \times 981}{125 \times (194 - 125)}$$

or $A_3 = 1510 \text{ ft}^2$

(b) Heat to condense = $W_3\lambda_3$

or heat to condense = 14300 × 1022 Btu/h

or heat to condense = 14710000 Btu/h

Water requirement = 14710 000 (120 – 85)

or water requirement = 420000 Btu/h

or water requirement = $\frac{420000}{500}$ gpm = 840 gpm

Thus, economy = $\frac{\text{Evaporation (lb)}}{\text{Steam (lb)}} = \frac{40000}{191000} = 2.09$

Question 2 A solution of cellular colloids in water is concentrated from 10% to 50% solids in single effect evaporation. Steam is available at a guage pressure of 15 lb force/in^2 (249°F). A pressure of 4 in Hg absolute is maintained in the vapour space, which corresponds to boiling point of water of 125°F. The feed rate to the evaporator is 55000 lb/hour and the overall heat transfer coefficient is 500 Btu/ft^2-hour-°F. The solution has a negligible elevation in boiling point and a negligible heat of dilution.

Calculate the steam consumption, economy, and heating surface required if the temperature of the feed is (a) 125°F, (b) 70°F.

The specific heat of the feed solution is 0.90, and the latent heat of vaporization of the solution is equal to that of water. Radiation losses may be neglected.

Answer We have, feed W_f = 55000 lb/h

Total solid in feed = 55000 × 0.10 = 5500 Btu

Total product = $\frac{5500}{0.5}$ = 11000 lbs

Thus, the flow rate of thick liquor is 11000 Btu. Now,

the total evaporation = 55000 – 11000 lb/h = 44000 lb/h

Thus, flow rate of thick liquor,

W = 55000 – 44000 = 11000 lb/h

(a) By enthalpy balance, we have

$W_1\,\lambda_1 = W_s\,\lambda_s + W_f \times C_F(t\rho_1 - t_1)$

Now, at 15 lb force/in^2 and 249°F, λ_s = 946 Btu/lb, λ_1 = 1023 Btu/lb, $t = t_f$ = 125°F

We also know that

$W_s\,\lambda_s = (W_f - W)\lambda + W_f C_p\,(t - t_f)$

or W_s × 946 = (55000 – 11000) × 1023 + 55000 × 0.9(125 – 125)

or $W_s = \frac{44000 \times 1023}{946}$ = 47580 lb/h

$$\text{Economy} = \frac{\text{Evaporatoin (lb)}}{\text{Steam (lb)}}$$

$$\text{or economy} = \frac{44000}{47580} = 0.925$$

Heating surface,

$$A = \frac{W_s\,\lambda_s}{\mu\,(t_s - t)}$$

$$\text{or } A = \frac{47580 \times 946}{500 \times (249 - 125)}$$

$$\text{or } A = \frac{47580 \times 946}{500 \times 124}$$

Thus, $A = 725 \text{ ft}^2$

(b) We have, $W_s\lambda_s = (W_f - W)\lambda + W_F C_F(t_f - t)$

or $W_s \times 946 = (55000 - 11000)\ 10243 + 55000 \times 0.90\ (70 - 125)$

or $W_s \times 946 = 44000 \times 1023 - 55000 \times 0.9 \times 55$

$$\text{or } W_s = \frac{44000 \times 1023 - 55000 \times 0.9 \times 55}{946}$$

or $W_s = 50460$ lb/h

$$\text{Economy} = \frac{44000}{50460} = 0.872$$

Heating surface,

$$A = \frac{W_s\lambda_s}{\mu\,(t_s - t)}$$

$$\text{or } A = \frac{50460 \times 946}{500\ (249 - 125)}$$

Thus, $A = 770 \text{ ft}^2$

4.17 REFRIGERATION

Cooling is a fundamental operation in many biochemical processes. When heat has to be removed continuously from a system at low temperatures and rejected to the atmosphere at higher temperatures, a mechanical refrigeration system is required. A mechanical refrigeration system acts as a pump that extracts heat at low temperatures and transfers it to the surroundings of the system at higher temperatures, which is against the first law of thermodynamics.

4.17.1 Basic Components of Refrigeration System

For accomplishing the first law of thermodynamics, a refrigeration cycle is required. The basic components of a refrigeration system are shown in Figure 4.39. The major components of the system are evaporator, compressor, condenser, liquid reservoir, and expansion valve.

4.17.2 Refrigeration Cycle

Whenever there is a demand, the refrigerant flows through the evaporator and evaporates by picking up latent heat of vaporization. Gas kept at low pressure then enters the compressor, where it is adiabatically compressed to a higher pressure. This gas can be condensed under high pressure at the ambient temperature, which releases latent heat of condensation. The high pressure liquid is then expanded to low pressure by means of a valve and enters the evaporator. The refrigeration cycle can be shown by means of the pressure–enthalpy diagram (Figure 4.40).

4.17.2.1 Enthalpy change and work done during adiabatic compression

For adiabatic compression,

$$PV = P_1 V_1^{\gamma} \tag{4.209}$$

$$\text{or } V = V_1 (P_1/P)^{1/\gamma} \tag{4.210}$$

Also,

$$dH = Vfp$$

Substituting the value of V from Equation 4.210 and values of f and p, we get

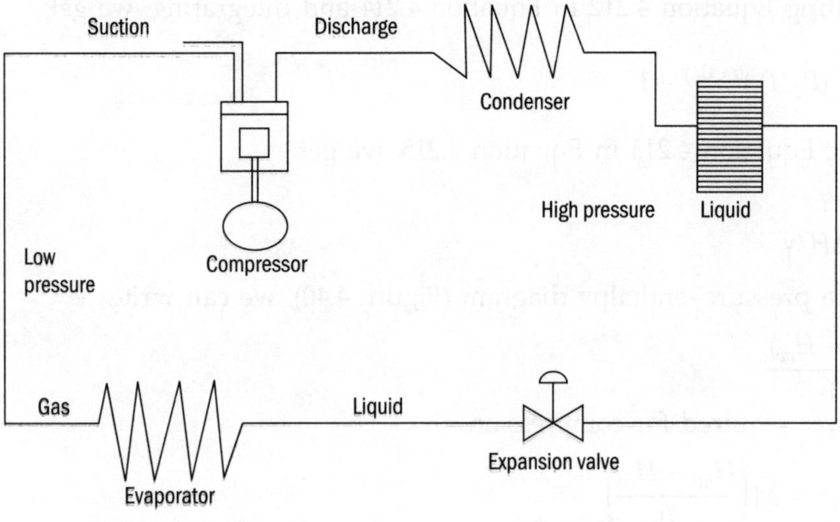

Figure 4.39 Schematic diagram of a mechanical refrigeration system

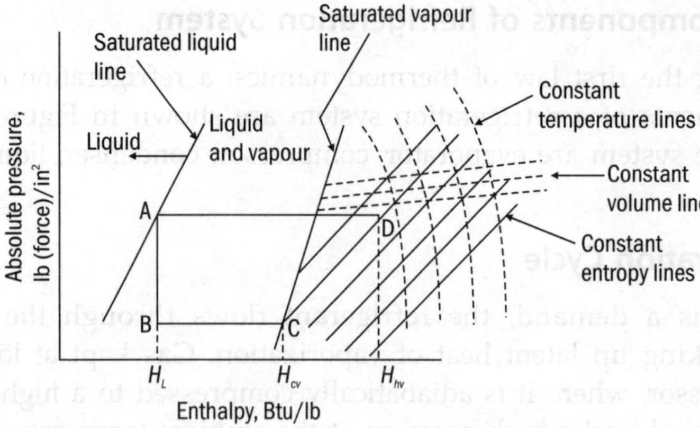

Figure 4.40 Refrigeration cycle on a pressure–enthalpy diagram

$$dH = V_1(P_1/P)^{1/\gamma}(P)^{-1/\gamma}dp \tag{4.211}$$

Now,

$$V_1 = (RT_1)/P_1 \tag{4.212}$$

Substituting Equation 4.212 in Equation 4.211 and integrating, we get

$$\Delta H = \left(\frac{\gamma}{\gamma^{-1}}\right) (RT_1) \left[\left(\frac{P_1}{P_2}\right)^{\frac{1-\gamma}{\gamma}} - 1\right] \tag{4.213}$$

Similarly,

$$dW = -1/\gamma(V_1)\ (P_1)^{1/\gamma}\ (P)^{\ 1/\gamma}\ dp \tag{4.214}$$

Substituting Equation 4.212 in Equation 4.214 and integrating, we get

$$W = \frac{RT_1}{1-\gamma}\ (P_2/P_1)^{(r-1)/\gamma} - 1 \tag{4.215}$$

Dividing Equation 4.213 in Equation 4.215, we get

$$\Delta H/W = -\gamma \tag{4.216}$$

$$\text{or } W = -\Delta H/\gamma \tag{4.217}$$

From the pressure–enthalpy diagram (Figure 4.40), we can write

$$W = \frac{(H_{hv} - H_{cv})}{\gamma} \tag{4.218}$$

The power required for compression,

$$P = W/\text{time} = M\left(\frac{H_{hv} - H_{cv}}{\gamma}\right)$$

where M is the mass of the refrigerant recirculated through the system for unit time. The refrigeration capacity can be expressed in tonnes of refrigeration. One tonne of refrigeration is the rate of removal of heat sufficient to freeze 1 tonne (2000 lb) of water in 24 h. Therefore,

$$(\text{tonnes})_{\gamma} = \frac{(H_{cv} - H_1)M}{12000}$$

4.18 HEAT EXCHANGERS

The transfer of heat is a common occurrence in industrial processing. Heat exchangers are equipment used for transferring heat from one fluid to another. A copper tube cooled by transferring heat to ambient air is an example of the simplest form of a heat exchanger.

4.18.1 Design Shapes of Some Conventional Heat Exchangers

The design shapes of some conventional exchangers include (1) double pipe heat exchanger (Figure 4.41[a]), (2) shell and tube heat exchanger (Figure 4.41[b]), (3) swept surface heat exchanger (Figure 4.41[c]), and (4) plate heat exchanger.

4.18.1.1 Double pipe heat exchanger

A double pipe heat exchanger consists of one pipe inside another. The walls of the pipe act as the heat transfer surface. An advantage of this type of heat exchanger is that it can be built and installed in the field itself.

4.18.1.2 Shell and tube heat exchanger

A shell and tube heat exchanger consists of a bundle of tubes enclosed in a shell. This type of head arrangement allows for single or multiple passes for the product. In a single-pass arrangement, the product enters at one end and exits at the opposite side.

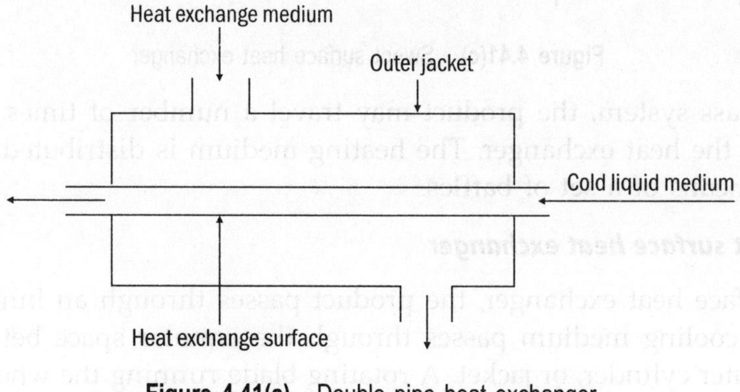

Figure 4.41(a) Double pipe heat exchanger

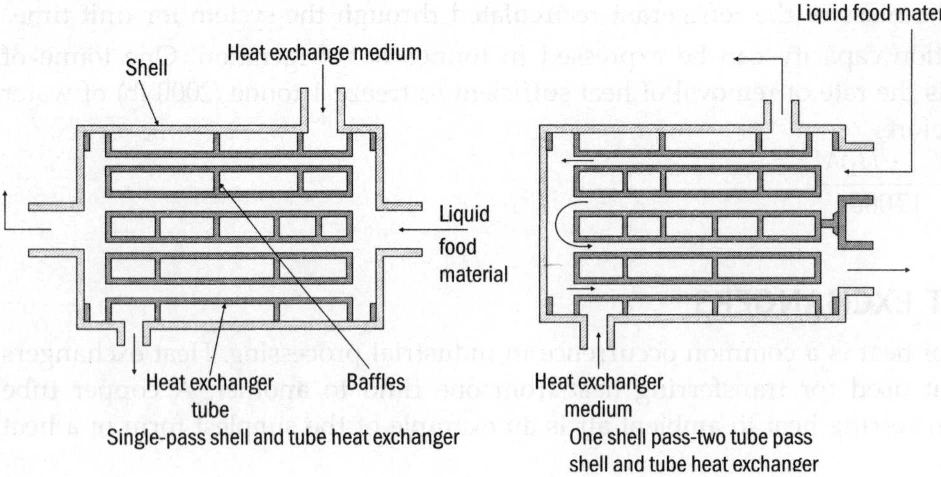

Figure 4.41(b) Shell and tube heat exchangers

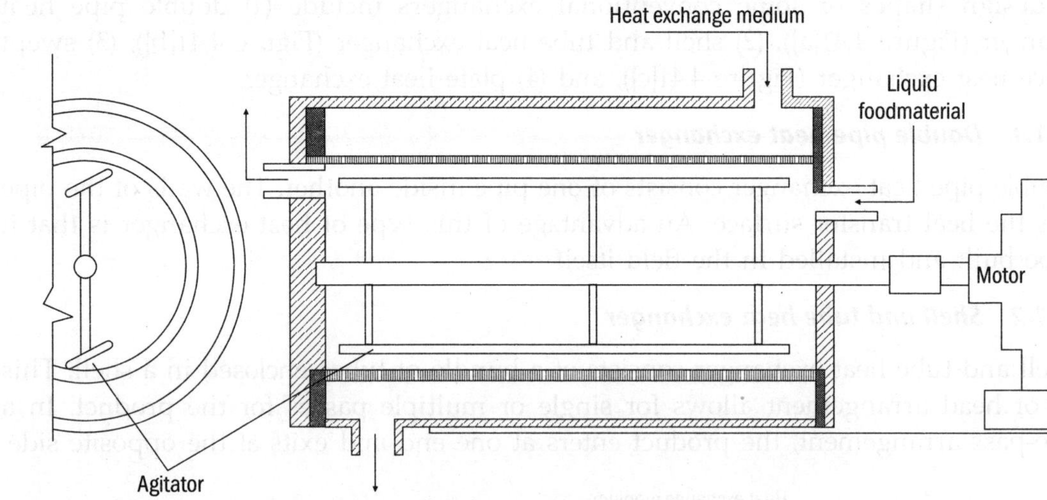

Figure 4.41(c) Swept surface heat exchanger

In a multiple-pass system, the product may travel a number of times back and forth before it leaves the heat exchanger. The heating medium is distributed inside the heat exchanger by means of a set of baffles.

4.18.1.3 Swept surface heat exchanger

In a swept surface heat exchanger, the product passes through an inner cylinder and the heating or cooling medium passes through the annular space between the inner cylinder and outer cylinder, or jacket. A rotating blade running the whole length of the cylinder continuously agitates and scrapes the wall through which heat exchange takes place. This type of heat exchanger is most suitable for concentrating viscous materials or materials containing insoluble substances.

4.18.1.4 Plate heat exchanger

Plate heat exchangers consist of a series of plates clamped together on a frame. The product and heating medium pass through alternative channels. The channels are narrow and the product and heating medium pass through them at high velocity. Due to this, high heat transfer rates per unit area can be achieved. A major disadvantage of this type of exchanger is its inability to withstand high pressure and temperatures. Plate heat exchangers are mostly used for heating fluids at temperatures below the boiling point of water at atmospheric pressure.

4.18.2 Energy Balances

The designs of heat exchangers are based on energy balances and the estimation of rates of heat transfer.

4.18.2.1 Enthalpy balances

Neglecting mechanical, potential, and kinetic energy terms (as they are negligible), the energy balance equation in a heat exchanger can be given by Equation 4.219.

$$m(H_b - H_a) = q \tag{4.219}$$

where m is the average mass flow rate of the stream, q is the rate of heat transfer into the stream, H_a and H_b are enthalpies per unit mass of stream at entrance and exit, respectively.

Equation 4.219 can be written for each stream flowing through the heat exchanger. If the loss of heat to surroundings is negligible, the heat balance equation can be given by Equation 4.220.

$$q = m_h(H_{ha} - H_{hb}) = m_c(H_{cb} - H_{ca}) \tag{4.220}$$

where m_c and m_h are the mass flow rates of cold and warm fluids, respectively, and H_{ca} and H_{ha} are the enthalpies per unit mass of entering cold and warm fluids, respectively.

If constant specific heats are assumed, Equation 4.220 can be written as

$$m_hC_{ph}(T_{ha} - T_{hb}) = m_cC_{pc}(T_{cb} - T_{ca}) \tag{4.221}$$

where C_{ph} is the specific heat of hot fluid and C_{pc} is the specific heat of cold fluid.

For a condenser, the enthalpy balance equation can be given by Equation 4.222.

$$m_h\lambda = m_cC_{pc}\ (T_{cb} - T_{ca}) = q \tag{4.222}$$

where m_h is the rate of condensation of vapour and λ is the latent heat of condensation.

Equation 4.222 is applicable only if the vapour leaves the condenser in a saturated state (that is, as a saturated vapour). If it loses its sensible heat, then Equation 4.222 has to be modified as given by Equation 4.223.

$$m_h \lambda C_{ph}(T_h - T_{hb}) = m_c C_{pc}(T_{cb} - T_{ca}) \quad (4.223)$$

where C_{ph} is the specific heat of condensate.

4.18.2.2 Overall heat transfer coefficient

The driving force for heat flux is the difference between the average temperature of hot fluid (T_h) and cold fluid (T_c). The heat flux is proportional to the driving force. The quantity $T_h - T_c$ is the overall local temperature difference, denoted by ΔT. Heat flux varies from position to position in the heat exchanger since ΔT also varies significantly. The local flux can be given by Equation 4.224.

$$dg/dA = U\Delta T = U(T_h - T_c) \quad (4.224)$$

where U is the overall heat transfer coefficient.

If A is taken as the outside area, then U is denoted by U_0. Likewise, if the area chosen is inside the tube (A_i), then the coefficient is also based on that area and is denoted by U_i. Since ΔT and q are independent of area, it follows that

$$U_0/U_1 = dA_i/dA_o = D_i/D_o \quad (4.225)$$

where D_i and D_o are the diameters of the tube inside and outside, respectively.

To use Equation 4.225, it must be integrated over the entire area through which heat transfer takes place. This can be done if (1) the overall heat transfer coefficient is constant, (2) C_p is constant, (3) heat loss to ambient is negligible, (4) flow is steady and either parallel or countercurrent. Under these circumstances, straight lines are obtained if T_c and T_h are plotted against q (Figure 4.42). Thus,

$$\frac{d(\Delta T)}{dq} = \frac{\Delta T_2 - \Delta T_1}{q_T} \quad (4.226)$$

$$\text{or } \frac{d(\Delta T)}{U\Delta T dA} = \frac{\Delta T_2 - \Delta T_1}{q_T} \quad (4.227)$$

Integrating Equation 4.227 between limits ΔT_1 to ΔT_2, we get

$$\int_{\Delta T_1}^{\Delta T_2} \frac{d(\Delta T)}{\Delta T} = \frac{U(\Delta T_2 - \Delta T_1)}{q_T} \int_0^{AT} dA \quad (4.228)$$

$$\text{or } \ln \frac{\Delta T_2}{\Delta T_1} = \frac{U(\Delta T_2 - \Delta T_1)}{q_T} AT \quad (4.229)$$

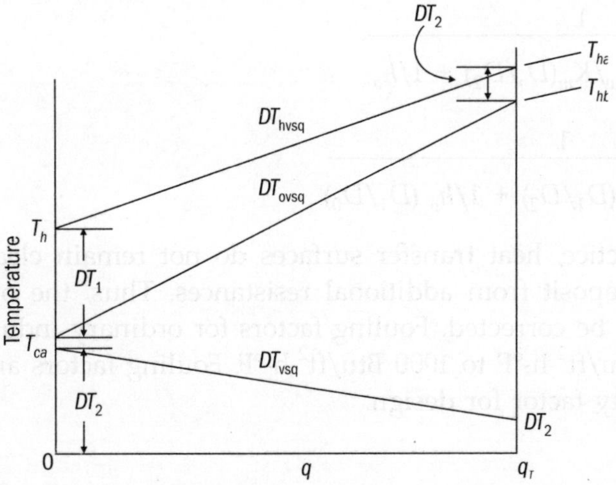

Figure 4.42 Temperature versus heat flow rate countercurrent flow

Therefore,

$$q_T = UA_T \frac{\Delta T_2 - \Delta T_1}{\ln(\Delta T_2/\Delta T_1)} \tag{4.230}$$

or $$q_T = UA_T \Delta T_{LM} \tag{4.231}$$

where $$\Delta T_{LM} = \frac{\Delta T_2 - \Delta T_1}{\ln(\Delta T_2/\Delta T_1)} \tag{4.232}$$

The area A_T required for heat transfer can be calculated if the values of ΔT_{LM}, U, and q are known.

In the derivation of design Equation 4.232, it is assumed that the overall heat transfer equation does not vary along the heat exchanger. If U varies linearly with temperature drop over the entire heating surface, then q_T is expressed by Equation 4.233.

$$q_T = A_T \frac{U_2\Delta T_1 - U_1\Delta T_2}{\ln(U_2\Delta T_1/U_1\Delta T_2)} \tag{4.233}$$

where U_1 and U_2 are overall heat transfer coefficients at the ends of the heat exchanger.

In most general case, where D^*T and U varies significantly from point to point, Equation 4.224 has to be integrated either graphically or numerically to obtain the area A_T required for heat transfer.

4.18.2.3 Calculation of overall coefficient from individual coefficients

The overall heat transfer coefficient can be calculated from individual coefficients and the resistance of the tube wall from Equations 4.234 and 4.235 given below.

$$U_o = \frac{1}{\frac{1}{h_1}(D_o/D_i) + X_w/K_m(D_o/D_2) + 1/h_o} \quad (4.234)$$

$$U_i = \frac{1}{\frac{1}{h_i} + X_w/K_m + (D_1/D_2) + 1/h_o\,(D_1/D_0)} \quad (4.235)$$

However, in practice, heat transfer surfaces do not remain clean since scale, dirt, and other solids deposit from additional resistances. Thus, the overall heat transfer coefficient needs to be corrected. Fouling factors for ordinary industrial liquids fall in the range of 300 Btu/ft^2-h-°F to 1000 Btu/ft^2-h-°F. Fouling factors are set at values that also provide a safety factor for design.

4.18.2.4 *Tutorial*

Question 1 A heat exchanger is constructed out of 12 gauge brass tube not exceeding 8 ft in length. It heats 15 000 lb/h of water entering at 70°F and flowing at 1.4 ft/s. The water is heated by a saturated stream entering at 35 psia. Calculate the number of tubes per pass, number of passes, and length of the tube if the desired outlet temperature is 150°F. (Data: 12 gauge = 0.109 in; K brass = 60 Btu/h-ft-°F; ρ water = 62.3 lb/ft^3; m = 1 cP; outer diameter = 1 in; internal diameter = [1 – 0.218] in = 0.782 in).

Answer t_s = 259°F corresponding to 35 psia

x_m = 0.109°F

We have

W = 15 000 lb/h

$\Delta t_1 = 259 - 70 = 189°F$

$\Delta t_2 = 259 - 150 = 109°F$

$$\Delta t_m = \frac{189 + 109}{2} = \frac{298}{2} = 149°F$$

$$N_{Re} = \frac{Dvp}{\mu}$$

$$\text{or } N_{Re} = \frac{\frac{0.782}{12} \times 1.4\,62.3}{1 \times 6.72 \times 10^{-4}}$$

$$\text{or } N_{Re} = \frac{0.782 \times 1.4 \times 62.3 \times 10^4}{12 \times 6.72}$$

$$\text{or } N_{Re} = 8457$$

This implies that the flow is turbulent.

Also,

$$\frac{L_i D}{K} = 0.023 \left(\frac{cp\mu}{k}\right)^{0.33} (N_{Re})^{0.8}$$

$$\text{or } h_i = \frac{\dfrac{0.0230 \times 0.346}{0.782}}{L-2} \times \left(\frac{86 \times 0.721 \times 10^{-4}}{0.346}\right)^{0.33} \times (8457)^{0.8}$$

or $h_i = 60$

The value of h_0 is to be assumed.

Now, we have

$$u = \frac{1}{\dfrac{A_i}{L_0 A_0} \times \dfrac{K_m A_i}{K_B A_n} \times \dfrac{1}{L_i}}$$

$q = u \times A \times \Delta t_m$

or $q = u \times n \times \pi \times D \times L \times \Delta t_m$

or $q = 15000 \times 1(150 - 70)$

or $q = 15000 \times 80$

or $q = 1200000$ Btu/h

Now, $A = x$ (say)

Thus, total length $L = \dfrac{x}{\pi_0}$

$$\text{or } L = \frac{x}{\pi \times \dfrac{0.782}{12}}$$

Thus,

number of tubes $= \dfrac{L}{8} = \text{n}$

Velocity = 1.4 fps = 1.4 × 3600 ft/h

Thus,

velocity = 504010 ft/h

We have

$A = \pi \times D \times n \times L \times m$

where the number of passes $m \approx 10$ and n is the number of tubes.

If $\dfrac{Lm}{8} = 9.5$, we have

$$\text{or } \frac{x}{\pi D n 10} = L$$

Now, $q = 15000 \times 80$ Btu/h $= 1200000$ Btu/h

$\Delta t_1 = 259 - 70 = 189°F$

$\Delta t_2 = 259 - 150 = 109°F$

Thus,

$$\Delta t_m = \frac{189 + 109}{2} = 149°F$$

Taking h_0 for steam = 2000 Btu/h-ft-°F

We have,

$$h_i = P_r = 7,\ N_{Re} = 8470$$

$$h_i = \frac{0.346 \times 0.023}{\frac{0.782}{12}} \times (8470)^{0.8}\ (7)^{0.33}$$

or h_i = 325 Btu/h-ft^2-°F

$$\frac{1}{L_i} = \frac{1}{L_i} + \frac{1}{h_0 D_0} + \frac{D_i l_u \dfrac{D_0}{D_1}}{K_m^2} = 0.003604$$

Thus,

u_1 = 277.5 Btu/h-ft^2-°F

We have,

$$n \times \frac{\pi D^2}{\mu} \times \nu = \frac{15000}{60.3}$$

Thus,

$$n = 15$$

We have,

$$A_i = \frac{q}{n\ (\Delta t) Lm}$$

or $A_i = \pi \times D_i\ L \times n \times m$

or A_i = 29.08 ft^2

Thus,

$$L \times m = \frac{12 \times 29.08}{\pi \times 0.762 \times 15} = 9.5$$

$$\frac{L \times m}{8} = \frac{9.5}{8} = 1.19$$

If m = 2, then

$$L = \frac{1.19 \times 8}{m}$$

or $L = 1.19 \times 4$ ft

or $L = 4.26$ ft $= 4\frac{3}{4}$ ft

Colour Plates

The skin of an average-sized adult has over 3000 in^2 of surface area, weighs about 6 lb and varies in thickness from 1/50th of an inch on the eyelids to as much as 1/3rd of an inch on the palms and soles. The skin is composed of three distinct layers—epidermis, dermis, and subcutaneous fatty tissue.

The skin protects against the invasion of bacteria, receives and uses radiation from the sun, acts as a temperature regulator and cooling mechanism, serves as an organ of perception for the nervous system through its many specialized nerve endings, and keeps blood and tissue fluids from escaping. One-third of all blood circulating throughout the body is received by the skin.

Plate 1.1 shows the subcutaneous layer as seen from the underside, revealing some of the networks of veins. The larger veins are cut at points where they penetrate to deeper levels to join other veins returning blood to the heart.

2. Abdominal venous plexus
9. Basilic vein
23. Cephalic (upper arm) vein
37. Cubital vein
41. Epigastric artery and vein, superficial
82. Mammillary venous plexus
129. Saphenous vein, great
153. Thoracoepigastric vein

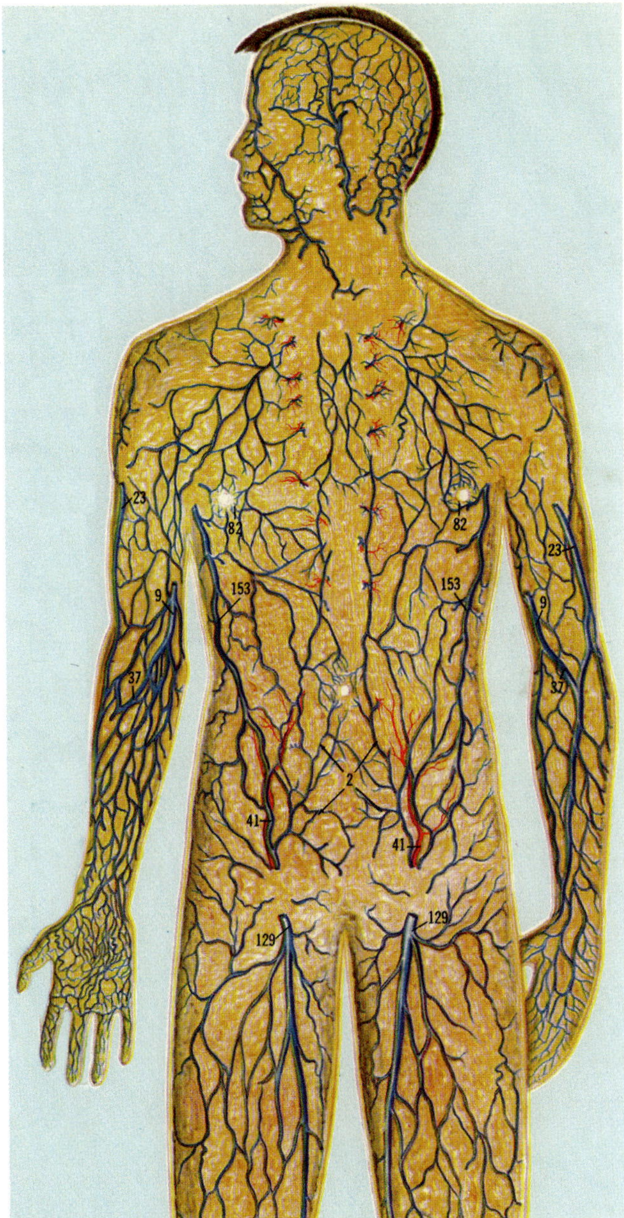

Plate 1.1 NADE- and RADE-based living human body design product (man): network of subcutaneous layers and skin layers (back view)

Source Bauer (1965)

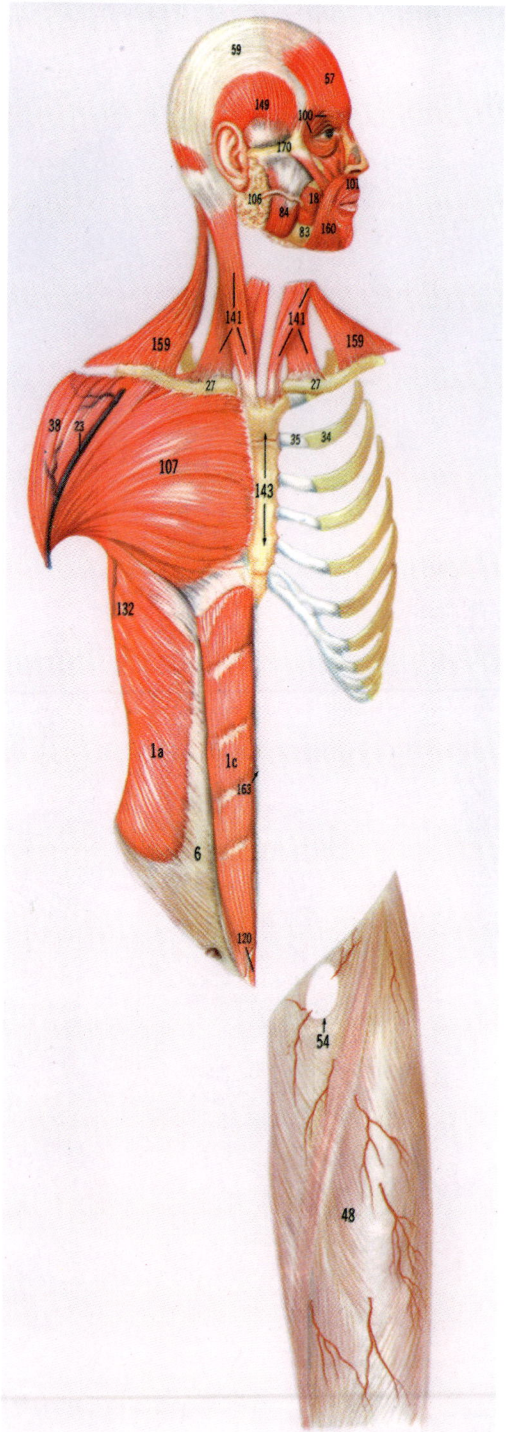

1 Abdominal muscle
  a. Abdominal oblique (flank) muscle, external
  c. Rectus abdominis (belly) muscle
6 Aponeurosis (strong tendosheath) of external abdominal oblique muscle
18 Buccinator (cheek) muscle
23 Cephalic (upper arm) vein
27 Clavicle (collarbone)
34 Costae (ribs)
35 Costal (rib) cartilages
38 Deltoid (shoulder–arm action) muscle
48 Fascia (sheath of the thing)
54 Fossa ovalis (opening for saphenous vein)
57 Frontalis muscle
59 Galea aponeurotica (strong tendon sheath–scalp)
83 Mandible (jawbone)
84 Masseter (chewing) muscle
100 Orbicularis oculi (eye) muscle
101 Orbicularis oris (mouth) muscle
106 Parotid (salivary) gland and duct
107 Pectoralis major (chest) muscle
120 Pyramidalis muscle
132 Serratus anterior (chest, side) muscle
141 Sternocleidomastoid (head rotating) muscle
143 Sternum (breastbone)
149 Temporalis muscle
159 Trapezius (neck–shoulder action) muscle
160 Triangularis (chin) muscle
163 Umbilicus (navel)
170 Zygomatic bone (cheekbone)

Plate 1.2 NADE- and RADE-based living human body design product (man): mainly muscles (front view)

Source Bauer (1965)

1 Abdominal muscle
   a. Abdominal oblique (flank) muscle, external
   b. Abdominal oblique, internal
   c. Rectus abdominis (belly) muscle
   d. Transversus abdominis
27 Clavicle (collarbone)
34 Costae (ribs)
35 Costal (rib) cartilages
36 Cranial (brain) cavity
38 Deltoid (shoulder—arm action) muscle
39 Diaphragm (breathing muscle)
41 Epigastric artery and vein deep inferior
48 Fascia (sheath of the thing)
54 Fossa ovalis (opening for saphenous vein)
55 Frontal (forehead) bone
73 Intercostal (rib) muscles, internal
78 Lacrimal (tear) gland
83 Mandible (jawbone)
85 Maxilla (cheekbone)
86 Maxillary (cheek) sinus
96 Occipital bone (skull base)
99 Omohyoid (tongue depressor) muscle
102 Orbit (eye socket)
105 Parietal bone (skull dome)
106 Parotid (salivary) gland and duct
107 Pectoralis major (chest) muscle
108 Pectoralis (chest) minor muscle
115 Pterygoid muscle, medial (closes the jaw)
125 Round ligament of liver
130 Sartorius (leg–crossing) muscle
141 Sternocleidomastoid (head rotating) muscle
142 Sternothyroid (moves larynx) muscle
143 Sternum (breastbone)
146 Sublingual (salivary) gland
147 Submandibular (salivary) gland
151 Thoracic artery and vein, internal
152 Thoracic (breathing) muscle, transverse
159 Trapezius (neck–shoulder action) muscle
163 Umbilicus (navel)

Plate 1.3 NADE- and RADE-based living human body design product (man): muscles and ribs (back view)

Source Bauer (1965)

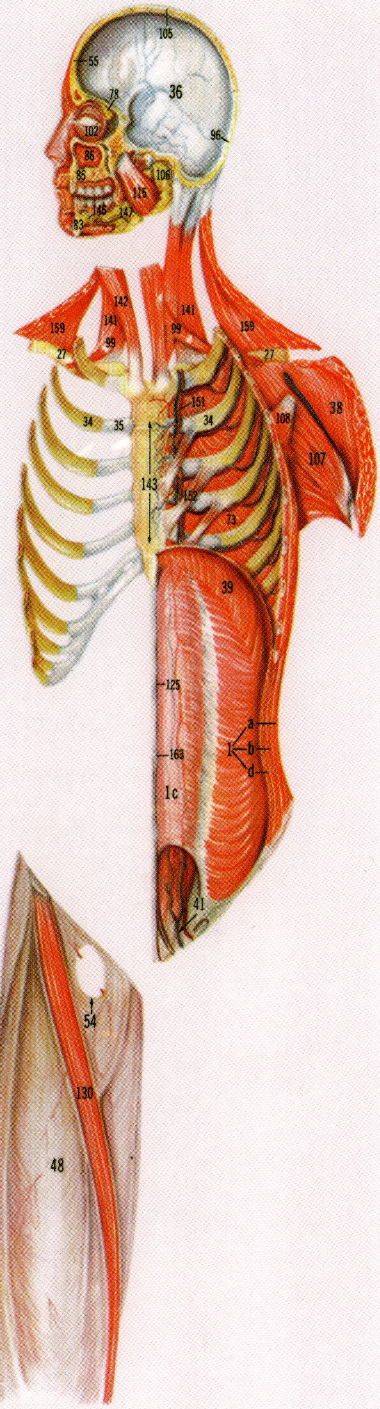

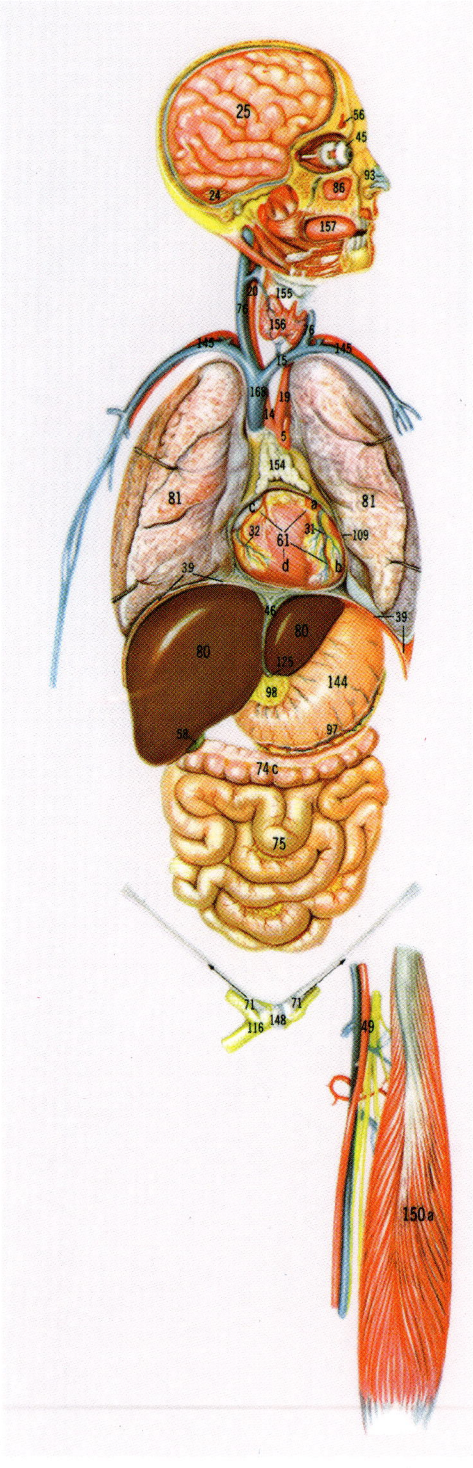

5 Aorta (great artery)
14 Brachiocephalic trunk (artery)
15 Brachiocephalic vein
19 Carotid artery, left common
20 Carotid artery, right common
24 Cerebellum (hindbrain)
25 Cerebrum (forebrain)
31 Coronary (heart) artery and vein, left
32 Coronary (heart) artery and vein, right
39 Diaphragm (breathing muscle)
45 Eyeball
46 Falciform ligament (supporting tissue for the liver)
49 Femoral artery, vein, nerve
56 Frontal sinus (airspace)
58 Gall bladder
61 Hearts
- a. Left auricle
- b. Left ventricle
- c. Right auricle
- d. Right ventricle

71 Inguinal (groin) ligament
74 Intestine, large (colon)
f. Transverse colon
75 Intestine, small (jejunum and ileum)
76 Jugular vein, internal
80 Liver
81 Lung
86 Maxillary (cheek) sinus
93 Nasal cartilages
97 Omentum, greater (cut)
98 Omentum, lesser
109 Pericardium (heart sac cut)
116 Pubic bone
125 Round ligament of liver
144 Stomach
145 Subclavian artery and vein
148 Syphysis pubis
150 Thigh muscle
- a. Rectus femoris

154 Thymus gland
155 Thyroid (voice box) cartilage
156 Thyroid gland
157 Tongue
168 Vena cava (great vein), superior

Plate 1.4 NADE- and RADE-based living human body design product (man): locations of heart and liver and location of rectum for the removal of excreta (front view)

Source Bauer (1965)

5 Aorta
11 Bile duct
20 Carotid artery
24 Cerebellum (hindbrain)
25 Cerebrum (forebrain)
33 Corpus callosum (connecting the brain hemispheres)
39 Diaphragm (breathing muscle)
44 Eustachian tube opening
49 Femoral artery, vein, nerve
56 Frontal sinus
58 Gall bladder
61 Heart
  b. Left ventricle
  d. Right ventricle
  e. Left atrium
  f. Right atrium
63 Hepatic (liver) artery
64 Hepatic (liver) veins
71 Inguinal (groin) ligament
74 Intestine, large (colon)
  c. Transverse colon
75 Intestine, small (jejunum and ileum)
76 Jugular vein, internal
79 Larynx (voice box)
80 Liver
  a. Bare area—not covered by peritoneum
81 Lung
87 Medulla oblongata (upper enlarged part of spinal cord)
91 Mesentery (supporting membrane) of intestine
95 Nasal turbinates
98 Omentum (apron of fat), lesser
111 Pituitary (master) gland
112 Pons (bridge, connecting upper and lower brain centres)
113 Portal vein
116 Pubic bone
117 Pulmonary artery
118 Pulmonary veins
126 Root of mesentery
135 Sphenoid sinus
144 Stomach
145 Subclavian artery and vein
148 Symphysis pubis
150 Thigh muscle
  a. Rectus femoris
157 Tongue
167 Vena cava (great vein), inferior
168 Vena cava, superior

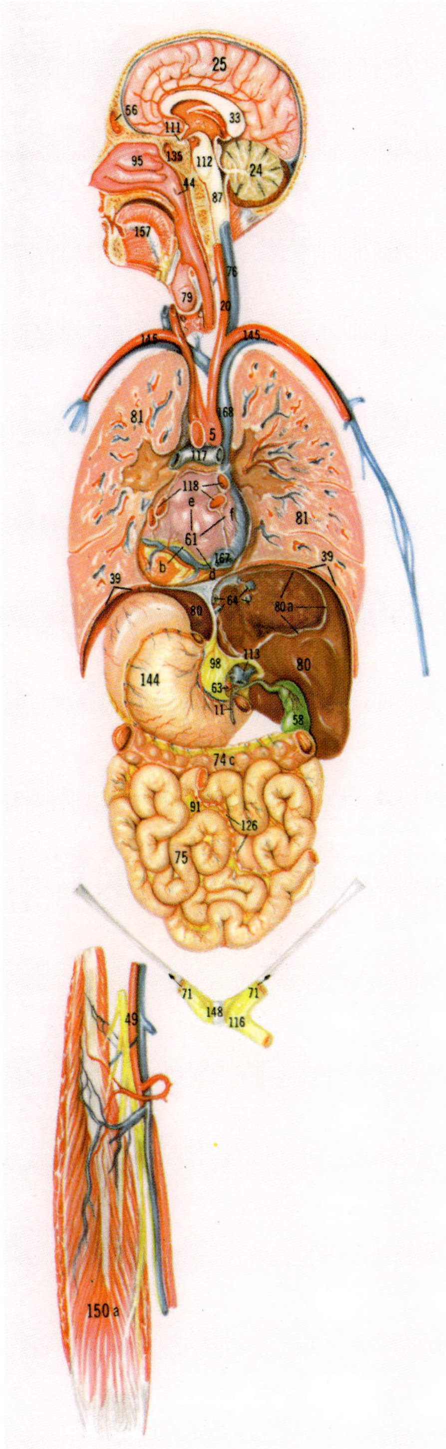

Plate 1.5 NADE- and RADE-based living human body design product (man): location of brain and position of mouth (back view)

Source Bauer (1965)

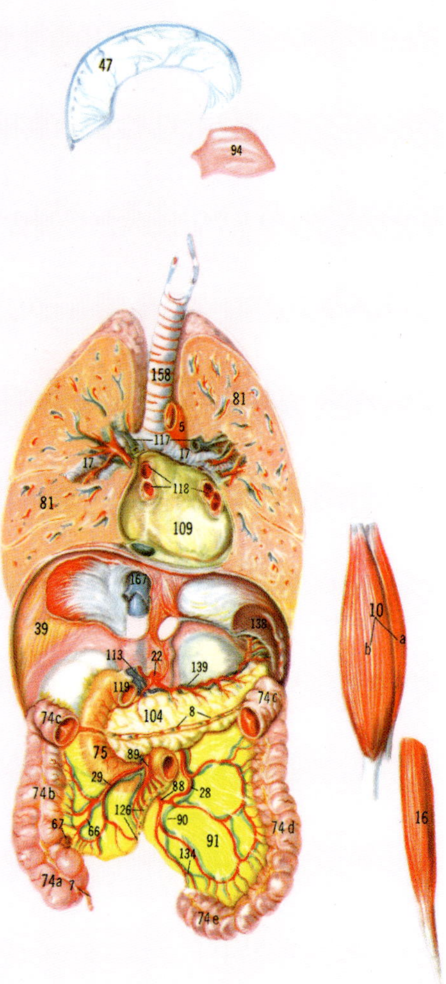

5 Aorta (great artery)
7 Appendix, vermiform
8 Attachment of transverse mesocolon (mesentery)
10 Bicepa brachil muscle flexes arm and forearm, supinates forearm)
  a. Long head
  b. Short head
16 Brachioradialis muscle (flexes and supinates forearm)
17 Bronchus (branch of windpipe)
22 Celiac artery
28 Colic artery and vein, left
29 Colic artery and vein, right
39 Diaphragm (breathing muscle)
47 Falx cerebri (membrane separating cerebral hemispheres)
66 Ileocolic artery and vein
67 Ileocolic junction
74 Intestine, large (colon)
  a. Cecum
  b. Ascending
  c. Transverse
  d. Descending
  e. Sigmoid
75 Intestine (duodenum), small
81 Lung
88 Mesenteric artery and vein, inferior
89 Mesenteric vein, inferior
90 Mesenteric vein, superior
91 Mesentery (supporting membrane)
94 Nasal septum
104 Pancreas
109 Pericardium (heart sac, cut)
113 Portal vein
117 Pulmonary (lung) artery
118 Pulmonary veins
119 Pylorus (exit) of stomach
126 Root of mesentery
134 Sigmoid artery and vein
138 Spleen
139 Splenic artery and vein
158 Trachea (windpipe)
167 Vena cava (great vein), inferior

Plate 1.6 NADE-and RADE-based living human body design product (man): various organs and muscles used for breathing (front view)

Source Bauer (1965)

5 Aorta (great artery)
7 Appendix, vermiform
10 Bicepa brachil muscle flexes arm and forearm, supinates forearm)
  a. Long head
  b. Short head
11 Bile duct, common
16 Brachioradialia muscle (flexes and supinates forearm)
28 Colic artery and vein, left
29 Colic artery and vein, right
39 Diaphragm (breathing muscle)
47 Falx cerebri (membrane separating cerebral hemispheres)
62 Hepatic flexure of colour
66 Ileocolic artery and vein
67 Ileocolic junction
74 Intestine, large (colon)
  a. Cecum
  b. Ascending
  d. Descending
  e. Sigmoid
75 Intestine (duodenum), small
81 Lung
88 Mesenteric artery and vein, inferior
89 Mesenteric vein, inferior
90 Mesenteric vein, superior
91 Mesentery (supporting membrane)
94 Nasal septum
104 Pancreas
110 Peritoneum, parietal
113 Portal vein
133 Sigmoid artery and vein
138 Spleen
139 Splenic artery and vein
140 Splenic flexure of colon
158 Trachea (windpipe)

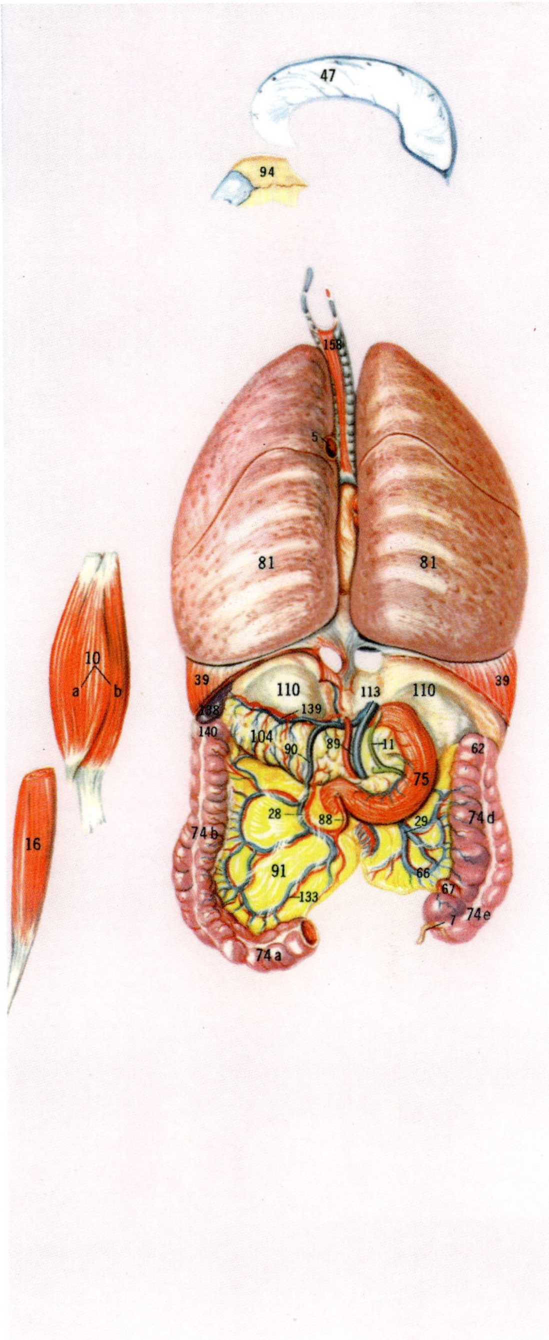

Plate 1.7 NADE- and RADE-based living human body design product (man): location of lungs (front view)

Source Bauer (1965)

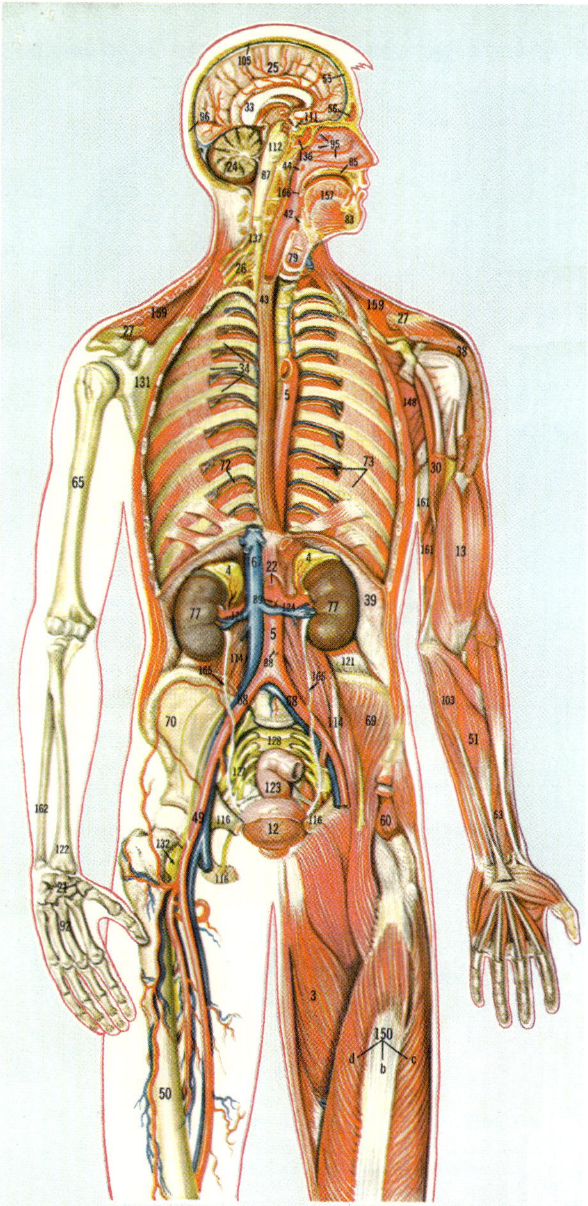

3 Adductor muscles of thigh
4 Adrenal (suprarenal) gland
5 Aorta (great artery)
12 Bladder, urinary
13 Brachialis muscle
21 Carpal (wrist) bones
22 Celiac artery
24 Cerebellum (hindbrain)
25 Cerebrum (forebrain)
26 Cervical nerves
27 Clavicle (collarbone)
30 Coracobrachialis muscle
33 Corpus callosum
34 Costae (ribs)
38 Deltoid (shoulder–arm action) muscle
39 Diaphragm (breathing muscle)
42 Epiglottis (windpipe valve)
43 Oesophagus (gullet)
44 Eustachian tube opening
49 Femoral artery, vein, nerve
50 Femur (thigh bone)
51 Flexor carpi radialis muscle
53 Flexor digitorum sublimes muscle (bends fingers)
55 Frontal (forehead) bone
56 Frontal sinus (airspace)
60 Gluteus medius muscle
65 Humerus (upper arm bone)
68 Iliac artery and vein, common
69 Iliacus muscle (bends thigh)
70 Ilium (hip bone)
72 Intercostals (rib) artery, vein, and nerve
73 Intercostals (rib) muscles, internal
77 Kidney
79 Larynx (voice box)
83 Mandible (jawbone)
85 Maxilla (cheekbone)
87 Medulla oblongata
88 Mesenteric artery, inferior
89 Mesenteric artery, superior
92 Metacarpal (hand) bone
95 Nasal turbinates
96 Occipital bone (skull base)
103 Palmaris longus muscle
105 Parietal bone (skull dome)
111 Pituitary (master) gland
112 Pons (connecting bridge)
114 Psoas muscles
116 Pubic bone
121 Quadratus lumborum muscle
122 Radius (lower arm bone)
123 Rectum
128 Sacrum
136 Sphenoid sinus
150 Thigh muscle
c. Vastus lateralis
159 Trapezius (nick–shoulder action) muscle
165 Ureter
124 Renal artery and vein
131 Scapula (shoulder blade)
137 Spinal cord
d. Vastus medialis
161 Triceps brachii muscle
166 Uvula
167 Vena cava (great vein) inferior
127 Sacral nerve plexus
132 Sciatic nerve
148 Subscapularis muscle
b. Vastus intermedius
157 Tongue
162 Ulna (lower arm bone)

Plate 1.8 NADE- and RADE-based living human body design product (man): skeletal structure (front view)

Source Bauer (1965)

1 Transverse carpal ligament
2 Abductor pollicis brevis muscle
3 Flexor pollicis brevis muscle
4 Abductor pollicis muscle
5 Lumbrical muscles
6 Flexor pollicis longus muscle
7 Tendons of flexor digitorum sublimes muscle
8 Abductor digiti V muscle
9 Flexor digiti V brevis muscle
10 Opponnens digiti V muscle
11 Digital ligaments
12 Annular ligaments
13 Pisiform bone
14 Tendon of flexor digitorum profundus muscle

Plate 1.9 Palmar view of the right hand

Source Bauer (1965)

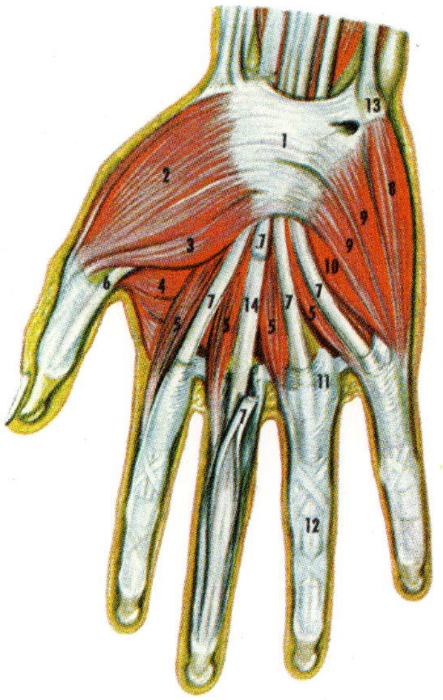

1 Dorsal carpal ligament
2 Tendon of extensor digitorum communis muscle
3 Dorsal interosseus muscle
4 Tendon of extensor pollicis longus muscle
5 Tendon of extensor pollicis brevis muscle
6 Abductor digiti V muscle
7 Digital ligaments

Plate 1.10 Back of the right hand

Source Bauer (1965)

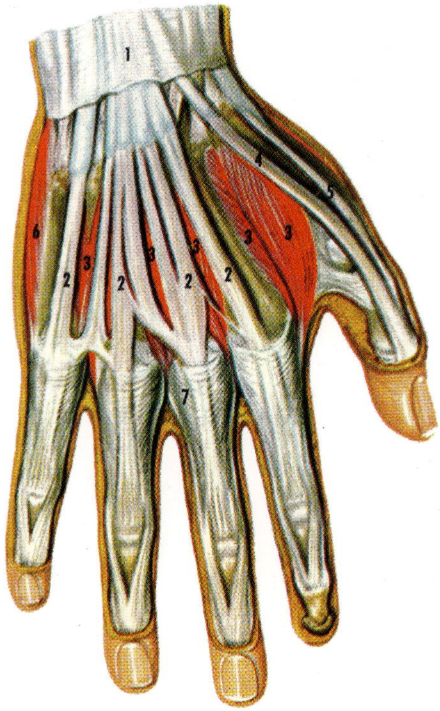

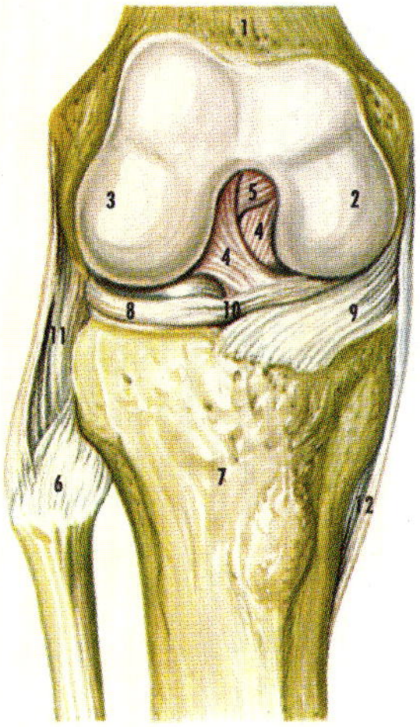

1 Femur
2 Medial condyle
3 Lateral condyle
4 Anterior cruciate ligament
5 Posterior cruciate ligament
6 Head of fibula and ligament
7 Tibia
8 Lateral meniscus
9 Medial meniscus
10 Transverse ligament
11 Collateral fibular ligament
12 Collateral fibular ligament

Plate 1.11 Right knee
Source Bauer (1965)

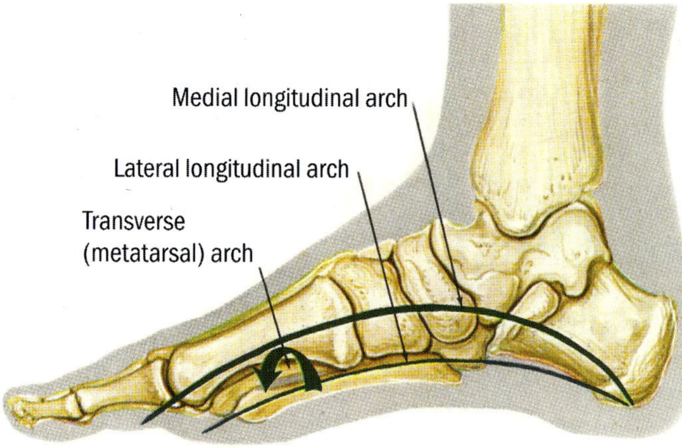

Plate 1.12 Arches of the foot
Source Bauer (1965)

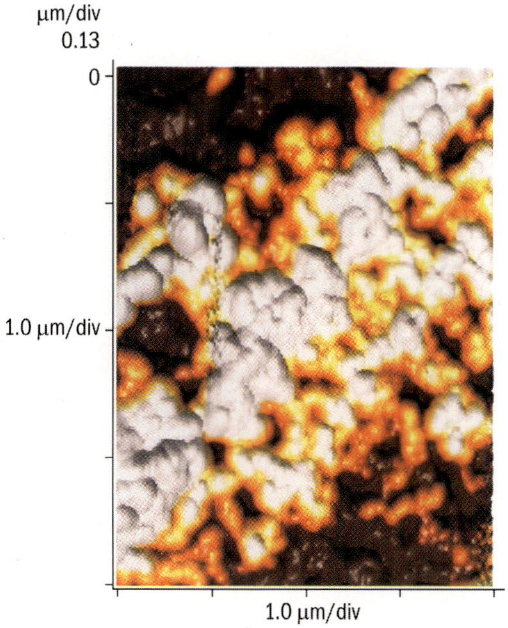

Plate 4.1 AFM image of polymeric drug RISUG

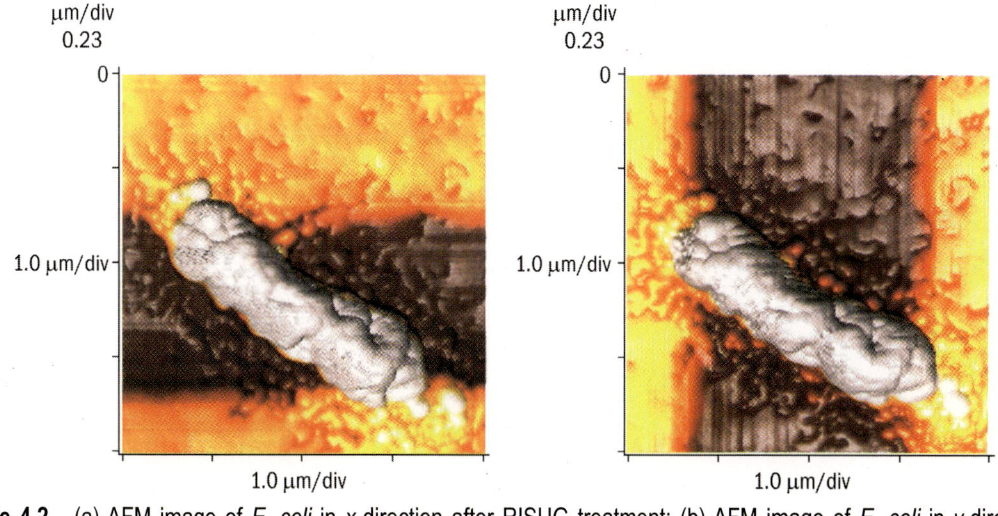

Plate 4.2 (a) AFM image of *E. coli* in *x*-direction after RISUG treatment; (b) AFM image of *E. coli* in *y*-direction after RISUG treatment

Plate 4.3 AFM image of a cluster of *E. coli* cells after RISUG exposure

4.19 ADSORPTION

4.19.1 Theory of Adsorption

Every surface takes up some amount of gas or vapour with which it is in contact relative to the extent and type of surface exposed to the gas phase. This phenomenon of concentration of gas or vapour molecules at a solid surface is known as adsorption. The surface of a solid, which is the most important factor in practical adsorption applications, is not the external surface, which is visible to the naked eye or through a microscope, but the internal surface. In order for the internal surface to be available for adsorption, it must be permeable to the gas or vapour. Adsorbents are porous solids; porosity plays a key role in determining the way in which gases or vapours are adsorbed under different conditions of concentration, temperature, and so on. However, it is not necessarily the most porous adsorbent that adsorbs most gas; the structure of the pores, their size uniformity, and arrangement as well as other factors concerning both the solid and material adsorbed have to be taken into account.

A solid exhibits adsorbent properties if the atoms at the surface do not exist in a homogeneous force field as those in the interior. There are residual unsatisfied forces in the atoms just outside the surface and if not adsorbed, these would otherwise interact with the nearest atoms. Therefore, a state of strain exists at the surface and the free energy tends to decrease. In case of solids in contact with the gas phase, this is achieved by adsorption of gases or vapours by the surface.

The extent of adsorption by a solid surface under particular conditions is mainly determined by the type and arrangement of atoms of the surface within the pores of the adsorbent (that is, the material from which the surface is constructed) and the physical and chemical characteristics of the gas or vapour adsorbed.

4.19.2 Types of Adsorption

There are two main types of adsorption processes. The gas or vapour may either interact with the solid surface in a physical or chemical manner or the two processes may take place simultaneously within the same system. Adsorption processes are categorized as follows.

4.19.2.1 Physical adsorption (physisorption)

Physical adsorption occurs when the forces of attraction is active between the molecules of the adsorbate and adsorbent are weak van der Waals type forces. Thus, physical adsorption is also known as van der Waals adsorption. In such cases, the gas or vapour is in a condensed state. The equilibrium between the solid and gas is rapidly set up and readily reversible; a new equilibrium condition is quickly established when the gas pressure or temperature is altered. For a given system, the increase in temperature results in a reduction of the amount of adsorption. As expected, heat involved in this process is of the same order as heat of condensation.

4.19.2.2 Chemical adsorption (chemisorption)

Chemical adsorption involves forces of a chemical nature, the bond formed between the gas and the solid surface is similar in strength to a bond joining separate atoms in a chemical compound. For a majority of cases, the equilibrium between solid and gas is not readily reversible or rapidly set up as with physisorption. In general, the rate and amount of adsorption increases with increase in temperature, and the heat liberated during the process is of the order of the heat of a chemical reaction.

Other types of adsorption are activated adsorption, persorption, and sorption.

4.19.2.3 Activated adsorption

The term activated adsorption is used in cases where the rate of chemisorption is considerable at low temperature and the activation energy of the process is high. For a given gas and surface, a transition from physical to activated adsorption may be observed when the temperature is raised.

4.19.2.4 Persorption

From the X-ray analysis of various materials, it is apparent that molecules can be fitted into the lattice structure of certain compounds. For example, water can be taken up by certain aluminosilicates such as chabazite. Chabazite readily takes up water into the vacant spaces that exist between the aluminium, silicon, and oxygen atoms in the lattice. Methyl and ethyl alcohol are also taken up but not ether and acetone vapours. The size of the molecule adsorbed should be such that it is able to fit in the lattice. Thus, persorption may be thought of as a molecular sieve effect.

4.19.2.5 Sorption

It is sometimes difficult to experimentally distinguish between the different types of vapour or gas picked up by solids. In 1909, McBain suggested the term "sorption", which covers all types of vapours and gases. This term is now sometimes used to refer to the phenomenon if it is intended to be non-committal.

4.19.2.6 Distinction between absorption and adsorption

Confusion in terminology exists due to the existence of the word "absorption", which refers to a phenomenon different from "absorption". In practice, the words absorption and adsorption are used somewhat indiscriminately to mean the same thing. However, there is a distinction between the two terms. When water vapour is taken up by silica gel, it "condenses" on the surface of the material but can easily be evaporated by increasing the temperature; this is adsorption. On the other hand, when water vapour is taken up by a desiccant such as calcium chloride, the material is converted into a hydrate. In this case, there is a change in the composition of the material. Furthermore, addition of more moisture to the desiccant would cause it to lose its crystalline shape and eventually dissolve in the water absorbed. Water thus permeates through the whole of the material and finally a phase change occurs; this is absorption.

4.19.2.7 Adsorption equilibria and isotherms

In practical application, the most important sorption phenomenon is physical adsorption, which has been widely studied in the static way. In such static or equilibria studies, a gas or vapour (the adsorbate) is maintained at a constant pressure in contact with a given amount of solid (the absorbent). The whole system is maintained at a specified temperature. The amount of gas or vapour taken up is measured volumetrically at various partial pressures of the adsorbate at a fixed temperature. By this means, an adsorption isotherm for the gas–solid system can be expressed in the form of a graph. The adsorption isotherm is the most important and by far the most often used of the various equilibria data that can be measured. The other expressions of adsorption equilibria are (1) the adsorption isobar, which is a plot of the amount of gas or vapour absorbed against temperature at constant pressure and (2) the adsorption isostere, which is the variation of equilibrium pressure with temperature corresponding to a constant amount of gas adsorbed.

Various types of isotherms are observed, and the shape of the graph varies according to the adsorbate-adsorbent system. According to Braunauer, the five types of adsorption isotherms exist on the adsorption of gases (Figure 4.43).

In principle, the five types of adsorption isotherms can be obtained on any one adsorbate, the variable factor being the physical structure of the adsorbent concerned. For example, a range of silica gels can be prepared and from these, it is possible to select materials that, when using water vapour as the adsorbate, would give isotherms similar to each of the five types listed by Braunauer.

4.19.2.8 Specific adsorption theories

Various equations have been derived to fit the isotherms observed by Braunauer (Figure 4.43). In order of increasing theoretical and mathematical complexity, these are the Freundlich, Langmuir, and the Braunauer, Emmett, and Teller equations. Purely statistical approaches have also been made to the adsorption theory.

4.19.2.8.1 Monomolecular layer theory

Langmuir proposed that the forces active in adsorption are similar to those that act in chemical combinations. Furthermore, during adsorption, a molecular layer of thickness

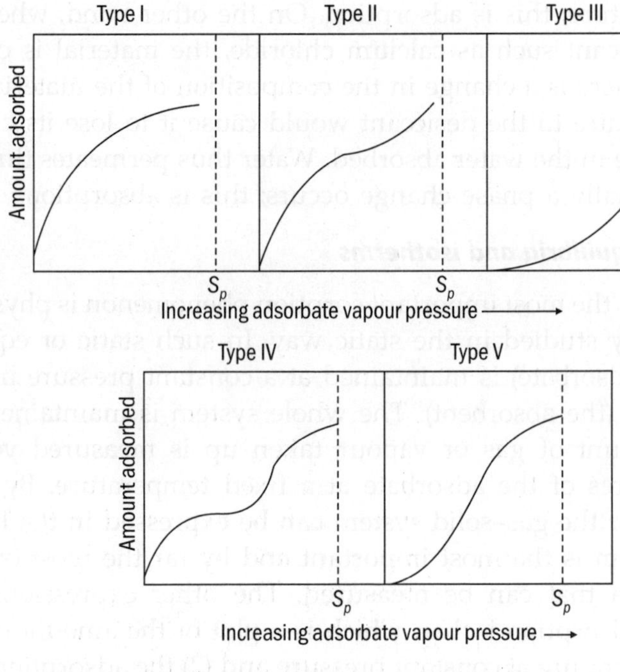

Figure 4.43 Types of adsorption isotherms

1 mol forms on the solid surface. This concept satisfactorily explains observations on the initial amount of adsorbent taken up by a solid at fixed temperature, whether or not the forces involved are chemical or physical. However, on progressively increasing the pressure of the adsorbate to the saturation pressure, the Langmuir treatment produced results that did not meet expectations and, therefore, more rigorous analyses were required.

4.19.2.8.2 Multimolecular layer theory

By assuming that adsorption on solid surfaces takes place with the formation of secondary, tertiary, and finally multilayers upon the primary monomolecular layer, Langmuir, Braunauer, and others derived an equation that fitted the intermediate and higher pressure regions of the various types of adsorption isotherms.

4.19.2.8.3 Capillary condensation theory

All adsorbents of practical importance have a porous structure. The pores within the adsorbent can be visualized as cylindrical in shape or having a narrow entrance and bulbous interior. Carter-type pores also exist. Amorphous adsorbents such as silica-activated alumina and activated carbon contain pores of various shapes and sizes.

Since saturation vapour pressure within a small capillary is lower than that for a vapour in a large space, some approaches to the theoretical treatment of adsorption use capillary condensation as the basis for explaining adsorption in a porous adsorbent.

4.19.2.8.4 Representation of adsorption isotherms

Isotherm data are expressed in the form of (1) a graphical or tabular record based on measurements, (2) empirical algebraic expressions fitted to the data and usually selected for their generality and simplicity of calculation use, and (3) equations reflecting the molecular statistics of the process.

4.19.2.9 Concentration variables

The quantity of each solute can be expressed in units of mass or number of moles (for adsorption from gas phase), volume (for adsorption from liquid phase) or number of ion equivalents (for ion exchange). The fluid phase concentration C is the number of quantity units per volume unit of fluid phase; for a gas, the partial pressure p may replace concentration. The stationary phase or solid phase concentration q is the number of quantity units obtained per mass unit of solid phase.

For generality, the isotherm equation will be expressed in terms of dimensionless concentrations, formed as the ratio of the actual C or q to a reference value of C or q in the same units. Hence, for each solute species,

$$x = \frac{C}{C_{ref}} \text{ and } y = \frac{q}{q_{ref}}$$

For single-solute adsorption, C_{ref} is usually the highest fluid phase concentration and q_{ref} is the equilibrium solid phase concentration coexisting with C_{ref}.

The isotherms for two-component systems are single curves of q versus C or p, or of y versus x, expressed for the species taken up by the solid. In operations where the concentration ranges only over a part of the isotherm, between C', q' and C'', q'' or between x' y' and x'', y'', the following definitions of fractional fluid concentration change x and fractional solid concentration change y are also useful.

$$x = \frac{c - c'}{c'' - c'} = \frac{x - x'}{x'' - x'} \tag{4.236}$$

$$y = \frac{q - q'}{q'' - q'} = \frac{y - y'}{y'' - y'} \tag{4.237}$$

Both x and y begin at 0 and end at 1; x and y also usually lie in the range of 0 to 1. The relation of a partial range transition to a full range transition is shown in Figure 4.44.

Representative isotherms are shown in Figure 4.43, as classified by Brunauer and coworkers (Brunauer 1943). Curves that are convex upwards throughout (type I) are

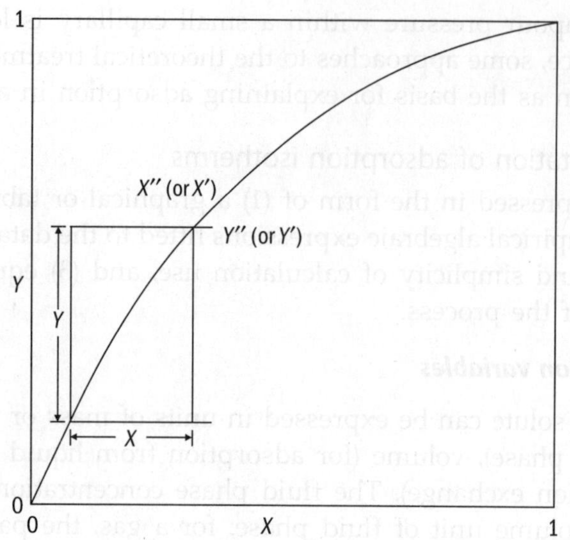

Figure 4.44 Construction of the isotherm for a partial change in equilibrium concentration between $(x'\ y')$ and (x'', y'')

designated as "favourable" to uptake of solute, while those that are concave upwards throughout (type III) are designated as "unfavourable" to uptake of solute. The terms favourable and unfavourable refer to the uptake step for the component to which the isotherm applies. The opposite designation holds for the discharge step. Thus, elution is "unfavourable" for a component having favourable isotherm. For non-inflected isotherms, the part near the origin may be linear as in Langmuir curves or it may lean towards the vertical or horizontal as in Freundlich curves (Table 4.21). Other isotherms (types II, IV, and V) have one or more inflection points.

The simplest algebraic equations to describe shapes of isotherms are given in Table 4.21. One of these, the trinomial, can be used to fit all except type IV of the curves shown in Figure 4.43. The constants of the equations in Table 4.21 must be within defined ranges to fit a specific isotherm type.

The bilinear isotherms approximate an equilibrium curve by two straight lines. The Langmuir isotherm applies to adsorption on completely homogeneous surfaces, with negligible interaction between adsorbed molecules. The Langmuir isotherm can be mathematically expressed as shown in Equation 4.238.

$$W = \frac{bC}{1 + KC} \tag{4.238}$$

where W is the adsorbate loading, C is the concentration in the fluid, and b and K are constants.

Unfortunately, the Langmuir isotherm has a simple theoretical basis and is not a good fit for most physical adsorption systems. The empirical Freundlich equation is often a better fit, particularly for adsorption from liquids (Equation 4.239).

$$W = bC^m \tag{4.239}$$

where $m < 1$

4.19.2.10 Principles of adsorption in fixed beds

4.19.2.10.1 Concentration patterns in fixed beds

In fixed-bed adsorption, the concentrations in the fluid phase and solid phase change with time as well as with the position in the bed. At first, most of the mass transfer takes place near the inlet of the bed, where the fluid contacts fresh adsorbent. If the solid contains no adsorbate at the start, the concentration in the fluid drops exponentially with distance, essentially to zero before the end of the bed is reached. This concentration profile is shown by curve t_1 in Figure 4.45(a), where C/C_0 is the concentration in the fluid relative to that in the feed. After a few minutes, the solid near the inlet is nearly saturated, and most of the mass transfer takes place further from it; the concentration gradient becomes S-shaped as shown by curve t_2. The region where most of the change in concentration occurs is called the mass transfer zone, and the limits are often taken as C/C_0 = 0.95 to 0.50. With time, the mass transfer zone moves down the bed, as shown by profiles t_3 and t_4. Similar profiles could be drawn for the average concentration of adsorbate on the solid showing nearly saturated solid at the inlet, a large change in the region of the mass transfer zone, and zero concentration at the end of the bed. Instead of plotting the actual concentration on the solid, the concentration in the fluid phase in equilibrium with the solid is shown as a dashed line for time t_2. This concentration must always be less than the actual fluid concentration, and the difference in concentrations, or driving force, is large where the concentration profile is steep and mass transfer is rapid.

4.19.2.10.2 Breakthrough curves

Few fixed beds have internal probes that would permit measurement of profiles such as those shown in Figure 4.45(a). However, these profiles can be predicted and used in calculating the curve of concentration versus time for fluid leaving the bed. The

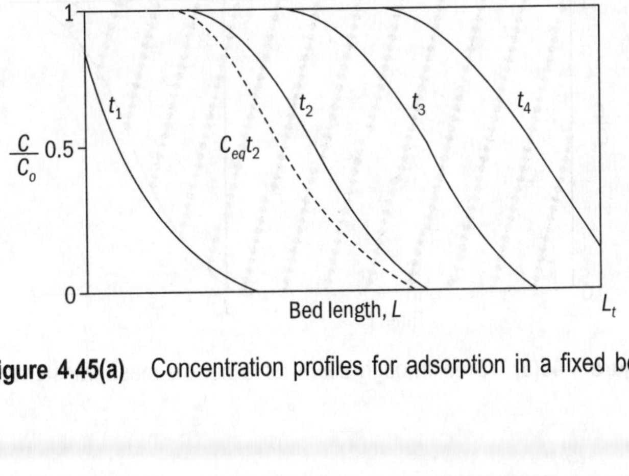

Figure 4.45(a) Concentration profiles for adsorption in a fixed bed

curve shown in Figure 4.45(b) is called a breakthrough curve. At times t_1 and t_2, the exit concentration is almost zero, also shown in Figure 4.45(a). When the concentration reaches some limiting permissible value or break point, the flow is stopped or diverted to a fresh adsorbent bed. The break point is often taken as a relative concentration of 0.05 or 0.10. Since only the last portion of fluid processed has such a high concentration, the average fraction of solute removed from the start to the break point is often 0.99 or higher.

If adsorption is continued beyond the break point, the concentration would rise rapidly to about 0.5 and then slowly approach 1.0 as shown in Figure 4.45(b). This S-shaped curve is similar to those for internal concentration profiles. By material balance, it can be shown that the area between the curve and a line at $C/C_0 = 1$ is proportional to the total solute adsorbed if the entire bed is in equilibrium with the feed. The area up to the break point time t_b represents the actual amount adsorbed. If the mass transfer zone is narrow relative to the bed length, the breakthrough curve will be rather steep, as shown in Figure 4.46(a), and most of the capacity of the solid will be utilized at the break point. When the mass transfer zone is almost as long as the bed, the breakthrough curve is greatly extended, as shown in Figure 4.46(b), and less than half of the bed capacity is utilized. A narrow mass transfer zone is desirable

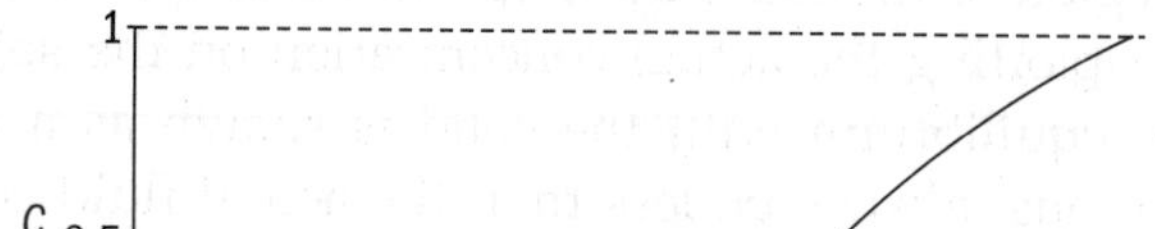

Figure 4.45(b) Breakthrough curve for adsorption in a fixed bed

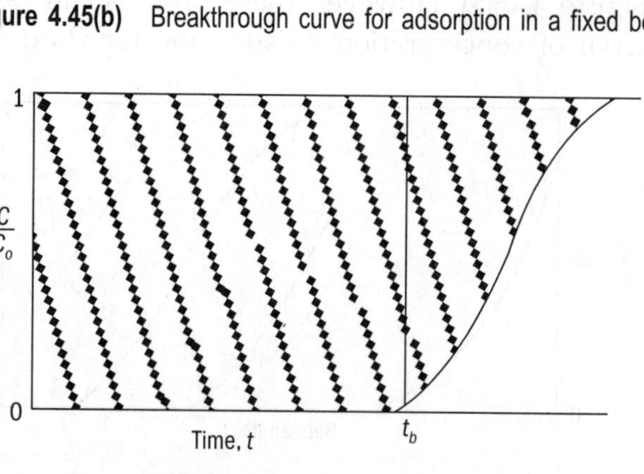

Figure 4.46(a) Breakthrough curve for a narrow mass transfer zone

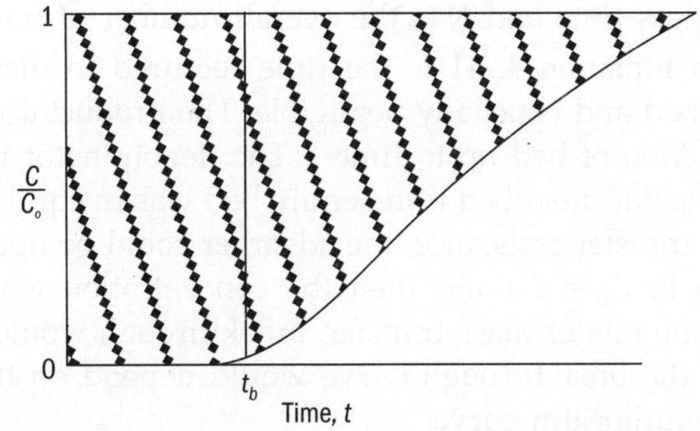

Figure 4.46(b) Breakthrough curve for a wide mass transfer zone

to make efficient use of the adsorbent and to reduce the energy costs in regeneration. In the ideal case of zero mass transfer resistance and no axial dispersion, the mass transfer zone would be of infinitesimal width, and the breakthrough curve would be a vertical line from 0 to 1.0 when the solid is saturated.

4.19.2.11 Basic equations for adsorption

Although adsorbers are generally designed from laboratory data, their approximate performance can sometimes be predicted from equilibrium data and mass transfer calculations.

The rate of mass transfer process for isothermal adsorption in fixed beds is approximated using an overall volumetric coefficient and an overall driving force (Equation 4.240).

$$\rho_p(1 - \varepsilon)\frac{dW}{dt} = K_C a(C - C^*) \tag{4.240}$$

where ρ_p is the density of particle, ε is the external void fraction of bed, W is the adsorbate loading, K_C is the overall mass transfer coefficient, a is the external surface area per unit volume of bed (given by $a = \varepsilon[1 - e]/D_p$ [for sphere], where D_p is the particle diameter), C is the concentration of adsorbate in fluid, and C^* is the concentration of adsorbate in equilibrium with concentration in solid.

The solutions to Equation 4.240 for different isotherms and controlling steps are given by Equations 4.241 and 4.242.

$$\tau \equiv \frac{u_0 C_0\left(t - \dfrac{L_T\varepsilon}{d_0}\right)}{\rho_p(1 - \varepsilon)L_T W} \tag{4.241}$$

$$N \equiv \frac{K_a L_T}{d_0} \tag{4.242}$$

where τ is dimensionless time and N is the overall number of transfer units.

The term $L_T\varepsilon/d_0$ in Equation 4.241 is the time required to displace the fluid from external voids in the bed and is usually negligible. The product d_0C_0t is the total solute fed to a unit cross section of bed up to time t. The denominator is the capacity of the bed or the amount of solute adsorbed if the entire bed was in equilibrium with the feed. If there was no mass transfer resistance, the adsorber could be operated with complete removal of solute up to C_0 = 1.0 and then the concentration would jump from 0 to C/C_0 = 1.0. With a finite rate of mass transfer, breakthrough would occur at C/C_0 = 1.0 and the steepness of the breakthrough curve would depend on the parameter N and on the shape of the equilibrium curve.

4.20 EXTRACTION

Out of the different applications of mass transfer in chemical engineering and absorption of gases, which has been covered separately, the subsequent sections focus on the process of extraction and leaching.

Removal of a soluble constituent from a solid or a liquid by the use of a solvent is termed extraction. It may be liquid–liquid extraction, also called solvent extraction, or solid–liquid extraction, also called leaching or lixiviation. After extraction or leaching, the target component is separated from its mixture with the solvent by distillation or rectification.

4.20.1 Liquid–Liquid Extraction

Liquid–liquid extraction has special fields of application such as separation of acetic acid and water by benzene or toluene (Figure 4.47). It usually involves three steps.

Table 4.21 Simplest algebraic forms of equations describing shape of isotherms

Equation type	Relative solid concentration	Relative fluid concentration
Linear	x	y
Constant separation factor	$\frac{x}{r + x - rx}$	$\frac{ry}{1 + ry - y}$
Langmuir	$\frac{(1 + K_L C_{ref})\, x}{1 + K_L C_{ref}\, x}$	$\frac{y}{1 + K_L C_{ref}(1 - y)}$
Freundlich	x^{β}	$y^{1/\beta}$
Trinomial y [$-2 < a_2 < 1 - a_1$]	$x(1 - a_1 - a_2) + a_1x^2 + a_2x^3$	–
Trinomial x [$-2 < b_2 < 1 - b_1$]	–	$y(1 - b_1 - b_2) + b_1y^2 + b_2y^3$

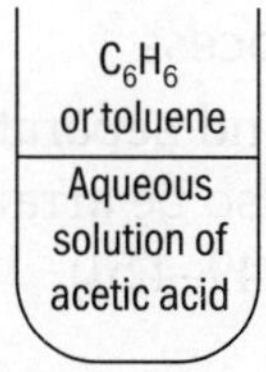

Figure 4.47 Separation of acetic acid and water by benzene or toluene

1. Intimate contact of the original solution with the solvent.
2. Separation of the two solution phases with the solvent phase.
3. Recovery of the solute from the solvent phase.

In liquid–liquid extractions, the solution to be extracted is called the feed and the liquid with which the feed is contacted is the solvent. The solvent-rich product is called extract and the residual liquid from which the solute has to be removed is raffinate.

4.20.1.1 Stepwise equipment

The equipment required for the first step of extraction process are agitated vessels, jet mixers, baffle plate mixers, bubble-cap plates, and packed towers. For the second step, the equipment required are decantation, centrifuging or packed towers, and for the third step, the necessary equipment for distillation process are required.

4.20.1.2 Methods of operation

Extraction may be carried out either as a batch (single or multiple contact) or a continuous process.

4.20.1.2.1 Single-contact batch process

In a single-stage batch process (Figure 4.48), the solvent and solution are mixed together in a mixer and then separated into two phases. The extract E contains the required solute in the added solvent, and the raffinate R contains the weaker solution with some of the solvent.

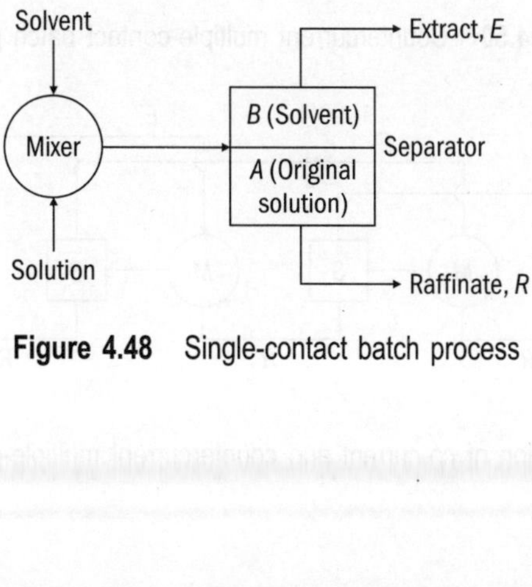

Figure 4.48 Single-contact batch process

4.20.1.2.2 Multiple-contact batch process

In the single-contact process, mixing and separation occur in the same vessel. A series of mixers (*M*) and separators (*S*) may also be arranged to give a multiple-contact system, which may be "co-current" (Figures 4.49–4.51).

4.20.1.2.3 Continuous process

Continuous countercurrent operation is obtained by using a column apparatus such as spray or packed towers (Figure 4.52).

4.20.2 Equilibrium Conditions

The mechanism of transfer of the solute from the initial solvent to a new solvent is one of diffusion (the entire process is similar to that of absorption, with the gas phase

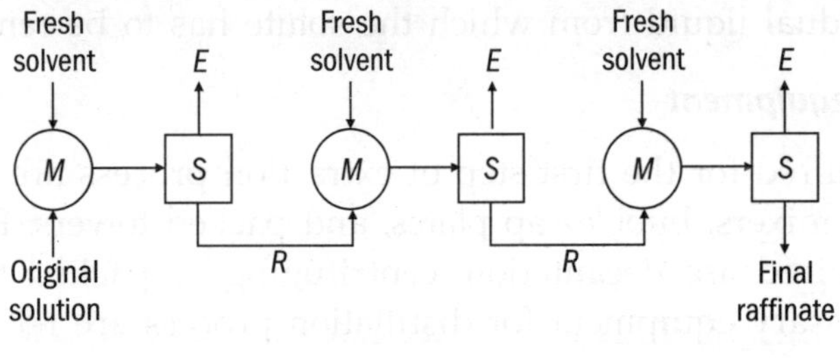

Figure 4.49 Co-current multiple-contact batch process

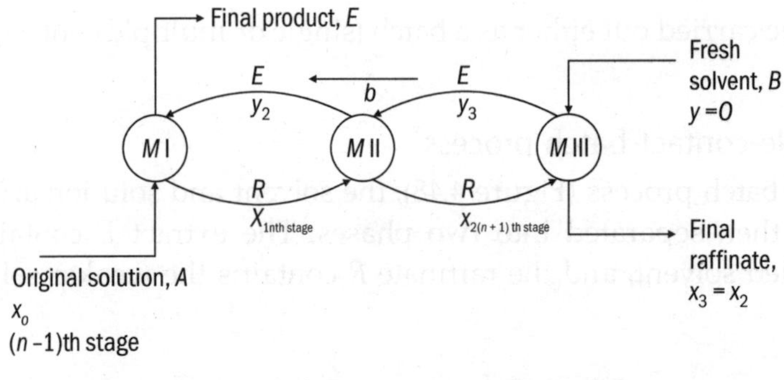

Figure 4.50 Countercurrent multiple-contact batch process

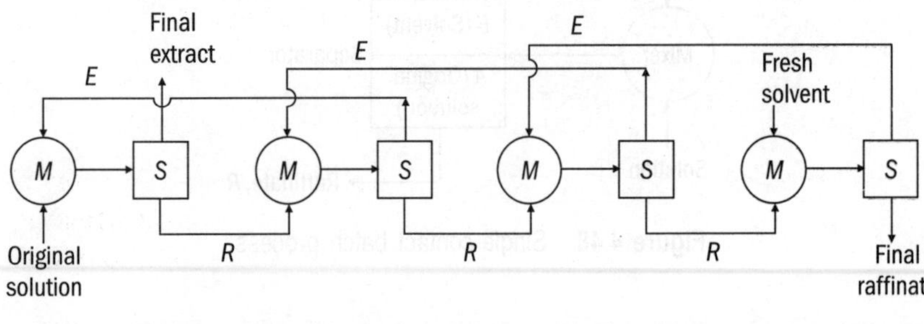

Figure 4.51 Combination of co-current and countercurrent multiple-contact batch processes

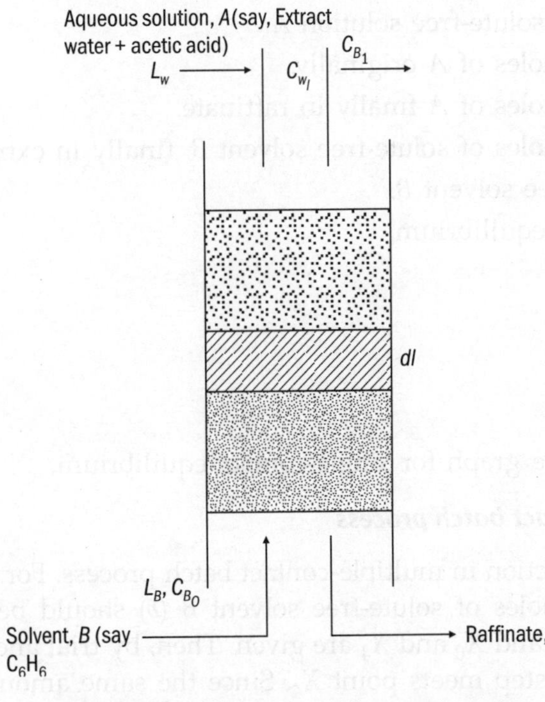

Figure 4.52 Continuous countercurrent process

replaced by a second liquid phase). While calculating the capacity of the extraction equipment, two significant factors to be taken into consideration are (1) the equilibrium of the solute between the two phases and (2) the resistance to diffusion of the solute. On the basis of the two film theory, it is assumed that there are two liquid films, one on each side of the interface.

The equilibrium condition for distribution of one solute between two liquid phases is considered in terms of the distribution law. At equilibrium, the ratio of the concentration of the solute in the two phases is given by Equation 4.243.

$$\frac{C_1}{C_2} = K \tag{4.243}$$

where C_1 is the concentration of the solute in the solvent phase, C_2 is the concentration of the solute in the aqueous phase, and K is the distribution coefficient.

4.20.3 Mathematics of Operation Cases

4.20.3.1 Immiscible solvents, single contact (K is constant)

Figure 4.48 shows extraction of immiscible solvents in single-contact batch process. Assuming ideal contact, that is, extract and raffinate are in equilibrium with each other. Let

a = moles of original solute-free solution A.
x_0 = moles of solute/moles of A originally.
x_1 = moles of solute/moles of A finally in raffinate.
y_1 = moles of solute/moles of solute-free solvent B, finally in extract.
b = moles of solute-free solvent B.

By material balance equilibrium,

$$ax_0 = ax_1 + by$$

$$\text{or } a(x_1 - x_0) = -by$$

$$\text{or } \frac{y_1}{x_1 - x_0} = -\frac{a}{b}$$

Figure 4.53 shows the graph for single-contact equilibrium.

4.20.3.2 Multiple-contact batch process

Figure 4.49 shows extraction in multiple-contact batch process. For maximum efficiency, the same amount of moles of solute-free solvent B (b) should be used. Suppose four extracts are to be made and X_0 and X_4 are given. Then, by trial and error, four steps are drawn so that the last step meets point X_4. Since the same amount of solvent is used each time, slope = $-\frac{a}{b}$. Figure 4.54 shows the graph for multiple-contact equilibrium.

4.20.3.3 Countercurrent multiple-contact batch process

Figure 4.55 shows extraction in countercurrent multiple-contact batch process. Assuming ideal stage, the material balance of solute for the first stage is given as follows.

$$by_2 + ax_0 + by_1 + ax_1$$

The subscripts indicate the stage from where solvents A and B are leaving. In general, the material between the top and the nth stage is given as follows.

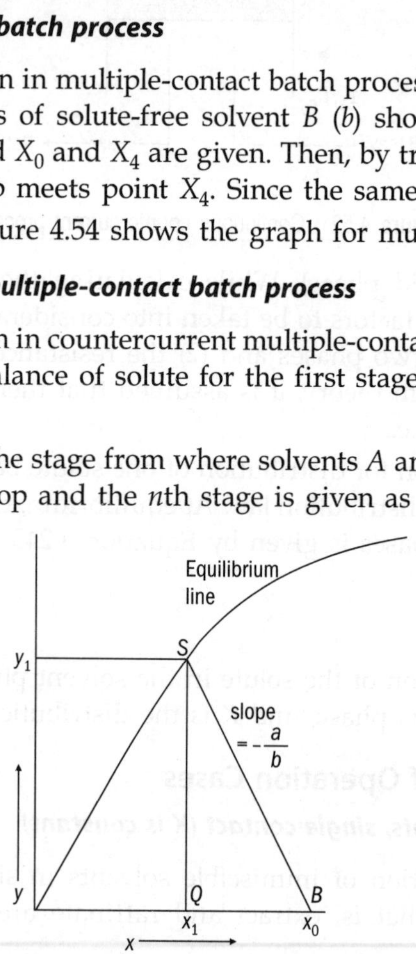

Figure 4.53 Single-contact equilibrium

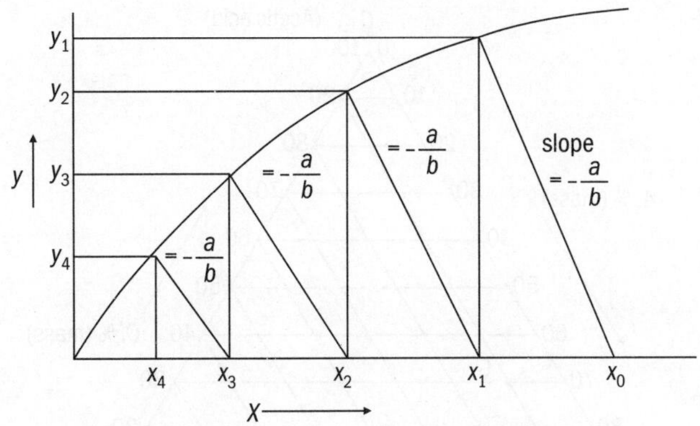

Figure 4.54 Multiple-contact equilibrium

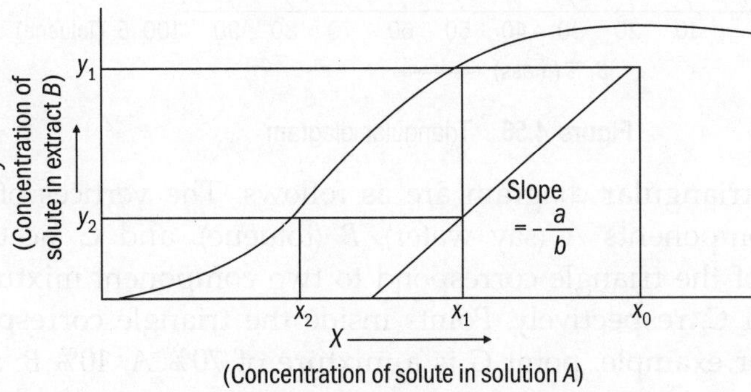

Figure 4.55 Countercurrent multiple-contact equilibrium

$$by_{n+1} + ax_0 = ax_n + by_1$$

or $y_{n+1} = \frac{a}{b}(x_n - x_0) + y_1$

The values of x_0, y, and so on are fixed. Thus, a straight line is obtained if y_{n+1} is plotted against x_n, having a slope of $\frac{a}{b}$.

In case of countercurrent multiple-contact equilibrium, the operating line is on the right of the equilibrium line, otherwise all other procedures are similar. If the equilibrium line and x_0 are given, slope $\frac{a}{b}$ is known, and if it is given that three stages are to be drawn, the trial and error method is used. By assuming a value for y, stages are made and on obtaining three stages, the calculation is correct (Figure 4.55).

4.20.3.4 Two partially miscible solvents (K is constant)

In extraction, problems of this nature are usually solved by using triangular or rectangular phase diagram (Figure 4.56).

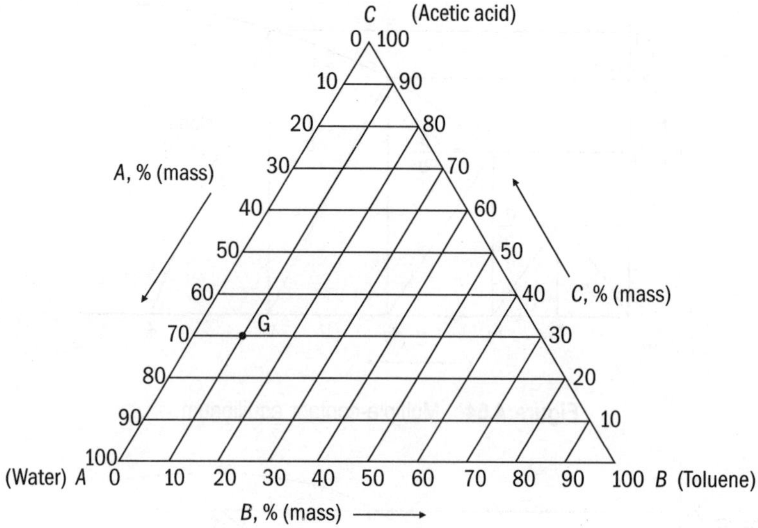

Figure 4.56 Triangular diagram

The features of the triangular diagram are as follows. The vertices of the triangle correspond to pure components *A* (say water), *B* (toluene), and *C* (acetic acid). The sides *AB*, *BC*, and *AC* of the triangle correspond to two component mixtures of *A* and *B*, *B* and *C*, and *A* and *C*, respectively. Points inside the triangle correspond to three component systems. For example, point G is a mixture of 70% *A*, 10% *B*, and 20% *C*.

1. Component *C* is miscible in all proportions with *A* and *B*, but *A* and *B* are only partially miscible (Figure 4.57[a]).

   MNOPQ is the phase boundary line or solubility curve. Outside this curve, the two components are completely miscible, that is, there is only one phase. However, within the curve the two components are partially miscible and the compositions at any point are given by lines (say, *xy*).

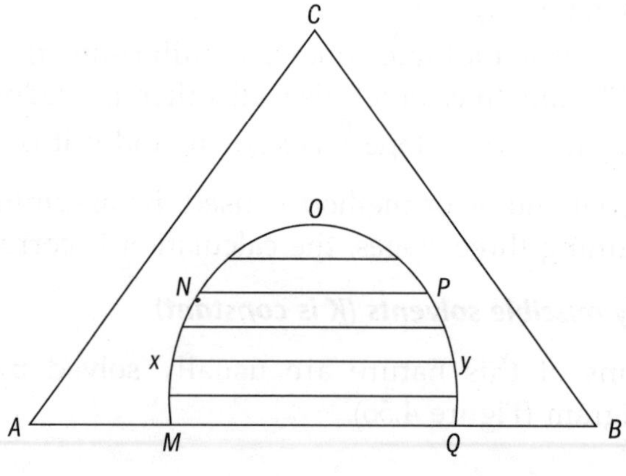

Figure 4.57(a) *C* completely miscible with *A* and *B*, while *A* and *B* are partially miscible

2. *C* is miscible with *A* in all proportions, but *A* and *B* and *C* and *B* are partially miscible. Outside *MNOP*, they are miscible in all proportions (Figure 4.57[b]).
3. Each component is only partially miscible with another, that is, *C* and *A*, *A* and *B*, *B* and *C* (Figure 4.57[c]).

a = single-phase region; *b* = two-phase region; *c* = three-phase region

In these three cases, all the components are liquids and case (a) is mostly used.

4.21 LEACHING AND BIOLEACHING

Leaching is a solid–liquid extraction; it is also called lixiviation. Leaching is the process of separating soluble matter (solute) from its mixture with an insoluble solid. If the solute is uniformly dispersed in the solid, the material close to the surface will dissolve first, leaving a porous structure in the solid residue. The process can be considered to take place in three parts.

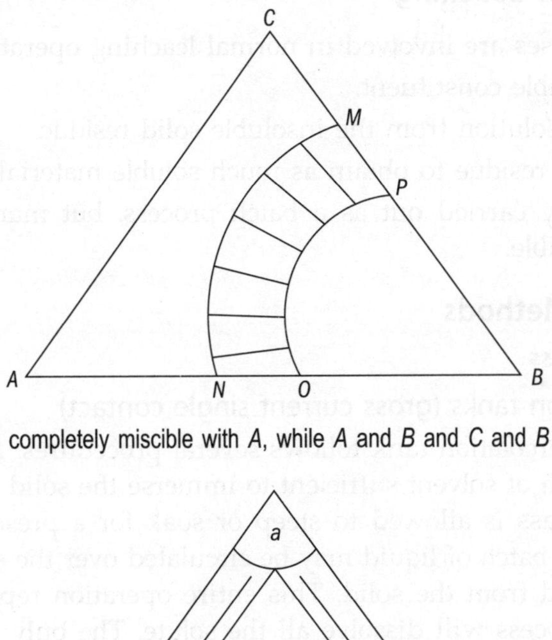

Figure 4.57(b) *C* is completely miscible with *A*, while *A* and *B* and *C* and *B* are partially miscible

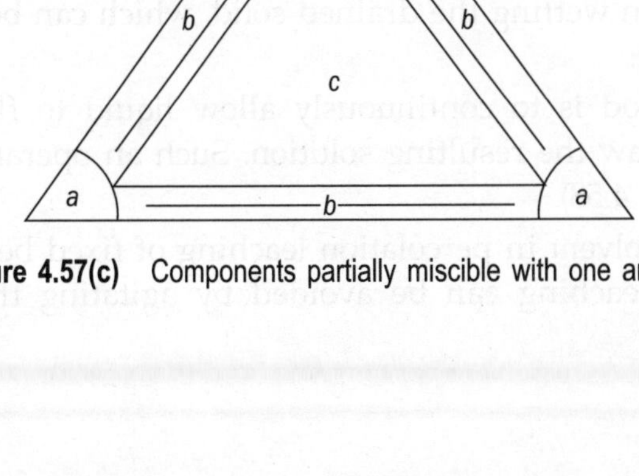

Figure 4.57(c) Components partially miscible with one another

1. Change of phase of the solute as it dissolves in the solvent.
2. Diffusion of the solute through the solvent in the pores of the solid.
3. Transfer of the solute from the solution in contact with the particles to the main bulk of the solution.

Any of these processes may be responsible for limiting the rate of dissolution. In some cases, the soluble material is distributed in small isolated pockets in a material that is impermeable to the solvent (example, gold in rock). In such cases, the material is crushed so that all soluble matter is exposed to the solvent. If the solid has a cellular structure, the rate of dissolution will be low because the cell walls provide additional resistance (example, extraction of sugar from beet). In the leaching of oil from seeds, the solute itself is liquid and may diffuse towards the solvent. Major factors influencing the rate of leaching include particle size, type of solvent, temperature, and agitation of the fluid.

4.21.1 Methods of Leaching

Three distinct processes are involved in normal leaching operation.

1. Dissolution of soluble constituent.
2. Separation of the solution from the insoluble solid residue.
3. Washing the solid residue to obtain as much soluble material as possible.

Leaching is mainly carried out as a batch process, but many continuous process plants are now available.

4.21.2 Mode of Methods

4.21.2.1 Batch process

4.21.2.1.1 Percolation tanks (gross current single contact)

The operation of a percolation tank follows several procedures. After the tank is filled with the solid, a batch of solvent sufficient to immerse the solid completely is pumped into it. The entire mass is allowed to steep or soak for a prescribed period of time. During this time, the batch of liquid may be circulated over the solid by pumping. The liquid is then drained from the solid. This entire operation represents a single stage. Repetition of this process will dissolve all the solute. The only solute retained is that dissolved in the solution wetting the drained solid, which can be washed by filling the tank with fresh solvent.

An alternative method is to continuously allow liquid to flow into the tank and continuously to withdraw the resulting solution. Such an operation may be equivalent to many stages (Figure 4.58).

Channelling of the solvent in percolation leaching of fixed beds with its consequent slow and incomplete leaching can be avoided by agitating the liquid and solid in

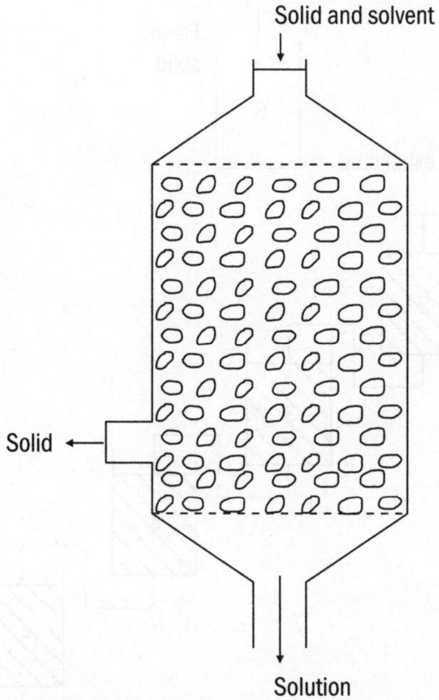

Figure 4.58 Batch leaching vessel

leaching vessels. Coarse solids can be leached by passing the solvent through a bed of the material, but fine solids offer high resistance to the flow. Finely divided solids are, therefore, suspended in leaching solvents by agitation.

4.21.2.1.2 Countercurrent multiple-contact leaching vessels

Countercurrent multiple-contact leaching vessels are called the Shank system. Leaching and washing of the leached solute by the crosscurrent single-contact method invariably result in weak solutions of the solute. The strongest solution will result if a countercurrent scheme is used, where the final withdrawn solution is taken from contact with fresh solid. Fresh solvent is added to this solid, from which most of the solute has already been leached. To avoid moving the solids from one tank to another, the process is operated following the steps listed below (Figure 4.59).

- Tanks 1 to 5 are filled with solid and liquid and are in operation for some time. Tank 6 is empty.
- Strong solution from tank 5 is collected. The liquid is transferred from tank 4 to 5, from 3 to 4, from 2 to 3, and from 1 to 2. Fresh solid is added to tank 6.
- Spent solid from tank 1 is discarded. The liquid is transferred from tank 5 to 6, from 4 to 5, from 3 to 4, and from 2 to 3. Fresh solvent is added to tank 2. At this stage, the process is the same as that at the start except that the tank numbers are each advanced by one.
- The operation is continued to make it a countercurrent multiple-contact process.

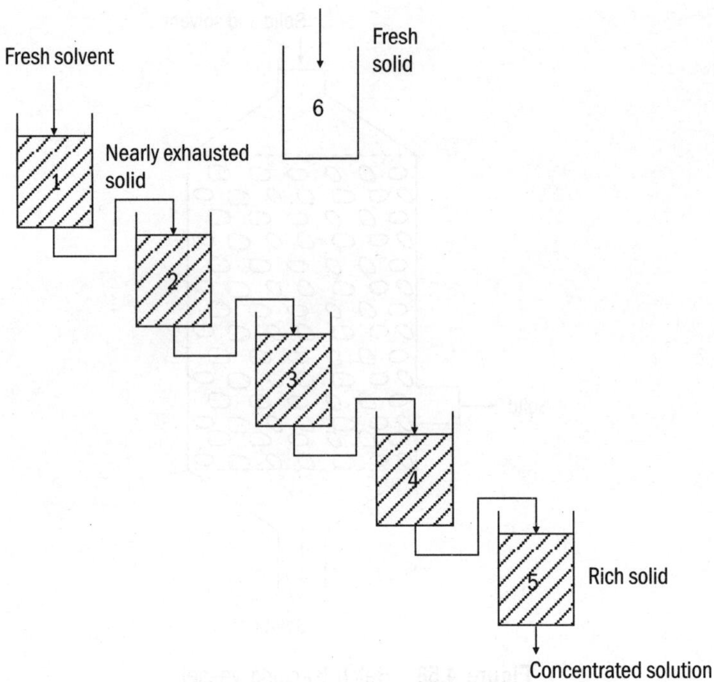

Figure 4.59 Countercurrent multiple-contact leaching (Shank system)

The flow is under gravity, although it may also be maintained by a pressure flow. The diffusion battery operates under pressure and is also provided with heaters (if the liquid is slightly heated, its viscosity decreases and solubility increases). As in the Shank system, the diffusion battery (Figure 4.60) in the system when completely exhausted and washed is changed with the solid. The process is thus made continuous and countercurrent.

4.21.2.2 Continuous process

4.21.2.2.1 Continuous countercurrent decantation

When the solid is very fine (Figure 4.61) and is suspended in liquor, it may be pumped in a countercurrent treatment. Solids to be leached together with solution from the

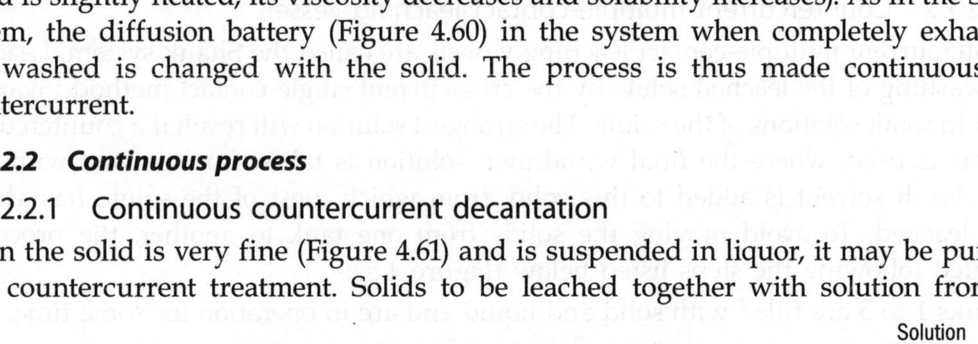

Figure 4.60 Diffusion battery (countercurrent multiple-contact leaching)

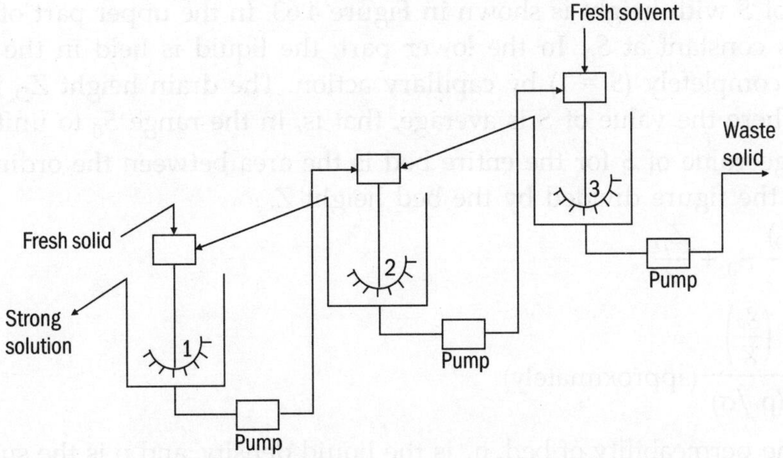

Figure 4.61 Continuous countercurrent decantation

second thickener are introduced into leaching agitators. The strong solution is decanted from the solids by the first thickener. This is a single-stage process although more than three stages are shown.

4.21.2.2.2 Continuous countercurrent classifier

When the solid is not very fine, it can be moved in a direction countercurrent to the solvent by means of hydraulic classifier such as a Dorr classifier (Figure 4.62).

4.21.3 Mathematical Treatment

4.21.3.1 Ordinary batch leaching vessel

Let us consider a bed of granular solids whose void spaces are completely filled with liquid, as shown in Figure 4.58. When the liquid is allowed to drain under the influence of gravity, with the admission of air to the voids from the top of the bed, the rate of liquid flow is at first very rapid. It gradually falls and after some time, no drainage occurs although the bed still contains liquid. The fraction of the void volume occupied by this liquid is called residual saturation *S*.

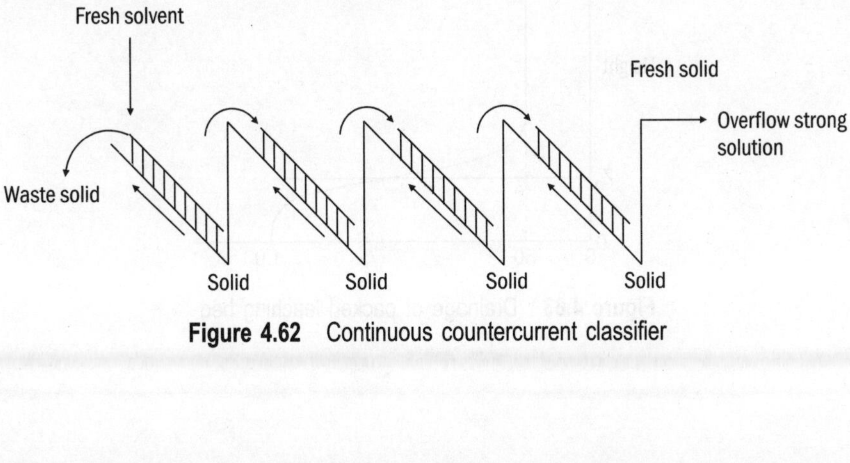

Figure 4.62 Continuous countercurrent classifier

Variation of S with height is shown in Figure 4.63. In the upper part of the bed, the value of S is constant at S_0. In the lower part, the liquid is held in the void spaces, filling them completely (S = 1) by capillary action. The drain height Z_D is defined as the height where the value of S is average, that is, in the range S_0 to unity.

The average value of S for the entire bed is the area between the ordinate axis and the curve of the figure divided by the bed height Z.

$$S_{av} = \frac{(Z - Z_D)}{Z} S_0 + \frac{Z_D}{Z} \tag{4.244}$$

$$Z_D = \frac{0.275 \left(\frac{g_c}{g}\right)}{(K/g)^{0.5}(\rho_L/\sigma)} \text{(approximately)} \tag{4.245}$$

where K is the permeability of bed, ρ_L is the liquid density, and σ is the surface tension of the liquid.

The value of S_0 depends on $\left(\frac{K\rho_L}{g\sigma}\right)\frac{g}{g_c}$, which is called capillary number.

S_0 = 0.075, when the capillary number is less than 0.02.

and $S_0 = \frac{0.0018}{\frac{K\rho_L}{g\sigma}\left(\frac{g}{g_c}\right)}$, when the capillary number is greater than 0.02.

K is the proportionality constant in the flow equation for laminar flow through the bed.

$$G = \frac{K\rho_L \, \Delta P \, g}{\mu_L \, Z \, g_c} \tag{4.246}$$

where ΔP is the pressure drop across the bed and G' is the mass velocity of flow based on the entire cross section of the bed.

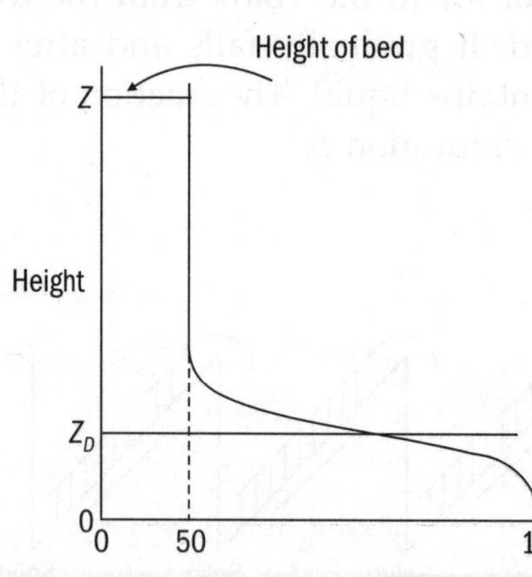

Figure 4.63 Drainage of packed leaching bed

For laminar flow through a bed of packed solid, P is correlated by the Ergun equation (Equation 4.247).

$$\frac{\Delta P}{z}\,\frac{g_c\, e^3\, d_p \rho_g}{(1-e)\,G^2} = \frac{150(1-e)}{R_e} \tag{4.247}$$

where d_p is the effective diameter of the particles given by

$$d_p = \frac{\text{Diameter of a space of the same surface}}{\text{Volume ratio as the packing in place}}$$

and e is the fraction void volume.

Now, if $\frac{\Delta P}{z}$ in Equation 4.246 is substituted in Equation 4.247 with replacement of R_e by $\frac{d_p G}{\mu_L}$, we get

$$K = \frac{d_p{}^2\, e^3\, g}{150(1-e)^2} \tag{4.248}$$

4.21.4 Bioleaching

From the beginning of creation, nature and natural phenomena have furnished the earth's crust with most of the metals and non-metals of economic and social importance. In some areas of the earth's crust, many of the metals are amassed in abundance in the form of high-grade ore deposits whereas in other areas, they are present in minute amounts in the form of low-grade ores. Through scientific excavations and with the help of geochemical and geophysical surveys, the locations of such ore deposits are being traced through the ages. Metallurgical and chemical techniques are most useful in recovering these metals, mostly required in daily life, from their high-grade ores. However, most of these methods are uneconomic as well as inefficient in extracting metals from their low-grade ores, which are lost in large amounts at the time of extraction due to improper mining methods. Mining of many sulphide deposits has led to the accumulation of a large amount of low-grade rocks in the form of dumps, amounting to millions of tonnes. These dumps also contain non-ferrous and rare metals. Such losses amount to 5%–20% of the total ore deposits. At present, improvement of the national economy by preventing huge losses during extraction of metals is a burning issue. In the near future, it will be necessary to carefully consider reworking low-grade and abandoned ores when the supply of high-grade ores is exhausted.

Aware of the novel activities of microorganisms, metallurgists and chemical and biochemical engineers and scientists are trying to determine how microbial metabolic activities can be best exploited in metallurgy. Only in the last decade have people realized the significant role that microbes could play in hydrometallurgical industry for extraction of metals from their ores. The discovery of Bryner, Beck, Davis, *et al.* (1954), and Bryner and Jameson (1958) of *Thiobacillus ferrooxidans, Thiobacillus thiooxidans,* and *Ferrobacillus ferrooxidans* in acid mine drainage that accelerates the extraction of many metal sulphides, particularly copper sulphide, is exploited industrially for hydrometallurgical extraction of metals from their ores. Hydrometallurgical methods for leaching metals from low-grade ores consist of selective separation of metal ions in solutions after dissolving the ore by chemical or other means. Various techniques, such as ion exchange and froth floatation, are used for the separation of metal ions from their solutions. One technique utilizes microbial action for the dissolution of metallic ions from low-grade ores. This technique, known as microbial leaching, has opened a new discipline called biohydrometallurgy.

In normal hydrometallurgy for extraction of metals from their sulphide ores, metals are oxidized to corresponding sulphates and sulphuric acids until they pass into solution. This dissolution process is accelerated manifold through the metabolic activities of acidophilic iron oxidizing bacteria like Thiobacillus, depending on several factors governing in microbial leaching.

4.21.4.1 Factors governing microbial leaching

In microbial leaching, various types of microbes aid in increasing the rate at which metals are dissolved from ores. Thus, such leaching depends on several factors.

- Nature of microorganisms involved.
- Size and crystal structure of the ore from which the metal is to be extracted.
- Concentration of the metal present in the ore.
- Thermal environment during operation.
- Acid environment.
- Oxygen demand of the organism; extent of air supply and presence of CO_2.
- Inoculum size used.
- Exposure conditions to sunlight.
- Presence of surface agents and shaking arrangement.

Microorganisms that take part in the formation and degradation of minerals include bacteria, fungi, algae, and protozoa. However, organisms generally used for biological leaching belong to several species of bacteria such as *T. thiooxidans, T. ferroooxidans,* and *Ferrobacillus thiooxidans.* The primary characteristic of microorganisms used in microbial leaching is its strong affinity towards the metal to be extracted. Duncan, Trussell, and Walden (1964), Ehrlich and Fox (1967), Razzel and Trussell (1963a; b), and Bryner and Anderson (1957) reported that many of the microorganisms mentioned above are useful in leaching of copper, nickel, molybdenum, zinc, arsenic, antimony, and germanium from their sulphide ores. Malouf and Prater (1961) recommended proper selection of iron-oxidizing bacteria (Ferrobacillus or Thiobacillus) with different experimental conditions for leaching of uranium, vanadium, and manganese from their oxides.

The effect of particle size on metal sulphide oxidation has been examined in batch as well as in continuous flow experiments. Razzel and Trussell (1963a; b) showed that altering the surface area of chalcopyrite from 53–74 μ to that smaller than 44 μ doubled the rate of leaching in stationary cultures. Ehrlich and Fox (1967), Silverman, Rogoff, and Wenden (1961), and Temple and Delchamps (1953) noticed that increase in surface area by decreasing particle size significantly increases the rate of leaching of metals by microorganisms. It has also been observed that if the microorganisms attack the crystal lattice or the minerals directly, the rate of leaching is increased by providing close contact between the microorganism and the surface of minerals.

Similar to particle size, the concentration of minerals (that is, mass of ores per unit volume of liquid) is also a key factor in increasing the rate of microbial leaching. During operation, this concentration must remain at an optimum level, one which is tolerable by the microorganism. Zimmerly, Wilson, and Prater (1959) reported that in ZnS oxidation reaction, *T. ferrooxidans* can tolerate a zinc concentration of as high as 17 g/L. Marcheloritz and Schwartz (1961) observed that after adoption, organisms of Thiobacillus and Ferrobacillus groups grew well in solutions having zinc concentration of 25 g/L. There are also reports on the tolerance of zinc concentration as high as 56.9 g/L by *T. ferrooxidans.*

Thermal environmental effects on growth and leaching reaction by appropriate thiobacilli have been studied by Leathen, Kinsel, and Braley (1966), Kinsel (1960), Silverman and Ehrlich (1964), and Beck (1960). For chalcopyrite oxidation by *T. ferrooxidans,* Razzel and Trussell (1963a; b) reported an optimum temperature of 35°C. Ehrlich and Fox (1967) observed that synthetic CuS is readily oxidized bacteriologically at 30°C than at 25°C and not at all at 35°C. The temperature within a reacting ore body or leach dump of CuS or pyrite mineral may reach well above 50°C and may even reach the kindling point of the ore. This heat originates from microbial and chemical oxidations of the ore. The heat accumulates because of the poor conducting property of the leach dump. Excessive rise in temperature will obviously stop microbial activity. To avoid such a situation, proper cooling arrangements must be provided in addition to stirring of the dump so that the leaching reaction rate proceeds smoothly.

Regarding acidic environments, the optimal pH requirement for oxidation of iron and metal sulphides by Thiobacillus and Ferrobacillus groups of bacteria is 2.5–4.5 (Beck 1960). Not much variation of this range has been reported with different metal sulphides. The optimal pH for microbial oxidation depends on the energy source and associated gangue. The following ore properties or ore reactions affect acidic environments.

- Acid-consuming component in the gangue like limestone, aluminosilicates, and silicates.
- Oxidative reactions such as the following.

$$Cu_2S + (5/2)O_2 + 2H^+ \rightarrow 2Cu^{2+} + SO_4^{2-} + H_2O$$

Although these properties favour acidic environments, certain ore properties and ore reactions counteract acid consumption. These ore properties include the following.

- Acidic gangue
- Pyrite oxidation and ferric hydrolysis

$$4FeS_2 + 14H_2O + 15O_2 \rightarrow 4Fe(OH)_3 + 8H_2SO_4$$

- S^o–oxidation

$$S^o + H_2O + (3/2)O_2 \rightarrow H_2SO_4$$

Considering these reactions, scientists concluded that pH has an influencing effect on the surface structure of the cells and minerals for microbial cell attachment.

The bacteria involved in facilitating leaching require O_2 and CO_2 in their metabolism (Landesman, Duncan, and Walden 1966; Schnaitman and Lundgren 1965). Duncan, Trussell, and Walden (1964) and Ehrlich and Fox (1967) studied the effect of increasing oxygen supply by shaking and reported that shaking accelerated autoxidation. Malouf and Prater (1961) studied the effect of varying oxygen supply in air lift columns through which ferrous solution was cycled over bacteria bearing crushed stone and found that the rate of aeration and circulation affected bacterial leaching rate. However, existing data is insufficient and much information has to be revealed before generalization can be made about whether oxygen in the air limits microbial activity in leaching. Also, engineering data are not available regarding quantitative requirements of oxygen in the leaching system. This information is much needed in the technological advancement of biohydrometallurgy.

Since the Ferrobacillus and Thiobacillus groups of bacteria, which are commonly employed in microbial leaching, depend on CO_2 for their supply of carbon substrate. The bacteria should be supplied with requisite amount of CO_2 during mineral oxidation through their metabolic activities; otherwise, the metabolic pool of the organisms may be hindered, which will result in a decrease in the leaching rate. Schnaitman and Lundgren (1965) increased the yield of *F. ferrooxidans* growing on FeS by 30% by supplying additional CO_2. Published information regarding stimulating effects of added CO_2 on cell growth with mineral substrate is not sufficient. Definite data are also not available and have to be generated through investigations for better operating conditions of microbial leaching.

Torma, Walden, and Branion (1970), in a study considering leaching of ZnS concentrate, provided useful data regarding the values of pH, pulp density (solids concentration, that is, mass of solids per unit volume of liquid), specific surface of solids, and nutrient concentration that will lead to maximum rate of leaching. The study also delineated the factors and conditions under which extraction of zinc is limited.

Inoculum size, presence of sunlight, and incorporation of surface active agents significantly affect the extraction rate of metals from their minerals. Considering all these factors, it has been found that a particle size of 200–300 mesh (depending on the volume of rock), temperature between 30°C and 35°C, pH between 2.5 and 4.5, rapid aeration, large inoculum size, intervention of suitable surfactant, proper shaking, and protection from sunlight are suitable requirements for successful microbial leaching.

4.21.4.2 Microbial leaching method

The actual method of microbial leaching utilizes an air lift percolator (Figure 4.64). The mineral samples are held in a large tube by a perforated disc covered with a layer of glass wool resting in a constricted bottom. An exterior tube fitted to a source of compressed air serves as the lifting channel. An opening in the bottom of the percolator with a pinch clamp serves as a means of drainage. The compressed air is bubbled through water and filtered through a sterile air filter before use. The top of the percolator is plugged with a porous material like cotton, which prevents contamination. In large-scale operations, a number of such percolators are connected in series, forming a battery.

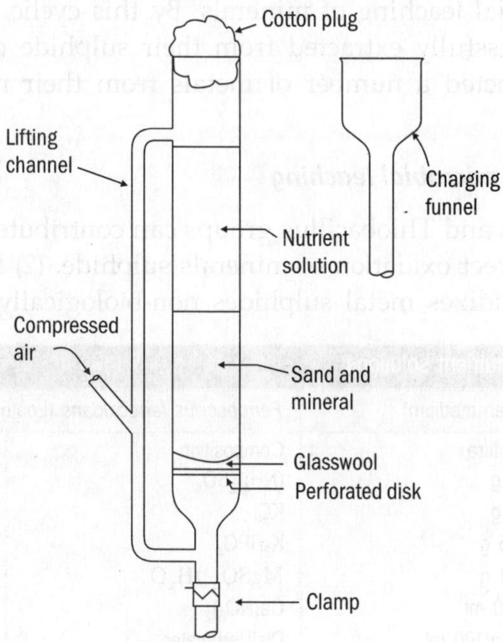

Figure 4.64 Percolator

The percolators are charged with a mixture of sand and desired sulphide ore, which is mixed with a small amount of distilled water, with the help of a charging funnel. Nutrient solution that acts as a leaching solution is then added to the percolator to allow growth of bacteria. The composition of this solution varies according to the organism utilized. For example, Waksman nutrient solution is most suitable with *T. thoooxidans*, while Leathen's nutrient solution provides the best results for *F. ferrooxidans*. The composition of these two solutions is given in Table 4.22. The filled percolators are sterilized at 140°C for 30 min and then cooled to a temperature of about 30°C. The first percolator in the battery is inoculated with about 10% inoculum (organism grown in nutrient solution for 24 h) under sterile conditions. The percolator is drained weekly and analysed for desired soluble metal.

When the rate of dissolution reaches steady state, the enriched culture from the percolator is used for inoculation of other percolators in the battery. The activity of microorganisms increases with repeated transfers. The bacteria involved in the process, being autotrophic in nature, obtain their energy for growth from the oxidation of sulphide minerals. Their needs for CO_2 and O_2 are met by air supplied to the percolators. To thrive, the bacteria force reactions to take place at rapid rates, which in turn affects the actual microbial leaching process.

Although the foregoing discussion focuses on batch process of microbial leaching, it can be operated continuously as well. Figure 4.65 shows the common steps adopted in a continuous microbial leaching of minerals. By this cyclic process, Mo, Zn, Cr, Ti, and U have been successfully extracted from their sulphide ores. Malouf and Prater (1961) successfully extracted a number of metals from their respective sulphide ores (Table 4.23).

4.21.4.3 Mechanism of microbial leaching

Bacteria of Ferrobacillus and Thiobacillus groups can contribute to leaching of minerals in three ways: (1) by direct oxidation of minerals sulphide, (2) by regeneration of ferric lixivant, which then oxidizes metal sulphides non-biologically, and (3) by depressing

Table 4.22 Composition of nutrient solutions

Thiobacillus thiooxidans (Waksman medium)		*Ferrobacillus ferrooxidans* (Leathen's medium)	
Composition	Per litre	Composition	Per litre
$(NH_4)_2SO_4$	0.2 g	$(NH_4)_2SO_4$	0.15 g
$MgSO_4{\cdot}7H_2O$	0.5 g	KCl	0.05 g
$CaCl_2{\cdot}2H_2O$	0.25 g	K_2HPO_4	0.05 g
$Fe_3O_4{\cdot}7H_2O$	0.01 g	$MgSO_4{\cdot}7H_2O$	0.50 g
Distilled water	1000 ml	$Ca(NO_3)_2$	0.01 g
Colloidal sulphur	1.0 g/100 ml	Distilled water	1000 ml
		$FeSO_4{\cdot}7H_2O$	10% w/v solution, 10 ml

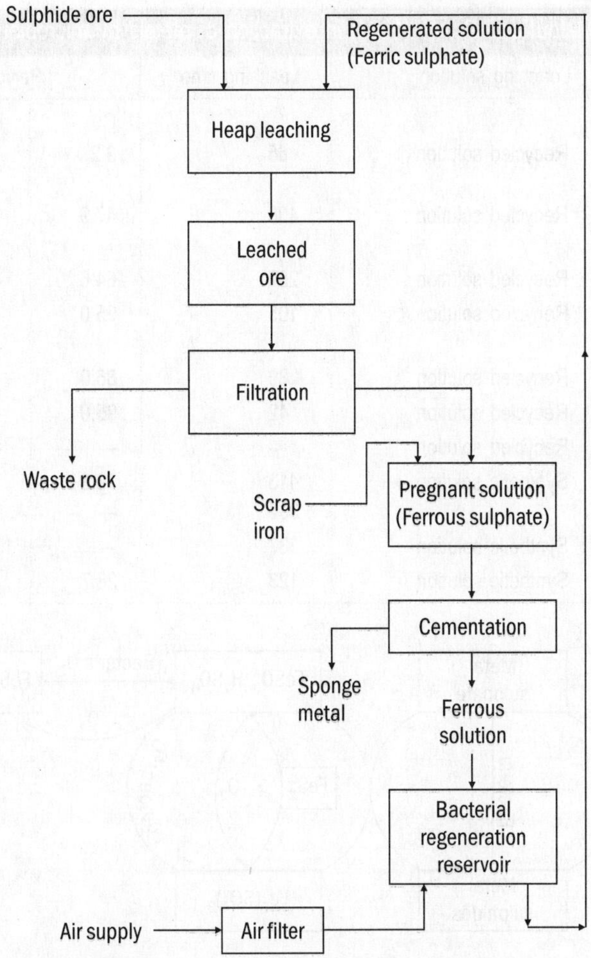

Figure 4.65 Cyclic process of microbial leaching

iron release from ore in the process of being leached. Involvement of these organisms in leaching mechanism is discussed in the subsequent sections.

Figure 4.66 explains the mechanism of microbial leaching. It is seen from the figure that the key agents responsible for oxidation of metals from minerals are iron-oxidizing bacteria, oxygen, and ferric sulphate. Generally, sulphide minerals of many metals contain a small amount of FeS_2, which helps in the formation of ferric sulphate. From Figure 4.66, it is understood that the mechanism of microbial leaching may be described under three distinct steps.

1. Conversion of small amounts of FeS_2 present in the sulphide ore to $FeSO_4$, in the presence of moisture, by an oxidation process. The representative reaction is given by Equation 4.249.

Table 4.23 Findings of Malouf and Prater (1961) in extraction of metals by microbial leaching

Minerals leached	Leaching solution	Leaching time	Per cent metal extracted			
			Cu	Fe	Zn	Mn
Chalcopyrite sample	Recycled solution	65	9.2	—	—	—
Chalcopyrite flotation concentrate	Recycled solution	408	47.9	—	—	—
Chalcopyrite-bearing mine waste	Recycled solution	287	64.6	—	—	—
Chalcocite-bearing ore	Recycled solution	105	95.0	—	—	—
Chalcocite-bearing flotation tailing	Recycled solution	89	86.0	—	—	—
Copper-bearing pyrite	Recycled solution	42	95.0	—	—	—
Copper-bearing pyrite	Recycled solution	—	—	—	—	—
Sphalerite	Synthetic solution	113	79.8	3.8	—	—
		337	—	—	19.6	—
Spahlerite and pyrite	Synthetic solution	337	—	100.0	48.6	—
Molyidemite concentrate	Synthetic solution	123	28.7	9.3	—	0.1

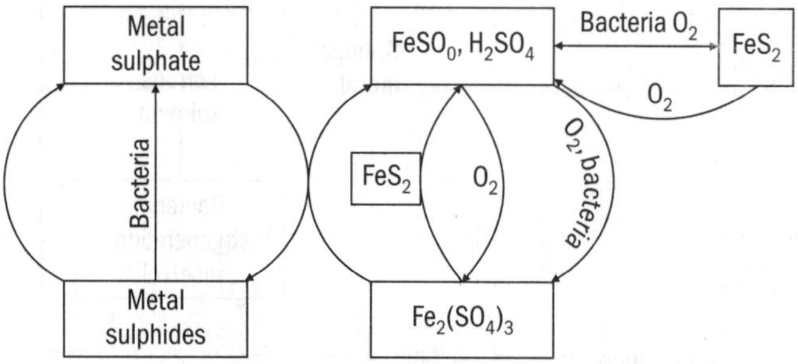

Figure 4.66 Mechanism of microbial leaching

$$2FeS_2 + 7O_2 + 2H_2O \rightarrow 2FeSO_4 + 2H_2SO_4 \quad (4.249)$$

2. The $FeSO_4$ formed is converted to $Fe_2(SO)_3$ by the following reaction.

$$4FeSO_4 + O_2 + 2H_2OSO_4 \xrightarrow{\text{Thiobacilli}} 2Fe_2(SO_4)_3 + H_2O \quad (4.250)$$

In Equation 4.250, the function of the bacteria is to transfer electrons from the metal sulphide to O_2 and CO_2 with the help of certain enzymes. The electron transfer aids the bacteria in generating useful energy, while at the same time reducing power for growth.

3. The $Fe_2(SO_4)_3$ produced reacts with the mineral ores, and the metal ions produced dissolve in solution (Equation 4.251).

$$MS + 2Fe^{3+} \rightarrow M^{2+} + 2Fe^{2+} + S^0 \quad (4.251)$$

For example,

$$CuS_2 + Fe_2(SO_4)_3 \rightarrow CuSO_4 + 2FeSO_4 + 2S$$

$$UO_2 + Fe_2(SO_4)_3 \rightarrow UO_2SO_4 + 2FeSO_4$$

$$FeS_2 + 7Fe_2(SO_4)_3 + 8H_2O \rightarrow 15FeSO_4 + 8H_2SO_4$$

The metal ions thus thrown into the solution are recovered by suitable separation processes like ion exchange and solvent extraction.

From these reactions it is clear that the main reagent that brings about leaching with the help of bacteria is ferric salt. This salt is usually produced slowly by chemical oxidation. Bacteria like Thiobacilli can tremendously increase the rate of conversion of Fe_2SO_4 to $Fe(SO_4)_3$. Once $Fe_2(SO_4)_3$ is used up in the reaction, it is again converted to $FeSO_4$ and bacterial reaction regenerates $Fe_2(SO_4)_3$ continuously (Figure 4.66). It is also evident that the net result of oxidation of metal sulphide by ferric sulphate is an increase in ferrous sulphate. Increase in $FeSO_4$ available as energy source results in an increase in the population of iron-oxidizing Thiobacilli and a subsequent increase in $Fe_2(SO_4)_3$. Increase in $Fe_2(SO_4)_3$ in turn can oxidize additional iron sulphides to produce greater amounts of $FeSO_4$. Thus, an ever-increasing cycle of oxidation and reduction reactions ensues, resulting in the oxidation of large quantities of metal sulphides. Iron-oxidizing bacteria regenerate $Fe_2(SO_4)_3$ from spent oxidants at a much faster rate than oxygen alone. Also, they are important in solubilization of metal sulphides, making them available for secondary deposition.

4.21.4.4 Merits of microbial leaching

The cost of chemical leaching by acidified ferric sulphate solution, which is commonly employed in the mining industry, is much higher compared to that of microbial leaching process. Also, the cost of replacing spent acid and ferrous sulphate solution or regenerating them by chemical means in microbial leaching plants is much less compared to chemical leaching plants. These merits encourage metallurgists to adopt chemical leaching process on a commercial scale.

4.21.4.5 Conclusions

As high-grade ore deposits are continuously becoming exhausted, increasing attention has been directed towards understanding the problems and possibilities of exploiting low-grade ore deposits. Available information in the literature indicates that leaching of metal sulphides using microbes, particularly bacteria, is technically feasible. From above discussions, it is apparent that the contribution of microbes towards the leaching of metals from their low-grade ores is of immense importance and is affected by a complicated array of enzymatic interactions. Control of some of these interactions is necessary for optimizing the bacterial assistance in the process. Although this process of low-cost biohydrometallurgy is well known in the USA and a few other countries, it is still not commercialized in many developing countries like India. These countries can easily exploit this economical process to benefit national economy. It is expected that in the near future, this process of leaching by virtue of its simplicity, economy, and other merits will be able to attract scientists and engineers to a greater extent. Since India is a rich source of many low-grade mineral ores, it is expected that engineers and scientists will pay due attention to this low-cost biohydrometallurgy.

4.21.5 Tutorial

Question 1 Nicotine (C) in a water solution (A) containing 1% nicotine is extracted with kerosene (B) at 20°C. Water and kerosene are insoluble.

(a) Determine the percentage extraction of nicotine if 100 kg of feed solution is extracted at one time with 150 kg solvent.

(b) What is the percentage extraction for three theoretical extractions using 50 kg solvent each?

The equilibrium data provided and expressed as kg nicotine/kg liquid is given in Table 4.24.

Table 4.24 Equilibrium data provided (in kg nicotine/kg liquid)

$x = \dfrac{\text{kg nicotine}}{\text{kg water}}$	0	0.001011	0.002460	0.00502	0.00751	0.00998	0.0204
$y = \dfrac{\text{kg nicotine}}{\text{kg kerosene}}$	0	0.000807	0.001961	0.00456	0.00686	0.00913	0.0187

Answer Equilibrium data are first plotted as shown in Figure 4.67.

$$x_F = \frac{1}{100} = 0.01 \text{ weight fraction of } C$$

$$x_F' = \frac{1}{10001} = 0.0101 \text{ kgC/kg } A$$

(a) In Figure 4.67, F represents the composition of the feed. FD is drawn with slope = -0.66 intersecting the equilibrium curve at D.

x_1 at D = 0.00425 and y_1 at D = 0.0038

Percentage of nicotine removed from water is

(0.0101 – 0.00425) × 99 = 0.58 kg × 100 = 58%

(b) For each stage,

$$\text{Slope } \frac{A}{B} = \frac{99}{58} = 1.98$$

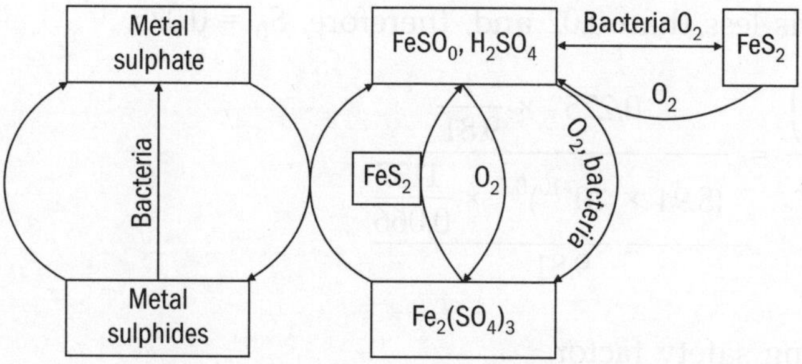

Figure 4.67 Plot of the equilibrium data

The construction began at *F*, with each operating line having a slope of –1.98. The final raffinate composition is $x_3 = 0.0034$.

Thus, nicotine extracted after three stages is

(0.0101 – 0.0034) × 99 = 0.663 kg

or 66.3% of that in feed

Question 2 Sugar remaining in a bed of bone char used for decolourization is leached by flooding the bed with water. The bed is then drained of the resulting sugar solution. The bed is of 1 m in diameter and 3.0 m in depth, and the temperature is 65°C. The sugar solution has a density of 1137 kg/m^3 and a surface tension of 0.66 N/m. The bulk density of the char is 960 kg/m^3 and the individual particle density is 1762 kg/m^3. The particles have a specific external surface of 16.4 m^2/kg. Estimate the mass of solution retained by the bed after dripping. Express it as mass solution/mass dry bone char.

Answer The fractional void volume, *e*, is given by Equation 4.252.

$$e = 1 - \frac{\text{Bulk density}}{\text{Particle density}} \tag{4.252}$$

$$\text{or } e = 1 - \frac{960}{1762} = 0.455 \text{ m}^3 \text{ void/m}^3 \text{ bed}$$

The particle surface, a_p, is given by Equation 4.253.

$$a_p = 16.4 \text{ m}^2/\text{kg} \times 960 \text{ kg/m}^3 \tag{4.253}$$

$$\text{or } a_p = 15\,744 \text{ m}^2/\text{m}^3 \text{ bed}$$

$$d_p = 6(1 - e)/a_p = 6(1 - 0.455)/15744 = 2.077 \times 10^{-4} \text{ m}$$

Now,

$$K = \frac{d_p^2 e^3 g}{150(1 - e)^2} = \frac{(2.077 \times 10^{-4})\,(0.455)^2\,(9.807)}{150(1 - 0.455)^2}$$

$$\text{or } K = 8.94 \times 10^{-10} \text{ m}^3/\text{s}$$

The value of K is less than 0.02 and, therefore, $S_0 = 0.075$

$$z_D = \frac{0.275 \times \left(\frac{1}{9.81}\right)}{\frac{(K)^{0.5}}{g} \times \frac{(PL)}{x}} = \frac{0.275 \times \frac{1}{9.81}}{\frac{(8.94 \times 10^{-10})^{0.5} \times \frac{1137}{0.066}}{9.81}}$$

or $z_D = 0.1705$ m

If $z = 3$ m, taking safety factor,

$$S_{av} = \frac{(z - z_D)S_O}{z} + \frac{z_D}{z}$$

$$\text{or } S_{av} = \frac{(3 - 0.1705)\,(0.075)}{3} + \frac{0.1705}{3}$$

or $S_{av} = 0.1276$

$$\frac{\text{Volume of liquid retained}}{\text{Volume of bed}} = 0.1276e = 0.0581 \text{ m}^3/\text{m}^3 \text{ bed}$$

$$\text{Mass of liquid in bed} = \frac{0.0581\pi(1)^2 3}{4} \times 1137 = 155.6 \text{ kg}$$

$$\frac{\text{Mass of liquid}}{\text{Mass of dry solid}} = \frac{0.0581\ (1137)}{1\ (960)} = 0.069 \text{ kg/kg}$$

4.22 DISTILLATION

Distillation is a method used to separate completely miscible liquid components of a mixture with minimal contamination. This method involves heating the liquid mixture to vaporize part of it and to condense the vapours. The composition of the condensate should be different from that of the residue left behind. To obtain one component of desired purity as condensate, vaporization and condensation (of part) of the condensate is necessary.

Let us consider distillation of a mixture of benzene and toluene. Let there be 1 mol of liquid mixture containing 0.5 mol of benzene (A) and 0.5 mol of toluene (B). If 0.2 mol of this mixture is vaporized and the vapours then condensed, the condensate will have more than 0.1 mol of benzene (A) and less than 0.1 mol of toluene (B).

A typical laboratory distillation unit has been shown in Figure 4.68. The schematic arrangement of an industrial distillation unit is shown in Figure 4.69.

4.22.1 Underlying Principle

A liquid mixture has two components: component A is of lower boiling point and component B is of higher boiling point. If part of the liquid mixture is vaporized, a

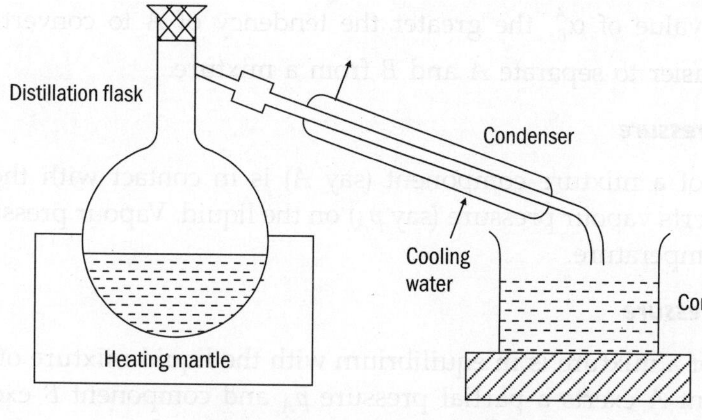

Figure 4.68 Laboratory distillation unit

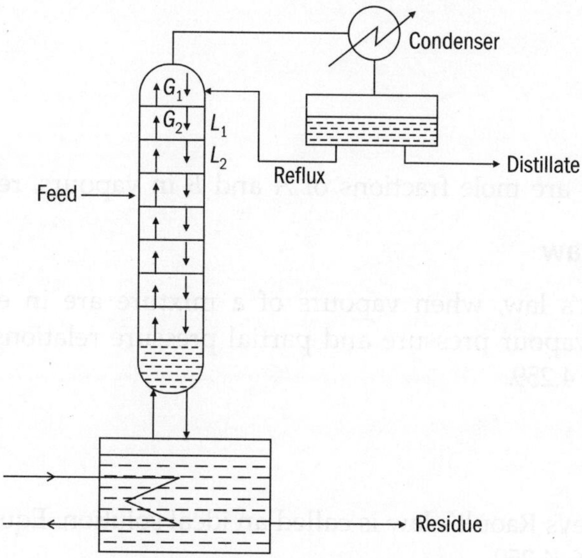

Figure 4.69 Fractionating column

larger fraction of component *A*, which is more volatile, is converted into vapour than component *B*. Let us consider a closed system where the vapour and liquid mixture of *A* and *B* are in equilibrium with each other. Let y^* be the mole fraction of component *A* in vapour phase and x be the mole fraction of *A* in liquid phase. Hence,

$1 - y^*$ = mole fraction of the less volatile component *B* in vapour phase

$1 - x$ = mole fraction of the component *A* in liquid phase

Relative volatility,

$$\alpha_B^A = \frac{y^*}{1 - y^*} \Big/ \frac{x}{1 - x} = \frac{y^*(1 - x)}{x(1 - y^*)} \tag{4.254}$$

The greater the value of α_B^A, the greater the tendency of A to convert into vapour phase, making it easier to separate A and B from a mixture.

4.22.1.1 *Vapour pressure*

When the vapour of a mixture component (say A) is in contact with the liquid in a closed system, it exerts vapour pressure (say p_A) on the liquid. Vapour pressure increases with increase in temperature.

4.22.1.2 *Partial pressure*

When the vapour of a mixture is in equilibrium with the liquid mixture of components A and B, component A exerts a partial pressure $\bar{p}_A$ and component B exerts a partial pressure $\bar{p}_B$ on the liquid. The total pressure is equal to the atmospheric pressure (Equation 4.255).

$$p_t = \bar{p}_A + \bar{p}_B \tag{4.255}$$

$$\bar{p}_A = p_t\, y^* \tag{4.256}$$

$$\bar{p}_B = p_t(1 - y^*) \tag{4.257}$$

where y and $(1 - y^*)$ are mole fractions of A and B in vapours, respectively.

4.22.2 Raoult's Law

According to Raoult's law, when vapours of a mixture are in equilibrium with the liquid mixture, the vapour pressure and partial pressure relationships are as given in Equations 4.258 and 4.259.

$$\bar{p}_A = p_A x \tag{4.258}$$

$$\bar{p}_B = p_B(1 - x) \tag{4.259}$$

A solution that obeys Raoult's law is called an ideal solution. Equation 4.260 is derived from Equations 4.256–4.259.

$$\alpha_B^A = \frac{y^*(1 - x)}{(1 - y^*)x} = \frac{\bar{p}_A}{\bar{p}_B} \tag{4.260}$$

The following three points can be concluded from the above discussion.

1. The composition of the vapour phase depends on the composition of the liquid phase.
2. The relative volatility depends on the vapour pressures of individual components.
3. Since vapour pressure depends on temperature, the relative volatility of individual components also depends on temperature.

Although all liquid mixtures do not obey Raoult's law, the above conclusions are still valid.

4.22.3 Vapour–Liquid Equilibrium

According to conclusion (1) above, for a given pair of mixture components (*A*–*B*) of value *x*, there is definite value of *y*. When $x = 0$, the pure component *B* would be in liquid phase and $y^* =$ zero. That is, only component *B* is in vapour phase. On the other hand, when $x = 1$, the pure component *A* would be in liquid phase and $y^* = 1$.

A typical equilibrium curve of acetone-water system is shown in Figure 4.70, and the data are also presented in Table 4.25. As is evident from Figure 4.70 and Table 4.25, the mole fraction of acetone goes on increasing as its mole fraction in liquid phase increases and the boiling point of the mixture (also called its bubble point) goes on decreasing. The total pressure is given in Equation 4.261.

$$p_t = \bar{p}_{\text{acetone}} + \bar{p}_{H_2O} = 1 \text{ atm} \tag{4.261}$$

At $x = 0.5$, $y^* = 0.849$

$$\alpha\frac{A}{B} = \frac{0.849 \times (1 - 0.5)}{(1 - 0.849)} \times 0.5 = 5.45$$

4.22.4 Single-Stage Operation: Flash Vaporization

Flash vaporization, or equilibrium distillation, is a single-stage operation wherein a liquid mixture is partially vaporized. The vapour is allowed to attain equilibrium

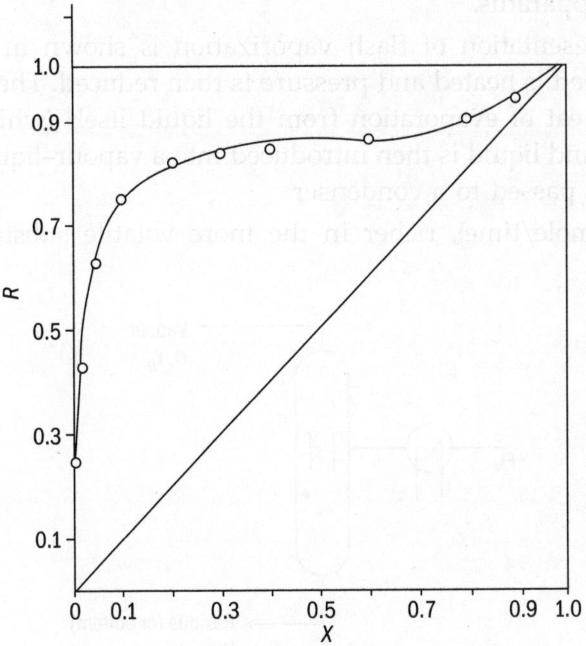

Figure 4.70 Acetone–water equilibrium data

Table 4.25 Equilibrium data for acetone–water system

X	Y	T (°C)
0.00	0.000	100.0
0.01	0.253	91.7
0.02	0.425	86.6
0.05	0.624	75.7
0.10	0.755	66.6
0.15	0.798	63.4
0.20	0.815	62.2
0.30	0.830	61.0
0.40	0.839	60.4
0.50	0.849	60.0
0.60	0.859	59.5
0.70	0.874	58.9
0.80	0.898	58.2
0.90	0.935	57.5
0.95	0.963	57.0
1.00	1.000	56.5

x – mole fraction of acetone in liquid; y^* – mole fraction of acetone in vapour; T – vapour–liquid temperature

Note Total pressure = 1 atm = 760 mm Hg

with the residual liquid, and the resulting vapour and liquid phases are separated and removed from the apparatus.

A schematic representation of flash vaporization is shown in Figure 4.71. In this process, the liquid feed is heated and pressure is then reduced. The vapours are formed taking their latent heat of evaporation from the liquid itself (which gets cooled). The mixture of vapours and liquid is then introduced into a vapour–liquid separating vessel, where the vapour is passed to a condenser.

The product D (mole/time), richer in the more volatile substance, is collected in vapour phase.

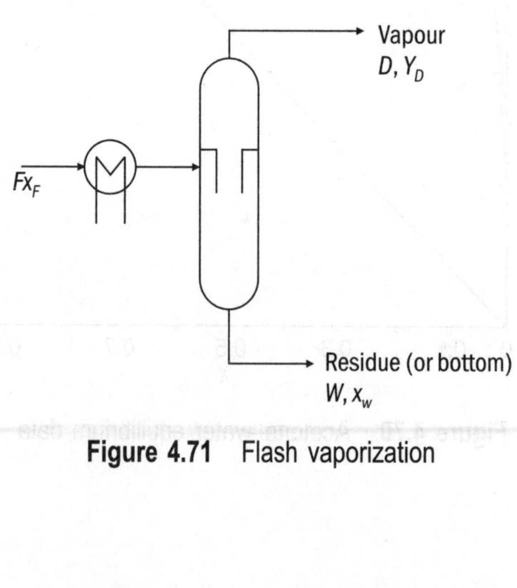

Figure 4.71 Flash vaporization

The material balances are given in Equations 4.262 and 4.263.

$$F = D + W \tag{4.262}$$

$$Fz_F = Dy_D + Wx_w \tag{4.263}$$

Solving Equations 4.262 and 4.263 simultaneously, we get

$$-\frac{W}{D} = \frac{y_D - z_F}{x_W - z_F} \tag{4.264}$$

where F is the feed moles, D is the distillate (moles), W is the residue (moles), and z_F, y_D, and x_W are the mole fractions.

If half of the acetone–water mixture containing 0.5 mole fraction of acetone is flash vaporized, the vapour phase will have 0.81 mole fraction and the liquid phase will have 0.19 mole fraction of acetone.

4.22.5 Differential or Simple Distillation

A batch of liquid mixture (A–B) is charged to a kettle and boiled slowly. The vapours are withdrawn rapidly and liquefied in a condenser. The condensate, or distillate, is collected in the receiver (Figure 4.68). The first portion of the distillate is richest in the more volatile component A, and as distillation proceeds, the vaporized product becomes cleaner. The distillate can, therefore, be collected in several separate batches called cuts to give a series of distilled products of various purities.

For a binary mixture, Equations 4.265 and 4.266 give the necessary relationships between feed, distillate, and residue.

$$\ln\frac{F}{W} = \int_{x_W}^{x_F} \frac{dx}{y^* - x} \tag{4.265}$$

$$Fx_F = Dy_{D,av} + Wx_W \tag{4.266}$$

where D is the moles of composited distillate and $y_{D,av}$ is the mole fraction of A in the composited distillate.

4.22.6 Continuous Rectification

Continuous rectification, or fractionate, is a multistage distillation operation. For a binary solution, with certain exceptions, it is ordinarily possible by this method to separate the solution into its components and recover each in any state of purity desired.

In this process, the feed is introduced more or less centrally into a vertical cascade of stages (Figure 4.69). Vapour rising in the section above the feed (called the absorption, enriching, or rectifying section) is washed with liquid to remove or absorb the less volatile component. Since no extraneous material is added, the washing liquid is provided by condensing the vapour issuing from the top, which is rich in component A.

The liquid returned to the top of the tower is called reflux, and the material permanently removed is the distillate, which is also rich in component *A*. In the section below the feed (stripping or exhausting section), the liquid is stripped of volatile component *A* by vapour produced at the bottom by the partial vaporization of the bottom liquid in the reboiler. The liquid removed is the residue, or bottoms, which is rich in less volatile component *B*. Inside the tower, the liquids are always at boiling temperature (called bubble point) and vapours are always at condensation temperature (called dew point). The highest temperature is at the bottom and the lowest is at the top.

The purities obtained for the two withdrawn products depend on the liquid/gas ratios used and the number of ideal or theoretical stages provided in the two sections.

4.22.6.1 *Calculations*

The overall material balance is given by Equations 4.267 and 4.268.

$$F = D + W \tag{4.267}$$

$$Fx_F = Dx_D + Wx_W \tag{4.268}$$

where *F* is the moles of feed per unit time, *D* is the moles of distillate per unit time, and *W* is the moles of residue per unit time.

4.22.6.2 *Enriching section: total condenser reflux at bubble point*

Consider a section of the fractionator entirely above the point of introduction of the feed (Figure 4.72). The condenser removes all the latent heat from the overhead vapour but does not cool the resulting liquid further. The reflux and distillate product are, therefore, liquids at the bubble point.

$$y_1 = x_D = x_0$$

where y_1 is the mole fraction of the more volatile component in the vapours rising from the top plate and x_0 is the mole fraction of component *A* in the reflux.

Liquid falling and vapours rising from each tray, *L*, will be constant if the molar latent heat of condensation (or evaporation) of the pure components and mixtures are equal. Also, vapours and liquid leaving each plate are in equilibrium with each other.

$$G_1 = L_0 + D = D(R + 1) \tag{4.269}$$

$$G_1 = G_2 = G_3 \cdots = G_i$$

$$L_0 = L_1 = L_2 \cdots = L_i$$

where G_i is the moles of vapours leaving plate *i* per unit time, L_i is the moles of liquid leaving plate *i* per unit time, x_i and y_i are the mole fractions of *A* in L_i and y_i, respectively.

$$G_{n+1}y_{n+1} = L_n x_n + Dx_D \tag{4.270}$$

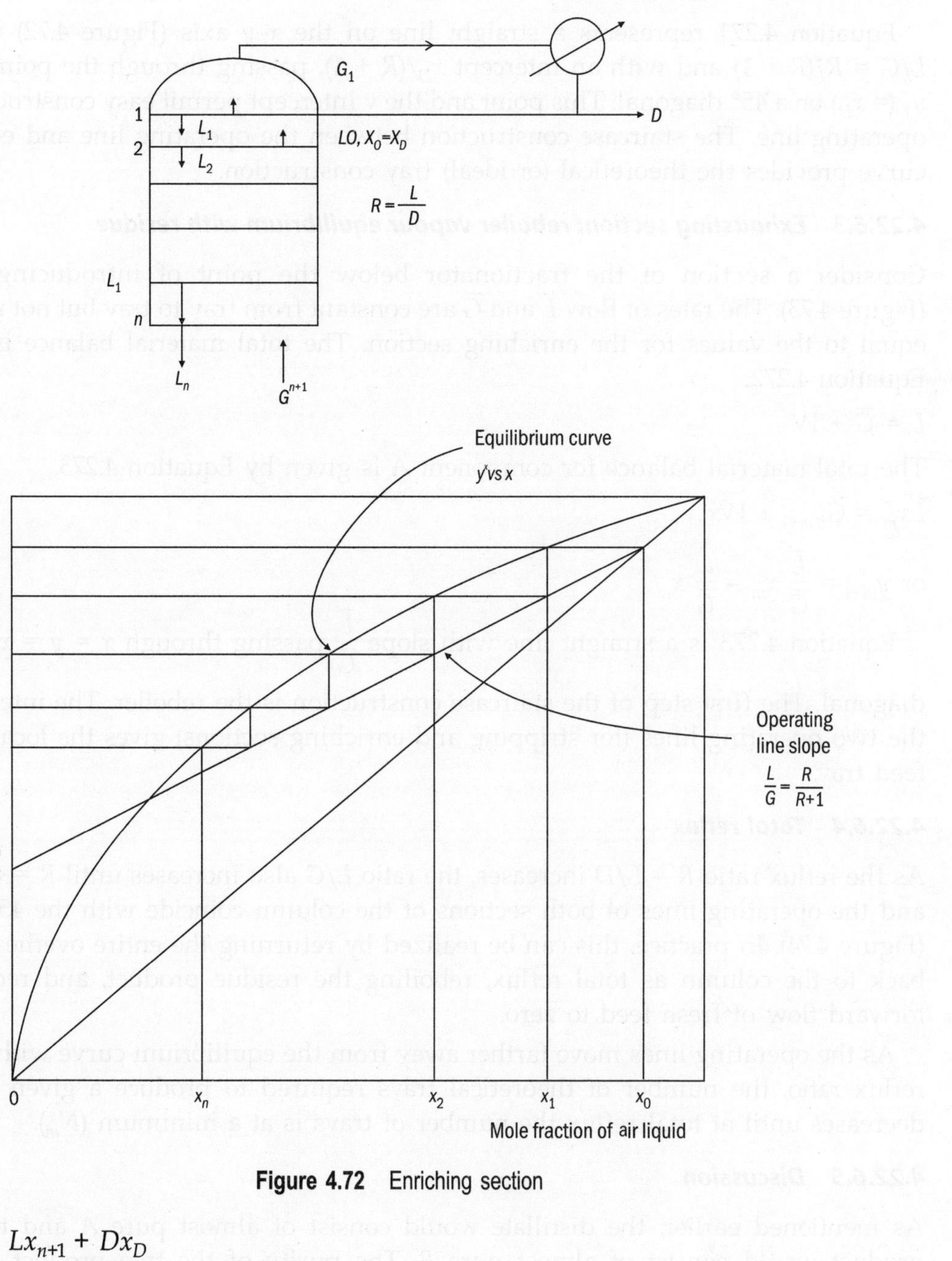

Figure 4.72 Enriching section

$$\text{or } Gy_{n+1} = Lx_{n+1} + Dx_D$$

$$\text{or } y_{n+1} = \frac{L}{G}\,x_n + \frac{D}{G}\,x_D$$

$$\text{or } y_{n+1} = \frac{R}{R+1}\,x_n + \frac{x_D}{R+1} \tag{4.271}$$

Equation 4.271 represents a straight line on the x–y axis (Figure 4.72) with slope $L/G = R/(R + 1)$ and with an intercept $x_D/(R + 1)$, passing through the point x_0 (= x_D), y_1 (= x_D) on a 45° diagonal. This point and the y intercept permit easy construction of the operating line. The staircase construction between the operating line and equilibrium curve provides the theoretical (or ideal) tray construction.

4.22.6.3 Exhausting section: reboiler vapour equilibrium with residue

Consider a section of the fractionator below the point of introducing the feed (Figure 4.73). The rates of flow $\overline{L}$ and $\overline{G}$ are constant from tray to tray but not necessarily equal to the values for the enriching section. The total material balance is given by Equation 4.272.

$$\overline{L} = \overline{G} + W \tag{4.272}$$

The total material balance for component A is given by Equation 4.273.

$$\overline{L}x_m = \overline{G}y_{m+1} + Wx_w$$

$$\text{or } y_{m+1} = \frac{\overline{L}}{\overline{G}}x_m - \frac{w}{\overline{G}}x_w \tag{4.273}$$

Equation 4.273 is a straight line with slope $\frac{\overline{L}}{\overline{G}}$ passing through $x = y = x_w$ on a 45° diagonal. The first step of the staircase construction is the reboiler. The intersection of the two operating lines (for stripping and enriching sections) gives the location of the feed tray.

4.22.6.4 Total reflux

As the reflux ratio $R = L/D$ increases, the ratio L/G also increases until $R = \infty$, $L/G = 1$, and the operating lines of both sections of the column coincide with the 45° diagonal (Figure 4.74). In practice, this can be realized by returning the entire overhead product back to the column as total reflux, reboiling the residue product, and reducing the forward flow of fresh feed to zero.

As the operating lines move farther away from the equilibrium curve with increased reflux ratio, the number of theoretical trays required to produce a given separation decreases until at total reflux the number of trays is at a minimum (N_m).

4.22.6.5 Discussion

As mentioned earlier, the distillate would consist of almost pure A and the bottom product would consist of almost pure B. The purity of the two product streams is determined by the reflux ratio and the number of theoretical trays. For a given x_F, x_D, and x_W, the theoretical number of trays required decreases with increasing reflux ratio. Distillation may also be performed in packet towers. Here, the change in the

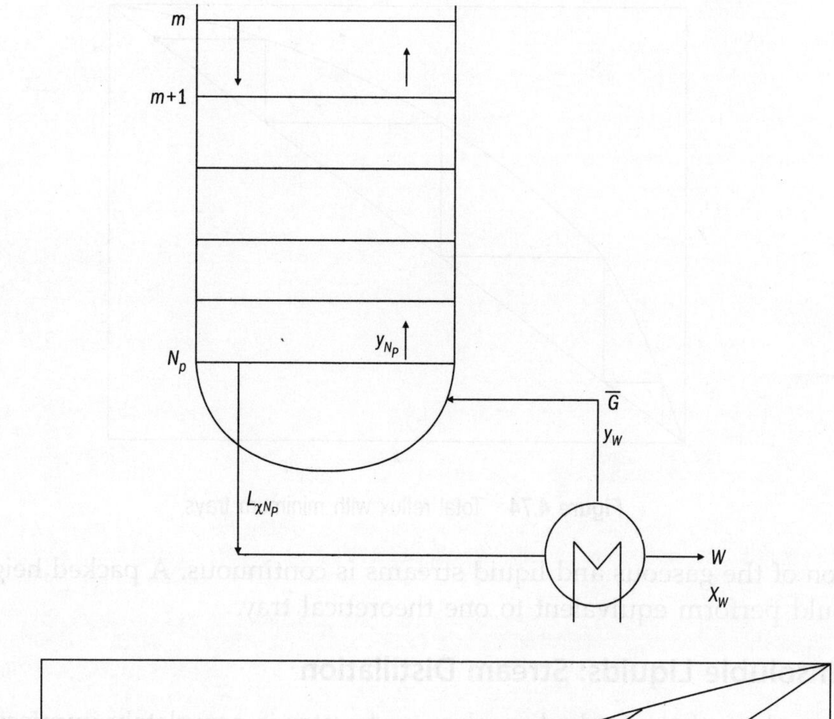

Figure 4.73 Exhausting section

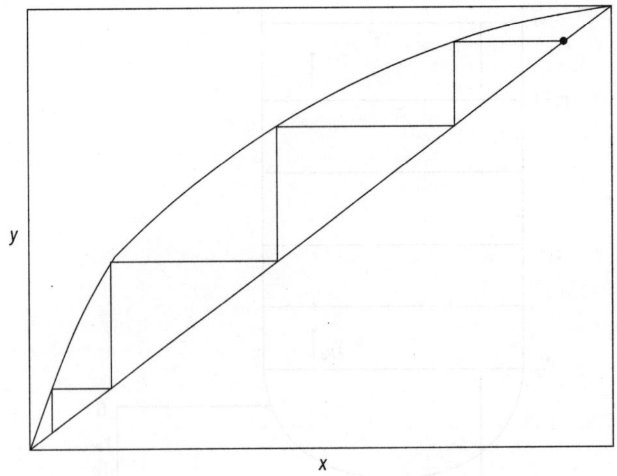

Figure 4.74 Total reflux with minimum trays

composition of the gaseous and liquid streams is continuous. A packed height of about 25 cm would perform equivalent to one theoretical tray.

4.22.7 Insoluble Liquids: Stream Distillation

A mixture such as that of a hydrocarbon and water is completely immiscible. In such cases, the vapour pressure of either component cannot be influenced by the presence of the other and each exerts its true vapour pressure at the prevailing temperature. When the sum of the separate vapour pressures equals to the total pressure, the mixture boils and the vapour composition is readily computed, assuming the applicability of the gas law.

$$p_A + p_B = p_t$$

$$y^* = \frac{p_A}{p_t} \tag{4.274}$$

So long as the two liquid phases are present, the mixture will boil at the same temperature and produce a vapour of constant composition.

For example, a mixture containing 50 g water and 50 g ethylamine, which is almost insoluble, is boiled at standard atmospheric pressure and 99.15°C.

p_A = 737.2 mm Hg

p_B = 22.8 mm Hg

Thus, $y^* = p_A/p_t = 737.2/760 = 0.97$ mole fraction water

$1 - y^* = p_B/p_t = 22.8/760 = 0.03$ mole fraction ethylamine

The original mixture contained 50/18.02 = 2.78 g mol water and 50/121.1 = 0.412 g mol ethylamine. The mixture will continue to boil at 99.15°C with an equilibrium vapour composition mentioned above until all the water has evaporated, along with 2.78(0.03)/0.97 = 0.086 g mol of ethylamine. The temperature will then rise to 204°C (boiling point of ethylamine), and ethylamine can be obtained at equilibrium.

4.23 DRYING OF BIOLOGICAL PRODUCTS

Drying of biological materials like foods and microbial biomass is often the last stage of a biochemical manufacturing process. It involves the final removal of water from a heat-sensitive material, ensuring minimum loss in nutritional value. Some of the important considerations for undertaking drying are as follows.

- Reduction in cost of transport.
- Proper handling and packing of the material.
- Convenient storage of the material.

The different types of driers are shown in Figures 4.75–4.79.

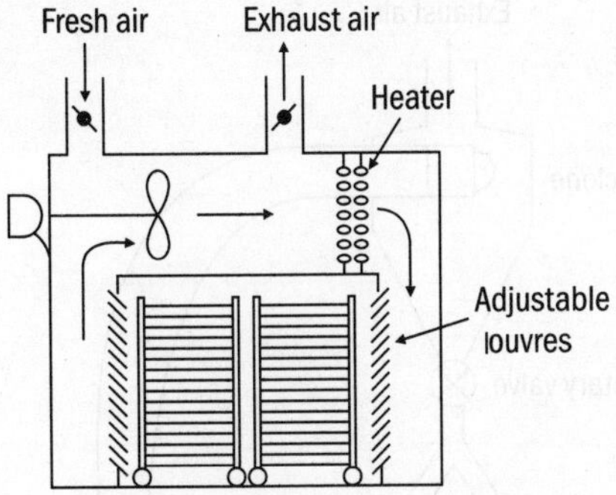

Figure 4.75 Cabinet drier

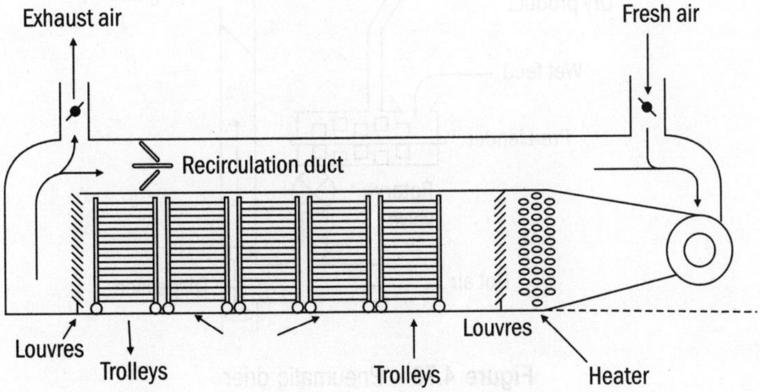

Figure 4.76 Tunnel drier

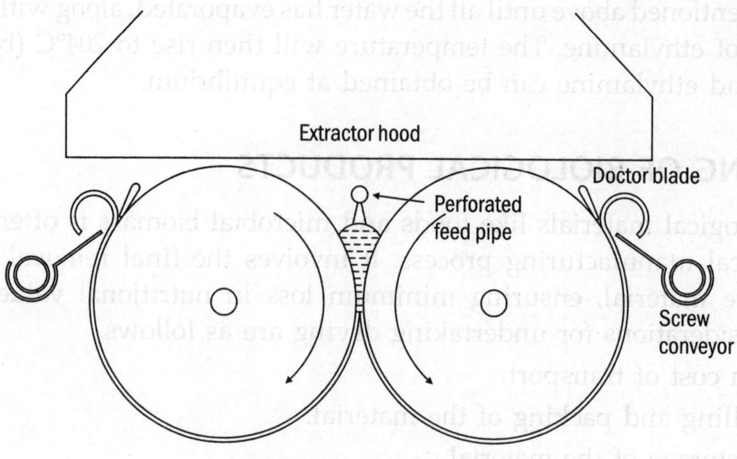

Figure 4.77 Drum drier

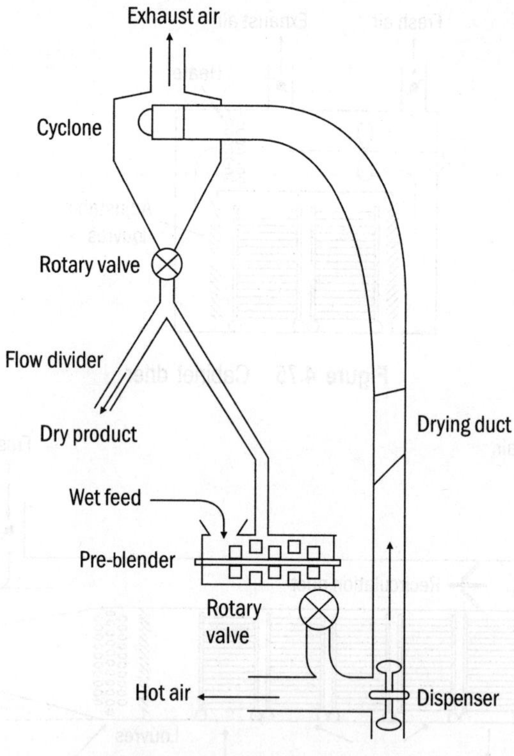

Figure 4.78 Pneumatic drier

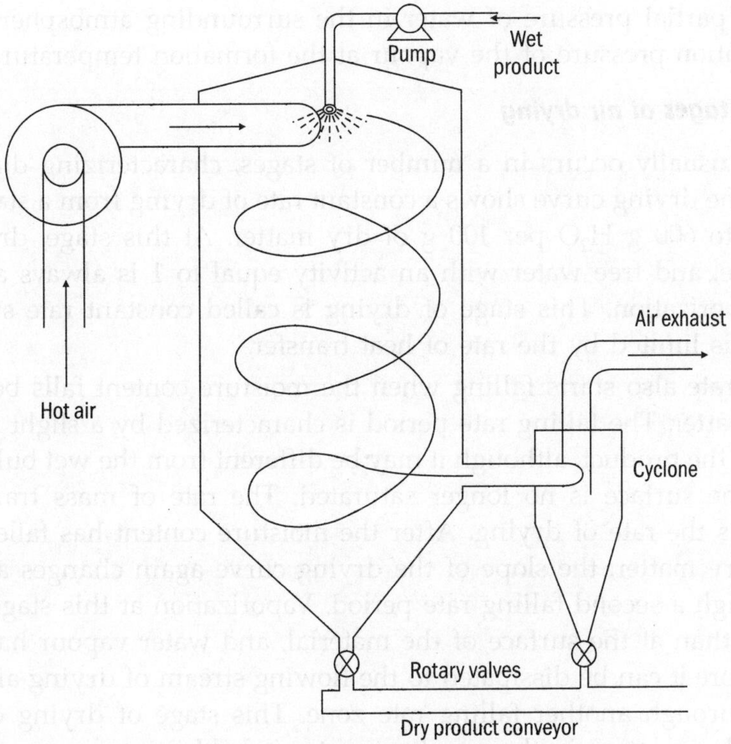

Figure 4.79 Spray drier

4.23.1 Drying Methods

During drying process, moisture is removed by the vaporization of liquid. The sensible heat of the air streams provides latent heat for the evaporation of water. Consequently, the air becomes cooler and more humid in the process and the product becomes warmer. Each mole of water vaporized requires heat transfer equivalent to the heat of vaporization. If the rate of mass transfer is too rapid for heat transfer to equal heat of vaporization, the material will drop in temperature. This process is called evaporative cooling.

4.23.2 Air Drying

A simple form of drier employs heated atmospheric air both for transferring necessary heat of evaporation and to remove the water vapour. However, when the operation is carried out under reduced pressure, the mass of air present is so low that the heat flux must be applied indirectly by means of conduction from heated surfaces, radiation or by a combination of the two. When the vapour is released from the material surface, it is removed as quickly as possible to maintain the drying rate. This is achieved by maintaining a partial pressure of water in the surrounding atmosphere that is lower than the saturation pressure of the vapour at the formation temperature.

4.23.2.1 The stages of air drying

Drying by air usually occurs in a number of stages, characterizing different rates of dehydration. The drying curve shows a constant rate of drying from a starting moisture of 700 g H_2O to 600 g H_2O per 100 g of dry matter. At this stage, drying occurs at the free surface, and free water with an activity equal to 1 is always available at the surface for vaporization. This stage of drying is called constant rate stage. Here, the rate of drying is limited by the rate of heat transfer.

The drying rate also starts falling when the moisture content falls below 600 g per 100 g of dry matter. The falling rate period is characterized by a slight increase in the temperature of the product, although it may be different from the wet bulb temperature. In this case, the surface is no longer saturated. The rate of mass transfer from the interior controls the rate of drying. After the moisture content has fallen below 120 g per 100 g of dry matter, the slope of the drying curve again changes and the drying rate goes through a second falling rate period. Vaporization at this stage occurs at the interior rather than at the surface of the material, and water vapour has to diffuse to the surface before it can be dissipated to the flowing stream of drying air. Dehydration again passes through another falling rate zone. This stage of drying corresponds to the region of the sorption isotherm where water is held in mono- or multi-molecular layers.

4.23.2.2 Theory of air drying

Transport of molecules from a surface to a fluid is analogous to the transport of heat by convection in heat transfer. The heat flux in convection is equivalent to the mass flux (Na/A) in mass transfer. The driving force for mass transfer is concentration gradient. The mass transfer rate at steady state is given by Equation 4.275.

$$dw/dt = Na/A = k_g(C_{Ai} - C_A) \tag{4.275}$$

where k_g is the mass transfer coefficient, and C_{Ai} and C_A are interface and bulk concentrations of component A, respectively.

The mass transfer equation for a material undergoing dehydration is given by Equation 4.276.

$$dw/dt = M_w/A = kg_w (H_i - H_a) \tag{4.276}$$

where M_w is the mass of water transferred per unit volume, H_i is the humidity at the interface where water is vaporized, and H_a is the humidity of the drying air.

H_e is the equilibrium humidity. H_i decreases when the temperature at the interface drops, resulting in a decrease in the rate of mass transfer.

In terms of temperature difference, the mass transfer equation for a material undergoing dehydration is given by Equation 4.277.

$$dw/dt = h_a/\lambda_w(t_i - t_a) \quad (4.277)$$

where t_a is the dry bulb temperature of the air, t_i is the temperature of the product surface, and λ_w is the enthalpy of vaporization. Therefore,

$$dw/dt = kg_w(H_i - H_a) = h_a/\lambda_w(t_i - t_a)$$

As long as there is adequate free water at the surface, the material will behave like a psychometric wet bulb. Thus the temperature at the surface is equal to that of the wet bulb (t_w). As long as the surface is at the wet bulb temperature, drying will proceed at a constant rate. The mass transfer and heat transfer coefficients are related to humid heat (C_S) as given by Equation 4.278.

$$\frac{h_s}{C_s} = \text{kg}\,\frac{\text{lb}}{(\text{h})(\text{ft}^2)} \quad (4.278)$$

where C_S = 0.24 + 0.0446 H_a Btu/lb dry air °F (H_a is taken as lb/lb of dry air).

4.23.2.3 *Prediction of drying times from drying rate data for material with one falling rate stage*

The drying rate curve for material with one falling rate stage shows a constant M_w from the initial moisture content X_0 to the critical moisture content X_c. The drying rate then falls in a linear relationship with decreasing moisture content until it becomes 0 at x = 0. The constant rate of drying is $R_C = -dx/d\theta$.

Therefore, the total time is

$$\theta_C = \frac{X_0 - X_c}{R_C}$$

At the falling rate period,

$$-\frac{dX}{d\theta} = \frac{R_C}{X_C}(X)$$

$$\text{or} \int_{\theta_C}^{\theta} d\theta = \frac{X_C}{R_C}\int_{X_C}^{X}\frac{dX}{X}$$

$$\text{or } \theta - \theta_C = (X_C/R_C)\ln(X_C/X)$$

$$\text{The total time } X_0 \text{ for } X = \frac{X_0 - X_C}{R_C} + \frac{X_C}{R_C}\ln\frac{X_C}{X}$$

The constant rate of drying (R_C) can be estimated by heat balance, that is, heat of vaporization = heat transferred.

Thus $\frac{dx}{d\theta}(M_S) = \frac{hA(T_a - T_s)}{\lambda}$

$$R_C = \frac{h(T_a - T_s)}{\lambda}\left(\frac{A}{M_S}\right) \tag{4.279}$$

v = surface × depth area = $A(d)$

$$\text{or } A(d) = M_S/\rho \tag{4.280}$$

Substituting Equation 4.280 in Equation 4.279, we get

$$R_C = \frac{h(T_a - T_s)}{\lambda\,(d)(\rho_s)} \tag{4.281}$$

Heat transfer can be calculated from any of the following correlations depending on the case.

- ***Air flow is parallel to the surface***
  In such a case, $h = 0.0128\ G^{0.8}$
  where h is the heat transfer coefficient (in Btu/h-ft^2-°F) and G is the mass rate of flow of air (in lb$_m$/h-ft^2).
  In SI units, $h = 0.0204\ G^{0.8}$
  where h is in W/m^{2}K and G is in kg/m^2-h.
- ***Air flow is perpendicular to the surface***
  In such a case, $h = 0.37\ G^{0.37}$ in FPS units, and $h = 0.429\ G^{0.37}$ in SI units.

4.23.3 Spray Drying

Spray drying is a process where a liquid droplet is rapidly dried as it comes in contact with a steam of hot air. Figure 4.79 is a schematic diagram of a spray drier where the atomized feed travels co-current with drying air. The small size of the liquid droplets allows rapid drying and the residence time of the material inside the spray drier is in the order of seconds. The dried material is separated in a cyclone separator. This material is continuously removed and cooled. While the droplet is drying, the temperature remains at the wet bulb temperature of the drying air. For this reason, high temperatures of the drying air can be tolerated in a spray drier with minimum damage to the heat sensitive components.

A major requirement for successful spray drying is reduction in the moisture content of a liquid droplet to a dryness level that would prevent it from sticking to a solid surface. Similar to air drying, the rate of drying passes through similar stages. For a spray drier, the constant rate drying time (θ_C) and falling rate drying time (θ_C) can be calculated by Equations 4.282 and 4.283.

$$\theta_C = \frac{4(X_0 - X_C)(r^2)(\rho_L)(h_{fg})}{3K_f(1 + X_0)(T_a - T_s)} \tag{4.282}$$

$$\theta_C = \frac{h_{fg}(\rho_s)(r_C^2)(X_C - X)}{3k_f(\overline{\Delta T})} \tag{4.283}$$

where X_0 is the initial moisture content, X_C is the critical moisture content, ρ_L is the density of liquid being dried, r_C^2 is the radius of droplet, h_{fg} is the heat of vaporization, k_f is the thermal conductivity of food material, and $\overline{\Delta T}$ is the mean temperature between the drying air and surface of the particle.

4.23.4 Drying Under Reduced Pressure

Vacuum-drying systems can be broadly divided into two categories.

1. In vacuum-drying systems, the major part of the drying process is carried out above the freezing point of the product. The water is then removed by a process of evaporation from liquid to vapour state. Typically, such equipment operate at about 20 Torr absolute pressure.
2. In freeze-drying systems, the earlier part of the drying process is carried out below the freezing point of the product. This is a process of sublimation, with water passing directly from solid to vapour state. The operating pressure must be below the triple point of ice, water, and vapour and corresponds to about 4 Torr at 0°C for pure water. This point is decreased by soluble solutes, which makes it possible for the system to operate at less than 1 Torr. In most freeze-drying operations, the temperature is allowed to rise well above 0°C during later stages of drying, so that the process resembles the later stages of air drying, except that the pressure is still kept at a lower level.

Some of the advantages of drying in a vacuum are listed below.

- The temperature of evaporation can be reduced theoretically to the equilibrium vapour pressure. This is strictly true for an agitated liquid and a wet surface, but may not be valid for a wet solid or a mass of frozen material.
- The concentration of oxygen can be reduced, often down to infinitesimally low levels, due to the sweeping out effect of water vapour.
- Under freeze-drying conditions, there is no opportunity for migration of salts and other solutes. This effect can be significant in the drying of wet solids above freezing point and can lead to protein denaturation.

The equipment necessary to dry solid and pasty materials under vacuum is more elaborate and expensive than that for atmospheric pressure drying. In most cases, the labour cost is also high. For these reasons, vacuum-drying process is reserved for relatively high-cost items and those where thermal or oxidation damage is unacceptable.

Freeze drying is even more expensive and until quite recently was limited to small-scale operations like drying of antibiotics, human blood, and other biological items.

4.23.5 Biodrying

Biodrying is an upcoming concept. In many areas such as tourist resorts and sites, pollution occurs owing to the accumulation of solid wastes. Rubbish and solid waste dumps may attract or act as breeding places for microorganisms, insects, and rodents, which increases the prevalence of certain diseases in such areas. These also reduce the aesthetic value of the tourist resorts. To avoid transportation and recycling cost of solid wastes, composting in the resort itself is useful. By composting/biomanuring in a heap composter, quick and easy drying of compost/biomanure by the metabolic heat of organisms is possible. A schematic of oxic–anoxic biotechnology principles involved in heap composting is shown in Figure 4.80. It indicates metabolic heat generation in forming hot zones, which in turn causes moisture removal in the compost by biodrying.

4.24 FREEZE DRYING OF BIOLOGICAL MATERIALS

Atmospheric driers are not effective in drying biological materials like pharmaceuticals, plasma, cultures or costly food with sensitive colour, flavour, and nutrients. Vacuum drying was introduced for producing goods of better quality. It is more satisfactory than atmospheric drying for the following reasons.

- Oxidation changes are reduced.
- Rate of drying is increased.
- Drying can be carried out at lower temperatures.

Vacuum drying may be carried out above or below the freezing point. When carried out above the freezing point, water is removed by the process of evaporation. On the other hand, when drying is carried out below the freezing point, water is removed by the process of sublimation. This is called freeze-drying, lyophilization, or sublimation drying.

Materials to be freeze dried are first frozen and then dried under high vacuum and restricted heat. Ice sublimes off without going through the liquid stage, and most of the

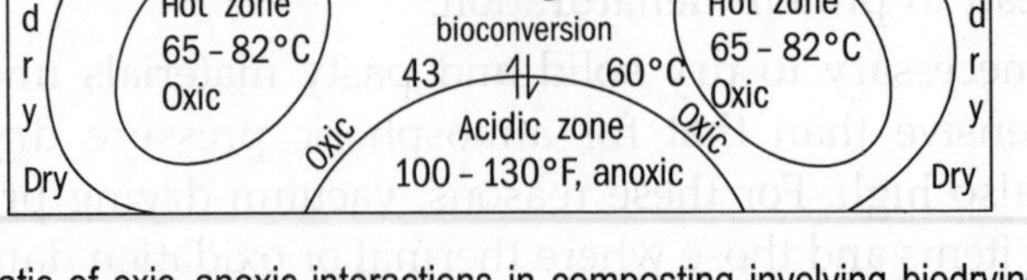

Figure 4.80 Schematic of oxic–anoxic interactions in composting involving biodrying by metabolic heat

defects in ordinary vacuum drying above freezing point are remedied. The following points justify the extra cost and complexity of freeze drying.

- Heat injury to products is completely eliminated since the temperature is very low.
- Solutes remain evenly and permanently dispersed without undergoing concentration since the frozen solvent sublimes.
- Case hardening is avoided since the evaporating ice layer of surface recedes, leaving porous residue of solute exposed.
- There is minimum coagulation since solutes are colloidally dispersed.
- Sterility is maintained since bacterial growth or enzyme changes cannot take place under frozen condition of drying.
- Foaming or bubbling is avoided since the produce is frozen solid and sublimation takes place.

4.24.1 Theoretical Considerations

Freeze drying depends on the creation and maintenance of a difference in vapour pressure between the immediate dry surroundings of a substance and the ice in its frozen interior, where it never melts. Figure 4.81 shows the heat and mass flow in freeze drying.

Ordinarily, the necessary vapour pressure gradient is produced by operating the equipment at a moderately high vacuum (0.1–2 mm Hg, absolute pressure), along with the following conditions.

- A low temperature condenser in the vapour space.
- An unobstructed passage for vapour entering from the subliming spaces of the frozen body.
- A steady flow of heat to the frozen body to keep its temperature higher than that of ice, but not high enough to cause melting anywhere within the body. Heat equal

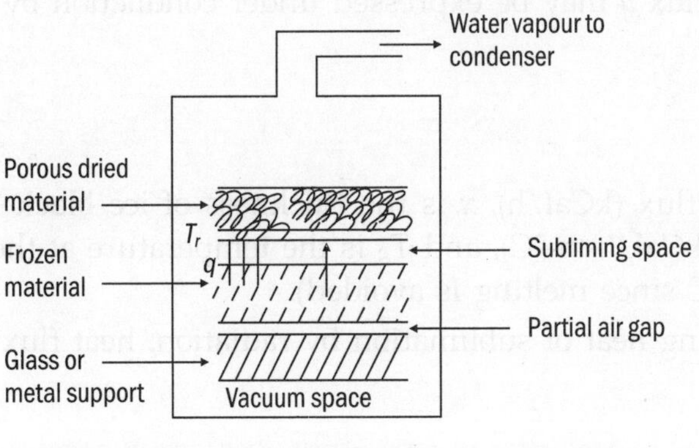

Figure 4.81 Heat and mass flow in freeze drying

to the latent heat of sublimation of ice must be continuously supplied to the zone of sublimation within the body to keep the temperature from falling, thus diminishing the vapour pressure gradient and rate of further drying.

4.24.2 Analysis of Mechanism in Heat Transfer and Mass Transfer in Freeze Drying

The physical mechanism involved in freeze drying includes the following.

- Transport of water vapour from the ice core to the surface, and finally through the porous outer layer of the frozen product.
- Transport of water vapour from the surface of the body to the vacuum pressure or low temperature condenser.
- Supply of the necessary amount of heat.

4.24.2.1 *Mass transfer*

In Figure 4.81, T_x is the constant temperature of the subliming surface of the ice (absolute temperature, °R), P_x is the corresponding vapour pressure (mm Hg Torr; 1 mm Hg = 1 Torr), T_e is the condenser temperature (very low that the corresponding vapour pressure, P_e, is assumed negligible), and A is the sublimation area (m^2). It is assumed that the resistance of the space between the ice and condenser is negligible. Under such ideal situations, sublimation achieves its maximum rate, G_{max} (Equation 4.284).

$$G_{max} = K_1 \frac{AP_x}{\sqrt{T_x}} \tag{4.284}$$

where K_1 is the constant dependent on the molecular weight of the subliming substance. G_{max} is expressed in kg/h.

4.24.2.2 *Heat transfer*

In freeze drying, the transfer of heat to the drying zone may proceed by conduction or radiation. The heat flux q may be expressed under conduction by Equation 4.285.

$$q = \frac{T_S - T_x}{\frac{Ax}{Ak}} \tag{4.285}$$

where q is the heat flux (kCal/h), x is the thickness of ice block (m), k is the thermal conductivity of ice (kCal/h-m-°C), and T_S is the temperature at the back face of the ice (does not exceed 0°C since melting is avoided).

In case of supplying heat of sublimation by radiation, heat flux q may be expressed by Equation 4.286.

$$q = A \times 0.173 \times \left[\left(\frac{T_r}{100}\right)^4 - \left(\frac{T_x}{100}\right)^4\right] \tag{4.286}$$

where T_r is the temperature of the radiator.

4.24.2.3 *Combined heat and mass transfer*

Similar to dehydration, freeze drying is a combination of mass transfer and heat transfer processes. Figure 4.82 shows a schematic representation of heat and mass transfer resistances in freeze drying. The rate of sublimation is given by Equation 4.287.

$$G = \frac{A(p_x - p_c)}{\overline{R}_d + R_s + K_1^{-1}} \tag{4.287}$$

where G is the rate of sublimation, $\overline{R}_d$ is the resistance of the "dry" layer, R_s is the resistance of the space between the substance and the condenser, and K_1 is the constant dependent on the molecular weight of the subliming substance.

At the same time, the heat of sublimation (ΔH_s) must be supplied. Therefore,

$$G' = \frac{q}{\Delta H_S} \tag{4.288}$$

Therefore, freeze drying represents coupled heat and mass transfers and both must be considered simultaneously in analysis of an operation.

4.24.3 Freeze-Drying Equipment for Liquid: Lyophilizer

Freeze driers are mainly used in bath operations and are similar to shelf driers (Figure 4.83). The operations in freeze and shelf driers are also similar except that in freeze drying, the materials are first frozen and then put on trays. The essential components of a freeze drier include the vacuum chamber, vacuum pump, and condenser.

The simplest method is to place frozen material in metal trays on a heated shelf. This has one disadvantage: a dried layer may be formed in the heated surface, reducing heat transfer. To overcome this problem, accelerated freeze-drying method can be developed.

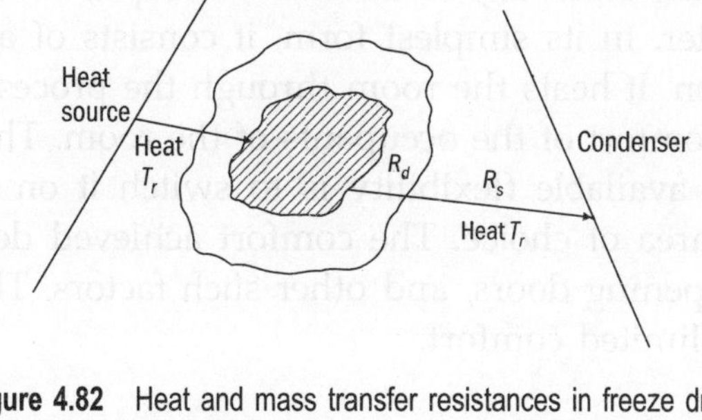

Figure 4.82 Heat and mass transfer resistances in freeze drying

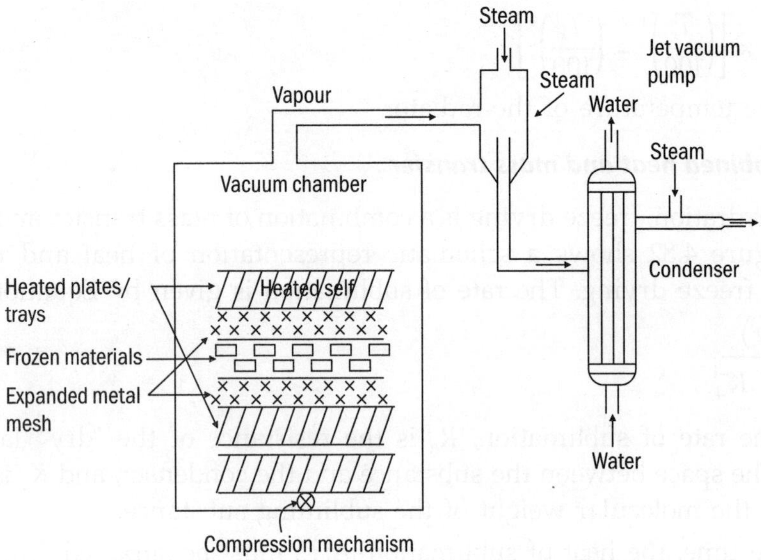

Figure 4.83 Freeze-drying cabinet

4.24.3.1 Accelerated freeze drying

In accelerated freeze drying, sheets of expanded metal mesh are inserted between frozen materials and heated plates to improve heat transfer and moisture removal. A refrigerated condenser may be used to condense the water vapour.

4.25 AUTOMATIC PROCESS CONTROL PRINCIPLES

The theory of automatic control has been extensively developed by electrical engineers and suitably adopted and applied by chemical and biochemical engineers to process industries. While the theory makes use of several advanced mathematical techniques, the following sections focus on the basic principles and concepts of automatic control.

The most convenient way to understand the concepts and basic features of automatic process control is to consider the operation of a familiar device and then think in terms of its manual control and then automatic control. For example, consider the use of an electric room heater in winter. In its simplest form, it consists of a single electrically heated rod when switched on. It heats the room through the process of radiation. The objective is to increase the comfort of the occupants of the room. The heater is usually manually operated and the available flexibility is to switch it on or off and turn it towards or away from the area of choice. The comfort achieved depends on the size of the room, frequency of opening doors, and other such factors. The device is cheap, has little control, and gives limited comfort.

A slightly improved and upgraded heater would consist of two rods where one or both or none of the rods may be switched on. This flexibility makes it more efficient but this version still has no control and gives limited comfort.

An improved and more efficient design is a convection or circulation heater. In such a device, the oil is heated by a source of heat (such as electric current). The heated oil circulates in the device making the surface hot and ultimately heating the surrounding air. Suitable design and operations can lead to warm air circulation in the room. On top of this, an automatic control can be incorporated in the device. One approach is that by a timer device the heating can be "on" for a chosen time period and "off" for the rest of the cycle. Further, flexibility in control design will alter the on-and-off portion of the cycle. Better control and greater efficiency can be attained at several levels of severity of winter as well as the demand of the user.

Alternative control strategy for the same heater is to make the switching of the unit dependent on the desired objective. The ideal would be to measure "comfort", or the specific amount of heat required from the device, and accordingly operate the switch. Unfortunately, there is no instrument that can measure "comfort". However, temperature is a reasonable measure of comfort and can be conveniently and continuously measured. The measured value of temperature can be compared with the desired value and the difference is used to decide whether the heater be switched on or off. This has all the ingredients of a commonly used control scheme. The situation can be expressed in terms of terminology used by control engineers.

4.25.1 Components in Control

The system consists of the process (heating of the room) and control system. The desired objective is to make the room comfortable. However, the quality of comfort cannot be measured and, therefore, a related variable that can be conveniently and reliably measured is chosen. The value given by the measuring instrument is compared with the desired value (the set point). The comparison gives the value of error, which is used by the controller to decide the corrective action (the controller action). The controller operates the switch (the valve) that regulates or changes the value of the manipulated variable (in this case, the current supply). The current and hence the heat supply change the temperature (the controlled, or measured, variable). This process is continuous and, if the system is well designed, the desired objective can be achieved within acceptable limits. Thus, the sequence of events can be divided into the following five points.

1. Measurement of the controlled variable (in this case, temperature) by respective measuring means.
2. Determination of error (difference between set point and measured value) by comparator.
3. Calculation of corrective action and giving the order (in this case, on or off) by the controller.

4. Operation of the value (in this case, switch) to change the manipulated variable (in this case, energy supply).
5. Effect of manipulated variable on the state of the process (in this case, temperature).

4.25.2 Representation of Control System

The control system used in bioprocessing can be a combination of fermenter and digital computer (PDP-II) control hardware. Its components include measuring element, comparator, controller, and valve. Each element receives and sends information in the form of a signal. For example, the comparator receives two pieces of information—the measured value and the desired value. It sends error as the signal to the controller. The measuring element receives the value of the controlled variable from the process and gives the measured value.

The approach discussed above is called feedback control. The control scheme works on the basis of error after it has existed. The value of the controlled (actually, the measured) variable is used to decide the control action. It is distinguished from open-loop control (like the proportion of on–off time period of the cycle described above) and feed-forward control (where control action is initiated on the basis of the anticipated, or expected, deviation of the controlled value from the desired value). Feedback control is most frequently used and is the only form of control discussed here.

Each element of the control system is continuously working or acting temperature is measured, error is calculated, and so on. The whole system is dynamic (not static or steady) all the time; for example, error is continuously changing (with controllers other than the on--off type, the error can be made to remain almost constant and zero, but the operation is analysed and viewed as a dynamic system).

4.25.2.1 On–off mode

The example illustrated above was the on–off control. In this case, the controller action is either off or on. As a result, the manipulated variable takes either a full value or a zero value. When the manipulated variable is at full value, the controlled variable, measured value, and error keep changing in one direction. When the error crosses zero, the controller action makes the manipulated variable take another position and the changes take place in the opposite direction. This leads to continuous cycling or oscillations of the controlled variable (around the desired value). Thus, with on–off control, the controlled variable cannot be maintained at a constant value. The minimum and maximum deviation (that is, the amplitude) may be small or large depending on the characteristics of the system. Similarly, the period of the cycle may be small or large. On–off control is a relatively simple and inexpensive device used typically in a domestic hot water tank having an immersion heater. Depending on the temperature of the water in the tank, the power supply is either connected to or disconnected from the heater.

Although such a system is inexpensive and extremely simple, the oscillatory nature of the control makes it suitable for those applications where it can be used alone and close control is not essential. Consequently, its use in the control of industrial processes is severely limited.

In many processes, it is desirable to eliminate cycling and maintain a constant value of the controlled variable. In particular, it is desirable that the manipulated variable should not take extreme values (like zero). For such cases and for more improved control, other types of controllers can be chosen, which are discussed below.

4.25.2.2 Proportional mode

The output of the controller is the input to the valve, that is, the controller indicates the course of action to the valve. The relation between the controller input (that is, the error) and controller output is characteristic of the type of controller. This is called the controller mode.

In the on–off control, the controller notes the sign of the error and instructs the valve on the basis of sign of error only. If error is positive, it will instruct the valve to close and vice versa (the operation in some processes is reverse).

Another controller mode is the proportional mode. In this case, the controller action is based on both the sign and magnitude of error. The control action can make the valve open or close or take any other intermediate position. Thus, when the error is zero, the valve position is, say, p_0. When the error is $-\varepsilon$, the controller action is to change the valve position by $-k_c$ to that the new valve position p. This can be expressed as given below (Equation 4.289).

$$p = p_0 + k_c\,(-\varepsilon) \tag{4.289}$$

where p is the output signal, k_c is the gain or sensitivity, and p_0 is the controller output when there is no error.

Such a controller is called a proportional controller, that is, the control action is proportional to the error, taking into account both the sign and magnitude of the error.

This seems to be a more reasonable form of control since it asks for greater change in manipulated variable only if the error is large. An important feature of proportional control is that it can maintain the variable at a constant value without the need for continuous oscillations. However, such a controller has a significant limitation, which is explained below.

An important question is the necessity of a control. Why will a system without a control not continuously keep operating at desired conditions? Why is adjustment and control required in most systems? Let us consider this by taking an example. Consider a steam-heated vessel used to heat a flowing stream of water. Under normal conditions—specified feed rate (100 L/min) and feed temperature (25°C)—a partially opened outlet valve (say, 50%) releases water at the desired temperature (say, 45°C). If the feed temperature or flow rate changes, the outlet temperature will not remain constant even if the valve position remains the same. The required (control) action to maintain the temperature at the desired value will be a new valve position. Thus, these are possible changes in process condition that require changes in manipulated variable to maintain the controlled variable. These variables in process condition are called load variables. In this example, the feed rate and feed temperature are load variables. Therefore, the need for control arises because of load variables.

A significant limitation of proportional control is that after load change, the controlled variable cannot be brought back to the desired value. That is, the error after load change cannot become zero although it can be made small. The deviation from the desired value after load change is called offset. The reason for offset is simple; after load change, a new valve position is required but is not possible if $\varepsilon = 0$. That is, a new valve position requires finite value of e^* and thus offset has to be accepted. Of course, for required change in valve position, offset (or e^*) can be small if k_c is large. It is an adjustable value called the proportional gain and can be changed by adjusting a knob on a proportional controller.

In a feedback control, the process settles down after going through some oscillation after a change in load. If the control system is not well devised, the system may even become unstable after a load change.

As k_c (the setting in the proportional controller) is increased, the response becomes more oscillatory (and the offset is reduced). However, the response may become unstable at a high value of k_c. At each oscillation, the value exceeds that obtained in previous oscillations. This emphasizes the importance of proper choice of the value of k_c. Thus, the greater the magnitude of the error, the larger the corrective action applied.

4.25.2.3 *Integral control*

The proportional controller acts according to the sign and magnitude of error and leads to offset. Another mode of control called the integral mode takes into account a combination of sign and magnitude of error as well as the time over, where the error has occurred. The particular combination is mathematically the integral of error over time (Equation 4.291).

$$p = p_0 + \frac{k_c}{T_i}\int \varepsilon dt \tag{4.290}$$

$$\text{or } p = p_0 + k_c\varepsilon + k_I\int \varepsilon dt \tag{4.291}$$

where k_I is a constant.

Thus, the valve position changes in proportion to the area under the error versus time curve. Large error over a short time calls for comparable change as small error over large time. Also, the controller will produce little action if error or time is small. This is a reasonable approach and possibly an improvement over the proportional controller.

Note that the value of p can be different from p_0 even after the value of ε changed to zero. Compared to the proportional controller, the offset of the integral controller after load change cannot be obtained, that is, even after load change error can be eliminated.

However, the integral mode takes time to act as against the proportional controller where $(p - p_0)$ is immediate. This makes the integral mode slow. Therefore, in practice proportional and integral modes are used in combination leading to a two-mode (or two-term) controller, the PI controller.

The PI controller tends to lead to greater oscillations and has a greater tendency to make the system unstable. The controller has two adjustments: proportional gain (k_c) and integral time (t). Setting, or tuning, the controller requires two adjustments. It is necessary to produce a reasonable or acceptable response. The problem of instability can be avoided by using lower k_c, but this leads to greater error during the first cycle. Thus, long-term response worsens by incorporating the integral mode (that is, the initial error is more).

4.25.2.4 Derivative mode

Another control mode is based on the principle that action should be taken depending on how fast the error is occurring and whether it is increasing or decreasing. Thus, control mode does not depend on the magnitude of error or its combination with time. More corrective action should be taken if the error is increasing rapidly and less if the error is increasing slowly (otherwise a large corrective action will be taken unnecessarily). In the derivative mode, action is taken in one direction (that is, start closing the valve) if the error is increasing and the correction is taken in the opposite direction (that is, start opening the valve) if the error is decreasing. It should be noted that in integral mode, the direction of corrective action is the same whether the error is increasing or decreasing. Mathematically, the derivative mode is expressed as a derivative (or rate) of ε (Equation 4.292) to define PD controller action.

$$p = p_0 + k_c e^* + k_D \frac{d\varepsilon}{dt} \tag{4.292}$$

where k_D is a constant.

The value of p can be different from p_0 after load change only if the error is changing, that is, an error must exist. Hence, the derivative mode will not eliminate offset. Also, if the error exists but is not changing with time, there is no control action regardless of how large or for how long the error existed. Thus, the derivative mode has limitations and is used in combination with P and I modes, leading to a three-term (or three-mode) PID controller.

The addition of derivative mode to the PI controller improves the short-term response, that is, the initial error after load change is reduced. This leads to three parameters: proportional gain, integral time, and derivative time. The values of these parameters need to be carefully selected to get acceptable or reasonable response from the system after load changes. Frequently, all three modes (PID control mode) are used. Mathematically, it is expressed by Equation 4.293.

$$p = p_0 + k_c\varepsilon + k_I\int\varepsilon dt + k_D\frac{de}{dt} \qquad (4.293)$$

In some operations, the controlled variable or the measured value of the controlled variable fluctuates around the average value. The magnitude of the fluctuations is small enough to be of little consequence to the process but rapid enough that the measured rate of change of error with time can be sizeable even when the error itself is small. This example is common in flow fluids where the pressure drop, which is a measure of flow rate, can have rapid small fluctuations. In such cases, the use of derivative mode should be avoided.

The preceding paragraphs have discussed the four controller modes—on-off mode, proportional mode, integral mode, and derivative mode—and the combined (PID) mode and their principal characteristics in a qualitative sense. The following sections will focus on other aspects of process control. The first concerns constructional and design features of the hardware of each of the elements of the control loop. The control instruments are classified as pneumatic (where the operating signal is air under pressure) or electronic (where the operating signals are electric direct current). Occasionally, the control loop consists of both electronic and pneumatic elements. In such cases, transducers that convert one type of signal to another are required. In the process, industrial pneumatic instruments are more frequently used.

Other key aspects are the application of control principles to evolving control strategy and selection of appropriate instruments for control of processes and operations. These two aspects relate to the quantitative predication of how a controlled system would behave under operating conditions. Following are a few points that are significant in this regard.

- Regarding feedback control of a single controlled/measured variable, more than one variable needs to be controlled simultaneously in many processes. This problem is called multivariable control. In cases where there is little or no connection between the different variables, the concept can be applied independently to each loop.

However, in some processes the variables influence each other leading to interference, or interaction. In such cases, modifications to the approach are required.

- Maintaining a constant value of the controlled variable in the presence of load changes is called regulatory control. In some operations, it is required to change the controlled variable in a desired pattern over a period of time. This problem is called servo operations and is significant in batch systems and even in some continuous operations.
- There has been a growing trend towards more advanced control strategies. The conventional controller computes corrective action via P, I, and D modes based on error. Instead of this conventional method, any kind of calculation may be carried out based on one or more measured values incorporating a detailed mathematical description of the complete process and corrective action may then be taken. Such calculations may be computed by a computer or microprocessor, leading to a computer/microprocessor control concept. In such cases, any variations in the process can be accounted for leading to an adaptive process control. Here, the control or system adapts itself to changing situations.

There are also more commonly applied simpler schemes such as ratio, cascade, feed forward, and feedback feed-forward combination controls.

While the area of control is vast and growing, the preceding paragraphs focused on the most basic and elementary aspects regarding the approaches that form the basis of conventional process control. Figures 4.84(a)–4.84(e) show the graphs of signals at various levels of k_c.

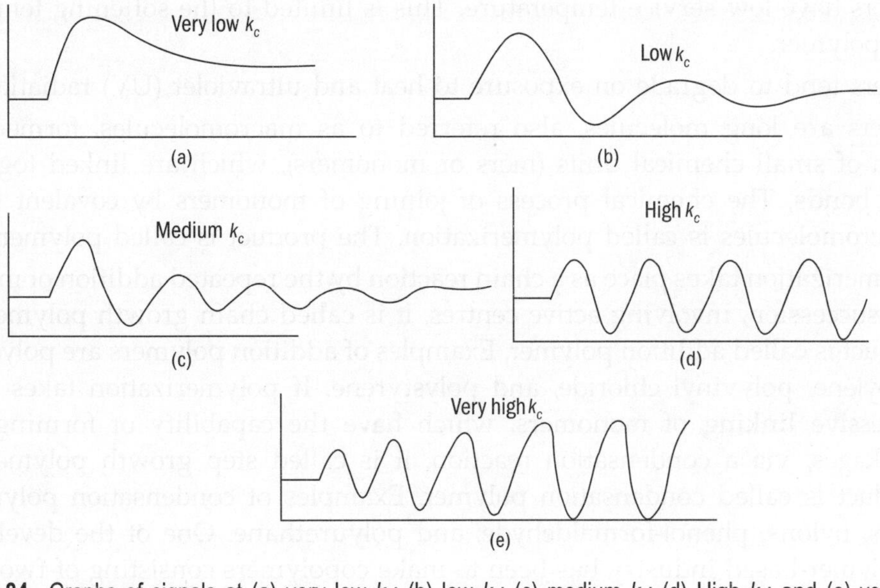

Figure 4.84 Graphs of signals at (a) very low k_c; (b) low k_c; (c) medium k_c; (d) High k_c; and (e) very high k_c

4.26 PROPERTIES AND FUNCTIONS OF POLYMERIC AND BIOPOLYMERIC ENGINEERING MATERIALS

The present time is sometimes referred to as the "plastics age" on account of the use of polymeric materials in a large variety of application areas. The rapid growth of these materials and related industries has taken place in the past three decades or so. The polymer-based industry can be broadly divided into (a) rubber/elastomer, (b) plastics, (c) synthetic fibres, (d) coating and paints, (e) adhesives, and (f) packing. There are several advantages of using polymers as engineering materials, which include the following.

- Many different types of polymers can be synthesized with widely varying properties. Some polymers are soft, flexible, and tough and others are rigid and strong.
- Polymers have high specific strength, that is, high strength-to-weight ratio.
- Polymers can be shaped into complex shapes with relative ease and at relatively low temperatures.
- Polymers operating in a variety of environments have useful range of deformability and durability, which can be exploited with careful design.
- Polymers have excellent chemical and corrosion resistance.
- Polymeric products are suitable for mass production technology at low cost.
- Overall material cost of polymers is low.

Some of the disadvantages of polymeric materials are listed below.

- Polymers have low dimensional stability. Polymeric products are inefficient in load-bearing applications.
- Polymers have low service temperature. This is limited to the softening temperature of the polymer.
- Polymers tend to degrade on exposure to heat and ultraviolet (UV) radiation.

Polymers are long molecules, also referred to as macromolecules, formed by the repetition of small chemical units (mers or monomers), which are linked together by chemical bonds. The chemical process of joining of monomers by covalent bonds to form macromolecules is called polymerization. The product is called polymer.

If polymerization takes place as a chain reaction by the repeated addition of monomers in quick succession, involving active centres, it is called chain growth polymerization. The product is called addition polymer. Examples of addition polymers are polyethylene, polypropylene, polyvinyl chloride, and polystyrene. If polymerization takes place by the successive linking of monomers, which have the capability of forming two or more linkages, via a condensation reaction, it is called step growth polymerization. The product is called condensation polymer. Examples of condensation polymers are polyesters, nylons, phenol-formaldehyde, and polyurethane. One of the developments in the polymer-based industry has been to make copolymers consisting of two or more repeat units and the modification of polymers to attain a wider range of properties. This has been successfully used to tailor-make materials for specific end uses.

4.26.1 Classification of Engineering Polymers

Polymeric materials can be classified in many ways.

- *Structural shape*: A polymer molecule may be linear, branched or cross-linked.
- *Physical state*: The physical state of polymer depends on the chemical structure and symmetry of the repeat units (monomers). These polymers can be further subdivided into two categories.

  Rubber and glass: A polymer may be rubbery or glassy depending on its glass transition temperature T_g. The macromolecules have sufficient internal motion above T_g, which makes the polymer soft and flexible. Such a polymer is called rubber. Below T_g, the molecular motion freezes causing the polymer to become hard and rigid. Such a polymer is called glass. Polymers with more rigid structures generally have a higher T_g.

  Amorphous and crystalline: In an amorphous polymer, the chains coexist in a random fashion without any microscopic order. In a crystalline polymer, the chains are arranged in some orderly fashion to form closely packed crystals. These crystals are considered physical cross-links that impart additional strength to the polymer. Crystallinity in a rubbery polymer makes it tough and less flexible, and in a glassy polymer makes it strong and less brittle. Only polymers with chemical symmetry will crystallize, and those with higher symmetry will crystallize to a greater extent. The temperature of use of crystalline polymers depends on their melting point, which is always higher than T_g.
- *Thermoplastics and thermosets*: This classification depends on how a material reacts when heated to processing temperatures. Thermoplastics soften when heated and harden when cooled, irrespective of the number of times the process is repeated. They flow with pressure and heat without any chemical changes on heating. The temperature of use is limited to the softening temperature (T_g for amorphous and T_m for crystalline). Thermoplastics can be reprocessed a number of times.

  Thermosets react irreversibly when heated to give strong intermolecular bonding, generally to form cross-linked (network) structures. They do not soften or flow on second heating, even on application of flow. Thermosets cannot be reprocessed. The use temperature limit of thermosets is higher than that of thermoplastics and is governed by degradation temperature.
- *Chemical family*: This depends on the family of monomers, which have a similar type of repeat unit or a similar type of linking unit. Examples are polyolefines, vinyl polymers, acrylic polymers, polyesters, polyamides, polyurethanes, epoxies, and fluoropolymers.

- *End* use: This depends on the final application of the polymers, which may cut across other classifications. These include plastics, rubbers, fibres, films adhesives, paints, membranes, and biomedical applications.

4.26.2 Polymers as Plastics

The use of polymers as plastics can be divided into two major categories.

1. *General purpose plastics*: They have low to medium mechanical properties and are thus meant for commodity items.
2. *Engineering plastics*: These have high mechanical strength and toughness and can be used in load-bearing applications.

4.26.2.1 General purpose plastics

Plastics for general use are normally crystalline polymers, which have a low T_g, or glassy amorphous polymers. The main five polymers in this category are low density polyethylene (LDPE), high density polyethylene (HDPE), polypropylene, polyvinyl chloride (PVC), and polystyrene. They account for about 90% of the total polymer production. A brief description of the physical and mechanical properties of these general purpose plastics, along with some applications, are given below.

4.26.2.1.1 Low density polyethylene

The repeat unit of LDPE (Figure 4.85) is a branched polyethylene prepared by free radical polymerization under high pressure. LDPE has a transluscent, wax-like appearance. It is a thermoplastic with a T_g of -100°C and T_m of 11°C. It has medium crytallinity (60%–75%), with a density of 0.91–0.94. LDPE has medium tensile strength with good impact resistance. It is flexible and tough over a wide temperature range. The surface is soft, with low scratch resistance. LDPE can be easily moulded, extruded, and machined.

Uses Blow moulded bottles and containers, gaskets, and flexible films for packaging.

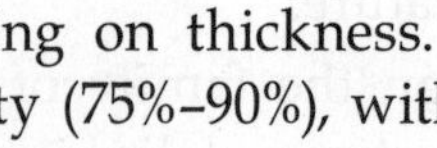

Figure 4.85 Repeat unit of polyethylene

4.26.2.1.2 High density polyethylene

The repeat unit of HDPE is a linear polyethylene prepared by coordination polymerization. HDPE also has a wax-like appearance but is harder than LDPE. It is translucent to opaque, depending on thickness. HDPE is a thermoplastic with a $T_g \approx -100°C$. It has high crystallinity (75%–90%), with a density of 0.94–0.96. HDPE has high tensile strength with moderate impact strength. The tensile strength increases with molecular weight. It is a tough and moderately rigid material.

Uses Wire insulation, dishpans, toys, buckets, containers, and film.

4.26.2.1.3 Polypropylene

The repeat unit of polypropylene (Figure 4.86) is an isotactic form of propylene, commercially useful and prepared by coordination polymerization. Polypropylene is harder and more rigid than polyethylene, with a T_g of ≈ 50°C. Polypropylene is highly crystalline and has a T_m of 176°C. It can withstand boiling water for long periods. Its tensile strength is higher, but impact strength lower, than that of polyethylene. Its chemical inertness is similar to that of polyethylene. Polypropylene has a unique property of a living hinge. One of its major drawbacks is its susceptibility to oxidation.

Uses Automotive interior parts, lamp housing, battery cases, cable covers, bobbins, refrigerator parts, and films.

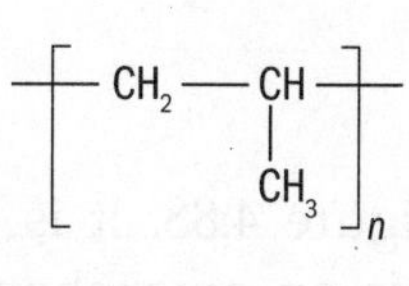

Figure 4.86 Repeat unit of polypropylene

4.26.2.1.4 Polystyrene

The repeat unit of polystyrene is shown in Figure 4.87. It is generally produced by free radical polymerization to give atactic polystyrene. Polystyrene is an amorphous glassy polymer, with a T_g of 100°C. It is transparent, hard, rigid, brittle, of moderate strength, and can be easily coloured. Polystyrene can be made impact resistance or chemical resistance by suitable modifications–generally by copolymerization. General purpose grade polystyrene is of low cost and low chemical resistance.

Uses Bottles, disposable glasses, toys, container lids, and other small household items.

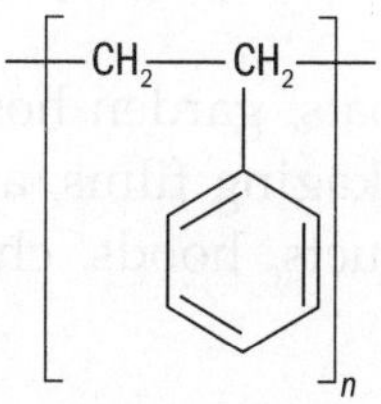

Figure 4.87 Repeat unit of polystyrene

4.26.2.1.5 High impact polystyrene

High impact polystyrene (HIPS) is rubber-modified polystyrene. A rubbery polymer (generally butadiene) is blended or graft polymerized with polystyrene to make HIPS. The rubber content of HIPS is 10%–30%. It is translucent with reduced rigidity. It has a much higher impact strength than polystyrene and poor chemical resistance.

Uses Refrigerator door liners, television and radio cabinets, stronger toys, and automotive interior parts such as the dashboard and control panel.

4.26.2.1.6 Styrene acrylonitrile copolymer

Styrene acrylonitrile copolymer (SAN) is made by copolymerization of styrene and acrylonitrile in a 70:30 ratio. It is transparent, hard, and rigid with slightly better impact strength. Its high tensile modulus makes it one of the stiffest plastics. It has excellent chemical resistance, making it resistant to acids, bases, and solvents.

Uses Similar to those of polystyrene (bottles, disposable glasses, toys, container lids, and other small household items) in which higher chemical resistance and rigidity is needed.

4.26.2.1.7 Polyvinyl chloride

The repeat unit of PVC is shown in Figure 4.88. It is generally made by free radical polymerization to give atactic PVC. It is an amorphous glassy polymer, with a T_g of 100°C. It is transparent, hard, and rigid. It has low impact strength, good chemical resistance, good electrical resistance, and is self-extinguishing. It has poor thermal and UV stability and, therefore, has to be stabilized carefully.

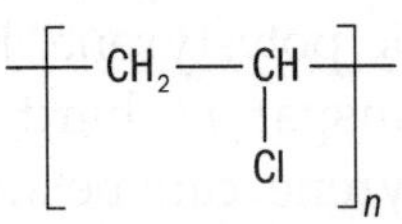

Figure 4.88 Repeat unit of polyvinyl chloride

PVCs are of two types: flexible PVC and high impact PVC. Flexible PVC is made by compounding with suitable plasticizers, while high impact PVC is made by blending with suitable rubbery polymers.

Uses Flexible PVC is used in raincoats, garden hoses, car upholstery leather-like brief cases and suitcases, pool liners, packaging films, and flexible shoes. High impact PVC is used in pipes in water works, ducts, hoods, chemical tanks, window frames, and house siding.

4.26.2.1.8 Engineering plastics

Engineering plastics have high mechanical strength and toughness. They are mainly used in load-bearing applications, generally to replace conventional materials such as metals and wood. The growth of engineering plastics has a greater potential than commodity plastics. It requires the knowledge of (1) plastics with high performance properties, (2) processing to make engineering products such as gears, (3) mould design and product design, and (4) structure property relations to optimize performance.

Engineering plastics have (1) high tensile strength, which makes them strong; (2) high modulus, which makes them stiff; (3) high impact strength, which makes them tough and non-brittle; (4) low creep under load, which makes them hard, with high T_g; and (5) high service temperature. They may be subdivided into (1) thermosets with high T_g, (2) thermoplastics with high T_g (crystalline or amorphous), (3) glass fibre reinforced, and (4) carbon fibre reinforced.

1 Engineering thermoplastics

Engineering thermoplastics include acrylonitrile butadiene styrene (ABS)/HIPS, polyamides (nylons), polyacetals, polycarbonate, polyesters, teflon, polysulphone, polyphenylene oxide, polyamide, polyurethane, and liquid crystalline polymers. Some of the above may be reinforces with glass/carbon fibres. Some blends also have engineering applications not attainable by individual components.

a. Acrylonitrile butadiene styrene

Acrylonitrile and butadiene impart chemical resistance and impact resistance, respectively, while styrene allows for easy processing. ABS can be tailored to yield an exact balance of properties. It can be utilized in low temperatures. It has a low creep for good load-bearing ability and can be easily blended with several other polymers.
Uses Helmets, pump, impellers, refrigerator liners, television and radio cabinets, luggage, extruded parts, and automotive moulded parts.

b. Polyamides (nylons)

There are two types of nylons: nylon 6 (Figure 4.89) and nylon 6,6 (Figure 4.90). Nylons are crystalline thermoplastics; nylon 6 has a T_m of 225°C and nylon 6,6 has a T_m of 260°C. They have high strength and are retained up to 150°C. They have high toughness, good impact resistance, high fatigue strength, and high moisture absorption. However, there is a loss in properties if exposed to moisture for a long time.
Uses Gears, valves, bearings, bobbins, surgical and pharmaceutical packaging, and high strength moulded parts.

$$\left[(CH_2)_5 - CO - NH \right]_n$$

Figure 4.89 Repeat unit of nylon 6

$$\left[OC - (CH_2)_5 - CO - NH - (CH_2)_6 - NH \right]_n$$

Figure 4.90 Repeat unit of nylon 6,6

c. Acetal resins

Acetal resins are also called polyoxymethylene (Figure 4.91). They are formed by the polymerization of formaldehyde or trioxane formaldehyde.

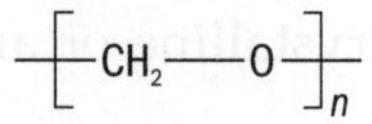

Figure 4.91 Repeat unit of polyoxymethane

The C–O bond in polyoxymethylene is short and hence the chain is packed more closely. Polyoxymethylene is highly crystalline, with a T_m of 175°C and T_g of –13°C. It has (1) high stiffness and high strength and toughness, (2) good fatigue endurance, (3) good resistance to creep, (4) good abrasion resistance and thermal stability, and (5) low coefficient of friction. It is resistant to solvents and has a good appearance.

Uses Bearing applications and gears; car parts including steering and clutch mechanism, emergency brake system, parts of the carburettor; fan blades, impellers, aerosol containers, blower wheels, and links in conveyor belts.

d. Polycarbonate

The repeat unit of a polycarbonate is shown in Figure 4.92. Polycarbonates are amorphous thermoplastics, with a T_g of 145°C. They are hard, rigid materials of high tensile strength. They have an outstanding impact resistance and low chemical resistance, but are susceptible to crazing under strain.

Uses Helmets, bulletproof protection covers, tool boxes, marine propellers, housing for small kitchen appliances, safety windows in aeroplanes, safety windows in prisons and jewellery shops, telephone parts, and baby bottles.

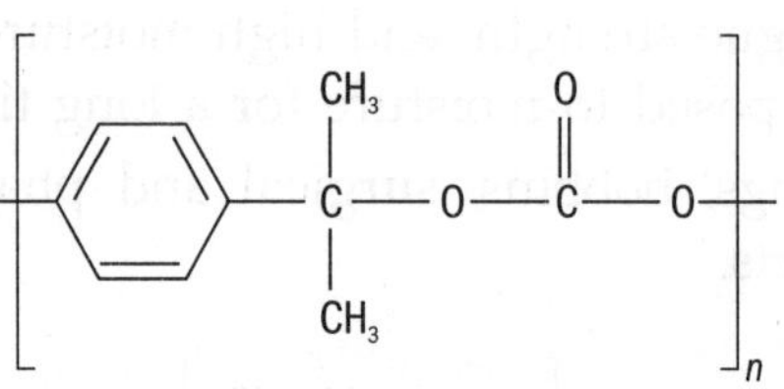

Figure 4.92 Repeat unit of polycarbonate

e. Polyesters

Polybutylene terephthalate

The repeat unit of polybutylene terephthalate (PBT) is shown in Figure 4.93. Polyethylene terephthalate (PET) has a $T_g \approx 30$°C and $T_m = 225$°C. It is made in a manner similar to PET. It has a faster crystallizing capacity than PET. PBT has high tensile strength and high flexural strength, good impact resistance, excellent dimensional stability. It is resistant to hydrocarbon oils and virtually free from stress cracking.

Uses Load-bearing applications, gears, mechanical parts, helmets, electronic components housing, and computer cabinets. It is also used as glass reinforcement to boost mechanical properties of materials. It blends with polycarbonate to boost impact strength.

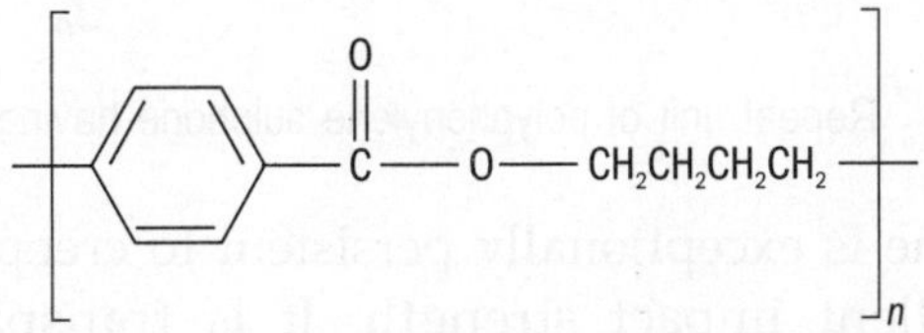

Figure 4.93 Repeat unit of polybutylene terephthalate

f. Teflon

The repeat unit of Teflon (polytetrafluoroethylene) is shown in Figure 4.94. Teflon is highly crystalline (T_m = 327°C), tough, flexible, non-resilient. Its tensile strength is moderate. The mechanical properties can be maintained over a large temperature range. It has an excellent impact resistance even at very low temperatures. It has an outstanding chemical resistance and low coefficient of friction, which is a self-lubricating property.

Uses Self-lubricating bearings, chemically resistant pipes, high temperature electronic components, tank linings, gaskets, packings, and biomedical applications.

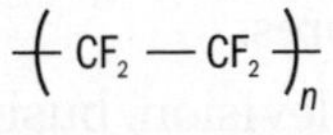

Figure 4.94 Repeat unit of Teflon

g. Polysulphones

Polyphenylene sulphone

Polyphenylene sulphone has two repeating units of T_g = 190°C (Figure 4.95) and T_g = 230°C (Figure 4.96).

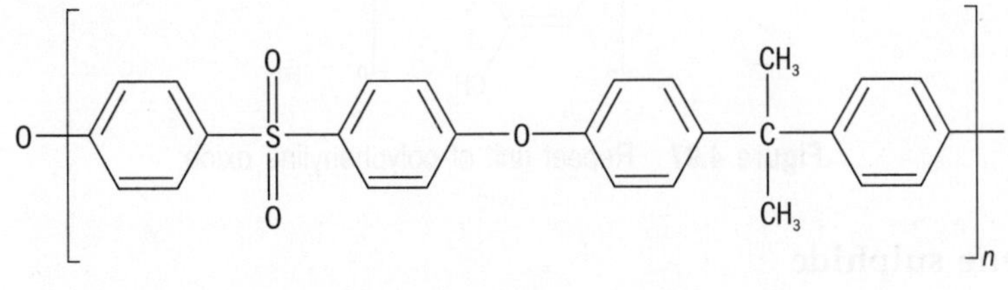

Figure 4.95 Repeat unit of polyphenylene sulphone having T_g = 190°C

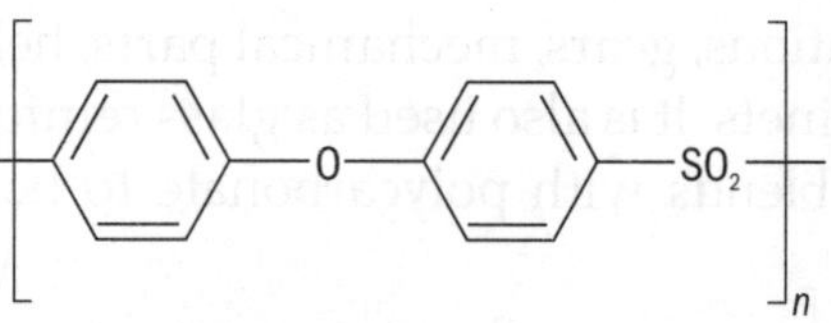

Figure 4.96 Repeat unit of polyphenylene sulphone having T_g = 230°C

Polyphenylene sulphone is exceptionally persistent to creep. It has high temperature resistance and an excellent impact strength. It is transparent and makes tough plastics.

Uses Polyphenylene sulphone is steadily replacing ceramics, thermosets, and metals. It is used as printed circuit boards and lamp bulb bases, in watches, television components, hair dryers, ovens, irons, fan heater components, car components (under the bonnet), coffee machines, and chemical plant pumps.

h. Polyphenyline oxide

The repeat unit of polyphenyline oxide (PPO) is shown in Figure 4.97. PPO, or poly(2,6-diethyl-p-phenylene) oxide, has a T_g of 208°C and T_m of 257°C. It has an exceptional dimensional stability due to high T_g, low coefficient of thermal expansion, low mould shrinkage, low water absorption, high strength and toughness, and good electrical properties. It can be blended with polystyrene to yield excellent properties. It can also be reinforced with glass fibres.

Uses Electrical industry, radio and television, business machines and computer housing, pumps and other plumbing applications, and in the automotive industry. It can also be utilized in instrument panels, steering column cladding, loudspeaker housings, ventilator grills, and shelves.

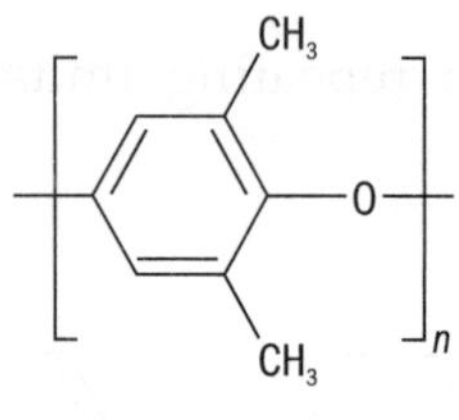

Figure 4.97 Repeat unit of polyphenyline oxide

i. Polyphenylene sulphide

The repeat unit of polyphenylene sulphide (PPS) is shown in Figure 4.98. PPS is crystalline, with a T_g of 88°C and T_m of 277°C. The mechanical properties of PPS are a little inferior to that of polysulphone, and its chemical resistance is less than that of polysulphone. It has excellent fire resistance and good electrical properties. PPS can withstand direct contact with metal at 250°C. It is heaper than polysulphones.

Uses Handles of irons, kettles, pressure cookers, and other cooking appliances, hair driers, fan heaters, laboratory equipment, and sterilizable medical components. In automobiles, it is used as exhaust gas return valve, carburettor parts, ignition plates, and distributor cap.

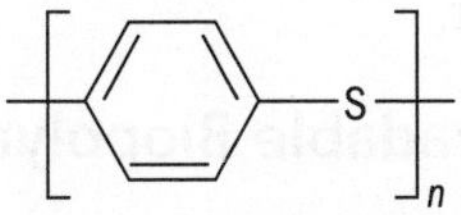

Figure 4.98 Repeat unit of polyphenylene sulphide

j. Polyurethane

Polyols, di-isocyanates, and glycols are reacted to produce block copolymers called polyurethane (Figure 4.99).

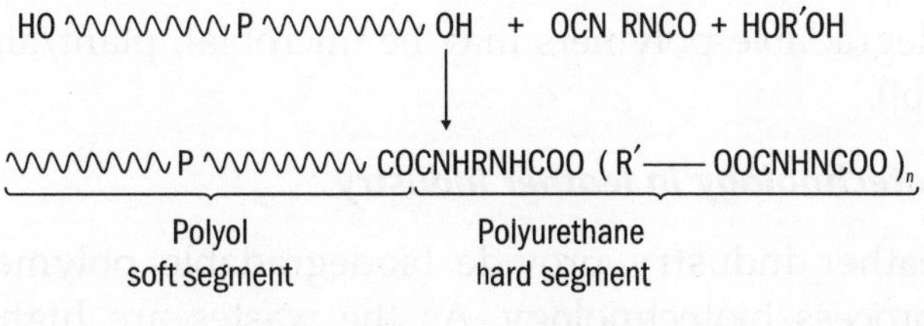

Figure 4.99 Reaction between polyol and polyurethane to produce segmented polyurethane

With proper choice of soft and hard segments, a large array of properties in polyurethane can be achieved. Soft polyurethane is rubbery in nature, while hard polyurethane is an engineering plastic.

k. Liquid crystalline materials

A liquid crystalline material exists in a liquid state and still retains a high degree of order. The state of liquid crystallinity is one in which the structural order is between a disordered liquid and an ordered crystalline solid.

Normal solids

$$\text{Solids} \xrightarrow{T_m} \text{Liquid}$$

Liquid crystalline material

$$\text{Solids} \xrightarrow{T_m} \text{Liquid crystal} \xrightarrow{T_i} \text{Liquid}$$

In the liquid crystalline state, a material can flow as if it was a fluid. The material has anisotropy due to the liquid crystals present. These liquid crystals can be made to orient by, say, electrical field, temperature or shear. Examples of uses of liquid crystals.

- In electronic watches and calculator displays, liquid crystals appear dark when oriented in the desired direction.
- In digital thermometers, liquid crystalline materials melt with temperature or change colour to become visible.
- In ultra-strong fibres like Kevlar.

4.26.3 Bioactive and Biodegradable Biopolymers

4.26.3.1 Classification

Polymers can be divided into synthetic non-biodegradable polymers and biosynthetic biodegradable polymers. Synthetic non-biodegradable polymers are further subdivided into engineering polymers and non-bioactive polymers. Biosynthetic biodegradable polymers are further subdivided into engineering polymers and bioactive polymers (Figure 4.100[a]).

Resources of biodegradable polymers may be microbial, plant/algal, animal or man-made (Figure 4.100[b]).

4.26.3.2 Process biotechnology in leather industry

Wastes from the leather industry provide biodegradable polymer and value-added products through process biotechnology. As the wastes are highly proteinaceous, a wide range of value-added products can be retrieved, making the leather industry an integrated business with value-creation opportunities.

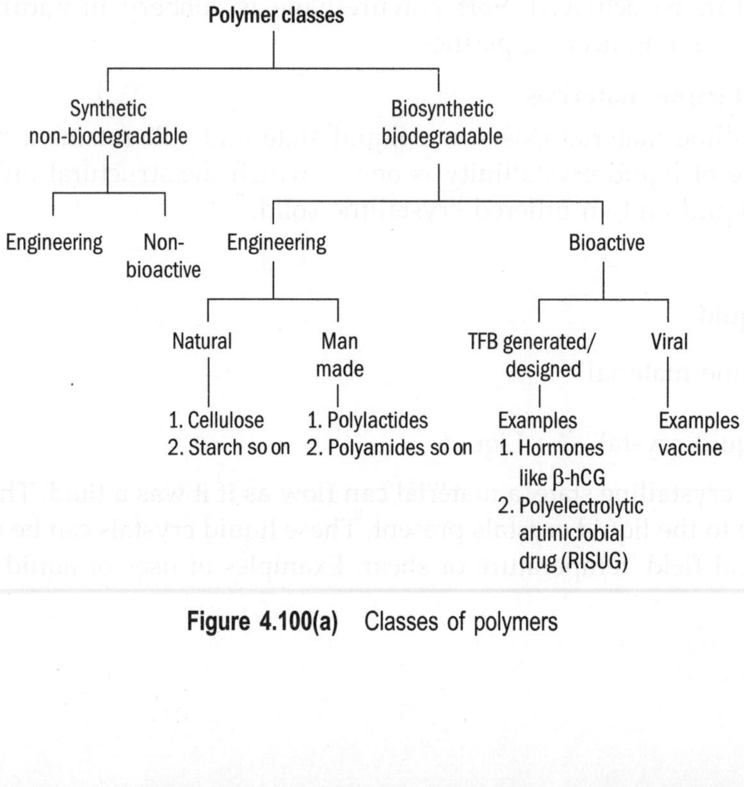

Figure 4.100(a) Classes of polymers

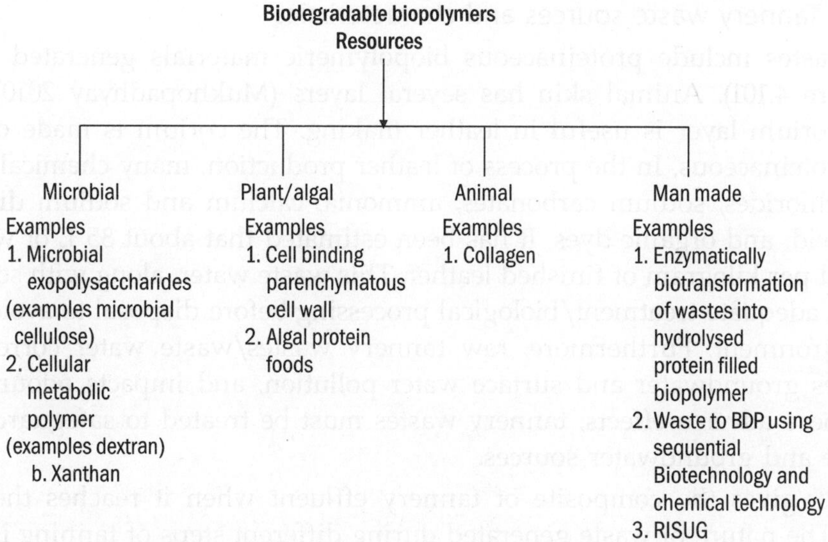

Figure 4.100(b) Resources of biodegradable polymers

The waste or waste water generated from the leather tanning industry is not ordinary. It has a treasure of value-added materials, which require retrieval. The industry generates substantial quantities of waste (Berry, Constantini, and Smart 2000; Taylor, Diefendorf, Fogila, *et al.* 1989; Taylor Cabeza, Marmer, *et al.* 1998; Taylor, Diefendorf, Brown, *et al.* 1992a; Taylor Cabeza, Marmer, *et al.* 1997; Taylor, Diefendorf, Thompson, *et al.* 1998; Chakraborty and Sarkar 1998; Cabeza, Taylor, DiMaio, *et al.* 1998). Waste water from different sections (Mukhopadhyay 2010) of leather tanning when mixed with its solid waste forms sludge with chrome shavings, which is greenish black in colour. Many possibilities of treating the chrome-tanned waste have been given in literature (Kolomaznik, Kupec, and Taylor 1997; Kolomaznik, Saha, and Taylor 1997; Kolomaznik, Mladek, Langmaier, *et al.* 1999; Parvathi, Suseela, and Nandy 1986). The major aim of the leather industry is abatement of environmental pollution and by-product retrieval. A scheme of the total recycle integration approach, which should economize the leather industry business and value-creation opportunities by biological process, is shown in Figure 4.101. It indicates many retrievable value-added products that can be obtained by various biotechnologies and provide higher economic returns. As shown in the figure, the by-products include biodegradable polymer and its derived materials, biogas, enzyme, inhibitor, recyclable chromium, and biofertilizer/compost. The availability of by-products largely depends on the biodegradability of the waste resource. Thus, an understanding of the fundamental of biochemical/chemical phenomena involved in biodegradation and biodeterioration is important.

4.26.3.2.1 Tannery waste sources and characteristics

Tannery wastes include proteinaceous biopolymeric materials generated in various steps (Figure 4.101). Animal skin has several layers (Mukhopadhyay 2010), but only the inner corium layer is useful in leather making. The corium is made of collagen, which is proteinaceous. In the process of leather production, many chemicals are used, including chlorides, sodium carbonates, ammonia, calcium and sodium dichromates, sulphuric acid, and organic dyes. It has been estimated that about 35 L of waste water is generated per kilogram of finished leather. This waste water, along with solid wastes, requires an adequate treatment/biological processing before disposal to avoid pollution of the environment. Furthermore, raw tannery wastes/waste water corrodes sewer lines, causes groundwater and surface water pollution, and impacts odour problems. To lower these adverse effects, tannery wastes must be treated to safeguard the sinks and surface and groundwater sources.

Table 4.26 gives the composite of tannery effluent when it reaches the plant for treatment. The nature of waste generated during different steps of tanning is shown in Table 4.27, and the quantity of biological composition of settled and filtered composite tannery waste is given in Table 4.28. From the tables, it is clear that tannery waste has high protein content and is suitable as substrate resource in recycling through process biotechnology (biological treatment methods) in value-added products recovery.

4.26.3.2.2 Fundamentals of biodegradation

Structurally and composition-wise, leather industry wastes are polymeric biodegradable materials. The fundamental concept for chemical/biochemical biodegradation of these waste materials under aerobic and anaerobic conditions is given below.

Under aerobic conditions

$$\text{Polymer/biopolymer} + O_2 \xrightarrow{\text{Biocells/bioconversion}} CO_2\uparrow + H_2O + \text{Biomass} + \text{Residue} \quad (4.294)$$

Under anaerobic conditions

$$\text{Polymer/biopolymer} \xrightarrow{\text{Biocells/bioconversion}} CH_4\uparrow + CO_2\uparrow + H_2O + \text{Biomass} + \text{Residue} \quad (4.295)$$

Based on the above approaches, investigators have conducted biodeterioration/biodegradation of various wastes/refuse materials for their reuse by recycling as well as treatment for cleaner environment (Taylor, Diefendorf, Thompson, *et al.* 1997; Saha and Mukhopadhyay 1998; Poulopoulou, Katakis, and Vrachnou 1998). Some of the wastes undergo photodegradation and biodegradation simultaneously but slowly. In such cases, the combined incremental degradation (*dD*) of polymeric materials has been described by Equation 4.296 (Wool 1995).

$$dD = \left(\frac{\delta D}{\delta B}\right)_{MPC} dB + \left(\frac{\delta D}{\delta M}\right)_{PCB} dM + \left(\frac{\delta D}{\delta P}\right)_{BMC} dP + \left(\frac{\delta D}{\delta C}\right)_{BMC} dC \quad (4.296)$$

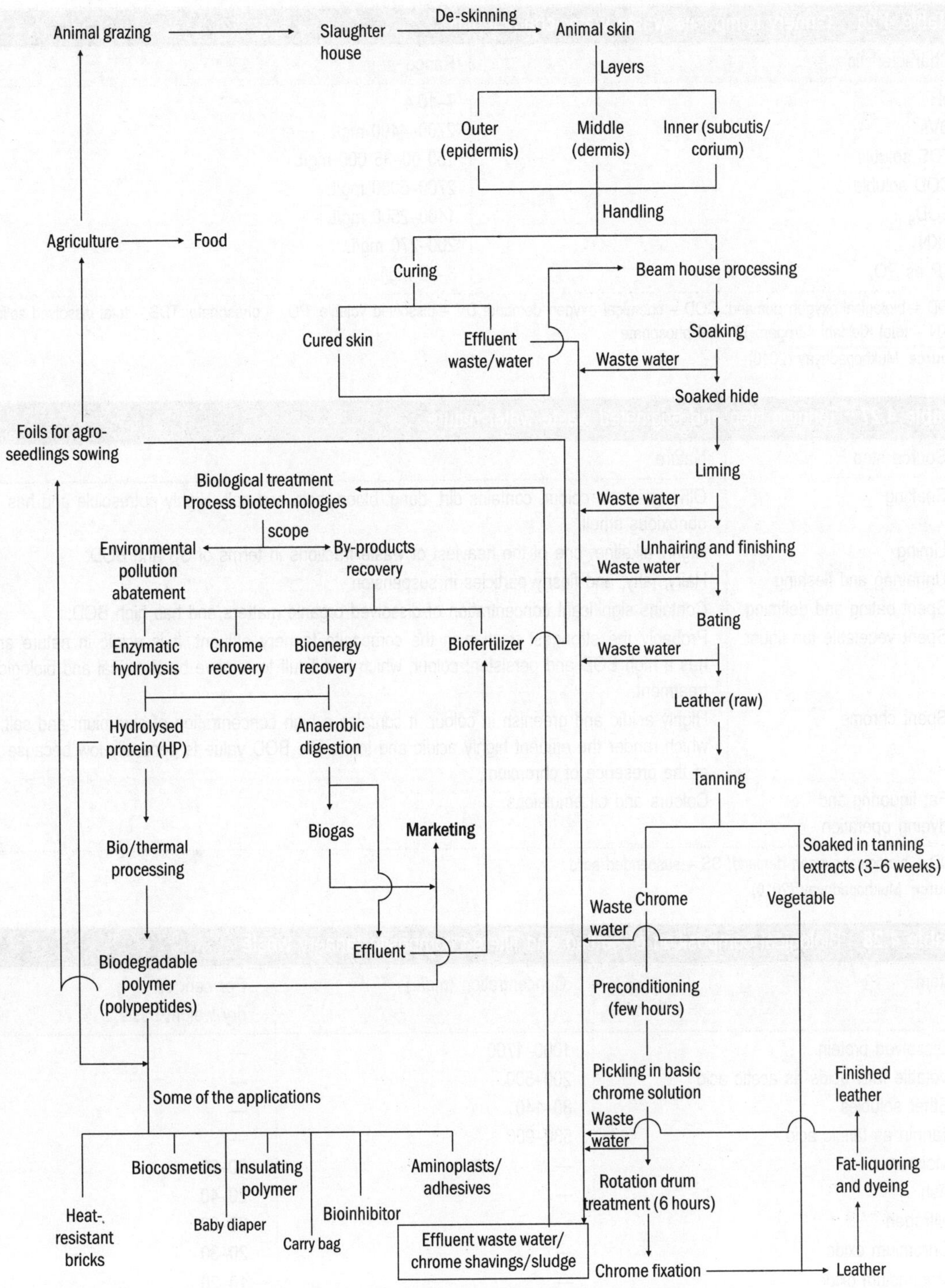

Figure 4.101 Total recycle integration approach

Table 4.26 Tannery composite waste water characteristics

Characteristic	Range
pH	7–10.4
DVs	2700–4400 mg/L
TDS soluble	150 00–35 000 mg/L
COD soluble	2700–5000 mg/L
BOD_5	1400–2500 mg/L
TKN	200–270 mg/L
TP as PO_4	3–22 mg/L

BOD – biological oxygen demand; COD – chemical oxygen demand; DV – dissolved volatile; PO_4 – phosphate; TDS – total dissolved solids; TKN – total Kjeldahl nitrogen; TP-total phosphate

Source Mukhopadhyay (2010)

Table 4.27 Tanning step and originated waste water nature

Source step	Nature
Soaking	Olive green in colour, contains dirt, dung, blood, hair, and salt. Highly putrescible and has obnoxious smell.
Liming	Highly alkaline, one of the heaviest of waste fractions in terms of SS and BOD.
Unhairing and fleshing	Hairy, fatty, and fleshy particles in suspension.
Spent bating and deliming	Contains significant concentration of dissolved organic matters and has high BOD.
Spent vegetable tan liquor	Probably the strongest fraction in the composite tannery effluent. It is acidic in nature and has a high BOD and persistent colour, which is difficult to remove by chemical and biological treatment.
Spent chrome	Highly acidic and greenish in colour. It contains a high concentration of chromium and salt, which render the effluent highly acidic and toxic. The BOD value is generally low because of the presence of chromium.
Fat liquoring and dyeing operation	Colours and oil emulsions.

BOD – biological oxygen demand; SS – suspended solid

Source Mukhopadhyay (2010)

Table 4.28 Biological composition of settled and filtered composite tannery waste

Item	Concentration (mg/L)	Per cent on dry weight basis
Dissolved protein	1000–1700	—
Volatile fatty acids as acetic acid	200–500	—
Ether solubles	80–140	—
Tannin as tannic acid	530–900	—
Moisture	—	80–90
Ash	—	20–40
Nitrogen	—	30–60
Chromium oxide	—	20–30
Magnesium oxide	—	10–20

Source Mukhopadhyay (2010)

where *B* is the total biomass, *M* is microorganism, *P* is photon, and *C* is chemical.

In these cases, the four differential terms in brackets interact and combine with each other to provide biodegradation end results. To observe the contributions of individual terms in Equation 4.296 in biodegradation ability test, the necessary conditions are stated.

- For $\delta D = \left(\frac{\delta D}{\delta B}\right)_{MPC}$, absence of sunlight, reactive chemicals, and microorganisms or photons are required.
- For $\delta D = \left(\frac{\delta D}{\delta P}\right)_{BMC}$, UV lamps are used in the absence of microbes, insects or autoxidants.

In such tests, it is usually presumed that one of the degradation mechanisms dominates and a standard is utilized with the assumption. In biodegradation shown by Equations 4.294 and 4.295, one needs to keep in mind that all natural living polymeric materials are composed mostly of carbohydrate units that are catabolized by providing energy for a reaction (respiration) similar to combustion (Equation 4.297).

$$(CH_2O)_n + nO_2 \rightarrow nCO_2 + nH_2O + \text{energy} \tag{4.297}$$

This catabolic biodegradation is required by a naturally acquired design engineering (NADE) process, like photosynthesis, to provide energy to living cellular materials of plant origin from solar energy reaction to renew the biomaterial resource via Equation 4.298.

$$nCO_2 + nH_2O \rightarrow (CH_2O)_n + nO_2 \tag{4.298}$$

The residue composition in Equations 4.294 and 4.295 varies considerably depending on the extent of biodegradation and bioresource material. This demonstrates that biopolymer-degraded residues are neither pollutants nor noxious to agricultural crops. That is why biodegradable film packages have been extensively used in foils for agro-seedling sowing as a solution to the disposal problem in plasticulture (Wool 1995).

4.26.3.2.3 Value-added products from waste

Figure 4.101 shows that chrome sludge is the major waste, which is connected to problems of real application in the tannery industry. The capital-intensive nature of the leather industry has been reported to be due to the high costs of raw materials, while the cost of labour is a relatively small component. From the cost structure break-up (Table 4.29), to achieve more economic return from this capital-intensive industry, experts suggest integration of its waste recycling for the retrieval of value-added products (Taylor, Cabeza, Marmer, *et al.* 1998; Kresalkova, Hnanickova, Kupec, *et al.* 2002).

Table 4.29 Relative cost break-up in the leather industry

Item	Cost component	Per cent share
Raw materials/supplies	• Raw hides	62–65
	• Chemicals	15–18
	• Labour	12–14
	• Environment	5–10
	• Energy	3–4
Overhead	• Organization/administration/selling	10
	• Capital	5–10

Note The cost break-up is likely to vary with location.

1. Improved biodegradable polymer

The production of improved biodegradable polymer (IBDP) is done by the following two steps.

a. Chrome sludge to hydrolysed protein (HP)

As the waste resource has high protein content, researchers have attempted to hydrolyse protein by enzyme (commercial grade) to produce hydrolysed protein (HP) and subsequently, to polymerize it by polycondensation reaction (Cabeza, Taylor, DiMaio, *et al.* 1998; Kolomaznik, Saha, and Taylor 1997; Cantera, Goya, and Mingo 1999; Simeonova and Dalev 1966) with some aldehydes/gluteraldehydes. In hydrolysing the proteinaceous waste, enzymatic process biotechnology was carried out in stages shown schematically in Figure 4.102. For enzymes bioreactor operation, the hydrolysis reactor is fed with 10 tonnes of water, 90 kg of cyclohexylamine or its equivalent amine compounds, and 60 kg of MgO. Another 3 tonnes of chrome shaving waste is gradually added under constant stirring to homogenize the mixture, which is then heated to carry out hydrolysis. The hydrolysis is run at an optimal temperature of 70°C and pH 9; organic base is added to adjust pH. After 3–5 h of the reaction, the hot heterogeneous mixture is filtered. The filtrate is collected in a metering storage tank and re-pumped into a continuously working three-stage vacuum evaporator. The filter cake is transferred to an enzyme reactor situated under the vacuum filter. In the reactor, after pouring 3 tonnes of water, the pH is adjusted to 9 and 0.9 kg of an alkaline proteolytic enzyme (ALCALASE 2.5 LD X OF Novo Nordisk, Denmark) is added. The enzyme reaction mixture is maintained at 70°C under constant stirring for 2–3 h. It is then filtered to give HP and filter cake. The HP clear solution is passed through the metering storage tank to a vacuum evaporator. As shown in Figure 4.102, the filter cake is recycled to the reactor and diluted with water. The heterogeneous mixture is fed into a hot solution of $Na_2Cr_2O_7$ and H_2SO_4 in the reactor for producing tanning salt, either in liquid or solid state. HP-containing free amino acid fractions are transformed to IBDP in a blend with polyvinyl alcohol (PVA) or starch by heating in an extruder (Kolomaznik, Mladek, Langmaier, *et al.* 1999).

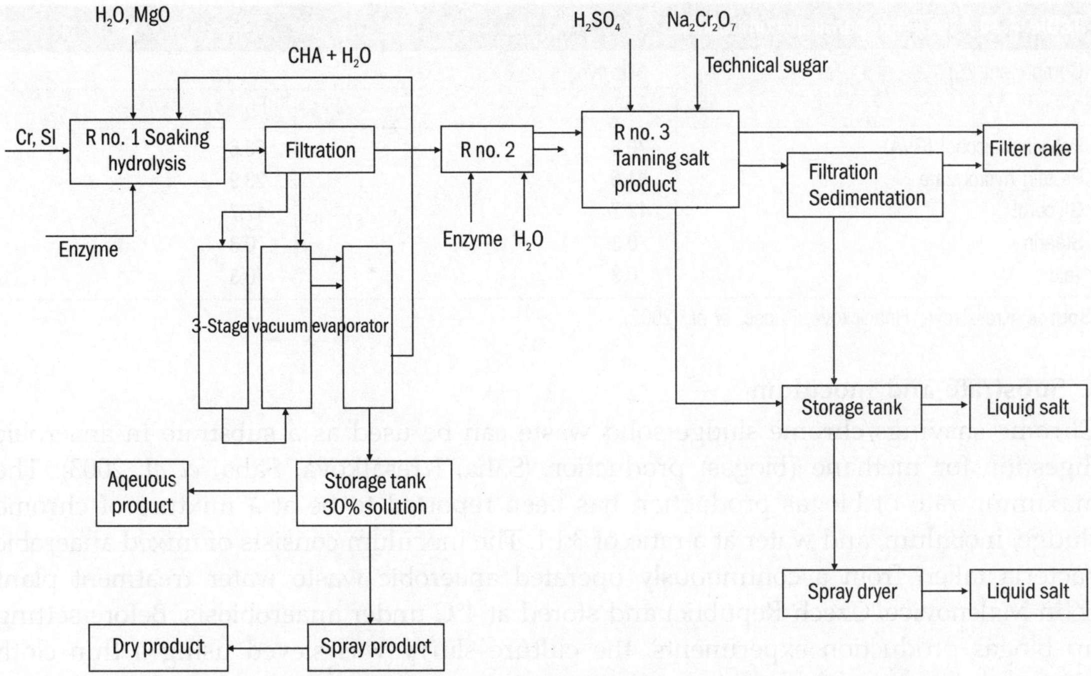

Figure 4.102 Enzymatic process biotechnology followed in Tannex Co. plant
Source Kolomaznik, Mladek, Langmaier, *et al.* (1999)

b. HP to IBDP

The HP obtained by the above enzymatic process is ready for producing biodegradable material based on suitable synthetic polymer like PVA. Although PVA is relatively poor in degradation under aerobic conditions, its modification with HP provides several advantages. HP-modified blow-extruded PVA film is a biodegradable plastic and has special advantages. HP in PVA not only functions as a filler, but also makes the material more economical. This is because HP powder is 50% cheaper than PVA. Moreover, in terms of biodegradability and mechanical strength, this HP-modified material has improved features.

In producing IBDP, blends were prepared by mechanical mixing of several components in different proportions. The components include PVA, HP, glycerol, and stearin. A typical proportion of these components in IBDP film production is given in Table 4.30. However, this composition needs optimization to obtain the best performance of IBDP.

2. Biogas

Biogas production using process biotechnology requires the following.

Table 4.30 IBDP preparation composition

Component (%)	Film type	
	F5	F6
Polyvinyl alcohol (PVA)	70.8	58.6
Protein hydrolyzate	11.5	23.9
Glycerol	17.7	17.7
Stearin	0.3	0.3
Talc	0.3	0.3

Source Kresalkova, Hnanickova, Kupec, *et al.* (2002)

a. Substrate and inoculum

Chrome shavings/chrome sludge solid waste can be used as a substrate in anaerobic digestion for methane (biogas) production (Saha, Kresalkova, Saha, *et al.* 2003). The maximum rate of biogas production has been reported to be at a mixture of chrome sludge, inoculum, and water at a ratio of 3:1:1. The inoculum consists of mixed anaerobic bacteria taken from a continuously operated anaerobic waste water treatment plant (Zlin-Malenovice, Czech Republic) and stored at 4°C under anaerobiosis. Before setting up biogas production experiments, the culture slurry was sieved using a thin cloth filter and the supernatant was used as inoculum.

b. Substrate treatment

The raw chrome sludge was denatured for 30 min by adding 1% (v/v) of MgO suspension and heating at 85°C. On cooling to room temperature, the pH of the suspension was adjusted to 8.5–9.5 using isopropic amine. One per cent (w/w) of a commercial proteolytic enzyme (Alcalase, Novonordisk, Denmark) was then added and enzymatic hydrolysis was conducted together with stirring at 70°C for 4 h. The hot mixture was then filtered. The grayish-green viscous residue was used as substrate. It contained 20.9% TS, 17.5% VS, and 3%–4% ash.

c. Anaerobic digestion

The treated substrate obtained above was used in a small scale, in different runs with various ratios of sludge, inoculum, and water (4:1:0, 3:1:1, 2:1:2, and 1:1:3). Anaerobic digestion (AD) was carried out at 50°C for 15 days, and the maximum biogas production was 1 ml/g substrate (17.5% VS). From these results, it is clear that pilot-scale studies for the production of biogas from chrome sludge can be carried out (Parvathi, Suseela, and Nandy 1986; Saha, Kresalkova, Saha, *et al.* 2003).

3. Enzyme

Enzyme treatment of leather industry waste to produce HP as described in the previous section has been reported to be a viable biotechnological process. However, to treat the large volumes of wastes produced in the industry, bulk quantities of enzyme are required. It is important that the allied enzyme production technology should not be high tech and expensive. Solid-state fermentation (SSF) process biotechnology has been reported to be suitable for the production of bulk enzyme at economical cost (Chakraborty and Sarkar 1998). Thus, in HP production from leather industry waste, the enzyme availability may be from two sources: recovering the used enzyme and reusing it or producing bulk enzyme by microorganisms through cheap SSF technology.

4. Recyclable chromium

The chrome tanning solution in the leather industry contains chromium (III) salts. In spite of introducing methods and agents that increase exhaustion of chrome salts, modern tanning technologies are not able to comply with the requirements of environmental protection. High levels of chrome in tannery effluents make the treatment difficult and yield tannery sludge that is hazardous and requires special disposal (Poulopoulou, Katakis, and Vrachnou 1998). Reported literature indicates that the tanning industry generates about 60 000 tonnes of solid chrome-tanned wastes each year worldwide. Also, processing 1 tonne of hide generates 200 kg chrome-tanned leather, 50 000 kg waste water containing 5 kg chromium, 250 kg non-tanned solid waste, and 200 kg tanned solid waste containing 3 kg of chromium (Kolomaznik, Zelinka, Langmaier, *et al.* 2002). The chromium in the effluent is in trivalent state and is non-toxic.

However, after using the process biotechnology of enzymatic hydrolysis of chrome sludge, the chrome cake obtained as residue is further chemically treated to give a recyclable chrome cake containing primarily trivalent chromium. The chrome cake contains protein or resins, which creates difficulties in the subsequent H_2SO_4 solution. It is thus removed by chemical treatment for chromium recycling process, as described by Kolomaznik, Mladek, Langmaier, *et al.* (1999). Figure 4.103 shows closed-loop chromium recycling in the beam house processing of leather.

5. Biofertilizer/compost

The simplest way of HP disposal is to utilize it in agriculture as an organo-nitrogenous fertilizer. With cooperation from the Agriculture Control and Testing Institute of the Czech Republic, Regional Department of Soil Agrochemistry and Plant Nutrition, the suitability of enzyme process biotechnology-generated HP and biodegradable polymer as biofertilizer was checked by conducting a vegetation test. The Czech Institute ran a comparative test on HP, commercial name EKO-N-HP, and available commercial fertilizer LOVODAN, which is a blend of ammonium nitrate and urea (1:1 ratio). The tested crop was lettuce. Both fertilizers were fed in such a way that the mass of nitrogen for all tested plants remained the same. A simultaneous comparative test on unfertilized soil was also conducted. The vegetation test results are given in Table 4.31

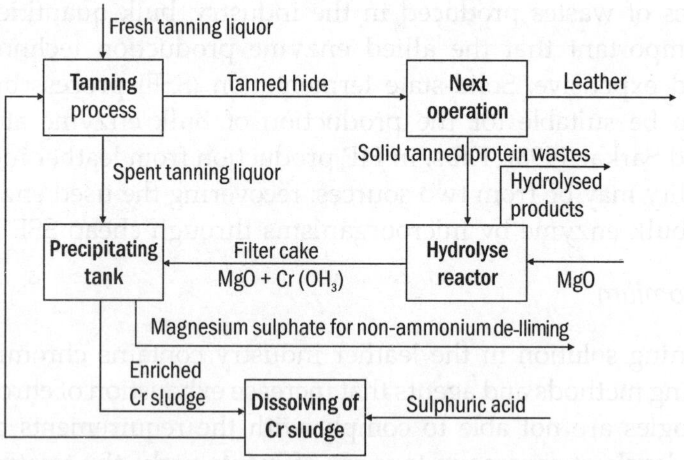

Figure 4.103 Closed-loop chromium recycling in beam house processing of leather

Table 4.31 Comparative vegetation tests on the suitability of HP as a biofertilizer

Type of fertilizer	Yield (g/test)	Nitrate content (ppm NO_3)
LOVODAN	187	519
EKO-N-HP	164	23
Unfertilized	105	52

Source Kolomaznik, Mladek, Langmaier, *et al.* (1999)

and indicate a substantial difference in the yields and nitrate contents in the consumable parts of lettuce.

The vegetation test showed positive effect of EKO-N-HP as biofertilizer on the growth and development of the agro-food product, and its use was extended in agriculture. The test for preventing the thinning of growing plants was also positive. This will help in sowing tapes bearing firmly fixed, specially treated seeds at optimal distance. Machine sowing requires the sowing tape to have good mechanical properties and meet soil conditions for optimal seed sprouting. PVA mixed with 20%–24% HP and processed by blow moulding can provide modified foil having higher mechanical strength and good composting characteristics. In addition, organic nitrogen slowly liberated from this biodegradable foil acts as a fertilizer and creates a favourable soil microclimate for the growth of sprouts. For this reason, the largest volume of HP is channelled into agriculture as an organic nitrogenous biofertilizer.

4.26.3.2.4 Conclusion

Process integration of leather-manufacturing industry has high potential for larger economic return through enzymatic retrieval of biodegradable polymer and other value-added products using process biotechnology. The retrieved HP and biodegradable polymer have already gained a lot of agro-industrial applications in different countries. Biogas generation by the anaerobic digestion of chrome sludge has also been shown to have high potential in terms of a low cost and eco-friendly process biotechnology. However, it needs further studies on scale up for industrial processing. Recyclable chromium, enzyme, and biotransformation of HP and IBDP as biofertilizer/compost also have great potential to provide higher economic return to the leather industry.

Finally, in enzymatic process biotechnology, the enzyme reaction provides advantage for the production of HP of a relatively good quality and chrome sludge. In turn, it has a place in the treatment of chromium-containing tannery waste by anaerobic digestion using microbial process biotechnology that gives a value-added product—biogas. Thus, this area of research, having brought satisfactory field results, definitely supports the idea that the process integration approach by total recycling of leather industry wastes using process biotechnology is a profitable business. However, its thorough economic evaluation is necessary.

4.27 BASICS OF COMPUTER AND ITS USE IN BIOSYSTEMS

A computer system helps in (1) solving mathematical problems from simplest to very complex in nature, (2) achieving online analysis of chemicals and biochemicals being either consumed or produced in a process, (3) controlling fermentators operation or those of immobilized enzyme reactors, (4) data acquisition, reduction, and analysis by making use of statistical techniques, and (5) making designs.

Although both analogue and digital computers have been used for scientific work, the ease of manipulation with digital computers has made it the choice of engineers and scientists in both academic work and R&D ventures.

4.27.1 Microcomputers

A microcomputer can be best understood with reference to Figure 4.104. The heart of the system is a very large-scale integrated chip (VLSIC) known as microprocessor. It coordinates the activities of all other devices attached to it by an interface bus. The interface bus is a cluster of data lines and control lines. From the microprocessor, another set of 16 lines known as address bus emanates. These lines are used by the processor to address any device with which it wants to interact.

Any problem whose answer is required is transformed into a set of instructions known as a program and a set of numbers or characters known as data. The program with data is termed user's software. The user's software can be entered into the microcomputer system with the help of an input device. Usually, a keyboard consisting of alphabets, numbers, and certain special function keys is used for this purpose. The

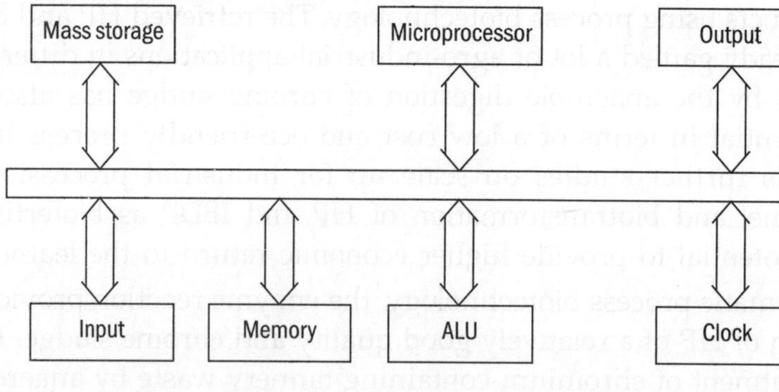

Figure 4.104 Essential features of a microcomputer

program and data lines so entered can be displayed or listed on output devices like cathode ray tube (CRT) or printers. The program and data once entered via a keyboard are stored in an area of the memory known as user's read/write memory. Also, during calculations, the memory stores the intermediate answers area in an internal read/write area of the memory just as we write intermediate answers on a scratch pad. The most important part of memory is the area known as the read only memory (ROM). The ROM contains a large number of set of instructions and data (operating systems software) that enables all the devices connected within the system to operate in a defined manner when power is switched on. A memory map of a typical microcomputer is shown in Figure 4.105.

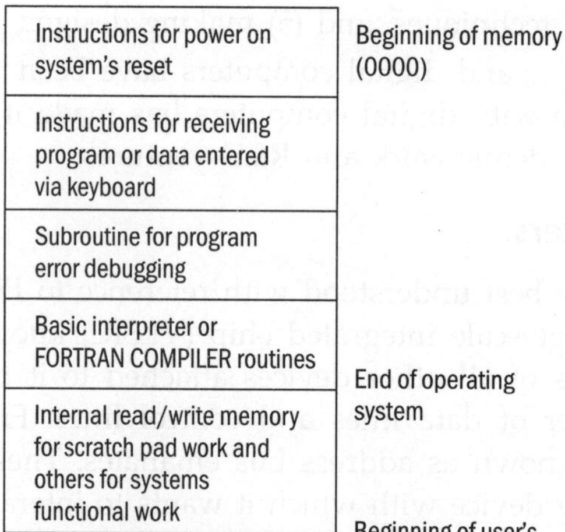

Figure 4.105 A typical memory map for a microcomputer

The arithmetic and logical unit (ALU) performs functions like multiplication, division, addition, and subtraction. This may not sound like much but is adequate for solving advanced mathematical problems because even the toughest problem can be reduced to arithmetic expressions. The ALU can also perform certain logical operations like $A < B$, $A > B$, and $A = B$.

An entered program when made to run is executed by the microprocessor line by line, beginning with the smallest line number. The microprocessor keeps track of the line number to be executed next. During execution of a program, whenever an output instruction is encountered, the microprocessor directs the output device to display or print the result.

Both the program and data developed by the user for the solution to any problem can be stored in mass storage devices. These include magnetic cassette tapes, floppy discs, and Winchester discs.

The cassette tape provides serial access to information stored, whereas the floppy and Winchester discs enable random access.

4.27.2 Applications in Biosystems

4.27.2.1 Solution to mathematical problems

Problem 1 A yeast growing aerobically in a shake flask on a YPG medium at 4.5 pH and 30°C has a specific growth rate of 0.2 per hour. How much time will it take for biomass concentration to reach 3 g/L and 5 g/L, if the initial concentration after inoculation was 0.1 g/L. Assume that the culture exhibits no lag time.

Answer The growth rate of the organism is proportional to biomass concentration (Equation 4.299).

$$\frac{dx}{dt} = \mu X \tag{4.299}$$

where x is the biomass concentration at any time t. The symbol μ represents the specific growth rate.

Integrating Equation 4.299 with initial condition $x = x_o$ at $t = 0$, we get

$$\ln(x/x_o) = \mu t$$

$$\text{or } t = (1/\mu) \ln(x/x_o) \tag{4.300}$$

Equation 4.300 is the model around which the computer program can be developed for obtaining an answer to a given problem.

A sample program written is BASIC is given below.

```
10 REM "EXPONENTIAL GROWTH"
20 DISP "FINAL BIOMASS CONCENTRATION"
```

```
30 INPT X1
40 READ XO, M
50 DATA 0.1, 0.2
60 T = (LOG (X1/XO))/M
70 DISP "TIME = "; T; "HOURS"; "XF = "; X1
80 GO TO 20
90 END
```

4.27.2.2 Online analysis

Problem 2 Develop a system for online analysis of protein from a fractionation column or from a fermenter.

Answer This problem requires two additional devices to be plugged into the microcomputer: (1) analogue-to-digital (ADC) convertor and (2) relay actuator. The other requirement is that of a continuous flow cell spectrophotometer adjusted to a wavelength of 280 nm. The configuration of the online analysis system is shown in Figure 4.106. The recorder output of the spectrophotometer's detector is connected to an ADC unit. Two solenoid pinch-type valves, one on the water line and the other on the sample line, are connected with the relay actuator. The relay actuator has four additional relays that can be programmed to perform any other switching task desired by the user.

To begin with, the online system has to be calibrated by placing samples of known strength and observing corresponding voltage signal from the recorder output. If the recorder output varies proportionally with protein concentration, a linear model can be accepted (Equation 4.301).

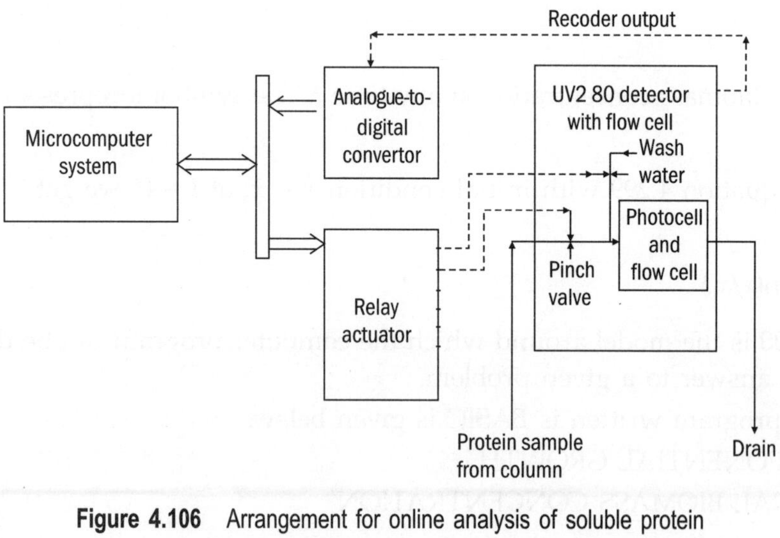

Figure 4.106 Arrangement for online analysis of soluble protein

$$P = mV \tag{4.301}$$

where V is the voltage output, P is protein concentration, and m is the slope of regression line passing through the V versus P data. If a regression line is not acceptable, a polynomial curve of higher order may give satisfactory correlation (for example, Equation 4.302).

$$P = aV + bV^2 \tag{4.302}$$

Either Equation 4.301 or Equation 4.302 is an essential requirement for online analysis of the protein because the computer can read only the recorder output voltage (V). It will compute the protein concentration of an unknown sample with the help of Equations 4.301 and 4.302. A sample program is given below.

```
10 REM "PROTEIN ANALYSIS"
20 DISP "WASH CYCLE"
30 GOSUB "WASH"
40 DISP "WASH CYCLE OVER"
50 DISP "PRESS RUN KEY TO RESUME"
60 STOP
70 DISP "ENGAGED IN ANALYSIS"
80 GOSUB "MEASUREMENT"
90 P = M*V
100 DISP "PROTEIN CONCENTRATION = "; P
110 T = TIME
120 DISP "TIME = "; T
130 WAIT 1000
140 ON KEY 1 GO TO 20
150 GO TO 80
160 END
170 "WASH"
180 CLOSE RELAY 2 AND OPEN RELAY 1
190 WAIT 1000
200 RETURN
210 "MEASURMENT"
220 READ RECORDER VOLTAGE AND STORE IT IN V
230 RETURN
```

4.28 GENETIC ENGINEERING PRINCIPLES AND CONSIDERATIONS IN HANDLING RECOMBINANT CELLS IN LARGE SCALE

In the late 1970s, a set of biochemical methodologies emerged, which made it possible to transfer genes of one organism into another. The organism transferred is completely unrelated to the donor organism. This technology, popularly known as genetic engineering, has revolutionized the potential of biotechnology for mankind's benefit. Through recombinant DNA technology, our knowledge of DNA function, gene structure and organization, gene regulation, and primary protein structure has increased tremendously. It has been possible to achieve high expression of human proteins like insulin, interferon, somatization in the common colon bacterium, *E. coli*.

While the techniques involved in the formation of recombinant cells are always carried out at laboratory scale, the end product (that is, transformed or recombinant cells) must be cultivated in large scale to get sufficient quantities of the protein of interest. During this shift in cultivation conditions, numerous problems arise, the most serious being that of plasmid shedding. Since the structural gene for the protein of interest is harboured on the plasmid DNA, its loss would result in lowered product yield.

4.28.1 Genetic Engineering Principles

The basic genetic engineering principles and its applications are discussed under the following three major headings.

4.28.1.1 General scheme for production of recombinant cells

The foreign DNA segment from a specific cell (microbial, plant or animal) or from a virus (in case of antigen-producing viruses) is obtained by cutting a large piece of DNA using certain enzymes called restriction endonucleases. These DNA segments are then ligated in appropriate fashion to vector molecules (plasmids, virus, and cosmids) using ligases. Chimeric plasmids are then introduced into suitable hosts. Cloning a DNA fragment ensures its production on a large scale in the cell for detailed analysis. In cases where the cloned gene expresses a product, cloning ensures production (or overproduction) of the corresponding polypeptide.

Both the choice of vectors and host cells will critically determine the amount of protein produced. Hence, its selection should be done with utmost care. In general, the properties the cloning vector must have are listed as follows (Bailey and Ollis 1986).

- Ability to replicate in the host cell.
- Ability to accommodate foreign DNA of various sizes without damage to replication functions.
- Easy insertion into cell after in vitro manipulations.

- Contains a selection marker to facilitate rapid positive selection of cells that contain the vector.
- Contains only one target site for one or more different restriction endonucleases.

Although *E. coli* is the most commonly used host for the production of proteins/enzymes using recombinant plasmids, it is not an industrially exploited microorganism. It is gram-negative, has toxic lipo-polysaccharides on its outer surface (*E. coli* infections in the blood stream have a 50% fatality rate), and it does not secrete proteins into the outer environment. Also, as a prokaryote, it cannot carry out RNA splicing and post-translational protein modifications. For these and other reasons, other cloning hosts have been developed (Table 4.32).

4.28.1.2 Mode of production of proteins

The course of production of proteins in living cells can be summarized as follows.

$$\text{DNA} \xrightarrow{\text{Transcription}} \text{RNA} \xrightarrow{\text{Translation}} \text{Protein}$$

The same sequence of events also governs production of proteins in recombinant cells. The "foreign DNA" must be transcribed into RNA by host RNA polymerases followed by the translation of mRNA into proteins. The host enzymes machinery must be geared to recognize foreign DNA as native and direct its protein synthesis. To maximize expression of the cloned gene, genetic engineering has specific DNA sequences that can be ligated at the *S′* end. These include strong promoters and strong ribosome-binding sites. With such combinations, it is possible to make *E. coli* produce 70% of the total cellular protein as the protein of interest.

4.28.1.3 General problems in overproduction by recombinant cells

4.28.1.3.1 Plasmid loss

Recombinant microorganisms, plant cells, and animal cells will sometimes lose the plasmids they harbour. If this plasmid contains the structural gene for the production of a commercially important protein or vaccine, it can cause specific problems at the time of scale up. The loss of the plasmid is thought to occur during cell division.

Table 4.32 Some cloning hosts and their useful properties

Host	Use
B. subtilis	Extracellular product, free from pyrogenic compounds, molecular switching possible.
Saccharomyces cerevisiae	• Extracellular product • RNA splicing • Post-translational modification possible • Known techniques for large-scale cultivation
Mammalian cells	Expression, stability, and secretion of animal-derived products like vaccines, hormones, and antibodies.

Even at low rates of plasmid loss, this can be a serious problem for cell harbouring, overproducing recombinant plasmids. Cells with plasmids have lower specific growth rates than those without plasmids.

To demonstrate the gravity of lowered growth rate on overall cell population, let us take the following example. Suppose the fraction of daughter cells shedding plasmid at any cell division is 0.01 and suppose that the growth rate of the resulting cell is twice that for cells with plasmid. After 15 generations of growth (for slower growing cells), the cells that have shed their plasmids will have overgrown their plasmid-containing counterpart by a ratio of 1000:1. Thus, the growth vessel would be full of cells not producing the desired product. This effect is more pronounced if the plasmid copy number in a recombinant cell is low.

A mathematical representation of this example is given for a yeast population growing in a chemostat in selective medium so that cells born without plasmids do not divide. The maximum possible dilution rate, that is, dilution rate at washout is given by Equation 4.303 (Bailey and Ollis 1986; Hartwell and Ungar 1977).

$$D_{\text{max}} = 1/P\,(\ln 2 - Q_D - Q_M) \tag{4.303}$$

where P denotes the time interval between mother cell birth and cell division, Q_D and Q_M are the fractions of daughter and mother cells that have no plasmids at birth, respectively.

If it is assumed that the dividing cell contains N plasmids that segregate randomly at birth, the fractions D and M may be calculated from Equations 4.304 and 4.305.

$$Q_D = f(N/N) \tag{4.304}$$

$$Q_M = f(O/N) \tag{4.305}$$

$$f(x/N) = \frac{N!}{x!\,(N-x!)}\alpha^x\,(1-\alpha)^{N-x} \tag{4.306}$$

The binomial distribution (Equation 4.306) gives the conditional probability that a newborn mother cell arising from the division of a cell with N plasmids will contain x plasmids. Here, α denotes the probability that the plasmid remains in the mother cell at birth and gives plasmid segregation properties.

If no cells without plasmid are born at cell division (that is, $Q_D = Q_M = 0$), the dilution rate at washout, D_{max}, is given by Equation 4.307.

$$D_{\text{max}} = \frac{\ln 2}{\text{p}} \tag{4.307}$$

On comparing Equations 4.303 and 4.307, it is seen that D_{max} (or μ_{max}) in a chemostat is considerably reduced because of plasmid instability. The magnitude of this effect, in which the dilution rate at washout is given by Equation 4.303, has been normalized by the dilution rate at washout for a completely stable plasmid D_{max}. It can also be seen that this ratio approaches unity if the plasmid copy number is high. If the copy number is low, the specific growth rate of the culture is reduced considerably.

Hence, at the time of scale up, it is necessary to ensure that recombinant cells do not lose their plasmids and also maintain a high copy number. Genetic engineering does provide potential solutions to this problem, some of which are as follows (Klotz 1982).

- *Constant selection of cells that do not lose their plasmids*: If the plasmid provides antibiotic resistance to the host cell, this antibiotic should be included in the fermentation medium. This will ensure that cells born without plasmids do not survive in the medium and hence do not overtake in number recombinant cells.
- *Controlled overproduction (molecular switches)*: In this mode of control, the recombinant cells are allowed to grow in a fermentation medium without expressing the product of interest. At the end of growth phase, product synthesis can be turned on by the use of special operator—repressor combinations, temperature-dependent promoters or sporulation promoters.
- *Integration of structural gene from the plasmid into the chromosomal DNA*: The chromosomal DNA is essential for the cell to function and survive. Procedures currently exist for stably integrating DNA regions of genetically engineered plasmids into the main chromosomal DNA of *E. coli*, yeast, and *B. subtilis*. In this way, the gene of interest is stable and can be expressed generation after generation.

4.28.1.3.2 Non-feasibility of large-scale isolation and purification of proteins from *E. coli*

A large number of proteins such as human insulin, human growth hormone, leucocyte, and interferon are being expressed and obtained from *E. coli*. Although high levels of expression are obtained, use of *E. coli* as a host cell presents the following disadvantages.

- Presence of proteases in *E. coli* results in protein degradation. The remedy is to use a protease inhibitor such a phenyl methyl sulphonyl fluoride in the cultivation broth or work with protease-deficient host cells.
- When *E. coli* fermentation is depleted of iron, an enzyme is derepressed, which catalyses the formation of an iron chelator, 2,3-dihydroxybenzoyl serine (Brot, Goodwin, and Fales 1966). This compound when complexed with iron is bright red and may cause problems during purification, such as binding to agarose gel.
- The presence of potent pyrogens such as lipopolysaccharide from the cell envelop causes fever response at levels 0.5 mg/kg body weight. Such contamination has been experienced in the purification of L-asparginase from *E. coli* (Grinnin 1981).

The preceding discussion shows that large-scale isolation of important protein products, especially those meant for human consumption, from *E. coli* is not feasible. Other hosts, such as *S. cerevisiae*, and established mammalian cell lines are being used for the expression of mammalian cell culture.

4.28.1.3.3 Cultivation of mammalian cells

Biologicals manufactured from mammalian cells in culture have considerable medical and clinical importance. The mammalian cell cultures discussed here are those that predominantly contain recombinant SV40, adenovirus or vaccinia virus molecules. These viruses contain information for the production of useful compounds (or vaccines) in mammalian cells. Some of the potential products from mammalian cell cultures are as follows.

- Growth factors like nerve growth factor and epidermal growth factor
- Blood factors
- Monoclonal antibodies
- Interferons
- Proteases like urokinase
- Vaccines like Hepatitis B
- Hormones including human growth hormone, insulin, calcitonin, and parathyroid hormone

The technology for cultivation of mammalian cells has till now been restricted to laboratory scale, which mostly involved use of rotating bottles. The rolling action ensures that the cells are alternatively exposed to growth medium and oxygen in the air space. This method is cumbersome and expensive for the production of a large quantity of cells (Glacken, Fleischaker, and Sinskey 1983).

Basically, there are two types of mammalian cells: anchorage-dependent cells (cells that require a surface for growth) and suspension cells (non-anchorage dependent). Barriers that need to be addressed to scale up mammalian cell systems are as follows.

- Oxygen transfer limitations
- Accumulation of toxic waste products
- Lack of regulatory hardware and software for process control
- High cost of serum

Barriers that have been addressed to some extent are listed below.

- Low surface-to-volume ratio of systems that cultivate anchorage-dependent cells
- Shear sensitivity of mammalian cells

Although many research laboratories and industries are engaged in the cultivation of mammalian cells, the process is not as cost effective. Recently developed methods to produce cells on a large scale, such as microcarriers, artificial capillaries, tubular spiral film, and microencapsulation, must be optimized and the problem of oxygen transfer limitation must be solved. Removal of toxic waste product must be ensured as these will inhibit growth and productivity. This effect can be reduced by change in cultivation conditions such as that of fed-batch culture, so that cell metabolism is shifted away from production of these compounds or by continuously perfusing fresh medium.

4.29 MODERN INSTRUMENTAL TECHNIQUES IN PROCESS BIOTECHNOLOGY

Modern instrumental techniques developed in physical science have been applied in process biotechnology in recent years. These techniques have facilitated rapid advancement of biological research. In 1982, Binnig and his associates developed the atomic force microscope (AFM), also called the scanning force microscope. With the help of the AFM, biochemical and physiological processes can be observed in real time at molecular and atomic resolution.

4.29.1 Atomic Force Microscopy

A major scientific breakthrough occurred in 1982 when scientists Gerd Binning and Heinrich Rohrer invented the scanning tunnelling microscope (STM), for which they were awarded the Nobel Prize for Physics (Binnig, Quate, and Greber 1986). In their technique, a fine sharpened tip, which is presumed to have a single atom at the apex, is scanned above the surface of the sample. A voltage (bias potential) is applied between the tip and the sample, and the small current (tunnelling current), which flows between the gap, is measured. The variation in the tunnelling current as it passes over the atomically corrugated surface is recorded. If this process is repeated across the entire sample, a three-dimensional "map" of the surface can be obtained (Rugar and Hansma 1990). However, the need for imaging insulating materials was soon realized and, in 1986, Binning invented the AFM. This instrument was similar to the STM, except that the tip and the sample surface were in contact and the inter-atomic van der Waals force acting between them provided the contrast mechanism, rather than tunnelling current. Hence, there was no longer a requirement for the sample to be conducting, which allowed for the study of other materials such as biological samples (Hansma, Elings, Marti, *et al.* 1988).

4.29.1.1 Principle and Modes of AFM

AFM probes the surface of a sample with a sharp tip, which is a couple of microns long and often less than 100 Å in diameter. The tip is located at the free end of a cantilever 100–200 μm long. Forces between the tip and the sample surface cause the cantilever to bend and deflect. A detector measures the cantilever deflection as the tip is scanned over the sample or the sample is scanned under the tip. The measured cantilever deflections allow a computer to generate a map of surface topography (Figure 4.107).

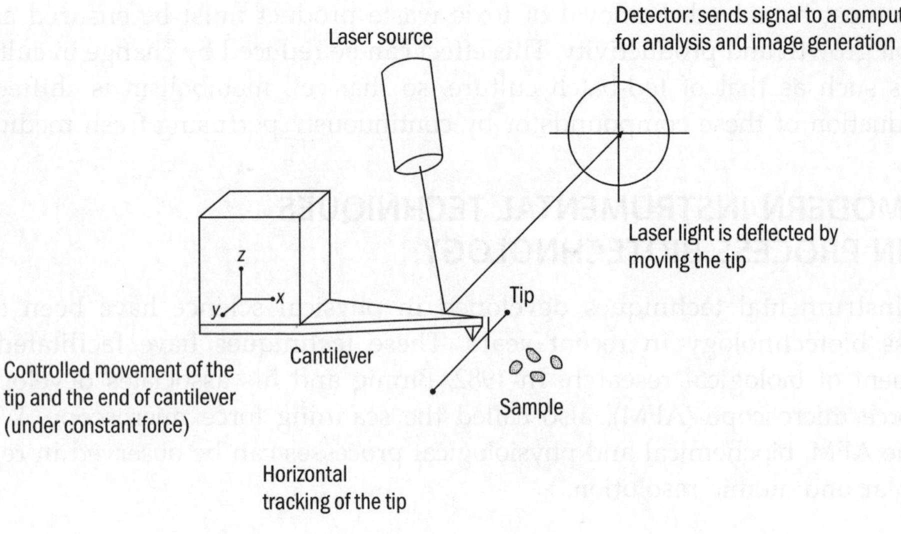

Figure 4.107 Method of operation of an atomic force microscope

Several forces contribute to the deflection of an AFM cantilever. The force most commonly associated is the inter-atomic force called the van der Waals force. The dependence of the van der Waals force on the distance between the tip and the sample is shown in Figure 4.108.

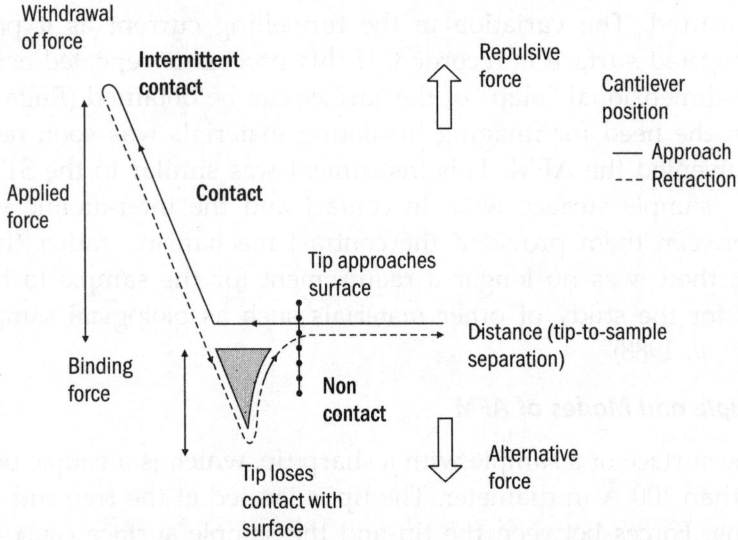

Figure 4.108 Representative force–distance curve generated using an AFM

4.29.1.1.1 Modes of AFM

1. Contact AFM

In the contact-mode AFM, also known as repulsive mode, the tip is mechanically contacted with the sample surface at an applied force. This applied force can be evaluated from the force–distance curve, measured when the tip is brought to and then retracted from the sample surface. A soft cantilever (spring constant 0.01–1 N/m) is usually used in contact AFM. The cantilever is so soft that it can be pulled onto the surface due to high attractive force between them. After a mechanical contact between the tip and sample, there is a repulsive force between them. This repulsive force is usually used as the feedback parameter (by maintaining a constant force through the adjustment of sample height while the tip is scanning the surface) to obtain AFM images. In the most common scheme, a laser beam bounces off the back of the cantilever onto a position-sensitive photodetector (PSPD). As the cantilever bends, the position of the laser beam on the detector shifts. The PSPD itself can measure displacements of light a small as 10 Å. The ratio of the path length between the cantilever and detector to the length of the cantilever itself produces a mechanical amplification. As a result, the system can detect sub-angstrom vertical movement of the cantilever tip. Once the AFM has detected the cantilever, it can generate the topographic data set by operating in one of the two modes: constant height and constant force mode (Kuk and Silverman 1989).

In constant force mode, the deflection of the cantilever can be used as input to a feedback circuit that moves the scanner up and down in the *z*-direction, responding to the topography constant by keeping the cantilever deflection constant. With the cantilever deflection held constant, the total force applied to the sample is constant. In constant force mode, the speed of scanning is limited by the response time of the feedback circuit, but the total force exerted on the sample by the tip is well controlled. This mode is generally preferred for most applications.

In constant height mode, the spatial variation of the cantilever deflection can be used directly to generate the topographic data set because the height of the scanner is fixed as it scans. The constant height mode is often used for taking atomic-scale images of atomically flat surfaces, where the cantilever deflections and thus variations in applied force are small. The mode is also essential for recording real-time images of changing surfaces, where high scan speed is essential. Hence, this type of mode is most suited to study surface characteristic alterations such as those induced by antibiotics or other anti-microbial agents in real time, that is, the mechanism of action can be investigated as the changes are taking place in biological cells.

4.29.1.2 Non-Contact AFM

In non-contact AFM, the cantilever is vibrated near the surface of a sample. Spacing between the tip and the sample for non-contact AFM is of the order of tens to hundreds of angstrom. This spacing is indicated on the van der Waals curve in Figure 4.108 as the non-contact regime.

A non-contact AFM is desirable because it provides a means for measuring sample topography with little or no contact between tip and sample. The system vibrates a stiff cantilever near its resonant frequency (100–400 kHz), and the change in the resonant frequency or vibrational amplitude as the tip approaches the sample is then detected. Since the force between tip and sample in a non-contact AFM is low, it is more difficult to measure than the force in contact mode, which can be several orders of magnitude greater.

In addition, cantilevers used for non-contact AFM must be stiffer than those used for contact mode because soft cantilevers can be pulled into contact with the sample surface. The small force values and greater stiffness of the cantilever used in non-contact AFM make the non-contact signal small; hence, this mode requires a sensitive AC detection scheme. However, the low force (total force) between tip and sample is low (10–12 N) gives a significant advantage to non-contact AFM for the study of soft biological samples such as cells, membranes, and biomolecules, which are otherwise damaged by the dragging force when subjected to numerous scans in contact mode.

4.29.1.3 Tapping-mode AFM

For biological samples, which are soft and susceptible to damage by the dragging of an AFM tip across the surface, another mode of AFM operation is available, called the tapping mode. It is similar to non-contact AFM, except that in tapping AFM, the vibrating cantilever tip is brought closer to the sample so that at the bottom of its travel, it "taps" the sample. The tapping AFM operating region is indicated on the van der Waals force curve (Figure 4.108). As for non-contact AFM and tapping-mode AFM, the cantilever oscillation amplitude changes in response to tip–sample spacing. An image representing surface topography is obtained by monitoring these changes.

Thus, this type of mode is most suitable for specific biological applications in comparison to contact or non-contact AFM. It is less likely to damage the sample than contact AFM because it eliminates lateral forces (friction or drag) between tip and sample. It is more effective than non-contact AFM for imaging larger scan sizes that may include greater variation in sample topography. Due to these features, this mode has evolved an important AFM technique as it overcomes some of the limitations of both contact and non-contact AFM (Howland and Benatar 1997).

4.29.1.4 AFM operating conditions

AFM pictures were taken employing the CP Research model of Thermo Microscopes, USA, operating in the non-contact mode. In this mode of operation, a silicon cantilever with a force constant of 17.2 N/m was vibrated near its resonance frequency of about 320 kHz. The deviation of the vibration frequency from resonance depends on the tip–sample separation and forms the data set providing the topography of the bacterial cells. For mounting the bacterial cells, glass, mica, and single crystalline silicon (100) substrates were employed. Of these, the glass substrates were found to be best suitable, and 10 µl of the bacterial suspension was mounted on a glass substrate. After air-drying, the glass was glued to a steel disk with the help of a double-sided adhesive tape and placed on the top of a microscope specimen holder cylinder, where the steel disk was magnetically held in place. The cells were imaged in air with AFM in tapping mode.

4.29.2 Scanning electron micrograph of *E. coli* in RISUG treatment

The scanning electron micrograph (SEM) results of *E. coli* are shown in Figures 4.109(a) and 4.109(b). The normal morphology of *E. coli* grown in the absence of RISUG shows the common of the bacillus structure with dispersed cytoplasm and cell walls with well-defined shape. The SEM of untreated *E. coli* shows normal smooth surface of these cells (Figure 4.109[a]).

The TEM results of RISUG of untreated and treated *E. coli* B37 cells are shown in Figures 4.110(a) and 4.110(b), respectively. Plate 4.1 shows the AFM image of polymeric drug RISUG. Plates 4.2(a) and 4.2(b) show the AFM image of *E. coli* in x-direction and y-direction after RISUG treatment, while Plate 4.3 shows the AFM image of a cluster of *E. coli* cells after RISUG exposure.

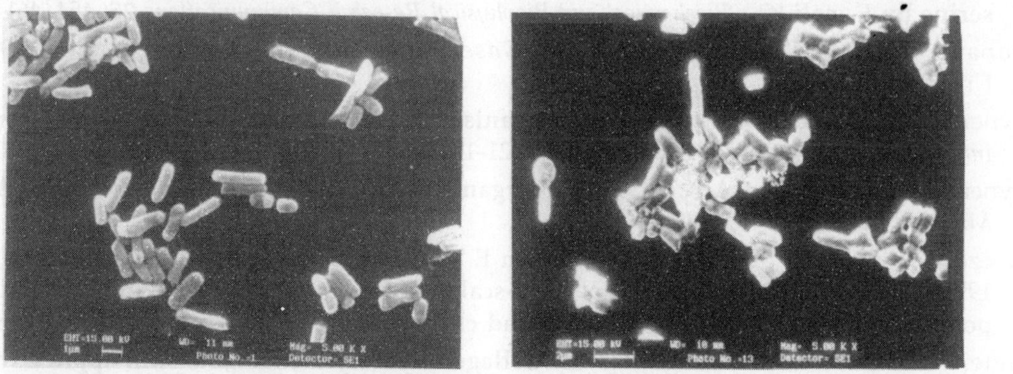

Figure 4.109 (a) SEM of RISUG of untreated *E. coli* B37 cells, (b) SEM of RISUG of treated *E. coli* B37 cells

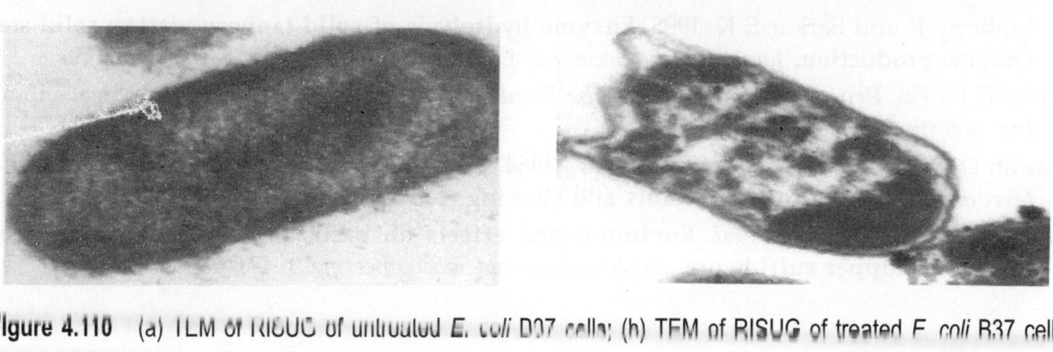

Figure 4.110 (a) TEM of RISUG of untreated *E. coli* B37 cells; (b) TEM of RISUG of treated *E. coli* B37 cells

REFERENCES

Aiba S, Humphrey A E, and Millis N F. 1973. ***Biochemical Engineering***, 2nd edn. Tokyo: University of Tokyo Press. 434 pp.

ASME (American Society of Mechanical Engineers). 1955. UPV. Code of 1952

Bailey J E and Ollis D F. 1986. ***Biochemical Engineering Fundamentals***, 2nd edn. New York: McGraw Hill Publishing Co. Ltd

Beck J V. 1960. **A ferrous-ion-oxidizing bacterium I: isolation and some general physiological characteristics**. *Journal of Bacteriology* **79**: 502–509

Berry F J, Constantini N, and Smart L E. **Recovery and reuse of chromium from leather shavings. Formation of chromium containing ceramic pigments.** *Technical paper presented at the 96th Annual Meeting of American Leather Chemists Association*, Oglebay Resort, Wheeling, West Virginia, 17-21 June 2000

Bryner L C, Beck J V, Davis D B, Wilson D G. 1954. **Microorganisms in leaching sulphide minerals**. *Journal of Industrial and Engineering Chemistry* **46**: 2587

Binnig G C, Quate C F, and Gerber Ch. 1986. **Atomic force microscope**. *Physics Review Letters* **56**: 930-933

Bolen P L and Slininger P J. 1984. **Continuous culture selection of mutant strains of Pachysolen tannophilus capable of rapid aerobic growth on D-xylose**. *Developments in Industrial Microbiology* **25**: 449–457

Brot N, Goodwin J, and Fales H. 1966. **In vivo and in vitro formation of 2,3-dihydroxybenzoyl serine by *E. coli* K_{12}**. *Biochemical and Biophysical Research Communications* **25**: 454–461

Brunauer S. 1943. ***The Adsorption of Gases and Vapours***, vol. 1. London: Oxford University Press

Bryner L C and Anderson R. 1957. **Microorganisms in leaching sulphide minerals**. *Journal of Industrial and Engineering Chemistry* **49**: 1721–1724

Bryner L C and Jameson A K. 1958. **Microorganisms in leaching sulphide minerals**. *Applied Microbiology* **6**: 281

Cabeza L F, Taylor M M, DiMaio G L, Brown E M, Marmer W N, Carrio R, Celma P J, Cot J. 1998. **Processing of leather waste: pilot-scale studies on chrome shavings. Isolation of potentially valuable protein products and chromium**. *Waste Management* **18**: 211–218

Cantera C S, Goya L, and Mingo R. 1999. **Collagen hydrolysate: soluble skin applied in post tanning processes Part 1: characterisation.** *Journal of the Society of Leather Technologists and Chemists* **84**: 29–37

Chakraborty R and Sarkar S K. 1998. **Enzyme hydrolysis of solid tannery wastes: solid state enzyme production.** *Journal of the Society of Leather Technologists and Chemists* **82**: 56

Clarke R J. 1958. ***Process Engineering in the Food Industries***. New York: Philosophical Library Inc. 360 pp.

Duncan D W, Trussell P C, and Walden C C. 1964. **Leaching of chalcopyrite with *Thiobacillus ferrooxidans*: effect of surfactants and shaking**. *Applied Microbiology* **12**(2): 122–126

Ehrlich H L and Fox S I. 1967. **Environmental effects on bacterial copper extraction from low-grade copper sulfide ore**. *Biotechnology and Bioengineering* **9**: 471–485

Glacken M W, Fleischaker R J, and Sinskey A J. 1983. **Large-scale production of mammalian cells and their products: engineering principles and barriers to scale-up**. *Annals of the New York Academy of Sciences* **413**: 355–372

Hartwell L H and Ungar M W. 1977. **Unequal division in *S. cerevisiae* and its implications for the control of cell division**. *Journal of Cell Biology* **75**: 422–435

Hansma P K, Elings V B, Marti O, Bracker C E. 1988. **Scanning tunneling microscopy and atomic force microscopy: applications to biology and technology**. *Science* **242**: 209–216

Howland R and Benatar L. 1997. ***A Practical Guide to Scanning Probe Microscopy***. Sunnyvale, CA: Park Scientific Instrument.

Keenan J H and Keyes F G. 1961. ***Thermodynamic Properties of Steam,***. New York: John Wiley and Sons

Kern D Q. 1950. ***Process Heat Transfer***. New York: McGraw Hill Publishing Co. Ltd

Kinsel N A. 1960. **New sulphur oxidizing iron bacterium: Ferrobacillus sulfooxidans**. *Journal of Bacteriology* **80**: 628–632

Klotz L C. 1982. **Overproduction of proteins in recombinant cells**. *Annals of the New York Academy of Sciences* **413**: 1-11

Kolomaznik K, Mladek M, Langmaier F, Janacova D, Taylor M M. 1999. **Practice of enzymatic dechromation of chrome shavings**. *Journal of the American Leather Chemists Association* **94**: 55

Kolomaznik K, Kupec J, and Taylor M M. 1997. **Engineering aspects of treatment of wastes from leather industry**. In *Proceedings of Environmental Engineering Conference*. [Alberta, Canada]

Kolomaznik K, Saha P, and Taylor M M. 1997. *R'97 Recovery, Recycling, Re-integration*. [Proceeding of the *3rd International Congress on Recovery, Recycling, Re-integration*, 4–7 February 1997, Geneva, Switzerland]

Kolomaznik K, Zelinka I, Langmaier F, Mladek M. 2002. *R'02 Recovery, Recycling, Re-integration*. [Proceeding of the *6th International Congress on Recovery, Recycling, Re-integration*, 12–15 February 2002, Geneva, Switzerland]

Kresalkova M, Hnanickova L, Kupec J, Kolomaznik K, Alexy P. 2002. ***Application of protein hydrolysate from chrome shavings for polyvinyl alcohol-based biodegradeable material.*** *Journal of the American Leather Chemists Association* **97**(4): 143–149

Kuk V and Silverman P J. 1989. ***Scanning tunneling microscope instrumentation***. *Review of Scientific Instruments* **60**(2): 165–181.

Landesman J, Duncan D W, and Walden C C. 1966. **Iron oxidation by washed cell suspensions of the chemoautotroph, *Thiobacillus ferrooxidans***. *Canadian Journal of Microbiology* **12**: 25–33

Leathen W W, Kinsel N A, and Braley S A. 1966. ***Ferrobacillus ferrooxidans*: a chemosynthetic autotrophic bacterium**. *Journal of Bacteriology*. **72**: 700

Malouf E E and Prater J D. 1961. **Role of bacteria in the alteration of sulphide minerals**. *Journal of Metals* **13**: 353

Mukhopadhyay S N. 2010. ***Process Biotechnology Fundamentals***. New Delhi: Viva Books. 247 pp.

Parvathi M S, Suseela K, and Nandy S C. 1986. **Chromium and leather wastes.** *Leather Science* **33**: 303–309

Pearson D. 1970. ***The Chemical Analysis of Foods,*** 6th edn. London: J&A Churchill Ltd. 604 pp.

Perry J H (ed). 1963. ***Chemical Engineers' Handbook,*** 4th edn. New York: McGraw Hill Publishing Co. Ltd. 52 pp.

Poulopoulou V S, Katakis D, and Vrachnou E. 1998. ***A method for the removal of chromium from tanned leather wastes.*** *Journal of the Air and Waste Management Association* ***48***(9): 846–852

Razzel W E and Trussell P C. 1963a. **Microbiological leaching of metallic sulphides.** *Applied Microbiology* **11**: 105–110

Razzel W E and Trussell P C. 1963b. **Isolation and properties of an iron- oxidizing Thiobacillus.** *Journal of Bacteriology* **85**: 595–603

Rugar D and Hansma P K. 1990. ***Atomic Force Microscope.*** *Physics Today* **43**: 23–30.

Saha N, Kresalkova M, Saha L, Kolomaznik K. 2003. ***Effect of anaerobic digestion on chrome sludge, a by-product of solid tanned leather waste.*** *Journal of the American Leather Chemists Association* **98**: 255

Saha N and Mukhopadhyay S N. 1998. ***Advances in Biotechnology.*** New Delhi: Education Publishers and Distributor

Schnaitman C and Lundgren D G. 1965. **Organic compounds in the spent medium of *Ferrobacillus ferrooxidans*.** *Canadian Journal of Microbiology* **11**: 23–27

Silverman M P and Ehrlich H L. 1964. **Microbial formation and degradation of minerals.** *Advanced Applied Microbiology* **6**: 153–206

Silverman M P, Rogoff M H, and Wenden I. 1961. **Bacterial oxidation of pyritic materials in coal.** *Applied Microbiology* **9**: 491–496

Simeonova L S and Dalev P G. 1996. ***Utilization of a leather industry waste.*** *Waste Management* ***16***(8): 765–769

Smiley K L and Bolen P L. 1982. **Demonstration of D-xylose reductase and Dxylitol dehydrogenase in *Pachysolen tannophilus*.** *Biotechnology Letters* **4**: 607–610

Taylor M M, Diefendorf E J, Thompson C J, Brown E M, Marmer W N, Cabeza L F. 1997. ***Extraction of value added by-products from the treatment of chromium containing collagenous leather industry waste.*** *Journal of the Society of Leather Technologists and Chemists* **81**: 5–13

Taylor M M, Cabeza L F, Marmer W N, Brown E M, Kolomaznik K. 1998. **Functional properties of hydrolysis products from collagen.** *Journal of the American Leather Chemists Association* **93**: 40–50

Taylor M M, Diefendorf E J, Brown E M, Marmer W N. 1992. **Characterisation of products isolated by enzyme treatment of chromium-containing leather waste.** *Journal of the American Leather Chemists Association* **87**: 380–388

Taylor M M, Diefendorf E J, Fogila T A, Bailey D G, Feairheller S H. 1989. **Enzymatic treatment of offal from fleshing machines.** *Journal of the American Leather Chemists Association* **84**: 71–78

Taylor M M, Diefendorf E J, Thompson C J, Brown E M, Marmer W N, Cabeza L F. 1998. **Extraction of value-added by-products from the treatment of chromium containing collagenous leather industry waste**. *Journal of the Society of Leather Technologists and Chemists* **81**: 5–13

Temple K L and Delchamps W E. 1953. **Autotrophic bacteria and the formation of acid in bituminous coal mines**. *Applied Microbiology* **1**: 255–258

Torma A E, Walden C C, and Branion R M R. 1970. **Microbiological leaching of a zinc sulfide concentrate**. *Biotech and Bioengineering* **12**(4): 501–517

Wickerham and Andreasen. 1942. ***The lyophil process: its use in the preservation of yeasts.*** *Wallerstein Labs Commun.* **5:** 165–169

Wool R P. 1995. **The Science and engineering of polymer composite degradation.** In *Degradable Polymers,* edited by G Scott, D Giled. London: Chapman and Hall

Zimmerly S R, Wilson D G, and Prater J D. 1959. US Patent, 2. 829, 964

BIBLIOGRAPHY

Badger W L and Banchero J T. 1955. ***Introduction to Chemical Engineering.*** New York: McGraw Hill Publishing Co. Ltd

Bailey J E, Hjorts M, Lee S B, Frienc S. 1983. **Kinetics of product formation and plasmid segregation in recombinant microbial populations**. *Annals of the New York Academy of Sciences* **413**: 71–87

Bird R B, Stewart W F, and Lightfoot E N. 1960. ***Transport Phenomena.*** New York: John Wiley and Sons

Brown C G. 1959. ***Unit Operations***. Mumbai: Asia Publishing House

Calderbank P H. 1967. **Mass transfer in fermentation equipment**. In *Biochemical and Biological Engineering Science*, vol. 1, p. 102, edited by N Blakebrought. New York: Academic Press Inc.

Danckwerts P V. 1970. ***Gas–liquid Reactions.*** New York: McGraw Hill Publishing Co. Ltd

DeBoer J H. 1953. ***The Dynamical Character of Adsorption***. London: Oxford University Press

Dodge B F. 1968. ***Chemical Engineering Thermodynamics***. New York: McGraw Hill Publishing Co. Ltd

Fabbri A. 1995. **The role of degradable polymers in agricultural systems**. In *Degardable Polymers,* edited by G Scott, D Giled. London: Chapman and Hall.

Green D W and Perry R H. 2008. ***Perry's Chemical Engineers' Handbook,*** 8th edn. New York: Tata McGraw Hill Publishing Co. Ltd

Grinnin E L. 1981. **L-asparginase: a case study of an *E. coli* fermentation product**. In *Insulins: growth hormone and recombinant DNA technology,* edited by J L Gueriginan. New York: Raver Press

Guha S K, Mukherjee B, Kapil B, Seth P. 2003. **RISUG as antimicrobial for reproduction health**. Presentation made at the *Seminar on Male Health and Annual Conference of the Society of Andrology,* New Delhi, India, 13–15 December 2003.

Heideman E. 1993. ***Fundamentals of Leather Manufacture***: Darmstadt: Eduard Roether K G.

Jeffries T W. 1982. ***A comparison of Candida tropicalis and Pachysoten tannophilus for conversion of xylose to ethanol.*** *Biotechnology and Bioengineering Symposium* **12**: 103-110

Lal R, John S A, and Arnsdorf M F. 1992. **Atomic force microscopy and force dissection of cardiac gap junctions**. *Journal of the Federation of American Societies for Experimental Biology* **6**: A409

Lord P G and Wheals A E. 1980. **Asymmetric division of *S. cerevisiae*.** *Journal of Bacteriology* **142**: 808–818

McCabe W L, Smith J C, and Harriott P. 1985. ***Unit Operations of Chemical Engineering***, 4th edn. New Delhi: Tata McGraw Hill Publishing Co. Ltd

Mukhopoadyay S N. 2007. ***Experimental Process Biotechnology Protocols***. New Delhi: Viva Books

Mukhopadyay S N and Saha V. 2005. ***Advanced Process Biotechnology***. New Delhi: Viva Books

Rettenberger G. 2008. Presentation abstracts. Paper presented at the *23rd international conference on Solid Waste Technology and Management (ICSWTM)*. Philadelphia, Pennsylvania, 30 March–2 April 2008

Sharam S. 2003. ***Biophysical, Biochemical, and Morphological Correlations of Action of the Contraceptive – RISUG^{TM} on E. coli***. New Delhi: IIT Delhi [Doctoral thesis submitted to BMRC, IIT Delhi]

Sharma S, Sen P, Mukhopadhyay S N, Guha S K. 2003. **Microbicidal male contraceptive-RISUG induced morphostructural damage in *E. coli***. *Colloids and Surfaces B: Biointerfaces* **32**(1): 43–50

Silverman M P and Lundgren D G. 1959. **Studies on the chemoautotrophic iron bacterium *Ferrobacillus ferrooxidans* II**. *Journal of Bacteriology* **78**: 326–331

Taylor M M, Diefendorf E J, Thompson C J, Brown E M, Marmer W N. 1993. **Effect of process variable on ash content of gelable and hydrolyzed protein products isolated from the treatment of chromium leather waste**. *Journal of the American Leather Chemists Association* **88**: 358

Taylor M M, Diefendorf E J, Brown E M, Marmer W N. 1994. **Effect of various alkalinity-inducing agents on chemical and physical properties of protein products isolated form chromium-containing leather waste.** *Journal of the American Leather Chemists Association* **89**: 221–228

Treybal R E. 1980. ***Mass Transfer Operations***, 3rd edn. New Delhi: Tata McGraw Hill

Section II

Bioprocess Engineering System: tutorials

5

Upstream Process Biotechnology: Solved Numerical Problems

5.1 PURE CULTURE MICROBIAL BIOTECHNOLOGY

5.1.1 Oxidation/Oxygen Related Questions

Question 1 NAD+/NADH is an important redox system in biocells. The measurement of the turnover of NAD+ and NADH in the biosynthesis of cell material is complex. Why?

Answer Measurement of the turnover of NAD+ and NADH in the biosynthesis of cell material is complex due to the rapid turnover of the ADP/ATP pool and the difficulty in rapid access to such pools.

Question 2 How can the net requirement or production of NADH be obtained, overcoming the difficulty in Question 1?

Answer Any net production of NADH would result in the uptake of oxygen without CO_2 production, while any net requirement for NADH would, in case of respiration, result in CO_2 production without O_2 uptake. Consequently, the redox state of the cell compared to the substrate reflects the net production or requirement of NADH. These effects may be qualitatively related to the overall composition of the cell as it reflects the redox state of the cell.

Question 3 How can one check the difference in oxidation states between the cell and substrate?

Answer The difference in oxidation states between the cell and substrate may be calculated by comparing with an arbitrary reference state to which both are theoretically oxidized. Either oxygen balance or electron balance can be used for comparison.

Question 4 Electron balance is identical to oxygen balance under aerobic/oxic conditions of cell cultivation. Why is electron balance and oxygen balance not identical under anaerobic/anoxic conditions of cell cultivation?

Answer Under aerobic/oxic conditions, substrate oxidation relies on oxygen as the terminal electron acceptor so that the electron balance becomes equivalent to oxygen balance. However, the overall redox balance is dependent only on the redox state of carbon in the substrate and in cells and not on any particular biochemical pathway of an organism. Thus, under anaerobic/anoxic conditions, since the cells do not rely on oxygen as the terminal electron acceptor, the electron balance and oxygen balance may not be equal.

Question 5 Why is the partial pressure of oxygen (pO_2) of air in lungs only about 100 mm Hg, while that of air in an open space is about 150 mm Hg?

Answer The pO_2 of air in an open space is much more (by about 50 mm Hg) than that of air present in lungs mainly due to its distribution in the alveoli.

Question 6 Oxygen plays a key role as a terminal electron acceptor in bioreaction systems where glucose is completely oxidized to CO_2 and H_2O. During these reactions, the released free energy is conserved in adenosine triphosphate (ATP) synthesized in oxidative phosphorylation reactions coupled to electron transport. The efficiency of oxidative phosphorylation is expressed either by P/O ratios or by measuring the time profile of oxygen concentration in the assay medium and adding the substrate during the course of the experiment. After glucose ingestion, the summary of bioreaction systems and efficiency of oxidative phosphorylation is as shown in Figures 5.1(a), 5.1(b), and 5.1(c).

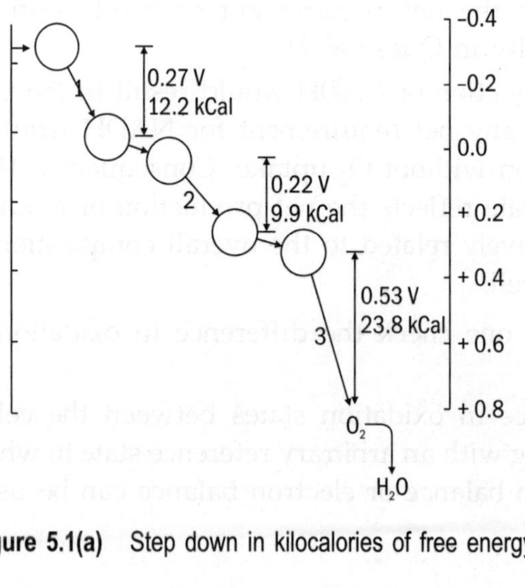

Figure 5.1(a) Step down in kilocalories of free energy

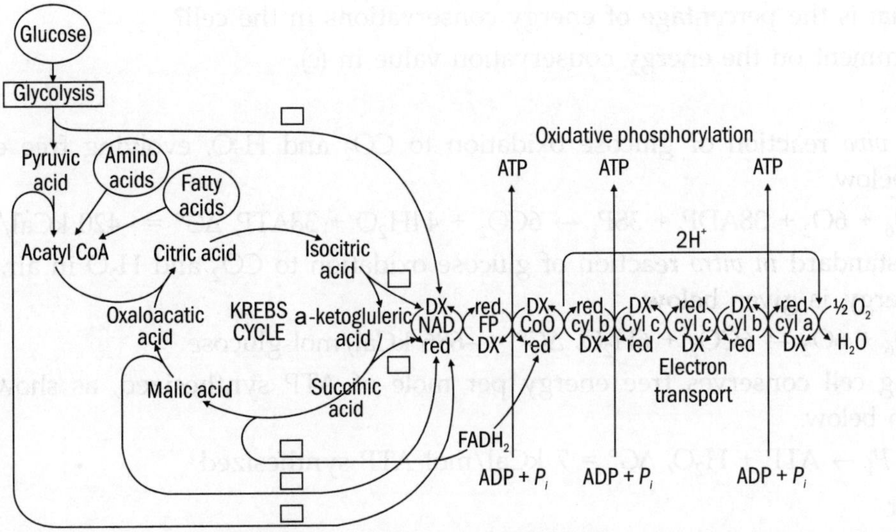

Figure 5.1(b) Summary of the reaction system

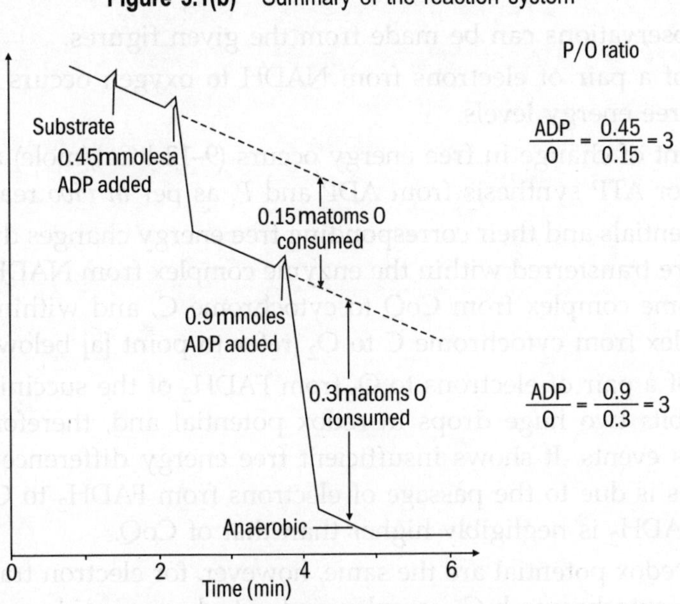

Figure 5.1(c) Oxidative phosphorylation efficiency

List the observations that can be made from these figures on bioreactions. Using the data given below, estimate the following.

(a) How much free energy is released as heat in a typical aerobic cell during the complete oxidation of 1 mol of glucose?

(b) How much free energy is conserved in the cell?

(c) What is the percentage of energy conservations in the cell?

(d) Comment on the energy conservation value in (c).

Data

The *in vivo* reaction of glucose oxidation to CO_2 and H_2O, evolving free energy, is given below.

$C_6H_{12}O_6 + 6O_2 + 38ADP + 38P_i \rightarrow 6CO_2 + 44H_2O + 38ATP, \Delta G° = -420$ kCal/mol ATP

The standard *in vitro* reaction of glucose oxidation to CO_2 and H_2O in air, evolving free energy, is given below.

$C_6H_{12}O_6 + 6O_2 \rightarrow 6CO_2 + 6H_2O, \Delta G° = -686$ kCal/mol glucose

A living cell conserves free energy per mole of ATP synthesized, as shown in the reaction below.

$ADP + P_i \rightarrow ATP + H_2O, \Delta G° = 7$ kCal/mol ATP synthesized

Answer

Observations

The following observations can be made from the given figures.

- The passage of a pair of electrons from NADH to oxygen occurs along a gradient of declining free energy levels.
- A large amount of change in free energy occurs (9–12 kCal/mole) at three different sites to sponsor ATP synthesis from ADP and P_i as per *in vivo* reaction.
- The redox potentials and their corresponding free energy changes drop in huge steps as electrons are transferred within the enzyme complex from NADH to CoQ, within b-C_1 cytochrome complex from CoQ to cytochrome C, and within the cytochrome oxidase complex from cytochrome C to O_2 (refer to point [a] below).
- The passage of a pair of electrons to O_2 from $FADH_2$ of the succinic dehydrogenase complex exhibits two huge drops in redox potential and, therefore, sponsors two ATP synthesis events. It shows insufficient free energy difference to sponsor ATP synthesis. This is due to the passage of electrons from $FADH_2$ to CoQ as the redox potential of $FADH_2$ is negligibly higher than that of CoQ.
- The drops in redox potential are the same, however, for electron transfers from CoQ to O_2 through cytochrome b-C_1 complex and cytochrome oxidase complex (c, a–a_3). This occurs when CoQ has accepted a pair of electrons from NADH or $FADH_2$.

(a) When glucose is oxidized to CO_2 and H_2O in air under *in vitro* conditions, the given data show free energy (*G*) difference, which is dissipated as heat.

$$G_{Glucose} - G_{Products} = -686 \text{ kCal/mol}$$

(b) A living cell conserves free energy of per mole of ATP synthesized. The free energy conserved is 7 kCal/mol and the ATP synthesized is 38 mol. Therefore, free energy conserved in a cell is 266 (= 38 × 7) kCal.

(c) Out of the *in vitro* free energy of 686 kCal/mol, 420 kCal/mol ATP is released *in vivo.* Therefore, *in vivo* free energy conserved in an aerobic cell is 686 – 420 = 266 kCal/mol glucose. Thus, the efficiency of energy conservation in cell is (266/686) × 100 = 39%.

(d) Energy conservation efficiency of 39% in a cell is quite high compared to a typical combustion engine having 10%–20% efficiency.

5.1.2 Engineering Biology in Cacterial Biotechnology

Question 7 In a host cell of *Escherichia coli,* the recombinant plasmid PBR322T leu, which contains β isopropyl malate dehydrogenase gene of *Thermus thermophilus* HB27, was cloned. For assessment of the stability of the plasmid in host cells, the following scheme (Equations 5.1 and 5.2) was proposed for a medium containing substrate concentration *S* mg/ml and inoculum containing different concentrations of plasmid-carrying cells, *P* (cells/ml), and cells without plasmid, *N* (cells/ml).

$$aS + P \rightarrow (2 - p)\ P + N;\ (N = pP) \tag{5.1}$$

$$bS + N \rightarrow 2\ N \tag{5.2}$$

The terms "*a*" and "*b*" in the scheme are coefficients having unit of cells/mg substrate, S is the substrate, *P* is the product, *N* is the plasmid number, *p* is the probability. What assumptions are required to formulate the above scheme and why? From this scheme, express the following.

(a) Concentration of plasmid-carrying cells.

(b) Concentration of cells without plasmid in relation to plasmid-carrying cells.

(c) Fraction of plasmid-carrying cells in total population.

Also arrive at the expressions from (c) in (1) a batch culture for nth generation and (2) chemostat. Explain its implications for high and small *n* values and state the considerations required.

Answer In the assessment of stability of plasmid in proliferating cells, the instability of plasmid is defined as the loss and/or deletion of phenotypic expressions. In such assessment, the following assumptions are required.

- Host cells grow exponentially.
- Cells are deprived of plasmid with a probability "*p*" per division of host cells.
- No consideration is needed on self-transmissible plasmid.

In such a case, the growth rate of both cells are given by Equations 5.3 and 5.4

$$\frac{dP}{dt} = (1 - p)\ \mu_1 P \tag{5.3}$$

$$\text{and } \frac{dP}{dt} = p\mu_1 P + \mu_2 N \tag{5.4}$$

where μ_1 is the specific growth rate (per hour) of cells with plasmid, μ_2 is the specific growth rate (per hour) of cells without plasmid, and t is the cultivation time (in hours).

Using the following initial conditions,

$$\left.\begin{array}{l} P = P_0 \\ N = N_0 \\ \text{at } t = 0 \end{array}\right\} \tag{5.5}$$

From Equations 5.3 and 5.5, we get

$$\int_{P_0}^{P} \frac{dP}{P} = \mu_1 (1 - p) \int_0^t dt$$

$$\text{or } \ln \frac{P}{P_0} = \mu_1(1 - p)t$$

$$\text{or } P = P_0 e^{(1-p)\mu_1 t} \tag{5.6}$$

From Equations 5.4, 5.5, and 5.6, we get

$$\frac{dN}{dt} + \mu_2 N = p\mu_1 P$$

$$\text{or } \frac{dN}{dt} + \mu_2 N = p\mu_1 P_0 e^{(1-p)\mu_1 t} \tag{5.7}$$

Solving Equation 5.7, we get

$$N = N_0 e^{\mu_1 t} + \frac{p\mu_1 P_0}{(1 - p)\mu_1 - \mu_2} \{e^{(1-p)\mu_1 t} - e^{\mu_2 t}\} \tag{5.8}$$

Fraction "F" of cells carrying plasmid in the total population is given in Equation 5.9.

$$F = \frac{P}{P + N} \tag{5.9}$$

For batch culture, assuming $N_0 = 0$, Equation 5.10 can be derived from Equations 5.6 and 5.8.

$$F = \frac{e^{(1-p)\mu_1 t}}{e^{(1-p)\mu_1 t} + \dfrac{p\mu_1 P_0}{(1 - p)\mu_1 - \mu_2} \{e^{(1-P)\mu_1 t} - e^{\mu_2 t}\}} \tag{5.10}$$

Then the number of generations, n, is as shown in Equation 5.11.

$$n = \frac{\mu_1 t}{\ln 2} \tag{5.11}$$

Considering $\alpha = \dfrac{\text{Specific growth rate of plasmid absent cells}}{\text{Specific growth rate of plasmid containing cells}} = \dfrac{\mu_2}{\mu_1}$ (5.12)

From Equations 5.10, 5.11, and 5.12 and at the nth generation,

$$F_n = \frac{1 - \alpha - p}{1 - \alpha - p\, 2^{n(\alpha+p-1)}} \tag{5.13}$$

where F_n is the fraction on the nth generation.

The following cases can be arrived at from Equation 5.13.

Case I. F_n versus α with P as parameter

For high n (say, $n = 25$), α usually ranges between 1 and 2. For $n = 25$, the plot of F_{25} versus α shows that as α increases, F_{25} decreases drastically.

Case II. F_n versus n with α as parameter

When n is small, F_n values for each α are nearly similar. As n and α increase, differences in F_n values increase. Thus,

$$\lim_{n\to\infty} F_n = 1 - p, \text{ when } \alpha = 0$$

$$\lim_{n\to\infty} F_n = 1 - p, \text{ provided } \alpha = 1, p \neq 0$$

Question 8 A syntrophism experiment was carried out using *Serratia marcescens* (wild W225) and its pigment mutants *pig* M, W592, M_2W622, and M_3W623 with a cell titre of 1×10^9/ml on PG agar plates. Control plates were prepared in order to recognize pigment formation under non-syntrophic conditions. Each PG agar plate with W225, M_1W592, M_2W622, and M_3W623 was labelled, and two parallel streaks were made on the plates when inoculated by a loop. Four bacterial suspensions were streaked 1 mm apart with a sterile inoculating loop onto three previously labelled PG agar plates (Table 5.1).

The observations made on the behaviour of the three *pig* mutants after two days are shown in Table 5.2.

Table 5.1 Experimental data on syntrophism test

Streak	PG plates		
	W225	M_1W592	M_2W622
a	225	592	225
b	622	622	592
c	623	623	—

Table 5.2 Experimental data on *pig* mutants

Acceptor	Colour of mutant on control plates	Inducer[a]			
		225	592	622	623
W225	Red violet	†	–	–	–
W592	Colourless (white)	±	†	–	–
W622	Yellow orange	–	+	†	+
W623	Grey–white	–	++	++	†

— indicates no pigment induction, + or ++ indicates weak or heavy pigment induction.

[a] An "inducer" strain releases a precursor into the medium and an acceptor strain can make use of it. A strain may be an inducer and an acceptor simultaneously.

(a) From the results, write the mutant on the following reaction step corresponding to the location of the genetic block in that mutant.

(b) Which of the two mutants, W622 and W623, can use a volatile pigment precursor?

(c) Explain the syntrophic behaviour of wild type W225.

(d) One of the pigment mutants does not form white or pale pink colonies but rather orange–yellow ones. Explain this observation.

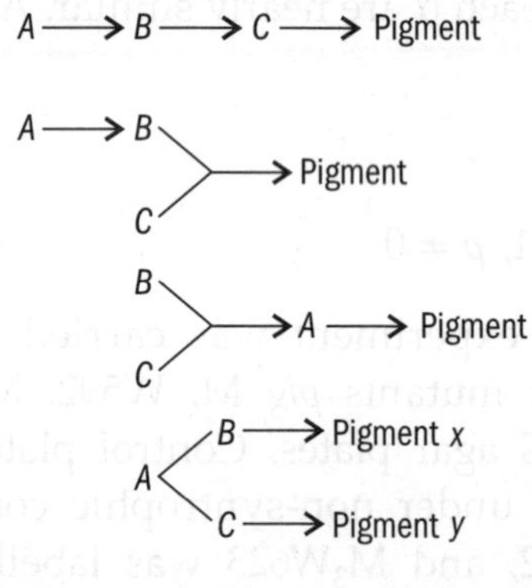

Reaction steps where *A*, *B*, and *C* are hypothetical pigment precursors.

Answer

(a) The schemes of the biosynthesis of the pigment that could apply are shown below.

The mutants W622 and W623 (in the formal schemes) are interchangeable. The correct scheme is the first one (Scheme A).

(b) Precursor B (MAP) is volatile as W623 is fed by W592 in the double plate experiment. Precursor C is not volatile. This scheme shows that mutant Y is by mutant X but not vice versa.

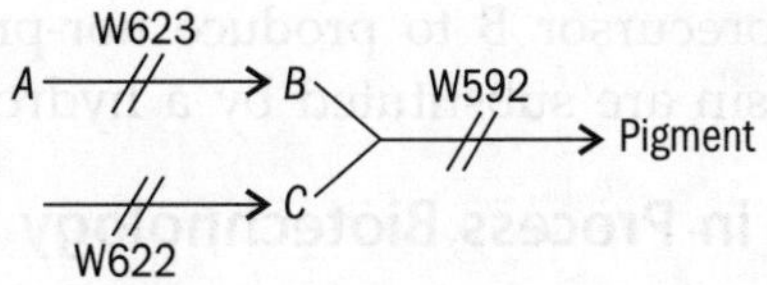

Scheme A

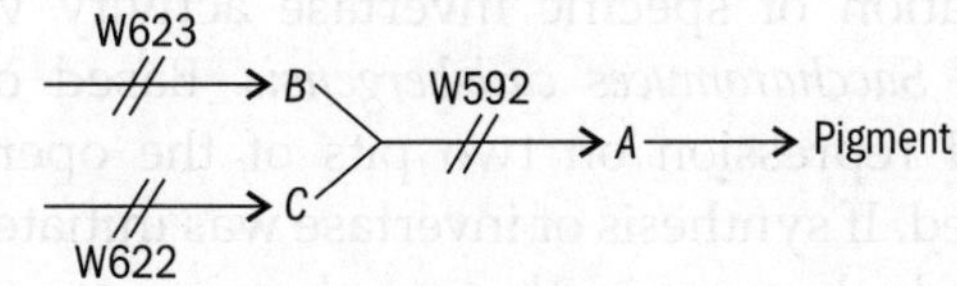

Scheme B

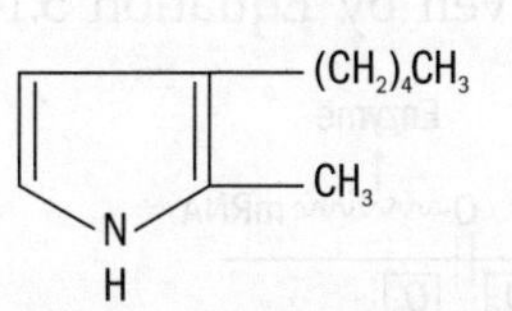

Precursor B: 2-methyl-3 amylpyrrol (MAP)

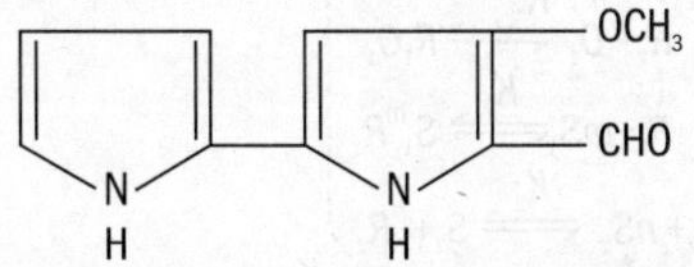

Precursor C: 4-methyl–2, 2–bipyrrol–5-carboxyl delyde (MBC)

(c) Wild type bacteria (W225) do not feed the *pig* mutant at all or do so only weakly, because the synthesis of prodigiosin might be regulated in such a manner that precursors B and C (at least) are not accumulated.

(d) A precursor of prodigiosin, which is still synthesized by the mutant W622, is yellow-orange or if it is colourless, it may be converted into a yellowish-orange pigment by a shunt reaction. In this case, the yellow-orange appearance of the inoculation streak of W622 (control plate) is due to the latter. The precursor that accumulates

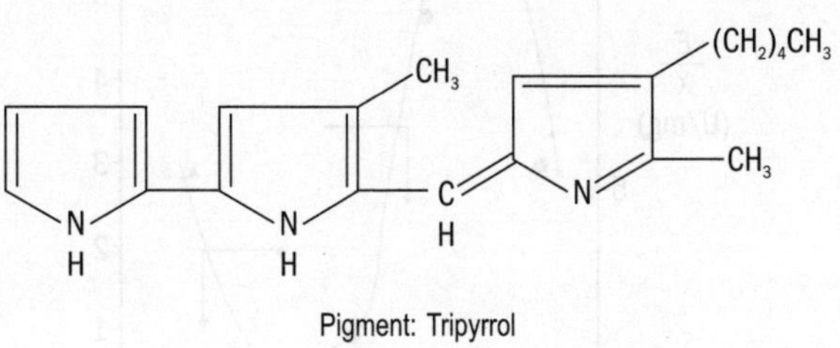

Pigment: Tripyrrol

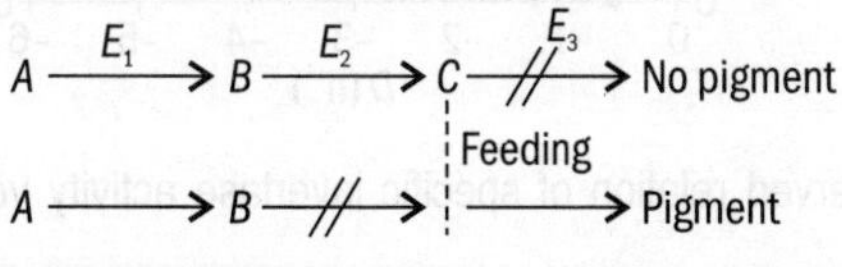

in strain W622 reacts with precursor B to produce nor-prodigiosin in which four methoxy groups of prodigiosin are substituted by a hydroxyl group.

5.1.3 Catabolic Repression in Process Biotechnology

Question 9 Figure 5.2 shows a catabolic repression molecular scheme and Figure 5.3 shows the observed relation of specific invertase activity versus dilution rate in a chemostat cultivation of *Saccharomyces carlsbergensis.* Based on the observation, dual control of induction and repression on two pits of the operator genes for invertase bio`synthesis was proposed. If synthesis of invertase was initiated only when functions of both O_1 and O_2 genes worked conveniently on a chromosome, the chemical equilibrium of the reaction shown in Figure 5.2 is given by Equation 5.14.

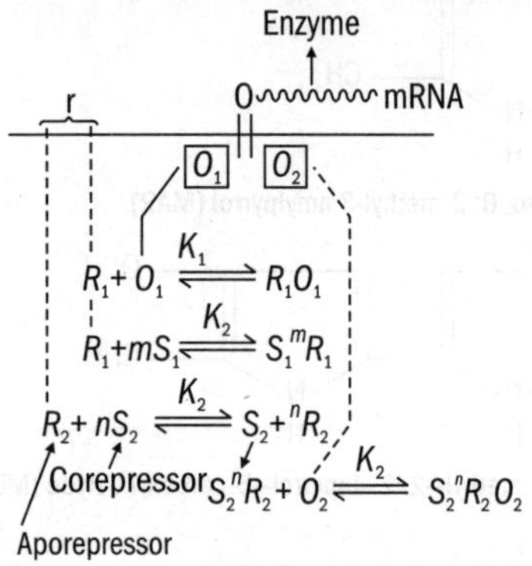

Figure 5.2 Catabolic repression molecule

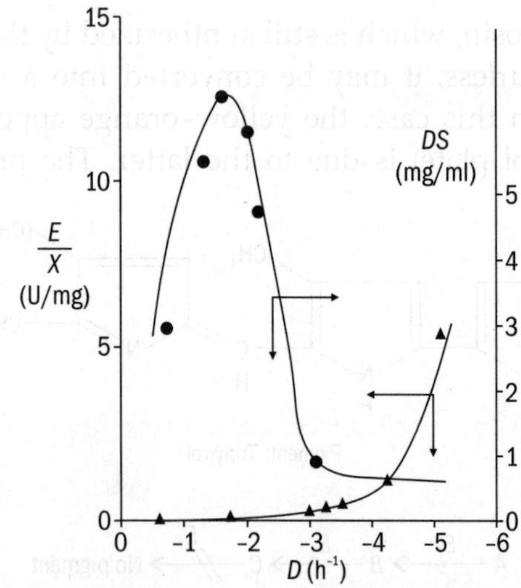

Figure 5.3 Observed relation of specific invertase activity versus dilution rate

$$Q_1 = \frac{[O_1]}{[O_1]_t} = \frac{1 + K_1\ S_1^m}{1 + K_1\ [R_1]_t + K_1\ S_1^m} \tag{5.14}$$

and

$$Q_2 = \frac{[O_2]}{[O_2]_t} = \frac{1 + K_2\ S_2^n}{1 + (1 + K_2\ [R_2]_t)\ K_2\ S_2^n} \tag{5.15}$$

where Q_1 and Q_2 are ratios, $[O_1]$ is the number of induced gene, $[O_2]$ is the number of non-repressed gene, and $[O_1]_t$ and $[O_2]_t$ are the total numbers of each operator gene.

The fraction of invertase specific activity (Q) is the actual ratio (E/X) value to the theoretical maximum value with full induction and no repression. Determine the values of Q_1 and Q_2 assuming values at lower dilution rate of m, n, K_1, K_2, k_1, and k_2, which are constants for the stoichiometry and adsorption equilibrium in the binding processes among molecules R_1, R_2, S_1, S_2, O_1, O_2, and $S_2^nR_2$ (assuming the value of S_1 same as glucose).

Answer $$Q_1 = \frac{1 + k_1\ S_1^m}{1 + k_1\ [R_1]_t + k_1\ S_1^m} \tag{5.16}$$

or $Q_1 + Q_1[k_1[R_1]_t + K_1S_1^m] = 1 + K_1S_1^m$

or $Q_1[k_1[R_1]_t + K_1S_1^m] = (1 - Q_1) + K_1S_1^m$

or $Q_1k_1[R_1]_t = (1 - Q_1) + (1 - Q_1)\ K_1S_1^m$

or $$\frac{Q_1}{1 - Q_1} = \frac{1 + K_1\ S_1^m}{(1 - Q_1)k_1\ [R_1]_t}$$

or $$\frac{Q_1}{1 - Q_1} = \frac{1}{(1 - Q_1)k_1\ [R_1]_t} + \frac{K_1}{(1 - Q_1)k_1\ [R_1]_t}\ S_1^m$$

For $m = 1$,

$$\frac{Q_1}{1 - Q_1} = \frac{1}{(1 - Q_1)k_1\ [R_1]_t} + \frac{k_1}{(1 - Q_1)k_1\ [R_1]_t}\ S_1$$

If $m = 1$, $K_1 = 470$, and at low dilution rates (Figure 5.3), using Equation 5.16, we get $k_1[R_1]_t = 10$

Assuming maximum possible specific activity of invertase = 25 U/mg cell mass from the highest E/X obtained with mutants,

$$Q = \frac{(E/X)_{exp}}{(E/X)_{\text{max theor}}} = Q_1 Q_2$$

$$Q_1 = \frac{1 + 470S}{1 + 10 + 470S}$$

From Figure 5.3, for $D = 0.4 = \mu$,

$\mu_m = 0.51$ per hour
$K_S = 0.056$ mg glucose/ml

$$\text{So } D = \mu = \frac{\mu_m S}{K_S + S}$$

$$\text{or } 0.4 = \frac{0.51\ S}{0.056 + S}$$

Solving S, we get

$$S = \frac{0.0224}{0.11} = 0.02$$

$$\therefore\ Q_1 = \frac{1 + 470 \times 0.02}{1 + 10 + 470 \times 0.02} = \frac{10.4}{20.4} = 0.5$$

From Figure 5.2, for dilution rate $D = 0.4$,

$E/X = 0.025$ U/mg
$(E/X)_{max} = 25$ U/mg

$$\text{Thus, } Q_2 = \frac{\left[\frac{(E/X)}{(E/X]_{max}}\right]}{Q_1}$$

$$\text{or } Q_2 = \frac{0.025/25}{0.5} \qquad \text{[from Figure 5.4]}$$

$$\text{or } Q_2 = \frac{0.025}{12.5} = 2 \times 10^{-3}$$

5.1.4 Homogenic Cell–cell Interaction

Question 10 In the production of cellulase by *Trichoderma reesei* in a stirred tank fermenter (STF), the organism formed pellets. The pellet diameter varied with cultivation time after its initiation. The diameters of bioreactor and impeller were 15 cm and 5 cm,

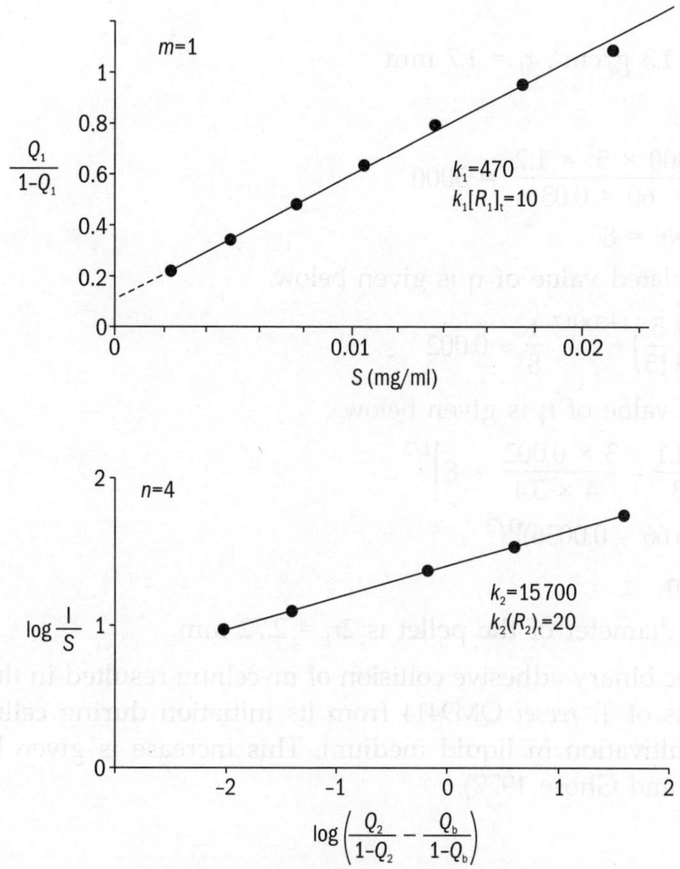

Figure 5.4 Estimates of the parameters

respectively, and the impeller rotation speed was 300 rpm. The pellet formation started on the first day of cultivation and the density of liquid was 1.1 g/cm^3. The average weight of a pellet, which was initially 0.03 g, increased to 0.09 g on the ninth day of cultivation when the density of liquid was 1.3 g/cm^3 and viscosity was 30 centipoise (cP). Considering the ratio of consistency index to liquid viscosity to be of Newtonian liquid, calculate the value of homogenization efficiency and diameter of the pellets of *T. reesei* on the ninth day of cultivation. During cultivation, the height of liquid in the fermenter was equal to the diameter of the fermenter. Assume α = 0.7, μ = 30, and cP = 0.03 g/cm/s.

Answer Given that N = 300 rpm, ρ_{bulk} = 1.2 g/cm^3

D_i = 5 cm, D_t = 15 cm, $t_2 - t_1 = 9 - 1 = 8$ days. Since the broth was Newtonian in character, we may assume $\left(\frac{K}{\mu}\right)^{n'}$ = 1 and K = 0.905. Also, at t = 1 (after pellet formation began), M_o = 0.03 g, ρ_o = 1.1 g/cm^3, r_o = 1.1 mm

At t = 9 days,

M_t = 0.09 g, ρ_t = 1.3 g/cm^3, r_t = 1.7 mm

Therefore,

$$N_{Re} = \frac{ND_i^2\rho}{\mu} = \frac{300 \times 5^2 \times 1.2}{60 \times 0.03} = 5000$$

For N_{Re} = 5000, Np = 8.

Thus, the calculated value of η is given below.

$$\eta = \left(\frac{3.14}{4 \times 0.905}\right)^3 \left(\frac{5}{15}\right)^{1+3\times0.7} \frac{1}{8} \approx 0.002$$

The calculated value of r_t is given below.

$$r_t = \left[\frac{0.09}{1.3} \times \frac{1 \times 1.1}{0.03} - \frac{3 \times 0.002}{4 \times 3.4} \times 8\right]^{1/3}$$

or r_t = [0.0692 (36.66 – 0.0036)]$^{1/3}$

or r_t = 1.3638 mm

Therefore, the diameter of the pellet is $2r_t$ = 2.72 mm.

Question 11 The binary adhesive collision of mycelium resulted in the increase in the number of pellets of *T. reesei* QM9414 from its initiation during cellulase production by submerged cultivation in liquid medium. This increase is given by Equation 5.17 (Mukhopadhyay and Ghose 1977).

$$\frac{dN}{dt} = \eta N^2 \tag{5.17}$$

where N is the number of pellets and η is the factor indicating the degree of homogeneity in the broth. Based on Equation 5.17, determine the number of pellets at any time from its initiation.

The value of η relating to fermenter parameters is given by Equation 5.18.

$$\eta = \left(\frac{\pi}{4K}\right)^3 \left(\frac{D_i}{D_t}\right)^{1+3\alpha} \frac{1}{N_p} \tag{5.18}$$

where K is consistency index, D_i is impeller diameter, D_t is fermenter diameter, N_p is power number, and α is an exponent depending on broth rheology.

Derive a relation to determine pellet radius at any time in terms of the mass and density of pellets and η.

Answer The pellet formation rate is given by Equation 5.19.

$$\frac{dN}{dt} = \eta N^2 \tag{5.19}$$

Integrating Equation 5.19 between limits $N = N_0$ at $t = t_1$ and $N = N_t$ at $t = t_2$, we have

$$\left[\frac{dN}{N^2}\right]_{N_0}^{N_t} = \eta \int_{t_1}^{t_2} dt$$

or $\left[-\frac{1}{N}\right]_{N_0}^{N_t} = \eta\ (t_2 - t_1)$

or $\left[-\frac{1}{N_t} + \frac{1}{N_0}\right] = \eta\ (t_2 - t_1)$

Thus,

$$N_t = N_0\ [1 - N_0\eta\ (t_2 - t_1)]^{-1} \tag{5.20}$$

According to Pirt,

$$N_t = \frac{M_t}{\frac{4}{3}\pi r_t^3\ \rho_t},\ N_o = \frac{M_0}{\frac{4}{3}\pi r_0^3\ \rho_o}$$

or $\dfrac{M_t}{\frac{4}{3}\pi r_t^3\ \rho_t} = \dfrac{M_0/\frac{4}{3}\pi\ r_0^3\ \rho_0}{1 - \dfrac{M_0\eta}{\frac{4}{3}\pi r_0^3\ \rho_0}\ (t_2 - t_1)}$

or $\dfrac{M_t}{M_0^{-1}}\ r_t^3\ \rho_t = \dfrac{1}{\frac{4}{3}\pi r_0^3\ \rho_0 - M_0\eta\ (t_2 - t_1)}$

or $r_t^3 = M_t\ \dfrac{\left[\frac{4}{3}\pi r_0^3\ \rho_0 - M_0\eta\ (t_2 - t_1)\right]}{M_0\ \rho_t}$

or $r_t^3 = \dfrac{M_t}{M_0}\dfrac{4}{3}\pi\ r_0^3\ \dfrac{\rho_0}{\rho_t} - \dfrac{M_t\ M_0\eta\ (t_2 - t_1)}{M_0\ \rho_t}$

or $r_t^3 = \dfrac{4}{3}\dfrac{M_t}{M_0}\dfrac{\rho_0}{\rho_t}\ \pi r_0^3 - \dfrac{M_t\eta}{\rho_t}\ (t_2 - t_1)$

$$\text{or } r_t = \sqrt[3]{\frac{4}{3}\frac{M_t}{M_0}\frac{\rho_0}{\rho_t}\ \pi\ r_0^3 - \frac{M_t}{\rho_t}\ \eta\ (t_2 - t_1)} \tag{5.21}$$

Equation 5.21 is the required relation.

Question 12 The dynamics of two species/components in a multi-interacting biosystem near steady state is given by the system of Equations 5.22 and 5.23 (Bailey and Ollis 1972).

$$\frac{d\lambda_1}{dt} = a_{11}\lambda_1 + a_{12}\lambda_2 \tag{5.22}$$

$$\frac{d\lambda_2}{dt} = a_{21}\lambda_1 + a_{22}\lambda_2 \tag{5.23}$$

(a) Show that this equation system has the following solution system given by Equations 5.24 and 5.25.

$$\lambda_1(t) = \frac{c_1 a_{12}}{\sigma_1 - a_{11}} e^{\sigma_1 (t-t_0)} + c_2\, e^{\sigma_2 (t-t_0)} \tag{5.24}$$

$$\lambda_2(t) = c_1\, e^{\sigma_1 (t-t_0)} + \frac{c_2 a_{21}}{\sigma_2 - a_{22}} e^{\sigma_2 (t-t_0)} \tag{5.25}$$

(b) Under what conditions does the system of Equations 5.24 and 5.25 show linearity?

Answer

(a) Given that $\frac{d\lambda_1}{dt} = a_{11}\lambda_1 + a_{12}\lambda_2$ and $\frac{d\lambda_2}{dt} = a_{21}\lambda_1 + a_{22}\lambda_2$. We have

$$\frac{d}{dt}\begin{bmatrix}\lambda_1\\ \lambda_2\end{bmatrix} = \begin{bmatrix}a_{11} & a_{12}\\ a_{21} & a_{22}\end{bmatrix}\begin{bmatrix}\lambda_1\\ \lambda_2\end{bmatrix}$$

Differentiating Equation 5.22, we get

$$\frac{d^2\lambda_1}{dt^2} = a_{11}\frac{d\lambda_1}{dt} + a_{12}\frac{d\lambda_2}{dt}$$

$$\text{or } \frac{d\lambda_2}{dt} = \frac{1}{a_{12}}\left[\frac{d^2\lambda_1}{dt^2} - a_{11}\frac{d\lambda_1}{dt}\right] \tag{5.26}$$

From 5.22, we get

$$\lambda_2 = \frac{1}{a_{12}}\left[\frac{d\lambda_1}{dt} - a_{11}\lambda_1\right] \tag{5.27}$$

Substituting Equations 5.26 and 5.27 in Equation 5.23, we get

$$\frac{1}{a_{12}}\left[\frac{d^2\lambda_1}{dt^2} - a_{11}\frac{d\lambda_1}{dt}\right] = a_{21}\lambda_1 + \frac{a_{22}}{a_{12}}\left(\frac{d\lambda_1}{dt} - a_{11}\lambda_1\right)$$

$$\text{or } \frac{d^2\lambda_1}{dt^2} - a_{11}\frac{d\lambda_1}{dt} = a_{12}a_{21}\,\lambda_1 + a_{22}\frac{d\lambda_1}{dt} - a_{22}\,a_{11}\lambda_1$$

$$\text{or } \frac{d^2\lambda_1}{dt^2} + (-a_{11} - a_{22})\frac{d\lambda_1}{dt} + (a_{22}\,a_{11} - a_{12}\,a_{21})\,\lambda_1 = 0 \tag{5.28}$$

Equation 5.28 changes to the form

$\sigma^2 + (-a_{11} - a_{22})\sigma + (a_{22}a_{11} - a_{12}a_{21}) = 0$

or $\sigma^2 - (a_{11} + a_{22})\sigma + (a_{11}a_{22} - a_{12}a_{21}) = 0$

$$\text{or, } \sigma^2 - (\text{tr } A)\sigma + \det A = 0 \tag{5.29}$$

where $\text{tr}A = a_{11} + a_{22}$; $\det A = a_{11}a_{22} - a_{12}a_{21}$ of the Jacobi matrix.

Equation 5.29 has roots σ_1 and σ_2, given as

$$\sigma_{1,2} = \frac{\text{tr}A}{2} \pm \frac{1}{2}\sqrt{(\text{tr}A)^2 - 4\det A}$$

For two distinct real roots of the characteristic equation,

$$\lambda_1 = A\,e^{\sigma_1 t} + Be^{\sigma_2 t}$$

$$\frac{d\lambda_1}{dt} = A\,\sigma_1\,e^{\sigma_1 t} + B\,\sigma_2 e^{\sigma_2 t}$$

$$\frac{d^2\lambda_1}{dt^2} = A\,\sigma_1^2\,e^{\sigma_1 t} + B\,\sigma_2^2\,e^{\sigma_2 t}$$

From 5.28, we get

$$A\,\sigma_1^2\,e^{\sigma_1 t} + B\,\sigma_2^2\,e^{\sigma_2 t} - (a_{11} + a_{22})(A\sigma_1\,e^{\sigma_1 t} + B\sigma_2 e^{\sigma_2 t}) + (a_{22}\,a_{11} - a_{12}a_{21})(Ae^{\sigma_1 t} + Be^{\sigma_2 t}) = 0$$

In order to evaluate the coefficients A and B, the initial condition for λ_1 is required.

At $t = t_0$, $\lambda_1 = \lambda_{10}$, $\dfrac{d\lambda_1}{dt} = 0$

$$\lambda_{10} = Ae^{\sigma_1 t_0} + Be^{\sigma_2 t_0} \tag{5.30}$$

Since $t = 0$, $\lambda_{10} = A + B$. Therefore, $\lambda_1 = Ae^{\sigma_1(t-t_0)} + Be^{\sigma_2(t-t_0)}$

The other condition required to solve A and B is given below.

$$\frac{d\lambda_1}{dt} = 0 \text{ at } t = t_0$$

Therefore,

$$A\sigma_1 e^{\sigma_1 t_0} + B\sigma_2 e^{\sigma_2 t_0} = 0$$

$$\text{or } \frac{A}{B} = -\frac{\sigma_2\,e^{\sigma_2 t_0}}{\sigma_1\,e^{\sigma_1 t_0}}$$

So from Equation 5.30,

$$\lambda_{10} = Ae^{\sigma_1 t_0} - \frac{\sigma_1\,e^{\sigma_1 t_0}}{\sigma_2\,e^{\sigma_2 t_0}}Ae^{\sigma_2 t_0}$$

Thus,

$$A = \frac{\sigma_2 \lambda_{10}}{\sigma_2\, e^{\sigma_1 t_0} - \sigma_1\, e^{\sigma_1 t_0}} = \frac{\sigma_2}{(\sigma_2 - \sigma_1)} \cdot \frac{\lambda_{10}}{e^{\sigma_2 t_0}}$$

and

$$B = \frac{\sigma_1\, e^{\sigma_1 t_0}}{-\sigma_2\, e^{\sigma_2 t_0}} \cdot \frac{\sigma_2}{(\sigma_2 - \sigma_1)} \cdot \frac{\lambda_{10}}{e^{\sigma_1 t_0}} = -\frac{\sigma_1}{(\sigma_2 - \sigma_1)} \cdot \frac{\lambda_{10}}{e^{\sigma_2 t_0}}$$

Thus,

$$\lambda_1 = \frac{\sigma_2}{(\sigma_2 - \sigma_1)} \lambda_{10}\, e^{\sigma_1 (t - t_0)} - \frac{\sigma_1}{(\sigma_2 - \sigma_1)} \lambda_{10}\, e^{\sigma_2 (t - t_0)}$$

Similarly, for λ_2

$$\frac{d\lambda_2}{dt} = a_{21} \left[A e^{\sigma_1 (t - t_0)} + B e^{\sigma_2 (t - t_0)} \right] + a_{22}\, \lambda_2$$

The solution to the above is given below.

$$\lambda_2 = C e^{\sigma_1 (t - t_0)} + D e^{\sigma_2 (t - t_0)}$$

$$C\sigma_1\, e^{\sigma_1 (t - t_0)} + D\sigma_2\, e^{\sigma_2 (t - t_0)} = a_{21} \left[A e^{\sigma_1 (t - t_0)} + B e^{\sigma_2 (t - t_0)} \right] + a_{22} \left[C e^{\sigma_1 (t - t_0)} + D e^{\sigma_2} (t - t_0) \right]$$

$$C\sigma_1 = a_{21} A + a_{22} C$$

$$\text{or } C = \frac{a_{21}}{a_{22}\, \sigma_1} A$$

$$D\sigma_1 = a_{21} B + a_{22} D$$

$$\text{or } D = \frac{\sigma_{21}}{\sigma_2 - a_{22}} B$$

Therefore,

$$C = \frac{\sigma_2 \lambda_{10}}{(\sigma_2 - \sigma_1)} \cdot \frac{a_2}{(a_{22} - \sigma_1)}$$

$$D = \frac{a_{21}}{(\sigma_2 - a_{22})} \cdot \frac{\sigma_1 \lambda_{10}}{(\sigma_2 - \sigma_1)}$$

Therefore,

$$\lambda_1(t) = A e^{\sigma_1 (t - t_0)} + B e^{\sigma_2 (t - t_0)} \tag{5.31}$$

$$\lambda_2(t) = C e^{\sigma_1 (t - t_0)} + D e^{\sigma_2 (t - t_0)} \tag{5.32}$$

The coefficient B/D has the ratio $(\sigma_2 - a_{22})/a_{21}$, which is written in the solution given in Equation 5.33.

$$\frac{A}{C} = \frac{a_{22} - \sigma_1}{a_{21}} = \frac{a_{12}}{\sigma_1 - a_{11}} \tag{5.33}$$

$[a_{11}a_{12} = -\sigma_1^2 + a_{11}\sigma_1 + a_{22}\sigma_1 - a_{11}a_{22}$, which arises from the equation $\sigma_1^2 - (a_{11} + a_{22})\,\sigma_1 + (a_{22}a_{11} + a_{12}a_{21}) = 0]$

Hence, Equations 5.31 and 5.32 can be written as follows.

$$\lambda_1(t) = \frac{C_1 a_{12}}{(\sigma_1 - a_{11})} e^{\sigma_1(t-t_0)} + C_2\, e^{\sigma_2(t-t_0)} \tag{5.34}$$

$$\lambda_2(t) = C_1\, e^{\sigma_1(t-t_0)} + \frac{C_2 a_{21}}{(\sigma_1 - a_{22})} e^{\sigma_2(t-t_0)} \tag{5.35}$$

where σ_1 and σ_2 are roots of the characteristic Equations 5.34 and 5.35 and C_1 and C_2 are constants.

(b) At $t = t_o$, the solutions to the equation systems are as given below.

$$\sigma_1\ (t) = \frac{C_1 a_{12}}{\lambda_1 - a_{11}} \times 1 + C_2 \times 1 \tag{5.36}$$

$$\sigma_2\ (t) = C_1 \times 1 + \frac{C_2 a_{21}}{\lambda_2 - a_{22}} \tag{5.37}$$

Equations 5.36 and 5.37 form a linear system for the determination of coefficients C_1 and C_2 from the initial perturbations $\sigma_1(t_o)$ and $\sigma_2(t_o)$.

Multiplying Equation 5.22 by a_{21} and Equation 5.23 by a_{11}, we get

$$a_{21}\frac{d\lambda_1}{dt} = a_{11}a_{21}\lambda_1 + a_{12}a_{21}\lambda_2 \tag{5.38}$$

and

$$a_{11}\frac{d\lambda_2}{dt} = a_{11}a_{21}\lambda_1 + a_{22}a_{11}\lambda_2 \tag{5.39}$$

Subtracting Equation 5.39 from Equation 5.39, we get

$$a_{21}\frac{d\lambda_1}{dt} - a_{11}\frac{d\lambda_2}{dt} = (a_{12}\,a_{21} - a_{22}\,a_{11})\lambda_2$$

Dividing Equation 5.22 by Equation 5.23, we get

$$\frac{d\lambda_1/dt}{d\lambda_2/dt} = \frac{a_{11}\,\lambda_1 + a_{12}\,\lambda_2}{a_{21}\,\lambda_1 + a_{22}\,\lambda_2}$$

$$\text{or } \frac{d\lambda_1}{d\lambda_2} = \frac{a_{11}\,\lambda_1 + a_{12}\,\lambda_z}{a_{21}\,\lambda_1 + a_{22}\,\lambda_2}$$

5.1.5 Process Biotechnology

5.1.5.1 *Batch culture*

Question 13 In a fermentation process, the following data were observed and recorded.

- Fermenter diameter, D_t = 3 m
- Impeller diameter, Di = 1.5 m
- Baffle plate width, W_b = 0.3 m
- Liquid depth, H_L = 5 m
- Liquid density, ρ = 1200 kg/m^3
- Liquid viscosity, μ = 0.02 kg/m-s
- Rotation speed of impeller, N = 60 rpm
- Aeration rate = 0.40 vvm

Calculate the following from the above data.

(a) Power requirement (P) for an ungassed system

(b) Power requirement (P_g) when aerated

(c) Volume coefficient (K_v) of oxygen transfer

(d) Hold up (H_o) of bubbles

Answer

(a) Given that D_t = 3, D_i = 1.5, H_L = 5 m. Therefore,

$$\frac{D_t}{D_i} = \frac{3}{1.5} = 2.0$$

$$\frac{H_2}{D_i} = \frac{5}{1.5} = 3.33$$

$$n = \frac{60}{60} = 1 \text{ revolution per second (rps)}$$

$$N_{Re} = \frac{nD_i^2\rho}{\mu} = \frac{1 \times 1.5^2 \times 1.2 \times 10^3}{2 \times 10^{-2}} = 1.35 \times 10^5$$

N_P is derived from the power curve graph.

Since g_L = 981,

$$P = \frac{\rho\, n^3\, D_i^5\, N_P}{g_c}$$

(b) $$Q = 0.4 \times \frac{\pi}{4} \times 3^2 \times 5 \times \frac{1}{60}$$

$$Q = \text{vvm} \times \frac{1}{60} \times \frac{\pi}{4} \times D_t^2 \times H_L$$

$$N_a = \frac{Q}{nD_i^3} = \frac{0.4 \times \frac{\pi}{4} \times 3^2 \times 5 \times \frac{1}{60}}{1 \times 1.5^3}$$

or $N_a = 6.95 \times 10^{-2}$

From power curve graph,

$$\frac{P_g}{P^{**}} = 0.65$$

Therefore,

$$P_g = P^{**} \times 0.65 = 99.1 \times 0.65 = 58 \text{ HP}$$

(c) $Q = \frac{2}{5} \times \frac{\pi}{4} \times 3^2 \times 5 = 14.1 \text{ m}^3/\text{min}$

$$V_s = \frac{14.1 \times 60}{\frac{\pi}{4} \times 3^2} = 119.3 \text{ m/h}$$

$$K_v = 0.0635 \left(\frac{58}{\frac{\pi}{4} \times 3^2 \times 5} \right)^{0.95} \times (119.7)^{0.67} = 2.50 \text{ kg mol/m}^3$$

$$\frac{H_L}{D_t} = \frac{5}{3} = 1.67$$

$$f_c = 1.3$$

$$K_v^* = 2.50 \times 1.3 = 3.25 \text{ kg mol/m}^3$$

(d) $\left(\frac{P}{V}\right)^{0.4} \times (V_S)^{0.5} = \left(\frac{89.1}{\frac{\pi}{4} \times 3^2 \times 5} \right)^{0.4} \times (119.7)^{0.5} = 15.9$

H_o = 18.5% (Aiba, Humphrey, and millis 1973; Figure 6.7)

Question 14 In hydrocarbon fermentation by *Candida tropicalis*, the following data were recorded.

- Maximum O_2 uptake rate = 12 mmol/h
- Cell concentration at the time of maximum O_2 uptake rate = 10 g/L (for 1% hydrocarbon source)
- Critical O_2 concentration for organism is 0.032 mmol/L
- Saturation O_2 concentration in the broth is 0.54 mmol/L

From the data given, determine the value of the minimum K_La required to supply sufficient dissolved O_2.

Answer At steady state, we have

Rate of O_2 demand = Rate of O_2 supply

Thus,

$Q_{O_2(max)}\ X = K_La\ (C^* - C_L)$

or $Q_{O_2(max)}\ X = K_La\ (C^* - C_{crit})$

Therefore,

$$K_La = \frac{Q_{O_2(max)}X}{C^* - C_{crit}}\quad \frac{\text{mmol/h} \times \text{g/L}}{\text{mmol/L}}$$

$$\text{or } K_La = \frac{12 \times 10}{0.544 - 0.032} = \frac{12 \times 10}{0.512}$$

$$\text{or } K_La = \frac{120}{0.512} \text{ per hour}$$

or $K_La \approx 235$ per hour

5.1.5.2 Continuous culture

Question 15 *Aerobacter aerogenes* was grown in a continuous culture in a single pilot plant having a culture vessel of 20 L capacity. Glycerol was the carbon source in the ammonium salts medium, which was at a concentration of 2.5 g/L. What would be the bacterial densities obtained if the flow rates of the incoming medium are 5, 10, and 20 L/h, and what would be the corresponding outputs of the system?

Data In a number of batch culture experiments, it was ascertained that 1 L of the medium yielded 1.325 g dry weight of bacterium and that the maximum-specific growth rate was 0.85 per hour. A value of $K_s = 1.35 \times 10^{-4}$ M was also observed (Herbert, Elsworth, and Telling 1956).

Answer The dilution rate is given as shown below.

$D = \frac{F}{V}$, where $V = 20$ L

Thus, the dilution rates corresponding to the flow rates of 5, 10, and 20 L/h are

- $D = \frac{5}{20} = 0.25$ per hour
- $D = \frac{10}{20} = 0.50$ per hour
- $D = \frac{20}{20} = 1$ per hour

The critical dilution rate (at which washout occurs) is attained when the substrate concentration of effluent is equal to the concentration of substrate in the incoming medium.

That is, when

$S_R = \tilde{S} = 2.5$ g/L

Thus,

$$\tilde{S} = K_S \left[\frac{D_c}{\mu_m - D_c}\right] = 2.5$$

and

$K_S = 1.35 \times 10^{-4}$ M

Molecular weight of glycerol ($C_3H_8O_3$) = 92

Thus,

1.35×10^{-4} M $= 92 \times 1.35 \times 10^{-4}$ g/L = 0.0124 g/L

That is,

$$0.0124 \left(\frac{D_c}{0.85 - D_c}\right) = 2.5$$

where D_c is the critical dilution rate.

Thus,

$$D_c = \frac{2.5 \times 0.85}{2.512} = 0.846 \text{ per hour}$$

Therefore, washout will occur when the dilution rate ≥ 0.846 per hour as is with the highest flow rates where 20 L/h ≡ D of 1.0 h-L. For the other two flow rates, the steady state bacterial concentration is obtained from Equation 5.40..

$$\tilde{X} = Y(S_R - \tilde{S}) = Y\left[S_R - K_S\left(\frac{D}{\mu_m - D}\right)\right] \quad (5.40)$$

where Y is the gram dry weight of bacteria produced per gram of glycerol consumed (Equation 5.41).

$$Y = \frac{1.325}{2.5} = 0.53, \text{ since } S_R = 2.5 \text{ g/L} \quad (5.41)$$

The steady state bacterial concentration for D = 0.25 per hour is thus

$$\tilde{X} = 0.53\left[2.5 - 0.0124\left(\frac{0.25}{0.85 - 0.25}\right)\right]$$

or $\tilde{X}$ = 0.53 [2.5 – 0.005] = 1.323 g/L

For D = 0.50 per hour,

$$\tilde{X} = 0.53\left[2.5 - \frac{0.0124 \times 0.50}{0.25}\right]$$

or $\tilde{X}$ = 0.53 [2.5 – 0.018] = 1.316 g/L

The outputs of bacteria at each flow rate will be $D\tilde{x}$ gram dry weight/L-h. Thus,

For D = 0.25 per hour

Output = 0.25 × 1.323 = 0.331 g/L-h

For D = 0.50 per hour

Output = 0.50 × 1.316 = 0.658 g/L-h

Thus, the volume of the culture vessel is 20 L. The total output of the apparatus per hour is 6.62 g and 13.16 g dry weight of cells, respectively.

Comment

As the maximum cell productivity is desired in the present system, it is evident from the solution that the productivity of the two-stage system ($N_1D_1 + N_2D_2$ = 0.2 + 2.309 = 2.523 UOD/ml–h) is much higher than the productivity of the single-stage system (ND = 1.136 UOD/ml–h). Thus, the two-stage system is preferable to the single-stage system.

Question 16 A two-stage system with fermenter volumes V_1 (100 L) and V_2 (150 L) is compared with a single-stage system with fermenter volume V_3 (250 L). The volumetric flow rate, F, is the same for both systems (F = 40 L/h). Comment critically on the system with higher productivity of cells. Calculate the cell concentration in the effluent stream for each stage. The growth curve data for batch process are given in Table 5.3.

Answer The cell concentration in the effluent stream for each stage is calculated and presented in Table 5.4.

From the batch data (t versus N), each value of dN/dt is calculated and plotted against N as shown in Figure 5.5.

We have,

$$N_n - N_{n\text{-}1} = \frac{\dfrac{dN_n}{dt}}{\dfrac{F}{V}}$$

Thus, the slope of N versus dN/dt as shown in Figure 5.5 will give the value of F/V.

From graphical analysis,

$$\text{Slope } \frac{F'}{V_1} = \left[\frac{2.09}{5.35}\right] = 0.40 \text{ per hour} = D_1$$

Also, from the graph, the cell concentration from first stage (N_1) of the two-stage system is

$$N_1 = 5.35 \times \frac{\text{UOD}}{\text{ml}}$$

The productivity of cell concentration in the first stage is

$$N_1 \times D_1 = 5.35 \times 0.40 = 0.214 \times \frac{\text{UOD}}{\text{ml-h}}$$

Table 5.3 Data of growth curve for batch process

t (hours)	$N\left(\frac{\text{UOD}}{\text{ml}}\right)$
1.00	0.12
1.50	0.14
2.00	0.17
2.50	0.22
3.00	0.28
3.50	0.35
4.00	0.45
4.50	0.57
5.00	0.72
5.50	0.91
6.00	1.15
6.50	1.39
7.00	1.76
7.50	2.21
8.00	2.79
8.50	3.46
9.00	5.35
9.50	5.35
10.00	6.30
11.00	7.60
11.50	8.14
12.00	8.51
12.50	8.84
13.00	9.13
13.50	9.40

N – number of cells in terms of UOD/ml

For the second stage, $F = 40$ L/h and $V_2 = 150$ L

Thus,

$$D_2 = \frac{F}{V_2} = \frac{40}{150} = 0.267 \text{ per hour}$$

From the point N_1 (as shown in Figure 5.5), a line is drawn to touch the curve $\left(\frac{dN}{dt} \text{ vs } N\right)$ with a slope of 0.267 per hour, and from that point, a vertical line is drawn onto the x-axis, which gives the value of N_2 for the second stage.

From the graph,

$$N_2 = 8.65 \times \frac{\text{UOD}}{\text{ml}}$$

Thus, the productivity in the second stage is

Table 5.4 Cell concentration in the effluent stream

t (hours)	$N\left(\frac{UOD}{ml}\right)$	$\frac{dN}{dt}\left(\frac{UOD}{ml\text{-}h}\right)$
1.00	0.12	0.04
1.50	0.14	0.05
2.00	0.17	0.06
2.50	0.22	0.09
3.00	0.28	0.13
3.50	0.35	0.16
4.00	0.45	0.21
4.50	0.57	0.27
5.00	0.72	0.35
5.50	0.91	0.44
6.00	1.15	0.55
6.50	1.39	0.67
7.00	1.76	0.85
7.50	2.21	1.06
8.00	2.79	1.33
8.50	3.46	1.63
9.00	5.35	2.09
10.00	6.30	1.70
10.50	7.00	1.30
11.00	7.60	1.06
11.50	8.14	0.89
12.00	8.51	0.74
12.50	8.84	0.62
13.00	9.13	0.54
13.50	9.40	0.44

N – number of cells in terms of UOD/ml

$$N_2 \times D_2 = 8.65 \times 0.267 = 2.309 \times \frac{\text{UOD}}{\text{ml-h}}$$

For the single stage system, $F = 40$ L/h and $V = 250$ L

Thus, $D = \frac{F}{V} = \frac{40}{250} = 0.16$ per hour. In Figure 5.5, the line touches the curve at p, and the intercept on the x-axis gives the value of the effluent cell concentration N. We find that

$$N = 7.1 \times \frac{\text{UOD}}{\text{ml}}$$

Therefore, the productivity of the single-stage system is

$$N \times D = 7.1 \times 0.16 = 1.136 \times \frac{\text{UOD}}{\text{ml-h}}$$

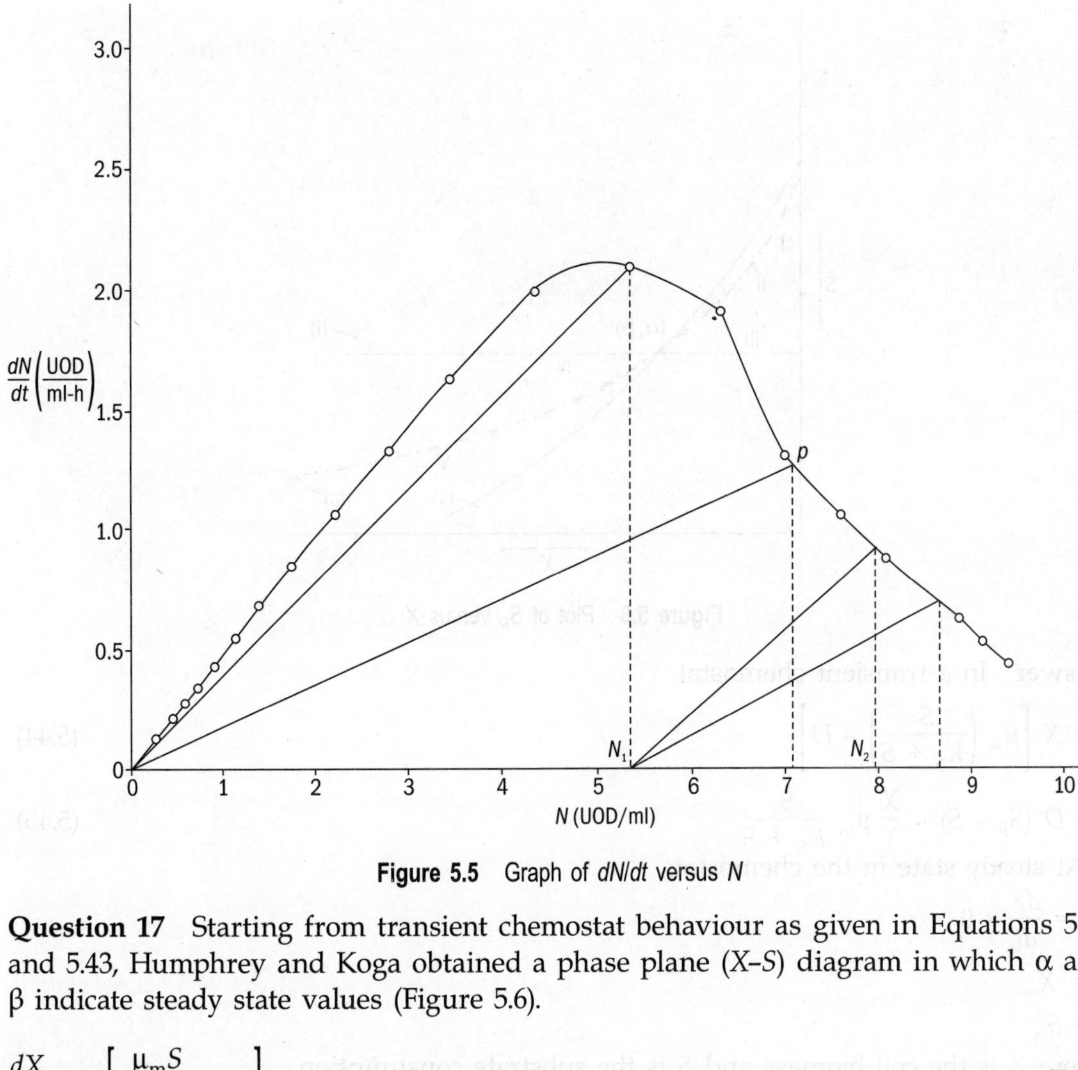

Figure 5.5 Graph of *dN*/*dt* versus *N*

Question 17 Starting from transient chemostat behaviour as given in Equations 5.42 and 5.43, Humphrey and Koga obtained a phase plane (*X*–*S*) diagram in which α and β indicate steady state values (Figure 5.6).

$$\frac{dX}{dt} = X\left[\frac{\mu_m S}{K_S + S} - D\right] \tag{5.42}$$

where *X* is the cell population, *S* is the substrate concentration, S_R is the inlet substrate concentration, *D* is the dilution rate, μ is the specific growth rate, *Y* is the yield coefficient, and K_S is a constant.

$$\frac{dS}{dt} = D\,(S_R - S) - \frac{X}{Y}\,\mu_m\left[\frac{S}{K_S + S}\right] \tag{5.43}$$

From Equations 5.42 and 5.43, considering steady state in the chemostat and stability by singular point, find out the match equations, giving steps of the profiles (i), (ii), and (iii) shown in Figure 5.6.

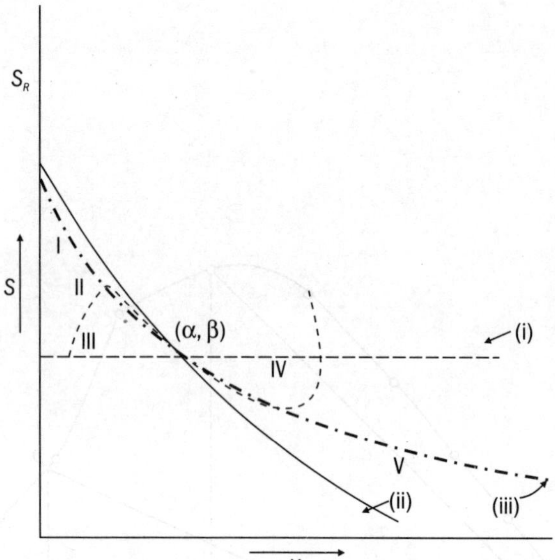

Figure 5.6 Plot of S_R versus X

Answer In a transient chemostat,

$$\dot{X} = X\left[\mu_m\left(\frac{S}{K_S + S}\right) - D\right] \tag{5.44}$$

$$\dot{S} = D\,(S_R - S) - \frac{X}{Y}\,\mu_m\,\frac{S}{K_S + S} \tag{5.45}$$

At steady state in the chemostat,

$$\frac{dX}{dt} = \frac{dS}{dt} = 0$$

$$X = X_\infty$$

$$S = S_\infty$$

where X is the cell biomass and S is the substrate consumption.

Thus, from Equation 5.44, we get

$$X_\infty\left[\mu_m\left(\frac{S_\infty}{K_S + S_\infty}\right) - D\right] = 0 \tag{5.46}$$

From Equation 5.45, we get

$$D(S_R - S_\infty) - \frac{X_\infty}{Y}\,\mu_m\left(\frac{S_\infty}{K_S + S_\infty}\right) = 0 \tag{5.47}$$

From Equation 5.46, since $X_\infty \neq 0$, we get

$$\mu_m\,\frac{S_\infty}{K_S + S_\infty} - D = 0$$

or $\mu_m S_\infty - D(K_S + S_\infty) = 0$

or $S_\infty(\mu_m - D) = DK_S$

$$\text{Thus, } \beta = S_\infty = \frac{DK_S}{\mu_m - D} \tag{5.48}$$

From Equation 5.47, we get

$$X_\infty = \frac{D(S_R - S_\infty)\,Y(K_S - S_\infty)}{S_\infty \mu_m} \tag{5.49}$$

Substituting Equation 5.46 into Equation 5.49, we get

$$X_\infty = \frac{YD\,(S_R - S_\infty)}{D}$$

Thus, $X_\infty = Y(S_R - S_\infty)$

or $X_\infty = Y(S_R - \beta)$ [from Equation 5.48]

Assuming

$$\alpha = X_\infty = Y(S_R - \beta) \tag{5.50}$$

Then,

$$\left.\begin{aligned} &\mu_m > D \\ &\text{and } S_R > \frac{K_S D}{\mu_m - D} \end{aligned}\right\} \tag{5.51}$$

The stability by singular point

$$X = \alpha + x \tag{5.52}$$

$$S = \beta + y$$

$$\frac{S}{K_S + S} = \frac{\beta\left(1 + \dfrac{y}{\beta}\right)}{(K_S + \beta)\left(1 + \dfrac{y}{K_S + \beta}\right)}$$

$$\text{or } \frac{S}{K_S + S} = \left(\frac{\beta}{K_S + \beta}\right)\left[1 + \left(\frac{1}{\beta} - \frac{1}{K_S + \beta}\right)y + \cdots\right] \tag{5.53}$$

$$\dot{x} = (\alpha + x)\left[\frac{\mu_m \beta}{K_s + \beta}\left\{1 + \left(\frac{1}{\beta} - \frac{1}{K_S + \beta}\right)y + \ldots\right\} - D\right]$$

$$\text{or } x = \left(\frac{\mu_m \beta}{K_S + \beta} - D\right) x + \frac{\mu_m \alpha \beta}{K_S + \beta}\left(\frac{1}{\beta} - \frac{1}{K_S + \beta}\right) y \tag{5.54}$$

Therefore,

$$\frac{dS}{dX} = 1 \frac{X\left[\mu_m \left(\frac{S}{K_S + S}\right) - D\right]}{D\,(S_R - S) - \frac{X}{Y}\,\mu_m \left(\frac{S}{K_S + S}\right)}$$

$$\text{or } \frac{dS}{dX} = \frac{D\,(S_R - S) - \frac{X}{Y}\,\mu_m \left(\frac{S}{K_S + S}\right)}{X\left[\mu_m \left(\frac{S}{K_S + S}\right) - D\right]}$$

Now,

$$D\,(S_R - S) - \frac{X}{Y}\,\mu_m \left(\frac{S}{K_S + S}\right) = 0$$

$$\text{or } D\,(S_R - S) = \frac{X}{Y}\,\mu_m \left(\frac{S}{K_S + S}\right) \tag{5.55}$$

$$\text{or } DS_R - DS - \frac{X}{Y}\,\mu_m \left(\frac{S}{K_S + S}\right) = 0$$

$$\text{or } DS_R - DS - \frac{X}{Y}\left[\frac{\mu_m S}{K_S + S} - D\right] - \frac{DX}{Y} = 0$$

$$\text{or } DS = DS_R - D\,\frac{X}{Y}$$

$$\text{or } S = S_R - \frac{X}{Y} = S_R - \frac{1}{Y} \cdot X \tag{5.56}$$

$$\text{Again, } S = \frac{K_S D}{\mu_m - D} \tag{5.57}$$

From Equation 5.55, we have

$$D(S_R - S)(K_S + S) - \frac{X}{Y}\,\mu_m\, S = 0$$

$$\text{or } - DS^2 + D(S_R - K_S)S + DS_R\, K_S - \frac{X}{Y}\,\mu_m\, S = 0$$

$$\text{or } - DS^2 + \left(DS_R - DK_S - \frac{X}{Y}\,\mu_m\right) S + DS_R K_S = 0$$

or $S^2 - \left(S_R - K_S - \frac{X\mu_m}{DY}\right) S - S_R K_S = 0$

or $S^2 - (S_R - K_S)\, S - S_R K_S = -\frac{X\mu_m}{DY} S$

where $X = f(S)$, $S = g(x)$, X is a constant, and $X \to \alpha/2$ and $S \to \frac{\beta}{2}$

The tracking function

$E = X_1 + YX_2$

Thus,

$$\frac{dE}{dt} = \frac{dX_1}{dt} + Y\frac{dX_2}{dt}$$

$$E = X_1\left[\frac{\mu X_1}{K + X_1} - D\right] + YD\,[S_R - X_2] - \mu\left[\frac{X_1 X_2}{K + X_2}\right] \tag{5.58}$$

Now,

$$\frac{\mu X_1^2}{K_1 X_1} - DX_1 + YDS_R - YDX_2 - \frac{\mu X_1 X_2}{K + X_2} = 0 \tag{5.59}$$

Rearranging Equations 5.58 and 5.59, we get

$$\frac{\mu K X_1\,(X_1 - X_2)}{(K + X_1)\,(K + X_2)} - D\,(X_1 + YX_2) - YDS_R = 0$$

Therefore,

$$E_{SS} = \frac{KD}{\mu - D} + \frac{Y}{2}\left[\left(S_R - K - \frac{\mu K}{Y(\mu - D)}\right) + \sqrt{\left[(S_R - D) - \frac{\mu D}{Y(\mu - D)}\right]^2 + 4\,S_R K}\right]$$

5.1.5.3 Scale up

Question 18 A glutamic acid fermentation yielded optimum conditions in a 5 L fermenter (d = 7.5 cm), where aeration of 2.5 L air/min and agitation at a rate of 500 rpm were applied. This is to be scaled up to a fermenter of 40 000 L (d = 112 cm) of similar geometry. If the power input per unit volume is to remain constant and if the aeration rate in the larger fermenter is to be only 10 000 L/min because of flooding considerations, recommend an agitator speed for the larger fermenter if the criteria of scale up is to

(a) Maintain constant volumetric oxygen mass transfer

(b) Maintain constant impeller tip speed

(c) Maintain constant mixing quality

Answer

(a) The constant volumetric oxygen transfer condition is

$$N_1 = N_2\left(\frac{D_2}{D_1}\right)^{73/63} \times \left(\frac{Q_1}{Q_2}\right)^{5/42}$$

Here, N_1 = 500 rpm, N_2 = ?, D_2 = 112 cm, D_1 = 7.5 cm, Q_1 = 2.5 L/min, and Q_2 = 10 000 L/min.

Therefore,

$$500 = N_2\left(\frac{112}{7.5}\right)^{73/63} \times \left(\frac{2.5}{10000}\right)^{5/42}$$

$$\text{or } 500 = N_2\,(15)^{73/63} \times \left(\frac{1}{4000}\right)^{5/42}$$

$$\text{or } N_2 = \frac{500}{(15)^{73/63}\left(\frac{1}{4000}\right)^{5/42}}$$

$$\text{or } N_2 = \frac{500 \times (4000)^{5/42}}{(15)^{73/63}}$$

Therefore, $N_2 = 58.21 \approx 59$ rpm

(b) Under constant impeller tip speed

$$N_1 = N_2\left(\frac{D_2}{D_1}\right)$$

Therefore,

$$500 = N_2\left(\frac{112}{7.5}\right)$$

$$\text{or } 500 = N_2 \times 15$$

$$\text{or } N_2 = \frac{500}{15} = 33.33 \approx 34 \text{ rpm}$$

(c) Under constant mixing quality

$$N_1 = N_2\left(\frac{D_1}{D_2}\right)^{1/4}$$

Therefore,

$$500 = N_2\left(\frac{7.5}{112}\right)^{1/4}$$

$$\text{or } 500 = N_2\left(\frac{1}{15}\right)^{1/4}$$

or $N_2 = 500 \times (15)^{1/4}$

Therefore,

$N_2 = 980$ rpm

5.1.6 Biothermodynamics, Bioenergetics, and Stoichiometry

Question 19 The conversion of glucose to lactic acid in an anaerobic organism is coupled to form 2 mol of ATP/mol of glucose.

(a) Calculate $\Delta G°$ of the overall reaction.

(b) What is the efficiency of energy conservation?

(c) How many moles of ATP/mol of glucose can be obtained in an aerobic organism at 40% efficiency of energy conservation?

Given the standard free energy change $\Delta G°'$,

Glucose → lactic acid (–52 000 cal/mol)

ATP → ADP + P_i (–7300 cal/mol)

Glucose → CO_2 + H_2O (–686 000 cal/mol)

Answer

(a) Glucose $\rightleftharpoons$ 2 lactic acid, $\Delta G_1°' = -52000$ cal/mol

2ADP + P_i $\rightleftharpoons$ 2ATP, $\Delta G_2°' = +7300 \times 2 = 14\,600$ cal/mol

Glucose + 2ADP + P_i $\rightleftharpoons$ 2 lactic acid + 2ATP

Therefore, $\Delta G°' = \Delta G_1°' + \Delta G_2°'$

or $\Delta G°' \cdots -52000 + 14\,600 = -37\,400$ cal/mol

(b) Conversion efficiency $= \dfrac{\Delta G_2^{\circ}}{\Delta G_1^{\circ}} \times 100$

Therefore, the conversion efficiency $= \dfrac{14600}{52000} \times 100 = 28.1\%$

(c) For complete combustion of glucose

$\Delta G°' = -686000$ cal/mol

The maximum number of ATP available at 100% efficiency is –686 000/–7300 = 93.97

Thus, at 40% of efficiency, the number of moles of ATP synthesized is

$93.97 \times 0.4 = 37.58 \approx 38$ mol

5.1.7 Stoichiometric and Energetic Calculation of Oxygen in Cells

Question 20 A typical cell composition of *Saccharomyces cerevisiae* is $CH_{0.78}O_{0.28}$. Taking water as a suitable reference state for comparison of redox state, check the oxygen and electron balance in the growth of yeast on ethanol as a substrate. (Consider N_2 of the cell to leave as NH_3, unaffected in redox state and in ash-free basis).

Answer

Basis: Conversion of 1 mol substrate to an amount of cell material with equivalent carbon content.

Oxygen balance:

Substrate	$C_2H_5OH + 3O_2 \rightarrow 2CO_2 + 3H_2O$
Cell	$2(CH_{0.78}\,O_{0.28}) + 2.11O_2 \rightarrow 2CO_2 + 0.78H_2O$
Difference	0.89 mol O_2

Electron balance:

Substrate	$C_2H_5OH + 3H_2O \rightarrow 2CO_2 + 12H^+ + 12\,e^-$
Cell	$2(CH_{0.78}\,O_{0.28}) + 3.44\,H_2O \rightarrow 2CO_2 + 8.44H^+ + 8.44e^-$
Difference	3.56 mol e^-

Thus, 3.56 mol $e^- \equiv (3.56/4)$ mol O_2 = 0.89 mol O_2 since

$O_2 + 4e^- = 2O_2^{2-}$

$2O_2^{2-} + 4H^+ = 2H_2O$

Question 21 From the redox differences between substrate and the cell in Question 20, calculate the NADH produced or required in the conversion of the substrate to cells. [Assume growth yield = 0.5 g cell/g ethanol, carbon content of the cell = 50%, and μ = specific growth rate of the cell = 0.25 per hour].

Answer The required amount of C_2H_5OH to yield 0.5 g cells is calculated as shown below.

$$\text{Carbon in cells} = 0.5 \text{ g cells} \times 0.5 \frac{\text{g C}}{\text{g cell}}$$

or carbon in cells = 0.25 gC

$$\text{or carbon in cells} = \frac{46 \text{ g Et OH/mol EtOH}}{24 \text{ g C/mol EtOH}} \times 0.25 \text{ g C}$$

or carbon in cells = 0.4792 g EtOH

The difference in oxidation in the formation of 0.5 g cells from 0.4792 g EtOH is calculated as shown below.

$$\text{mol } O_2 = \frac{0.4792 \text{ g EtOH}}{46 \text{ g EtOH/mol EtOH}} \times \frac{3.0 \text{ mol } O_2}{\text{mol EtOH}} - \frac{0.5 \text{ g cell}}{37.32 \text{ g cell/mol cell}} \times \frac{2.11 \text{ mol } O_2}{\text{mol cell}}$$

or mol O_2 = 0.0313 – 0.0283 = 0.003 mol O_2

Thus, the specific oxygen uptake rate (OUR) is

$$Q_{O_2} = \frac{0.003 \text{ mol } O_2}{0.5 \text{ g cell}} \times \frac{22400 \text{ mol } O_2}{\text{mol } O_2} \times 0.25 \text{ per hour}$$

or specific OUR = 33.60 mol O_2/g-h

This Q_{O_2} is necessary to balance the net rate of NADH production due to biosynthesis.

Question 22 A person wants to decrease 5 kg of body fat. The two routes available are (a) physical work and (b) dieting.

(a) For how long he has to work so that he burns 15 kCal/min to lose 5 kg of body fat?

(b) If his daily diet is 2500 kCal, how long he must stay on the diet at 2000 kCal/day?

(c) How much oxygen will be consumed by the person during the period of body weight loss?

Answer

(a) Let the person work for time *t*.

Thus, the time required to carry out an activity that burns 15 kCal/min to lose 5 kg of body fat is

$$t \times \frac{15 \text{ kCal}}{\text{min}} = 5 \times 10^3 \text{ g fat} \times \frac{9.3 \text{ kCal}}{\text{g fat}} = 46.5 \times 10^3 \text{ kCal}$$

$$\text{Thus, } t = \frac{46.5 \times 10^3}{15} \text{ min} = 3.1 \times 10^3 \text{ min} \approx 51.66 \text{ h}$$

(b) The time required by the person to go on a diet at 2000 kCal/day is

$$t = \frac{\text{Energy of 5 kg fat}}{\text{Energy deficit/day}} = \frac{46.5 \times 10^3 \text{ kCal}}{(2500 - 2000) \text{ kCal/day}} = 93 \text{ days}$$

(c) 4.7 kCal is released per gram of fat by 1 L of O_2.

Therefore, the total O_2 used during the weight lose period is

$$\frac{46.5 \times 10^3}{4.7} \text{ L} = 9.8 \times 10^3 \text{ L } O_2 = 980 \text{ L } O_2$$

5.1.8 Digital Computer Simulation

Question 23 Fermentative bioconversion of glucose to gluconic acid by *Pseudomonas ovalis* B1486 is given in Table 5.5. Based on the experimental results, the kinetic behaviour of the fermentative bioconversion can be expressed by Equations 5.60–5.64.

$$\frac{dC_s}{dt} = -\frac{1}{Y_s}\left(\frac{dC_x}{dt}\right) \tag{5.60}$$

Table 5.5 Batch fermentation data of gluconic acid production by *Pseudomonas ovalis* in horizontal rotatory fermenter (HRF)

t (hours)	C_X (UOD/ml)	C_S (mg/ml)	C_G (mg/ml)	C_P (mg/ml)	C_{GL} (mg/ml)
0	0.10	1.0	51.0	—	—
1.0	0.11	0.995	49.61	0.20	0.20
1.5	0.133	0.979	49.15	0.45	0.45
2.0	0.168	0.986	48.80	0.60	0.58
2.5	0.190	—	—	0.82	1.79
3.0	0.217	0.936	47.58	1.10	1.00
3.5	0.280	—	—	1.50	1.28
4.0	0.317	0.872	45.81	1.90	1.55
4.5	0.400	—	—	2.55	2.15
5.0	0.479	0.784	43.18	3.28	2.50
5.5	0.652	0.700	—	4.48	3.48
6.0	0.875	0.648	39.15	5.88	4.22
6.5	1.090	0.550	36.62	7.59	5.00
7.0	1.2400	0.448	34.15	9.63	6.58
7.5	1.491	0.330	30.72	12.82	—
8.0	1.732	0.280	26.71	15.08	9.25
8.5	2.00	0.123	22.08	19.02	10.35
9.0	2.200	0.094	17.20	22.02	11.50
9.5	2.200	—	—	—	—
10.0	2.200	—	10.00	30.86	10.60
10.5	—	—	—	34.20	9.50
11.0	—	—	6.25	37.05	8.52
11.5	—	—	—	39.68	7.20
12.0	—	—	3.33	42.11	5.90
12.5	—	—	–	44.00	5.08
13.0	—	—	1.00	45.30	3.24
13.5	—	—	—	46.20	2.65
14.0	—	—	—	46.82	2.85

t – fermentation time; C_X – cell concentration; C_S – ammonium acetate concentration; C_G – glucose concentration; C_P – gluconic acid concentration; C_{GL} – gluconolactone concentration

$$\text{or } \frac{dC_x}{dt} = \frac{\mu_m C_S}{K_S + C_S} C_X \tag{5.61}$$

$$-\frac{dC_G}{dt} = \frac{K_m(t) C_G}{K_\ell + C_G} C_X \tag{5.62}$$

$$\frac{dC_P}{dt} = a' C_{GL} \tag{5.63}$$

$$\frac{dC_{GL}}{dt} = \alpha' \frac{dC_G}{dt} - bC_{GL} \tag{5.64}$$

In Equations 5.60–5.64, C_G is the glucose concentration, C_P is the product (gluconic acid) concentration, C_{GL} is the intermediate (gluconolactone) concentration, K_m(t) is the Michaelis-Menten coefficient as a function of time, a' and b are stoichiometric coefficients, and α is a proportionality constant.

Develop a computer program giving a detailed flow chart for computer simulation of the above fermentative bioconversion kinetics and check the appropriateness of the kinetic models. Also examine the extent of accuracy for a chosen time interval of 0.01 s in computational iteration.

Answer Chisquare test indicated that the computational iteration in the time interval of 0.01s was accurate (Figures 5.7 and 3.79).

5.1.9 Immunobiotechnology

Question 24 When biological fluids such as blood serum or cells within broths are passed over a column containing antibodies attached to polyacrylamide beads, the antigen in the fluid binds to the antibody, while all other unrecognized molecules pass through undisturbed. The retention of the antigen on the column is proportional to the strength of binding as given in Equation 5.65.

$$V_e = [\varepsilon + (1 - \varepsilon) K] V \tag{5.65}$$

where V_e is the elution volume, ε is the column void fraction, $1 - \varepsilon$ is the volume fraction of column occupied by beads, V is the total column volume, and K is the apparent equilibrium constant for absorption.

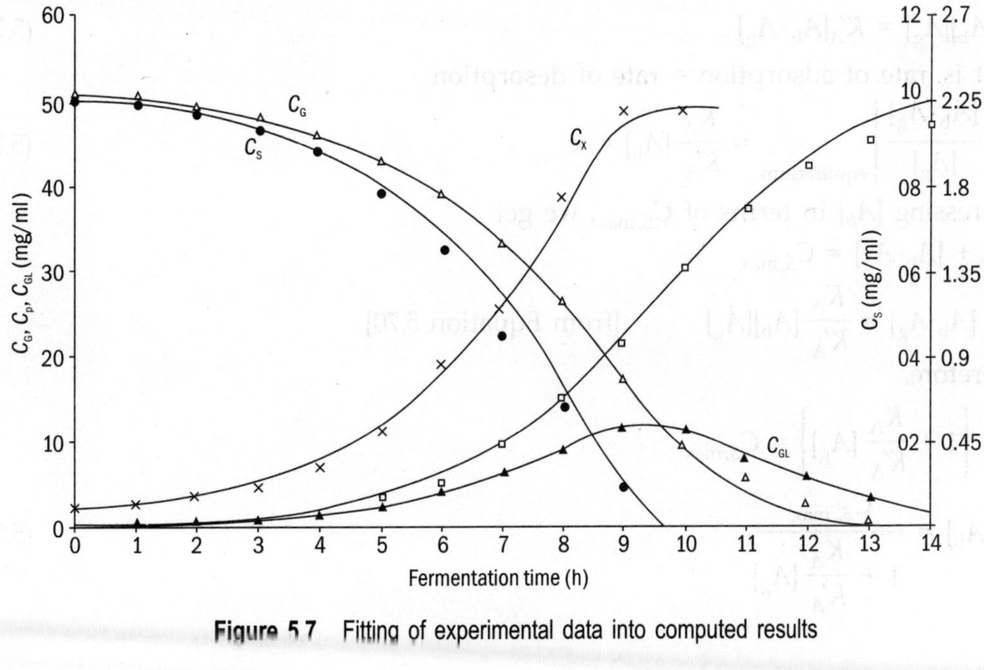

Figure 5.7 Fitting of experimental data into computed results

(a) Given the interaction kinetics between antibody and antigen, derive an expression for the elution volume of the antigen in terms of the antigen concentration in the bulk liquid and the maximum capacity for antigen binding, $C_{s,max}$.

(b) Given the presence of a constitutive inhibitor (I), derive an expression for the antigen elution time as a function of the inhibitor concentration.

Also given

$$K = \left.\frac{C_{S,S}}{C_{S,B}}\right|_{\text{equilibrium}} = \left.\frac{[A_b.A_g]}{[A_g]}\right|_{\text{equilibrium}} = \frac{\text{Concentration of bound antigen}}{\text{Concentration of antigen in bulk}} \quad (5.66)$$

$$A_b + A_g \underset{K'_A}{\overset{K_A}{\rightleftharpoons}} A_b \cdot A_g \quad (5.67)$$

$$A_b + I \underset{K'_{Ai}}{\overset{K_i}{\rightleftharpoons}} A_b \cdot I$$

Answer

(*a*) Let

$$\kappa_L = \frac{[A_b \cdot A_g]}{[A_b][A_g]} \quad (5.68)$$

$$\text{We have to find } \kappa = \left.\frac{[A_b \cdot A_g]}{[A_g]}\right|_{\text{equilibrium}} \quad (5.69)$$

From Equation 5.66, at equilibrium

$$K_A[A_b][A_g] = K'_A[A_b \cdot A_g] \quad (5.70)$$

That is, rate of adsorption = rate of desorption

$$\text{or } \left.\frac{[A_b \cdot A_g]}{[A_g]}\right|_{\text{equilibrium}} = \frac{K_A}{K'_A}[A_b] = \kappa \quad (5.71)$$

Expressing $[A_b]$ in terms of $C_{S,\max}$, we get

$$[A_b] + [A_b \cdot A_g] = C_{S,\max}$$

$$\text{But } [A_b \cdot A_g] = \frac{K_A}{K'_A}[A_b][A_g] \qquad \text{[from Equation 5.70]}$$

Therefore,

$$[A_b]\left\{1 + \frac{K_A}{K'_A}[A_g]\right\} = C_{S,\max}$$

$$\text{or } [A_b] = \frac{C_{S,\max}}{1 + \dfrac{K_A}{K'_A}[A_g]} \quad (5.72)$$

Using Equations 5.67 and 5.69 and expressing $\frac{K_A}{K'_A}$ as a function of κ_L, we get

$$\kappa_L = \frac{K_A}{K'_A} \tag{5.73}$$

From Equations 5.71, 5.72, and 5.73, we get

$$\kappa = \frac{\kappa_L \, C_{S,\,max}}{1 + \kappa_L \, [A_g]} \tag{5.74}$$

From Equations 5.65 and 5.74, we get

$$V_e = \left[\varepsilon + (1 - \varepsilon) \, \frac{\kappa_L \, C_{S,\,max}}{1 + \kappa_L \, [A_g]} \right] V \tag{5.75}$$

(b) Equation 5.71 is valid, but in expressing $[A_b]$ in terms of $C_{S,max}$, we get an additional term as shown below.

$$[A_b] + [A_b \cdot A_g] + [A_b \cdot I] = C_{S,max} \tag{5.76}$$

$$\text{Let } \kappa_i = \frac{[A_b \cdot I]}{[A_b][I]} \tag{5.77}$$

From Equation 5.67 and at equilibrium, we get

$$K_i[A_b][I] = K_i'[A_b \cdot I]$$

$$\text{or } [Ab \cdot I] = \frac{K_i}{K_i'} \, [A_b][I] \tag{5.78}$$

From Equations 5.77 and 5.78, we get

$$\frac{K_i}{K_i'} = \kappa_i \tag{5.79}$$

From Equations 5.70, 5.76, and 5.78, we get

$$[A_b] \left\{ 1 + \frac{K_A}{K'_A} \, [A_g] + \frac{K_i}{K'_i} \, [I] \right\} = C_{S,max} \tag{5.80}$$

Substituting Equations 5.73 and 5.79 in 5.80, we get

$$[A_b] = \frac{C_{S,max}}{1 + \kappa_L \, [A_g] + \kappa_i \, [I]} \tag{5.81}$$

From Equations 5.65, 5.71, and 5.81, we get

$$V_e = \left[\varepsilon + (1 - \varepsilon) \, \frac{\kappa_L \, C_{S,max}}{1 + \kappa_L \, [A_g] + \kappa_i \, [I]} \right] \tag{5.82}$$

5.2 MIXED CULTURE MICROBIAL BIOTECHNOLOGY

5.2.1 Waste Water and Water Treatment Biotechnology

Question 25 A portion of 5 ml sample of waste water was diluted to 300 ml in a biological oxygen demand (BOD) bottle. The dissolved oxygen content of the mixture was 8 mg/L. After five days of incubation at 20°C of waste water, the dissolved oxygen content of the mixture was 3 mg/L. Determine the BOD (20°C) of the waste water on the fifth day.

Answer Supplied volume of waste water = 5 ml

Volume of waste water after dilution = 300 ml

Thus, dilution ratio = 5/300 = 1/60

Water used for dilution = 300 – 5 = 295 ml

Initial dissolved oxygen content = 8 mg/L

Thus, the dissolved oxygen content after five days of incubation at 20°C is 3 mg/L.

Therefore, O_2 consumed by the organism is 8 – 3 = 5 mg/L.

Thus, the BOD of the original sample after 5 days of incubation at 20°C is

5 × 60 mg/L = 300 mg/L.

Question 26 For the same problem, if the BOD rate constant (*K*) of the waste water sample in Question 25 is 0.2 per day, determine the initial first stage BOD of the waste water.

Answer We know that

$$y = L_a(1 - 10^{-Kt})$$

where y is the five-day BOD at 20°C (= 300 mg/L), L_a is the initial first stage BOD, K = 0.2/day, t = 5 days

Therefore,

$$300 = L_a(1 - 10^{-0.2\times5})$$

$$\text{or } 300 = L_a(1 - 10^{-1}) = \frac{9}{10} L_a$$

$$\text{Thus, } L_a = \frac{300 \times 10}{9} = 333.33 \text{ mg/L}$$

Question 27 For the same problem, if the initial first-stage BOD of the waste water sample in Question 25 is 450 mg/L, evaluate the BOD rate constant (*K*).

Answer The initial first stage BOD, L_a = 450 mg/L

We know that y_5 = 300 mg/L and t = 5 days.

We also know that

$$y_5 = L_a (1 - 10^{-Kt})$$

That is, $300 = 450(1 - 10^{-5K})$

or $10^{-5K} = \dfrac{150}{450} = \dfrac{1}{3}$

or $K = \dfrac{\log (1/3)}{5} = 0.477/5 = 0.0954$ per day

Question 28 For Question 25, determine the five-day BOD at 30°C of waste water.

Answer We know that

$$\frac{y_5^{T}}{y_5^{T_0}} = 1 + C_L\,(T - T_0)$$

Now, from Question 25, we have

$y_5^{T_0} = y_5^{20°C} = 300$ mg/L

and $C_L = 0.02/°C$ in the range 10–30°C

Thus,

$$\frac{y_5^{30°C}}{300} = 1 + 0.02\,(30 - 20)$$

or $y_5^{30°C} = 300\,(1 + 0.2) = 360$ mg/L

Therefore, the five-day BOD at 30°C is 360 mg/L.

Question 29 The waste water in Question 27 is discharged into a stream. The volume of waste water is 5 million gallons per day (mgd). The volume of flow through the streams is 25 mgd. Determine the five-day BOD of the stream water after mixing if the temperature of the stream is (1) 20°C and (2) 25°C.

Answer

From Question 27,

Five-day BOD of waste water at 20°C = 300 mg/L

Volume of waste water flowing = 5 mgd

Stream water volume flow = 25 mgd

Therefore, the volume of the mixture = 25 + 5 mgd = 30 mgd

(a) Let the five-day BOD of mixture at 20°C be x mg/L.

From BOD balance,

$25 \times 0 + 5 \times 300 = 30 \times x$

Therefore, $x = 50$ mg/L

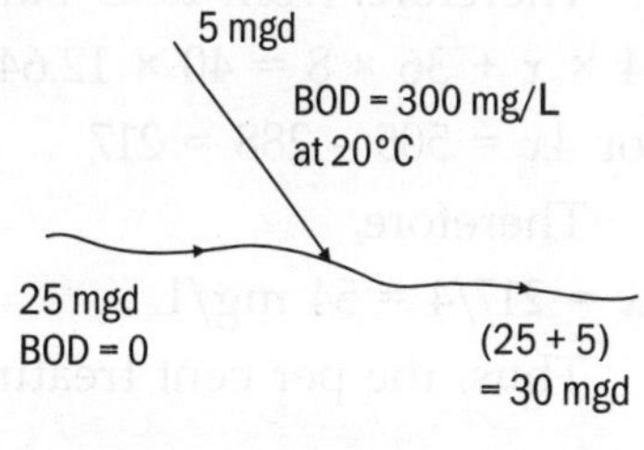

(b) At 25°C,

$$\frac{y_5^{25°C}}{y^{20°C}} = 1 + C_L\,(T - T_0)$$

or $\dfrac{y^{25°C}}{300} = 1 + 0.02\ (25 - 20)$

Therefore,

$y_5^{25°C} = 300 \times 1.1 = 330$ mg/L.

Thus, the five-day BOD of the waste water at 25°C is 330 mg/L.

Again, from BOD balance,

$25 \times 0 + 5 \times y_5^{25°C} = 30 \times x$

or $5 \times 330 = 30 \times x$

or $x = 55$ mg/L

Thus, the five-day BOD of the mixture at 30°C is 55 mg/L.

Question 30 A stream having a flow of 36 mgd was carrying a five-day BOD (20°C) of 8 mg/L. Waste water having a flow of 4 mgd and a five-day BOD (20°C) of 400 mg/L is discharged into the stream. The stream maintains a temperature of 20°C. If the initial stage BOD in the stream water is not more than 18 mg/L, estimate the percentage of BOD removal required by a treatment plant ($K = 0.1$ at 20°C).

Answer We know that

Stream flow = 36 mgd

Five-day BOD (20°C) of stream = 8 mg/L

Waste water flow = 4 mgd

Five-day BOD (20°C) of waste water = 400 mg/L

Thus, for the mixture,

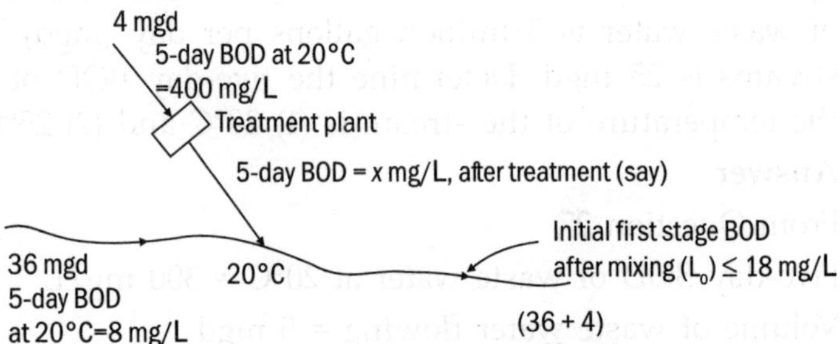

$y_5^{20°C} = L_0\ (1 - 10^{-Kt})$

or $y_5^{20°C} = 18\ (1 - 10^{-0.1 \times 5})$

or $y_5^{20°C} = 18(1 - 10^{-0.5})$

or $y_5^{20°C} = 18(1 - 0.32)$

or $y_5^{20°C} = 18 \times 0.68 = 12.64$ mg/L

Therefore, from BOD balance, we get

$4 \times x + 36 \times 8 = 40 \times 12.64$

or $4x = 505 - 288 = 217$

Therefore,

$x = 217/4 = 54$ mg/L

Thus, the per cent treatment in the plant is $[(400 - 54)/400] \times 100 = 346/4 = 86.5\%$

Question 31 The treated effluent of waste water was decided to be discharged into a stream having a flow of 36 mgd. The waste water before treatment and the stream waste were examined; the results are given in Table 5.6.

Determine the percentage of treatment if the initial stage BOD of the water is not to exceed 15 mg/L.

Answer

According to the problem,

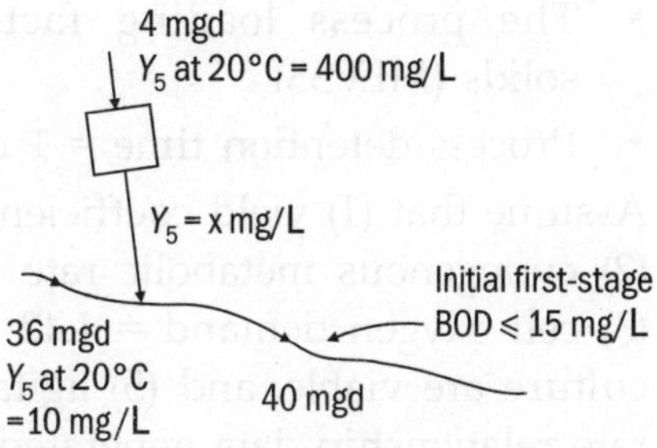

$$L_0^{25°C} = 15 \text{ mg/L}$$

From Arrhenius equation, we have

$$\frac{L^T}{L^{T_0}} = 1 + C_L (T - T_0)$$

$$\text{or } \frac{L^{25°C}}{L^{20°C}} = 1 + C_L (25 - 20)$$

$$\text{or } \frac{15}{L^{20°C}} = 1 + 0.02 \times 5$$

$$\text{or } L^{20°C} = \frac{15}{1.1} = 13.63 \text{ mg/L}$$

Thus,

$$y_5^{20°C} \text{ of the mixture} = L^{20°C} [1 - 10^{-Kt}]$$

$$\text{or } y_5^{20°C} \text{ of the mixture} = 13.63 [1 - 10^{-0.1 \times 5}]$$

$$\text{or } y_5^{20°C} \text{ of the mixture} = 13.63 \times 0.68 = 9.27 \text{ mg/L}$$

From BOD balance, we have

$$4x + 36 \times 10 = 40 \times 9.27$$

$$\text{or } 4x = 370.8 - 360 = 10.8$$

$$\text{or } x = \frac{10.8}{4} = 2.7 \text{ mg/L}$$

Thus, the per cent treatment in the plant is $\frac{400 - 2.7}{400} \times 100 = 99.3\%$

Table 5.6 Specifics of examined stream and waste water

Temperature	Stream water (25°C)	Waste water (25°C)
Five-day BOD (20°C)	10 mg/L	400 mg/L
K_1 (at 25°C)	0.1	0.1
K_2 (at 20°C)	0.25	—
Flow	36 mgd	4 mgd

5.2.2 Oxygenation Mixed Culture Process Biotechnology

Question 32 In the aerobic bioprocessing of waste liquid in a treatment plant, the following parameters were noted.

- Flow rate of waste liquid in the plant = 200 L/day
- Strength of waste liquid = 10 000 mg BOD_5 per litre
- The process loading factor = 0.4 kg/kg mixed liquor volatile suspended solids (MLVSS)
- Process detention time = 1 day

Assume that (1) yield coefficient = 0.55 g biocell synthesized per gram BOD removed, (2) endogenous metabolic rate = 0.05 g MLVSS destroyed per gram MLVSS per day, (3) cell oxygen demand = 1.42 g O_2 per gram cell, (4) all suspended materials in the culture are viable, and (5) agitator power input per unit volume and oxygen transfer rate relationship data generated by experiment is as shown in Table 5.7.

Calculate the following design parameters.

(a) Concentration of MLVSS

(b) Total O_2 requirement in the process

(c) Total power input when oxygen requirement is thrice the value obtained in (b)

Answer

(a) Given that the volume flow rate is 200 L/day.

$$\text{Thus, BOD loading rate} = 200 \times \frac{\text{L}}{\text{day}} \times 10 \ \frac{\text{g BOD}}{\text{L}}$$

or BOD loading rate = 2000 g BOD/day or 2 kg BOD/day

$$\text{Required MLVSS level} = \frac{\text{BOD loading}}{\text{Process loading factor}}$$

$$\text{or required MLVSS level} = \frac{2}{0.4} \times 1 = 50 \text{ kg MLVSS/day}$$

$$\text{Therefore, the concentration of MLVSS in waste liquid} = \frac{5}{200} \text{ kg/L}$$

or the concentration of MLVSS in waste liquid = 25 g/L

Table 5.7 Agitator power input per unit volume and oxygen transfer rate relationship data

Agitator power (W/L)	0	2	4	6	8	10	12	14	16
OTR $\left(\frac{\text{mmol } O_2}{\text{L-h}}\right)$	8.8	9.9	11.0	13.2	14.4	15.4	17.6	18.7	19.9

OTR – oxygen transfer rate

(b) The oxygen required in energy reaction for biocell synthesis is

$$\frac{(1 - 0.55)(\text{g cells})}{\text{g BOD}} \times \frac{1.42 \text{ g } O_2}{\text{g cells}} \times \frac{10 \text{ g BOD}}{\text{L–day}} = 2.2 \text{ g } O_2 \text{ per litre per day}$$

The oxygen required in an endogenous respiratory metabolism is

$$\frac{0.05 \text{ g MLVSS}}{\text{g MLVSS/day}} \times \frac{25 \text{ g cells}}{\text{L}} \times \frac{1.42 \text{ g } O_2}{\text{g cells}} = 1.8 \text{ g } O_2 \text{ per litre per day}$$

Thus, the total O_2 requirement in the process is

2.2 + 1.8 = 4.0 g O_2 per litre per day = 5.2 mmol O_2 per litre per day

(c) In this case, total O_2 required is

5.2 × 3 = 15.6 mmol O_2 per litre per hour

From Table 5.7, the plot of OTR (mmol O_2 per litre per hour versus agitator power/volume) shows that for OTR = 15.6 mmol O_2 per litre per hour, the power input per litre is 9.7 W.

Thus, for 200 L, the total power input is

9.7 × 200 W = 1940 W = 1.94 kW

5.3 IMMOBILIZED/ANCHORED BIOSYSTEMS

5.3.1 Microbial System

Question 33 Data collected from the performance of two bioreaction systems in immobilized whole cell (IWC) is shown in Table 5.8. Examine the nature of resistances encountered in the two systems. Does film resistance have any significant role for the two systems? If not, what resistance is a controlling one?

1. Glucose $\xrightarrow[\text{(Isomerase)}]{\textit{Actinoplane}\text{ sp.}}$ Fructose + Glucose

2. Cellobiose $\xrightarrow[\text{(Cellobiase)}]{\textit{Aspergillus wentii}}$ Glucose

Answer The following resistances are encountered in a packed bed bioreactor.

- External diffusion, that is, film diffusion
- Internal diffusion, that is, pore diffusion
- Back mixing

Film diffusion is characterized by the bed height required. When film diffusion is the controlling factor, the bed height required is given by

$$Z = \frac{\in (N_{Re})^{2/3}\ (N_{Sc})^{2/3}}{1.09\ a_v} \ln \frac{Y_1}{Y_2}$$

Table 5.8 Data of the performance of two immobilized whole cell bioreaction systems

Item	System 1	System 2
Bed height	22.5 cm	3 cm
Feed rate	42 ml/h	20 ml/h
Conversion efficiency	43%	50%
Feed concentration	36% glucose (w/v)	0.01 M cellobiose
Feed viscosity	0.94	0.8 cP
Feed density	1.2036 g/ml	1.0083 g/ml
Diffusivity	1.18×10^{-5} cm^2/s	2.5×10^{-5} cm^2/s at 45°C
Particle diameter	1.48 cm	0.009 cm
Effective diffusivity	1.11×10^{-5} cm^2/s	0.79×10^{-5} cm^2/s
Void fraction	0.44	0.45
Reaction order	First	First
Area/volume	4.03 cm^2/cm^3	333 cm^2/cm^3
Column diameter	5 cm	1 cm
Bulk concentration	2.00 M	0.01 M
Surface concentration at catalyst surface	1.0 M	–
Equilibrium constant	10.04	–
Intrinsic reaction rate	95.5 μM/min/g	108.9 μM/min/g
Product concentration	15.5% fructose	0.18% glucose
Thiele modulus	<2.8	~2.0

where ϵ is the void fraction, a_v is the surface area per unit volume, Y_1 is the inlet mole fraction of substrate, and Y_2 is the mole fraction of product.

$$N_{Re} = \frac{DG}{\mu}$$

where D is the particle diameter, G is the mean velocity, and μ is the liquid viscosity.

$$N_{Sc} = \frac{\mu}{\rho D_v}$$

where D_v is the effective diffusivity and ρ is the liquid density.

The pore diffusion is characterized by Thiele's modulus or effectiveness factor (η).

$$r = \eta \frac{k_2 E_0 S_s}{k'_m + S_s}$$

If the pore diffusion is the controlling factor in the packed bed bioreactor and the reaction is of first order, then the bed height is

$$Z = \frac{1}{A} \times F \times \frac{\tau}{\epsilon}$$

where F is the flow rate (cm^3/h), τ is the reactor space time (= V_R/Q min) given by,

$$\tau = -\ln(1 - \xi)/k_f \left[\frac{1}{\phi}\left(\frac{1}{\tanh 3\phi} - \frac{1}{3\phi}\right)\right]$$

where ϕ is the Thieles modulus, ξ is the fraction conversion, and k_f is the intrinsic dissociation constant.

For back mixing case,

$$\frac{D}{UL} = (N_{Pe})^{-1} \frac{dp}{\in L} \frac{\tau}{\tau_P}$$

$$\text{or } \frac{D}{UL} = 1 + k'\tau \frac{D}{UL}$$

where D/UL is the dispersion number and $N_{Pe} = \dfrac{Ud\rho}{D\in}$.

Checking for film diffusion

Case 1	**Case 2**
$G = \dfrac{42 \times 1.2036}{\frac{\pi}{4} \cdot 5^2 \times 3600}$	$G = \dfrac{20}{\frac{\pi}{4} \times 12} \times \dfrac{1.0083}{3600}$
$= \dfrac{1.4}{19.5} \times 10^{-2}\ \text{g/cm}^2\text{s}$	$= 0.712 \times 10^{-2}\ \text{g/cm}^2\ \text{s}$
$N_{Re} = \dfrac{DG}{\mu} = 1.48 \times \dfrac{1.4}{19.8} \times 10^{-2}$	$N_{Re} = \dfrac{DG}{\mu} = \dfrac{0.009 \times 0.712 \times 10^{-2}}{0.8 \times 10^{-2}}$
$\times \dfrac{1}{0.94 \times 10^{-2}} = 0.0895$	$= 0.801 \times 10^{-2}$
$N_{Sc} = \dfrac{\mu}{\rho D_v} = \dfrac{0.94 \times 10^{-2}}{1.2056 \times 1.11 \times 10^{-5}}$	$N_{Sc} = \dfrac{\mu}{\rho D_v} = \dfrac{0.8 \times 10^{-2}}{1.0083 \times 79 \times 10^{-5}}$
$= 7.1 \times 10^2$	$= 10.04 \times 10^2$
$\in = 0.44$	$\in = 0.45$
$a_v = 4.05\ \text{cm}^2/\text{cm}^3$	$a_v = 333\ \text{cm}^2/\text{cm}^3$
$\dfrac{y_1}{y_2} = \dfrac{1}{1 - 0.43} = \dfrac{1}{0.57}$	$\dfrac{y_1}{y_2} = \dfrac{1}{1 - 0.5} = \dfrac{1}{0.5}$
$Z = \dfrac{0.44 \times (0.0895)^{2/3} . (7.1 \times 10^2)^{2/3}}{1.09 \times 4.05}$	$Z = \dfrac{0.45\ (0.801 \times 10^{-2})^{2/3} \times (10.09 \times 10^2)^{2/3}}{1.09 \times 333}$
$\times \ln (1/0.57)$	$\times \ln \dfrac{1}{0.57}$
$= 0.89\ \text{cm}$	$= 0.00172\ \text{cm}$

In Case 1, since the actual bed height (22.5 cm) is greater than the calculated height, film diffusion is not limiting. In Case 2 also, the actual bed height (3 cm) is much higher than the calculated height and the film diffusion is not limiting.

Checking for pore diffusion

Case 1

$$k_f = 95.5 \frac{\mu\text{ mol}}{\text{min g}}$$

$$= 95.5 \times 10^{-6} \times 180 \text{ per minute}$$

[Molecular weight of glucose = 180]

$$\phi = 2.8$$

$$x = 0.43$$

$$\tau = -\frac{\ln(1-0.43)}{95.5 \times 10^{-6} \times 180} \times \left[\frac{1}{2.8}\left(\frac{1}{\tanh 8.4} - \frac{1}{8.4}\right)\right]$$

$$\left[\tanh 8.4 = \frac{e^{8.4} - e^{-8.4}}{e^{8.4} + e^{-8.4}} = \frac{4530 - \frac{1}{4530}}{4530 + \frac{1}{4530}} \approx 1\right]$$

Thus,

$$\tau = -\frac{\ln 0.57}{95.5 \times 10^{-6} \times 180}\left[\frac{1}{2.8}(1 - 0.117)\right]$$

$$\text{or } \tau = \frac{2.51 \times 2.3 \times 8.87}{9.55 \times 1.8 \times 2.8} \times 10$$

Therefore, $\tau = 10.4$ min

$$Z = \frac{1}{\frac{\pi}{4} \times 5^2} \times 42 \times \frac{10.4}{60} \times \frac{1}{0.44}$$

Therefore, $Z = 0.779$ cm

Case 2

$$k_f = 108 \times 10^{-6} \times 342 \text{ per minute}$$

[Molecular weight of cellobiose = 342]

$$\phi = 2.0$$

$$x = 0.5$$

$$\tau = -\frac{\ln(1-0.5)}{108 \times 10^{-6} \times 342} \times \left[\frac{1}{2}\left(\frac{1}{\tanh 6} - \frac{1}{6}\right)\right]$$

[tanh 6 = 1]

Thus,

$$\tau = \frac{-\ln 0.5}{108 \times 10^{-6} \times 342}\left[\frac{1}{2}\left(1 - \frac{1}{6}\right)\right]$$

$$= \frac{0.301 \times 2.303 \times 0.833}{108 \times 10^{-6} \times 342 \times 2}$$

or $\tau = 7.77$ min

$$Z = \frac{1}{\frac{\pi}{4} \times 1^2} \times \frac{7.77}{0.95} \times \frac{20}{60}$$

Therefore, $Z = 7.35$ cm

In Case 1, since the actual bed height (22.5 cm) is greater than the calculated height, film diffusion is not limiting. In Case 2, the actual bed height (3 cm) is less than the calculated height. Thus, pore diffusion is limiting in this case.

So in case of Glucose → Glucose + Fructose system, neither pore, nor film diffusion is limiting. However, in case of Cellobiose → Glucose system, pore diffusion is the limiting case but not film diffusion.

Back mixing case

$$\frac{D}{UL} = \frac{1}{N_{Pe}} \frac{\Delta P}{\in L} \frac{\tau}{\tau_p} \tag{5.83}$$

$$\tau = \frac{\in V}{F} = \frac{\in AL}{F} \tag{5.84}$$

$$\tau_p = \frac{d_p \, \mu}{N_{Re} \in D} N_{Pe} = \frac{d_p}{D} V \tag{5.85}$$

Substituting Equations 5.84 and 5.85 in Equation 5.83, we get

$$\frac{D}{UL} = \frac{D}{d_p U} \frac{\Delta P}{\in L} \frac{\in AL}{F} \frac{\rho D \in}{\mu}$$

$$\text{or} \quad \frac{d_p \, F\mu}{A\rho \, D\in \left(\dfrac{\Delta P}{L}\right)} = L^2$$

$\Delta P/L$ may be obtained from the Blake–Kozeny equation.

$$V_0 = \frac{\Delta P}{L} \frac{d_p^2}{150 \, \mu} \frac{\in^3}{(1 - \in)^2}$$

$$\therefore \quad \frac{\Delta P}{L} = \frac{V_0 . 150 \, \mu (1 - \in)2}{\in^3 d_p^2} = \frac{42}{3600 \times \dfrac{\pi}{4} \times 5^2} \times \frac{150 \times 0.94 \times 10^{-2} \, (0.56)^2}{(0.44)^3 \times (1.43)^2}$$

$$= 1.609 \times 10^{-3} \text{ g/cm-s}^2$$

5.3.2 Animal System

Question 34 The adsorption of vaccinia virus on cytodex 1 microcarrier culture of Vero cells shows a pattern as shown in Table 5.9 (at 37°C for 1.5 h). Examine what adsorption isotherm fits the data.

Table 5.9 Data of adsorption of vaccinia virus on Cytodex 1 microcarrier culture of Vero cells

Virus concentration (q), mg/L	0.01	0.02	0.03	0.06	0.10
Adsorption of Vero cells anchored on Cytodex (mg/cm^3)(W)	5.28	7.26	8.49	11.08	13.68

Answer Two adsorption isotherms, Langmuir and Freundlich, are tested on the given data set.

Langmuir isotherm

$$W = \frac{W_{max}\, q}{K + q}$$

$$\text{or } \frac{1}{W} = \frac{1}{W_{max}} + \frac{K}{W_{max}}\frac{1}{q}$$

In the plot of $(1/q)$ versus $(1/W)$, K/W_{max} is the slope and $1/W_{max}$ is the y-intercept.

The data are plotted in Figure 5.8.

Freundlich isotherm

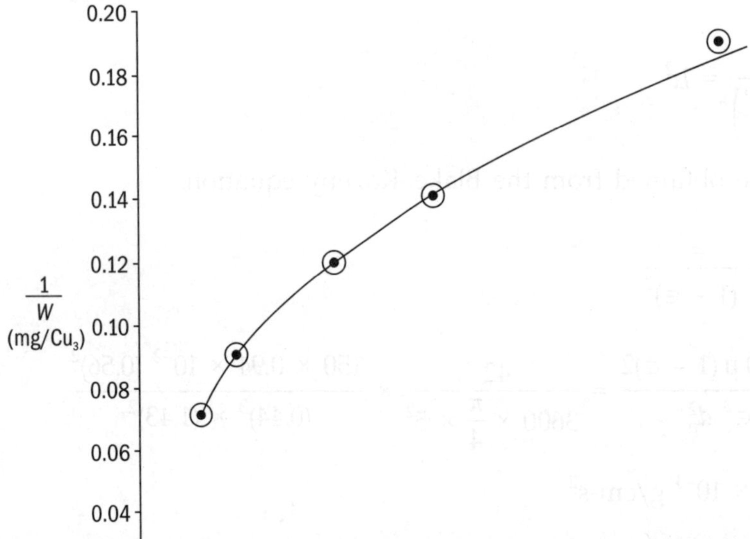

Figure 5.8 Langmuir curve

$W = Cq^n$

or $\log W = \log C + n \log q$

In the plot of $\log q$ versus $\log W$, n is the slope and $\log C$ is the y-intercept.

The data are plotted in Figure 5.9. The data are more consistent with the Freundlich isotherm. It can thus be concluded that the adsorption follows the Freundlich isotherm.

In the Freundlich isotherm,

$n = 0.4$

$C = 34.67\ (\text{g/cm}^3)$

Therefore, $W = 34.67\ q^{0.4}$

Question 35 The adsorption of vaccinia virus on Vero 76 cells at different cellular parameters (P) provided results as shown in Table 5.10.

What is the significance of P in this relation? Is there any linkage of P in terms of the data in Question 34? Comment on the profile (μ_t indicates specific growth rate).

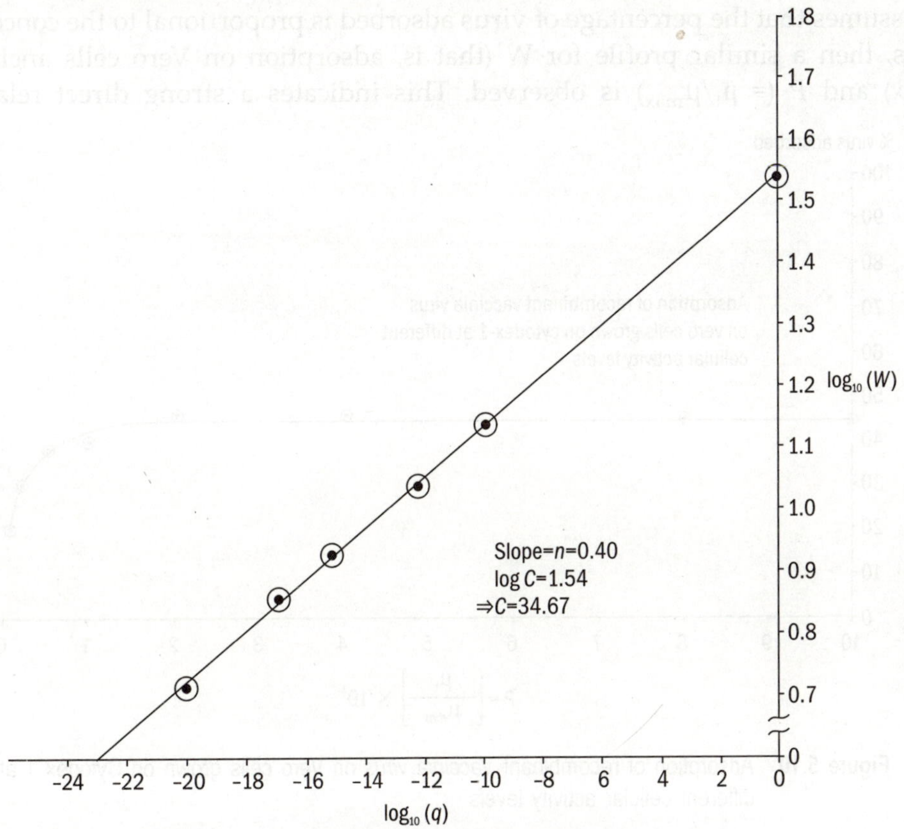

Figure 5.9 Freundlich isotherm

Table 5.10 Data of adsorption of vaccinia virus on Vero 76 cells at different cellular parameters

$P\left(=\frac{\mu_t}{\mu_{max}}\right) \times 10$	0.1	0.2	0.5	1.0	2.0	4.0	8.0	10
% virus adsorbed	20	30	38	40	46	46.8	45	44

Answer In this relation, *P* is the normalized cellular activity. A high cellular activity increases adsorption of virus up to a maximum level, while there is less chance of infection during low cellular activity.

It can be seen from the data and graph plotted (Figure 5.10) that increase in the adsorption of Vero cells on carrier results in higher cellular activity.

From Figure 5.10, we can see that virus adsorption increases with specific growth rate but saturates very fast, thus indicating a maximum level of infection. Also, the cellular activity must exceed 10% for effective virus adsorption to take place. High cellular activity does not increase the adsorption of virus but a lower one decreases virus adsorption drastically.

Linkage of P in terms of data in Question 34

If one assumes that the percentage of virus adsorbed is proportional to the concentration of virus, then a similar profile for W (that is, adsorption on Vero cells anchored on Cytodex) and *P* (= μ_t/μ_{max}) is observed. This indicates a strong direct relationship

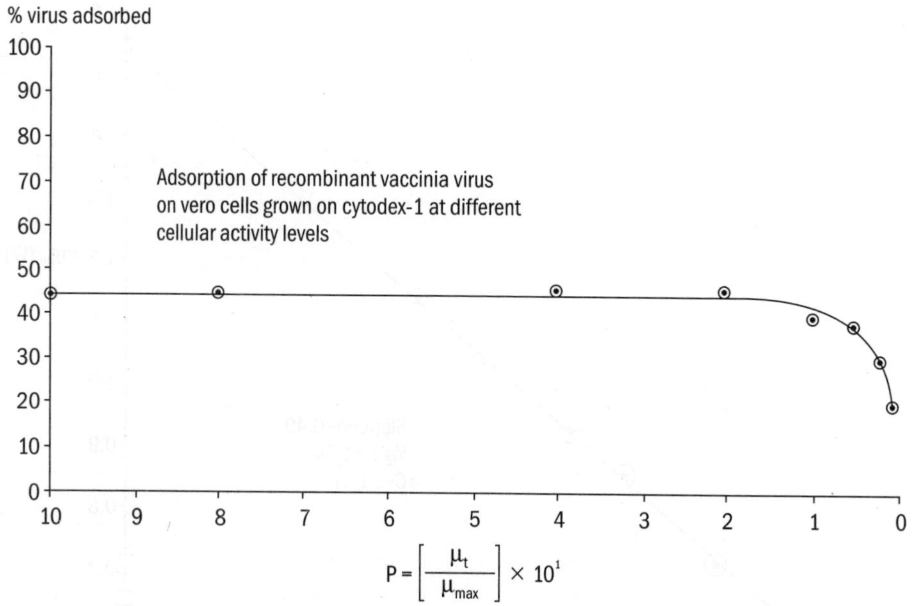

Figure 5.10 Adsorption of recombinant *Vaccinia virus* on Vero cells grown on Cytodex 1 at different cellular activity levels

between the adsorbed virus concentration and cellular activity. Thus, it can also be said that virus adsorption is a measure of cellular activity.

REFERENCES

Aiba S, Humphrey A E, and Millis N F. 1973. ***Biochemical Engineering,*** 2nd edn, Tokyo: University Tokyo Press

Bailey J E and Ollis D F. 1972. ***Biochemical Engineering Fundamentals,*** 2nd edn. New York: McGraw Hill Book Co. 646 pp.

Herbert D, Elsworth R, and Telling R C. 1956. **The continuous culture of bacteria; a theoretical and experimental study**. *Journal of General Microbiology* **14**: 601–622

Mukhopadhyay S N and Ghose T K. 1977. **Studies on cellulase production in relation to morphological variations of *Trichoderma reesei* QM9414**. 97 pp. [Proceedings of the *Bioconversion Symposium,* IIT Delhi, New Delhi]

BIBLIOGRAPHY

Atkinson B and Mavituna F. 1991. ***Biochemical Engineering and Biotechnology Handbook,*** 2nd edn. Hampshire, England: M Stocktor Press

Benedic R G, Schmidth W H, and Coghill R D. 1946. **The stability of penicillin in aqueous solution**. *Journal of Bacteriology* **51**(3): 291–292

Blakebrough N (ed.). 1969. ***Biochemical Engineering Science,*** vol. 1. London: Academic Press

Brownwell L E and Young E H. 1959. ***Process Equipment Design and Vessel Design,*** pp. 76–97. New York: John Wiley and Sons Inc.

Clark D P and Pandernik N J. 2009. ***Biotechnology.*** Amsterdam: Academic Press

Coulson J M and Richardson J F. 1962. ***Chemical Engineering,*** vol. I. Oxford: Pergramon Press

Flynn E H and Godzerki C W. 1967. **Pencillins and cephalosporis**. In *Antibiotics,* vol. 1, edited by D Gottlieb and P D Show, pp. 1–39. New York: Springer Verlag Inc.

Green D W and Perry R H. 2008. ***Perry's Chemical Engineers Hand Book,*** 8th edn. New York: McGraw Hill

Nelson W L. 1958. ***Petroleum Refinery Engineering,*** 4th edn. 960 pp. New York: McGraw-Hill

Hougen J A, Watson K M. 1962. ***Chemical Process Principles,*** part II. Mumbai: Asia Publishing House. 954 pp.

Kern D Q. 1950. ***Process Heat Transfer.*** New York: McGraw Hill

Kirk T K and Othmer D F. 1978. ***Encyclopaedia of Chemical Technology*** 2, 3rd edn. New York: John Wiley and Sons.

Kudrika A A, Grosh R J, and McFadden P W. 1965. **Heat transfer in two-phase flow of gas-liquid mixtures**. *Industrial and Engineering Chemistry Research* **4**(3): 339–344

Liddel J M. 1994. **Solvent extraction processes for biological products**. In *Engineering Processes for Bioseparations,* edited by L R Weatherly, pp. 166–201. Oxford: Butterworth-Heinemann

McCabe W L, Smith J C, and Harriot P. 1985. ***Unit Operations of Chemical Engineering.*** New York: McGraw Hill

Mukhopadhyay S N and Das D K. 1994. ***Oxygen Responses, Reactivities, and Measurements in Biosystems.*** Boca Raton, Florida: CRC Press

Mukhopadhyay S N (ed.). 2005. ***Advanced Process Biotechnology.*** New Delhi: Viva Books

Mukhopadhyay S N. 2007. ***Experimental Process Biotechnology Protocols.*** New Delhi: Viva Books

Mukhopadhyay S N. 2010. ***Process Biotechnology Fundamentals,*** 3rd edn. New Delhi: Viva Books

Perry R H and Chilton C H. 1963. ***Chemical Engineering Handbook,*** 4th edn. pp. 10–25. New York: McGraw-Hill

Prescott S C and Dunn C G. 1959. ***Industrial Microbiology,*** 3rd edn. New York: McGraw Hill

Solomons G L. 1969. ***Materials and Methods in Fermentation.*** New York: Academic Press

Steel R. 1958. ***Biochemical Engineering.*** London: Heywood and Co. Ltd.

Timoshenko S and MacCullough G H. 1949. ***Elements of Strength of Materials,*** 3rd edn. New Jersey: D Von Nostrand Princeton

Treybal R E. 1980. ***Mass Transfer Operations.*** New York: McGraw Hill

Van Winkle M. 1967. ***Distillation.*** New York: McGraw Hill

Van't Riet K and Tramper J. 1991. ***Basic Bioreactor Design.*** New York: Marel Dekkar Inc.

Walas S M. 1995. ***Chemical Reaction Engineering Handbook of Solved Problems,*** pp. 798–814. Austria: Gordon and Beach Publishers

6

Enzyme Engineering and Technology: Solved Numerical Problems

6.1 ENZYME KINETICS

6.1.1 Free Enzyme System

Question 1 Michaelis and Menten (1913) for the first time explained the general behaviour of an enzymatic reaction with the following mechanistic model.

$$A + E \underset{k_2}{\overset{k_1}{\rightleftharpoons}} (AE)^*$$

$$(AE)^* \xrightarrow{k_3} R + E$$

where A is the substrate, E is the enzyme, $(AE)^*$ is the intermediate complex or enzyme substrate complex, R is the product, k_1 is the reaction rate constant of forward reaction; k_2 is the reaction rate constant of backward reaction, and k_3 is the reaction rate constant.

This model is based on the assumption that the concentration of the intermediate $[(AE)^*]$ can be appreciable, and the total enzyme is distributed as shown in Equation 6.1.

$$[E_0] = [E] + (AE)^* \tag{6.1}$$

where $[E_0]$ is the concentration of the enzyme introduced into the mixture, which cannot be easily monitored.

Develop the rate equation for this reaction in terms of $[E_0]$ and $[A]$ and show that it explains the observed behaviour. Use steady state approximation in your development.[1]

[1] Problem taken from Levenspiel (2002)

Answer The Michaelis–Menten model is

$$A + E \underset{k_2}{\overset{k_1}{\rightleftharpoons}} (AE)^* \tag{6.2}$$

$$(AE)^* \xrightarrow{k_3} R + E \tag{6.3}$$

where A is the substrate, E is the enzyme, $(AE)^*$ is the intermediate complex or enzyme substrate complex, R is the product, k_1 is the reaction rate constant of forward reaction, k_2 is the reaction rate constant of backward reaction, and k_3 is the reaction rate constant.

Let us assume that $[E_0]$ is the concentration of total enzyme introduced, $[A]$ is the concentration of substrate introduced, and $(AE)^*$ is the concentration of the enzyme substrate complex. Applying law mass action for the rate of change of $[(AE)^*]$, we can write the rate of forward reaction, that is, rate of formation of $(AE)^*$ as

$$\gamma_f = k_1([E_0] - [(AE)^*])\,[A] \tag{6.4}$$

We can write the rate of backward reaction, that is, the rate of decomposition of $(A, E)^*$ as

$$\gamma_b = k_2[(AE)^*] \tag{6.5}$$

We can write the rate of $(AE)^*$ decomposition, that is, the rate of product formation as

$$\gamma_p = k_3[(AE)^*] \tag{6.6}$$

So the overall rate of change of $[(AE)^*]$ is

$$\frac{d[(AE)^*]}{dt} = \gamma_f - \gamma_b - \gamma_p$$

$$\text{or } \frac{d[(AE)^*]}{dt} = k_1([E_0] - [(AE)^*])\,[A]$$

$$\text{or } \frac{d[(AE)^*]}{dt} = k_2[(AE)^*] - k_3[(AE)^*] \tag{6.7}$$

As the steady state approximation has been made,

$$\frac{d[(AE)^*]}{dt} = 0$$

Thus, Equation 6.7 becomes

$$k_1([E_0] - [(AE)^*])\,[A]\,k_2\,[(AE)^*] - k_3\,[(AE)^*] = 0$$

$$\text{or } [(AE)^*]\,(k_1[A] + k_2 + k_3) = k_1[E_0]\,[A]$$

Thus,

$$[(AE)^*] = \frac{k_1[E_0]\,[A]}{k_1[A] + k_2 + k_3}$$

$$\text{or } [(AE)^*] = \frac{[E_0]\,[A]}{[A] + \left(k_2 + \dfrac{k_3}{k_1}\right)} \tag{6.8}$$

From Equation 6.6, we can calculate the rate of product formation in the enzyme reaction, v_p, as shown in Equation 6.3.

$$v_p = k_3[(AE)^*]$$

Substituting the value of $[(AE)^*]$ from Equation 6.8, we get

$$v_p = \frac{k_3[E_0]\,[A]}{[A] + \left(k_2 + \dfrac{k_3}{k_1}\right)} \tag{6.9}$$

$$\text{or } v_p = \frac{k_3[E_0]\,[A]}{[A] + \left(\dfrac{k_2}{k_1} + \dfrac{k_3}{k_1}\right)} \tag{6.10}$$

$$\text{But, } \frac{k_2}{k_1} = \frac{[(AE)^*]}{[A]\,([E_0] - [(AE)^*])} = K_s$$

where K_s is the equilibrium constant in the decomposition of enzyme-substrate couple $(AE)^*$. Now,

$$v_p = \frac{k_3[E_0]\,[A]}{[A] + K_s + \dfrac{k_3}{k_1}} \tag{6.11}$$

Again, $K_s + (k_3/k_1) = M$, which is a constant known as Michaelis constant. Thus

$$v_p = \frac{k_3[E_0]\,[A]}{[A] + M}$$

Substituting $[E_0] = C_{E0}[A] = C_A$, we get

$$v_p = \frac{k_3 C_A C_{E0}}{C_A + M} \tag{6.12}$$

where k_3C_{E0} represents a state at which all of the enzyme forms the enzyme–substrate complex. Therefore, $k_3C_{E0} = v_{max}$, which is the maximum rate of product formation. Thus, Equation 6.12 can be written as

$$v_p = \frac{v_{max} C_A}{C_A + M} \tag{6.13}$$

The following conclusions can be made from Equation 6.13.

- The limiting velocity or maximum velocity, v_{max}, is asymptotic to which the reaction velocity approaches as concentration of the substrate increases.
- The Michaelis constant corresponds to the concentration at which half of the limiting velocity is reached.

Proof

Substituting $v_p = \frac{v_{max}}{2}$ in Equation 6.13, we get

$$\frac{v_{max}}{2} = \frac{v_{max} C_A}{C_A + M}$$

or $2\ C_A = C_A + M$

Thus, $M = C_A$

Question 2 At room temperature, sucrose is hydrolysed by the catalytic action of the enzyme sucrase.

$$\text{Sucrose} \xrightarrow{\text{Sucrase}} \text{Products}$$

In a batch reactor, kinetic data were obtained starting with a sucrose concentration of C_{Ao} = 1.0 mmol/L and an enzyme concentration of C_{Eo} = 0.01 mmol/L (concentrations calculated from optical rotation measurement) (Table 6.1).

Determine whether the data can be reasonably fitted by the kinetic equation of the Michaelis–Menten type or

$$\gamma_A = \frac{k_3 C_A C_{E0}}{C_A + M}$$

where M is the Michaelis constant.

If the fit is responsible, evaluate the constants k_3 and M.[2]

Table 6.1 Kinetic data obtained from optical rotation measurement

C_A (mmol/L)	0.84	0.68	0.54	0.38	0.27	0.16	0.09	0.04	0.018	0.006	0.0025
Time (h)	1	2	3	4	5	6	7	8	9	10	11

[2] Problem taken from Levenspiel (2002)

Answer The time profile of C_A, $C_{A0}-C_A$, and velocity of reaction are given in the Table 6.2.

Plotting the velocity of reaction against the corresponding substrate concentration values, we get a hyperbolic curve, which clearly indicates that the data given in the problem fit with the Michaelis-Menten kinetic model. Plotting $1/v$ against the corresponding value of $1/C_A$, we get a straight line as shown in Figure 6.1. The calculated values of $1/v$ is given in Table 6.3.

Slope = $-1/K_m = -5.38$

Table 6.2 Time profile of C_A, $C_{A0} - C_A$, and velocity of reaction

Time (h)	C_A (mmol/L)	$C_{A0} - C_A$ (mmol/L)	Velocity of reaction, v_1 (mmol/t_1)
1	0.84	1 − 0.84 = 0.16	0.16
2	0.68	1 − 0.68 = 0.32	0.16
3	0.53	1 − 0.53 = 0.47	0.1567
4	0.38	1 − 0.38 = 0.62	0.155
5	0.27	1 − 0.27 = 0.73	0.146
6	0.16	1 − 0.16 = 0.84	0.140
7	0.09	1 − 0.09 = 0.91	0.13
8	0.04	1 − 0.04 = 0.96	0.12
9	0.018	1 − 0.018 = 0.982	0.1091
10	0.006	1 − 0.006 = 0.994	0.0994
11	0.0025	1 − 0.0025 = 0.9975	0.0907

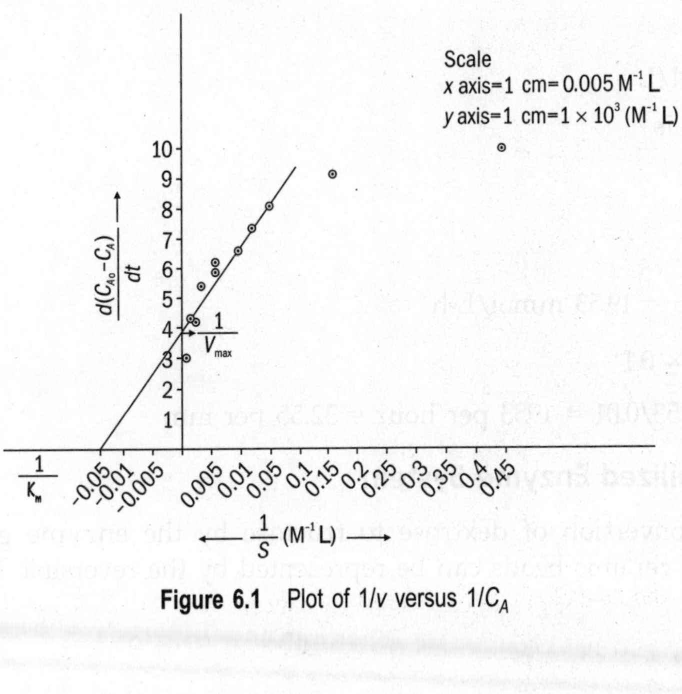

Figure 6.1 Plot of $1/v$ versus $1/C_A$

Table 6.3 Calculated values of 1/v

$1/C_A$ (L/mol)	$1/v$ (L-s/mol)
$\frac{1}{0.84} = 1.19$	$\frac{1}{0.16} = 6.25$
$\frac{1}{0.68} = 1.47$	$\frac{1}{0.16} = 6.25$
$\frac{1}{0.53} = 1.887$	$\frac{1}{0.1567} = 6.381$
$\frac{1}{0.38} = 2.631$	$\frac{1}{0.155} = 6.6451$
$\frac{1}{0.27} = 3.70$	$\frac{1}{0.146} = 6.853$
$\frac{1}{0.16} = 6.25$	$\frac{1}{0.14} = 7.14$
$\frac{1}{0.09} = 11.11$	$\frac{1}{0.13} = 7.7$
$\frac{1}{0.04} = 25$	$\frac{1}{0.12} = 8.333$
$\frac{1}{0.018} = 55.55$	$\frac{1}{0.1091} = 9.16$
$\frac{1}{0.006} = 166.66$	$\frac{1}{0.0994} = 10.06$
$\frac{1}{0.0025} = 400.00$	$\frac{1}{0.0907} = 11.02$

Thus,

$$K_m = 0.186 \text{ mmol/L}$$

The intercept is

$$\frac{V_m}{K_m} = 105$$

Thus,

$$V_m = 105 \times 0.186 = 19.53 \text{ mmol/L-h}$$

$$V_m = k_3[E] = k_3 \times 0.1$$

Thus, $k_3 = 19.53/0.01 = 1953$ per hour $= 32.55$ per mm

6.1.2 Immobilized Enzyme System

Question 3 Conversion of dextrose to fructose by the enzyme glucose isomerase immobilized on ceramic beads can be represented by the reversible Michaelis–Menten kinetics.

$$-r = \frac{V_m\left[S - \frac{P}{K}\right]}{K_m + S + \frac{K_m P}{K_F}}$$

where S is the substrate concentration of dextrose, P is the product concentration, K is the equilibrium constant, K_m is the Michaelis constant for the forward reaction, and K_F is the Michaelis constant for the reverse reaction.

(a) Obtain and integrated expression for the conversion in packed bed reactor (plug flow), assuming $K_m \approx K_F$.

(b) Compare the expression obtained in part (a) to the data given in Table 6.4

(c) Is it possible that the conversion may be mass transfer controlled? Suggest some experiments or calculations that one can carry out to determine if the overall reaction is controlled by kinetics or mass transfer.

(d) The activation energies for the enzyme in the soluble and bound forms are 16 kCal/mol and 15 kCal/mol, respectively. Does this give any clue to the probable rate controlling mechanism?

Answer

(a) We have

$$-r = \frac{V_m\left[S - \frac{P}{K}\right]}{K_m + S + \frac{K_m P}{K_F}}$$

Since $K_m \approx K_F$, we have

$$-r = \frac{V_m\left[S - \frac{P}{K}\right]}{K_m + S + P}$$

Table 6.4 Data for comparison

Per cent conversion to fructose	Normalized reactor residence time
8	0.01
15	0.02
28	0.04
36	0.06
42	0.08
45	0.10
48	0.12

Now,

$$S_0 = S + P$$

Also, if x is the conversion fraction, then

$$S = S_0(1 - x) \text{ and } P = S_0 x$$

$$r = \frac{V_m\left[S_0(1 - x) - \frac{S_0 x}{K}\right]}{K_m + S_0} \tag{6.14}$$

Now, for plug flow reactor,

$$\tau = S_0 \int_0^x \frac{dx}{r}$$

$$\text{or } \tau = S_0 \int_0^x \frac{K_m + S_0}{V_m\left[S_0(1 - x) - \frac{S_0 x}{K}\right]} dx$$

$$\text{or } \tau = \frac{S_0}{V_m} \int_0^x \frac{\left(\frac{K_m}{S_0} + 1\right)}{\left(1 - x - \frac{x}{K}\right)} dx$$

$$\text{or } \tau = S_0/V_m \left(\frac{K_m}{S_0} + 1\right) \int_0^x \frac{dx}{\frac{K - (K + 1)\, x}{K}}$$

$$\text{or } \tau = \left(\frac{K}{K + 1}\right) \frac{S_0}{V_m} \left(\frac{K_m}{S_0} + 1\right) \int_0^x \frac{dx}{\left(\frac{K}{K + 1} - x\right)}$$

$$= -\left(\frac{K}{K + 1}\right) \frac{S_0}{V_m} \left(\frac{K_m}{S_0} + 1\right) \ln \frac{\frac{K}{K + 1} - x}{\frac{K}{K + 1}}$$

Now,

$$\frac{P_{eq}}{K} = S_{eq}$$

$$\Rightarrow \frac{S_0 x_{eq}}{K} = S_0(1 - x_{eq})$$

$$\text{or } \frac{K}{K+1} = x_{eq}$$

$$\text{or } \tau = -\, x_{eq} \frac{S_0}{V_m}\left(\frac{K_m}{S_0} + 1\right) \ln \frac{x_{eq} - x}{x_{eq}}$$

Thus,

$$\tau = x_{eq} \frac{S_0}{V_m}\left(\frac{K_m}{S_0} + 1\right) \ln \frac{x_{eq}}{x_{eq} - x} \tag{6.15}$$

$$\text{or } \ln \frac{x_{eq}}{x_{eq} - x} = \frac{\tau}{x_{eq} \dfrac{S_0}{V_m}\left(\dfrac{K_m}{S_0 + 1}\right)}$$

$$\text{or } \ln \frac{x_{eq}}{x_{eq} - x} = \frac{\tau}{K}$$

$$\text{where } K = x_{eq} \frac{S_0}{V_m}\left(\frac{K_m}{S_0} + 1\right) = \text{constant}$$

$$\ln \frac{x_{eq} - x}{x_{eq}} = -\frac{\tau}{K}$$

$$\text{or } x_{eq} - x = x_{eq} e^{-\tau/K}$$

$$\text{or } x = x_{eq}\,(1 - e^{-\tau/K}) \tag{6.16}$$

(b) Equation 6.16 can be written as

$$x = x_{eq} - x_{eq} e^{-\tau/K}$$

Now, K is expressed in units of time. Hence, τ/K can be considered a normalized residence time. Thus, if we plot x versus $e^{-\tau_{\text{normalized}}}$, the plot should be linear. Hence, to compare the given data with Equation 6.16, we plot x versus $e^{-\tau_{\text{normalized}}}$ and see if it is linear. The values of $\tau_{\text{normalized}}$, and $e^{-\tau_{\text{normalized}}}$ for different values of x are listed in Table 6.5.

Table 6.5 Values of $\tau_{\text{normalized}}$ and $e^{-\tau_{\text{normalized}}}$ for different values of x

x	$\tau_{\text{normalized}}$	$e^{-\tau_{\text{normalized}}}$
0.08	0.01	0.0900
0.15	0.02	0.9801
0.28	0.04	0.9607
0.36	0.06	0.0417
0.42	0.08	0.9231
0.45	0.10	0.9048
0.40	0.12	0.8869

This data were processed on the linear regression analysis system of CTI SR SI-II calculator. The correlation coefficient was found to be 0.966. This corresponds to a degree of confidence of >99.9% (from statistical tables). Hence, the plot of x versus $e^{-\tau_{normalized}}$ is linear. Thus, the data compares well with Equation 6.16.

(c) From (b) we see that Equation 6.16 derived for kinetic reactor rate agrees well with the data. Hence, the reaction is kinetic rate controlled and the mass transfer effects seem to be negligible. However, to test whether the reaction is mass transfer controlled, we can carry out the following experiments.

- ***For external mass transfer***: Using different reactor lengths (in order to keep τ constant), determine the rate of reaction at different flow rates. If the rates of reaction are not the same (that is, outlet concentration is not the same) in each case, the reaction is controlled by external mass transfer control.
- ***For internal mass transfer***: Crush up the particle settling time enzyme activity is not affected, then determine the rate again. If the rate increases, then internal mass transfer is controlling.

(d) If mass transfer (either external or internal) was the controlling mechanism, then the activation energy for the bound enzyme should be higher than that of the soluble enzyme (since mass transfer rate is lower than kinetic rate). However, in this case, since activation energy is almost the same, the reaction rate appears to be the controlling mechanism. This is also shown by results in point (b). The slight decrease in activation energy may be attributed to a more favourable conformation of the enzyme in the immobilized state.

Question 4 In the kinetic studies of immobilized catalase reaction by H_2O_2 in a continuous stirred tank reactor (CSTR), dependence of free catalase inactivation by H_2O_2 was governed by Equation 6.17.

$$q(C_0 - C_1) = kE \tag{6.17}$$

And inactivation was governed by Equation 6.18.

$$-dE/dt = k_d E \tag{6.18}$$

The specific rate of catalase inactivation is given by Equation 6.19.

$$k_d(0°C) = \frac{aP}{b + P} + cP \tag{6.19}$$

Here, q is the flow rate (L/s), k is the first order reaction rate (per second), E is the amount of immobilized enzyme (mol), C_0 is the initial concentration (mol/L), C_1 is the outlet concentration, P is the concentration of H_2O_2 (mol/L), and a, b, and c are constants. Compute the outlet concentration if H_2O_2 enters the reactor at a concentration of 0.015 mol/L and the per cent conversion desired is 90%. What will be the specific rate of catalase inactivation? [Assume a = 0.072 per second, b = 0.15 mol/L, and c = 0.0185 L/s-mol]

Answer Since the desired per cent conversion is 90%, the per cent inactivation is 10%. Since $C_0 = 0.015$, C_1 = 10% of 0.015 = 0.0015 mol/L

Therefore,

$$k_d = \frac{0.072 \times 0.015}{0.15 + 0.015} + 0.0185 \times 0.015$$

or $k_d = 6.8 \times 10^{-3}$

6.2 ENZYME REACTORS

6.2.1 Free Enzyme

Question 5 Substrate (C_{A_0} = 2 mol/L) and enzyme (C_{t_o} = 0.001 mol/L) are introduced into a batch reactor. The conversion is 90% in 4.1 min, and the rate equation that represents this behaviour is given in Equation 6.20.

$$-\gamma_A = 10^3 \frac{C_{E0}C_A}{1 + C_A} \frac{\text{mol}}{(\text{L})(\text{min})} \tag{6.20}$$

There is a plan to build and operate a mixed flow reactor with a continuous feed of substrate (C_{A_0} = 2 mol/L) and enzyme (C_{E_0} = 0.001 mol/L).

(a) What should be the space time (τ) to achieve 90% conversion of the substrate? The reactor has been planned to be fabricated for time τ. In solution, it was found that this enzyme is not dispersed on a molecular scale as was expected but was present in finely dispersed state like distinct flocs or clumps. Although calculations are based on microfluid behaviour, the enzyme showed the same behaviour.

(b) On the basis of this new information, what conversion can be expected in the fabricated reactor system?

Answer

(a) For a mixed reactor and assuming microfluid behaviour.

$$\tau = \frac{C_{A0}x_A}{-\gamma_A}$$

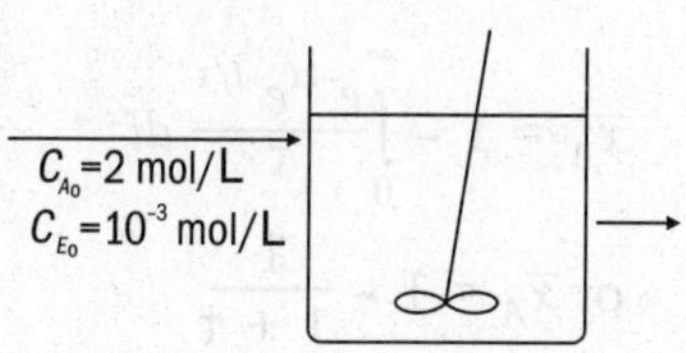

$$\tau = \frac{C_{A0}x_A(1 + C_A)}{10^3(0.001)\ C_A} \tag{6.21}$$

We have,

$$C_A = C_{A0}(1 - X_A) \tag{6.22}$$

Substituting for C_{A0} = 2 mol/L and X_A = 0.9 in Equation 6.22, we get

$C_A = 2(1 - 0.9)$

Thus, C_A = 0.2 mol/L

Substituting the values for C_A, C_{A0}, and X_A in Equation 6.21, we get

$$\tau = (2)\,\frac{(0.9)\,(1 + 0.2)}{0.2} = 10.8 \text{ min}$$

Thus, the space time required is 10.8 min for 90% conversion.

(b) For a mixed reactor and assuming macrofluid behaviour.

τ = 10.8 min (since reactor is already designed for this space time of 10.8 min)

Then, the conversion for macrofluid in a CSTR is given by Equation 6.23 (Levenspiel 2002).

$$\bar{x}_A = 1 - \int_0^\infty \left(\frac{C_A}{C_{A0}}\right)_{\text{batch}} \frac{e^{-t/\tau}}{\tau}\, dt \tag{6.23}$$

Now, the three cases that arise are as follows.

1. Reaction kinetics approximated as zero order.
2. Reaction kinetics approximated as first order.
3. Reaction kinetics followed as Monod kinetics.

First order

We have,

$-\gamma_A = -\,dC_A/dt = C_A/(1 + C_A)$

or $-\gamma_A = -\,dC_A/dt \approx C_A$ (when $C_A \ll 1$)

$$\text{Thus, } C_A/C_{A0} = e^{-kt} \tag{6.24}$$

From Equations 6.23 and 6.24, we get

$$\bar{x}_A = 1 - \int_0^\infty \frac{e^{-kt} e^{-t/\tau}}{\tau}\, dt$$

$$\text{or } \bar{x}_A = 1 - \frac{1}{1 + \tau}$$

$$\text{or } \bar{x}_A = \frac{\tau}{(1 + \tau)} = \frac{10.8}{11.8}$$

Thus, $\bar{x}_A$ = 0.91525

That is, 91.5% conversion for a macrofluid in CSTR assuming first order.

Zero order

We have,

$-\gamma_A = -dC_A/dt = C_A/(1 + C_A) \approx 1$

or $C_{A_0} - C_A = t$

or $C_A = C_{A_0} - t$ (6.25)

Thus, $C_A = C_{A_0}$ at $t = 0$

$C_A = 0$ at $t = 2$ min

$C_A = 0$ at $t > 2$ min

From Equations 6.23 and 6.25, we get

$$\bar{x}_A = 1 - \int_0^\infty \frac{(C_{A_0} - \tau)}{C_{A_0}} \frac{e^{-t/\tau}}{\tau} dt$$

$$\text{or } \bar{x}_A = 1 - \frac{1}{2\tau}\int_0^2 (2 - t)\, e^{-t/\tau}\, dt$$

$$\text{or } \bar{x}_A = 1 - \frac{1}{2\tau}\left[2\frac{e^{-t/\tau}}{(-1/\tau)} - \int te^{-t/\tau}\, dt\right]_0^2$$

$$\text{or } \bar{x}_A = 1 + \left[e^{-t/\tau} + \frac{1}{2\tau}\left(t\frac{e^{-t/\tau}}{(-1/\tau)} - \frac{e^{-t/\tau}}{(-1/\tau)^2}\right)\right]_0^2$$

$$\text{or } \bar{x}_A = 1 + \left[e^{-\tau/\tau} - \frac{t}{2}e^{-t/\tau} - \frac{\tau}{2}e^{-t/\tau}\right]_0^2$$

$$\text{or } \bar{x}_A = 1 + \left[e^{-t/\tau}\left(1 - \frac{t+\tau}{2}\right)\right]_0^2$$

$$\text{or } \bar{x}_A = 1 + \left[e^{-2/\tau}\left(1 - \frac{2+\tau}{2}\right) - 1\left(1 - \frac{\tau}{2}\right)\right]$$

$$\text{or } \bar{x}_A = 1 + \left[e^{-2/\tau}\left(-\frac{\tau}{2}\right) + \frac{\tau}{2} - 1\right]$$

$$\text{or } \bar{x}_A = \frac{10.8}{2}\left[1 - e^{-2/10.8}\right]$$

or $\bar{x}_A = 0.913$

That is, 91.3% conversion for a macrofluid in CSTR assuming zero order.

Mixed performance (Monod kinetics)

We have,

$$-\frac{dC_A}{dt} = \frac{C_A}{1 + C_A}$$

Therefore, considering

$$-\frac{dC_A}{dt} = C_A \text{ for } C_A < 1 \text{ mol/L}$$

that is,

$$-\frac{dC_A}{dt} = 1 \text{ for } C_A > 1 \text{ mol/L}$$

Thus,

$$C_A = C_{A0}e^{-t} \text{ for } C_A < 1 \text{ mol/L}$$

$$\text{or } C_A = (C_{A0} - t) \text{ for } C_A > 1 \text{ mol/L}$$

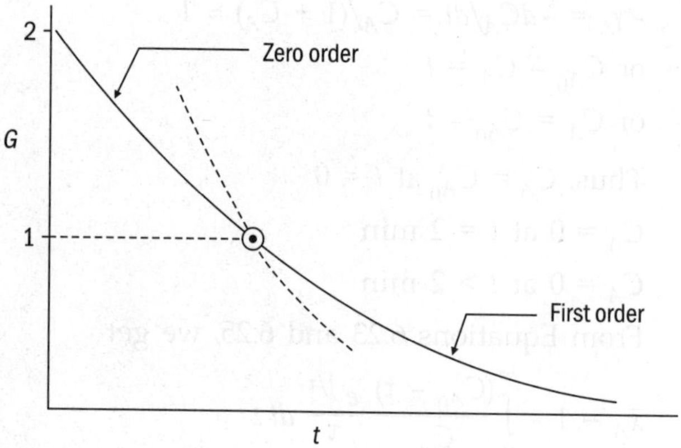

Now,

$$\bar{x}_A = 1 - \int_0^1 \frac{(2-t)}{2\tau} e^{-t/\tau}\, dt - \int_1^\infty \frac{e^{-t}\left(1+\frac{1}{\tau}\right)}{\tau}\, dt$$

$$\text{or } \bar{x}_A = 1 - \frac{1}{2\tau}\int_0^1 (2-t)\, e^{-t/\tau}\, dt - \frac{1}{\tau}\left[\frac{e^{-t}\left(1+\frac{1}{\tau}\right)}{-\left(1+\frac{1}{\tau}\right)}\right]_1^\infty$$

$$\text{or } \bar{x}_A = 1 - \frac{1}{2\tau}\left[\frac{2e^{-t/\tau}}{(-1/\tau)} - \frac{te^{-t/\tau}}{(-1/\tau)} + \frac{e^{-t/\tau}}{(-1/\tau)^2}\right]_0^1 + \frac{1}{1+\tau}\left[0 - e^{-\left(1+\frac{1}{\tau}\right)}\right]$$

$$\text{or } \bar{x}_A = 1 + \left[e^{-t/\tau}\left(1 - \frac{t+\tau}{2}\right)\right]_0^1 - \frac{1}{(1+\tau)}\, e^{-\left(1+\frac{1}{\tau}\right)}$$

$$\text{or } \bar{x}_A = 1 - 4.9\, e^{-\frac{1}{10.8}} + 4.4 - \frac{1}{11.8}\, e^{-\frac{11.8}{10.8}}$$

Thus, $\bar{x}_A = 0.9043$

That is, 90.5% conversion for Monod kinetics.

6.3 ENZYME PRODUCTION STRATEGIES

6.3.1 Batch

Question 6 In the production of cellulase enzyme by *Trichoderma reesei* QM9414 in a batch laboratory stirred tank bioreactor (STB), the organism was found to form pellets.

The pellet diameter varied with cultivation time after its initiation. The diameters of the bioreactor and the impeller were 15 cm and 5 cm, respectively, and the impeller rotation speed was 300 rpm. The pellet formation began on the first day of cultivation when the liquid density was 1.1 g/cm^3. The average weight of the pellets was 0.03 g initially; it increased to 0.09 g on the ninth day of cultivation, when the density of the liquid was 1.3 g/cm^3 and viscosity was approximately 30 centipoise (cP). Considering the ratio of consistency index to liquid viscosity to be of Newtonian type, calculate the value of the homogenization efficiency and diameter of *T. reesei* pellets on the ninth day of cultivation. During cultivation, the height of the liquid in the bioreactor was equal to the diameter of the bioreactor (assume $\alpha = 0.7$, $K = 0.903$).

Answer From the cellulase production system analysis, the following data were obtained: $\mu = 30$ cP = 0.03 g/cm/s; $n = 300$ rpm; $\rho_{(bulk)} = 1.2$ g/cm^3; $D_i = 5$ cm; $D_t = 15$ cm. Since the broth was Newtonian in character, we may assume

$(K/\mu)^{n'} = 1$ and $\alpha = 0.7$

Also, at $t = 1$ day (after initial pelletization),

$M_0 = 0.03$ g, $\rho_0 = 1.1$ g/cm^3, $r_0 = 1.1$ mm

And at $t = 9$ days (after initial pelletization),

$M_t = 0.09$ g, $\rho_t = 1.3$ g/cm^3, $r_t = 1.7$ mm

Using this data, we get

$$N_{Re} = \frac{nD_i^2\rho}{\mu} = \frac{300 \times 5^2 \times 1.2}{60 \times 0.3} = 500$$

For $N_{Re} = 500$, $N_p = 3$, the calculated value of η is

$$\frac{1}{\eta} = \left(\frac{4}{\pi}\right)^3 \left(\frac{K}{\mu}\right)^{n'} \times 3 \times \left(\frac{15}{3}\right)^{1+3\times 0.7} = 192$$

$$\therefore \eta = \frac{0.005}{h} = \frac{0.005}{3600} \text{ per second}$$

$$\text{Now, } r_t = 1.1\sqrt[3]{\frac{2.9}{0.39}}\,\{e^{-0.005(9-1)}\} = 1.98 \text{ mm}$$

Thus, $D_t = 3.96$ mm

The per cent deviation of calculated pellet diameter from the experimental value is

$$\frac{3.96 - 3.4}{3.96} \times 100 \approx 12\%$$

This is in a reasonable range.

Question 7 A continuous cellulase production study was carried out in a single-stage 5 l CeCa (Gallen kamp) laboratory fermenter fitted with control devices at 28°C and pH 5.0 using an aeration rate of 3 L/min at an agitation speed of 450 rpm. The density and viscosity of the broth were found to be 1.2 g/cm and 3.0 cP, respectively. The dilution rate in the vessel was 0.012–0.028 per hour. The fermenter diameter and working volume in the fermenter were 18.0 cm and 3 L, respectively. The variation of yield coefficient with dilution rate is given in Table 6.6.

(a) Calculate the ungassed power requirement.
(b) Calculate the gassed power requirement.
(c) Calculate the maintenance coefficient.
(d) Calculate the true yield coefficient.

Also, assuming an isothermal expression of air, determine the actual compressor horsepower if its efficiency is 90%.

Answer We have,

D_t = 18 cm

Thus,

$$D_i = \frac{1}{3} D_t = 18/3 = 6 \text{ cm}$$

(a) For an ungassed system

$$N_{Re} = \frac{NDi^2\rho}{\mu} = \frac{450 \times 6^2 \times 1.2}{60 \times (3/100)} = 1.08 \times 10^4$$

$$\text{Since } N_p = \frac{Pg}{\rho N^3 Di^5} = \frac{P \times 980}{1.2 \times \left(\frac{450}{60}\right)^3 \times (6)^5}$$

$$\text{or } 6 = \frac{P \times 980 \times (60)^3}{1.2 \times (450)^3 \times (6)^5}$$

Table 6.6 Variation of yield coefficient with dilution rate

D (per hour)	Y_0 (g cell/g cellulase)
0.012	0.212
0.016	0.238
0.020	0.280
0.024	0.277
0.028	0.285

D – dilution rate; Y_0 – yield coefficient

$$\text{or } P = \frac{6 \times 1.2 \times (450)^3 \times (6)^5}{980 \times (60)^3} = \frac{6^6 \times 1.2 \times (450)^3}{6^3 \times 10^3 \times 980}$$

$$\text{or } P = \frac{6^3 \times 1.2 \times 450 \times 450 \times 450}{980 \times 1000} = \frac{216 \times 1.2 \times 45^3}{980}$$

$$\text{or } P = \frac{216 \times 55 \times 350}{980} \text{ gcm/s}$$

$$\text{or } P = \frac{216 \times 5535 \times 1000}{98 \times 3.67 \times 10^5 \times 735} \text{ HP}$$

Thus, $P = 3.3$ HP

(b) For gassed power system,

$$\frac{P_g}{P} = 0.68$$

Thus, $P_g = 3.3 \times 0.68 = 2.24$ HP

(c) Maintenance coefficient

From the data given in Table 6.6, a plot of $1/D$ versus $1/Y_0$ is obtained (Figure 6.2). From the slope of this plot, the maintenance coefficient m is 0.031.

(d) From the intercept of this slope, the true yield coefficient Y_m is 0.45.

Finally, considering the isothermal expansion of air and compression efficiency 90%, actual compressor horsepower is $0.90 \times 2.24 \approx 2.21$ HP.

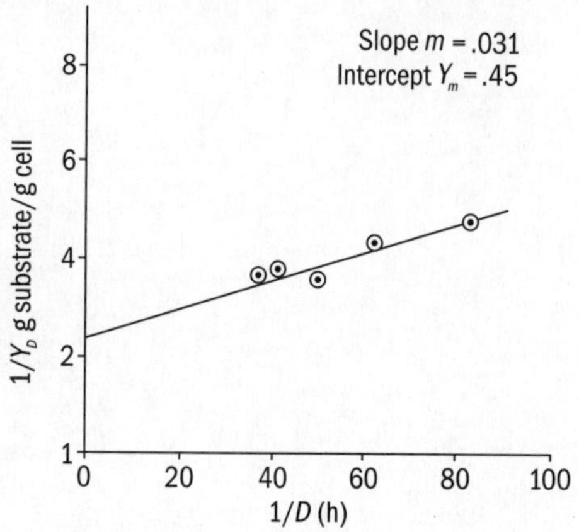

Figure 6.2 Steady state data: yield coefficient versus dilution rate

BIBLIOGRAPHY

Aiba S, Hamphrey A E, and Millis N F. 1973. ***Biochemical Engineering***, 2nd edn. Tokyo: University of Tokyo Press

Bailey J E and Ollis D F. 1986. ***Biochemical Engineering Fundamentals***, 2nd edition. New York: McGraw-Hill Book Company. 918 pp.

Green D W and Perry R H. 2008. ***Perry's Chemical Engineers' Handbook***, 8th edition, pp. 715–716. New York: McGraw Hill.

Levenspiel O. 2002. ***Chemical Reaction Engineering***, 3rd edition. New York: John Wiley and Sons

Mukhopadhyay S N (ed.). 2010. ***Process Biotechnology Fundamentals***, 3rd edition. New Delhi: Viva Books

Sahai V and Ghose T K. 1977. **Proceedings of the Bioconversion Symposium**, pp. 269–280. New Delhi: IIT Delhi

7

Product Recovery and Purification: Solved Numerical Problems

7.1 INTRODUCTION

The increasing demand for enzymes and biocellular products as highly specific and active biocatalysts for a variety of processing and analytical applications depends on the ready availability of reasonably pure and low-cost enzymes and biocells. This inevitably entails the development of large-scale processes and equipment. Progress in this direction depends not only on the refinement of individual separation techniques but also on engineering advances in the application of these procedures to large-scale continuous flow or batch processing. The approach used in technical evaluation of a situation is listed below.

- Identification of existing methods of purification.
- Establishment of criteria for scale up of techniques with a view to identifying their potential.
- Analysis of engineering problems involved in the scale up.
- Identification of methods most suitable for large-scale operation and comparison with existing processes if possible.

It is appreciated that these would be more meaningful if techno-economic studies were available. Unfortunately, meaningful economic data are scanty due to limited resources and other restraints.

Methods for separating and isolating components of a mixture have existed since the beginning of chemical science, and the progress of chemical process industry has been related to the development of such separation techniques to a great extent.

All separation processes are concerned with two forces: (1) the impelling forces (F_1) and (2) the retarding forces (F_2). Thus, the rate of movement of a solute molecule

A will depend on $F_1^A - F_2^A = \Delta F^A$, while that of solute molecule *B* will depend on $F_1^B - F_2^B = \Delta F^B$. Also, for a given distance in the apparatus, the separation achieved will depend on the ratio $\Delta F^A/\Delta F^B$. The ΔF values of all separation processes for various species should be different so that adequate resolution can be obtained for short distances. Mukhopadhyay (2010) showed the different conventional separation processes available for separation of components of a mixture, along with the types of forces (F_1 and F_2) involved in each process.

7.2 FILTRATION

Question 1 A biological slurry is to be classified by filtration. The results obtained from a leaf filter with 205 ft^2 of filtering area during constant filter at 50 psi are given in Table 7.1.

The slurry before filtration was found to contain 5 lb of solid per 100 lb of liquid. The densities of the solid and liquid were 78.5 lb/ft^3 and 62.4 lb/ft^3, respectively. The viscosity of the slurry was 2 cP, and the cake was found to be incompressible.

Determine the specific resistance, γ, of the cake and the equivalent length of the fitter cloth.

Answer From Carman's theory, we have

$$dV/d\theta = P\,A^2/\mu(\gamma\,C\,V + R_m\,A) \tag{7.1}$$

where *V* is the volume of filtrate obtained at time θ, γ is the specific resistance of the cake, R_m is the resistance of the filter medium, *A* is the filter area, *C* is the solid content in liquid before filtration, *P* is the filtration pressure, and μ is the viscosity.

Integration of Equation 7.1 of constant pressure gives

$$\theta = (\mu\,\gamma\,C/2PA^2)V^2 + (\mu R_m/P\,A)V \tag{7.2}$$

$$\text{or } \theta/V = (\mu\,\gamma\,C/2PA^2)V + (\mu R_m/P\,A) \tag{7.3}$$

$$\text{or } \theta/V = b\,V + a \tag{7.4}$$

where $b = \mu\,\gamma\,C/2PA^2$ and $a = \mu R_m/P\,A$

Plotting θ/V against *V* (Table 7.2) gives a straight line plot whose slope is $b = \mu\,\gamma\,C/2PA^2$ (Equation 7.5). The value of *r* can be determined since the values of μ, *C*, *P*, and *A* are known.

Table 7.1 Results obtained

Time (min)	Filtrate volume (ft^3)
15	100
30	163
45	213
60	258
90	235

Table 7.2 Calculated values of θ/V from the corresponding values of V

θ/V (min/ft^3)	V (ft^3)
15/100 = 0.150	100
30/163 = 0.184	163
45/213 = 0.211	213
60/258 = 0.232	258
90/235 = 0.385	235

$$b = 7.5 \times 10^{-4} \text{ min/ft} = 7.5 \times 10^{-4} \times 60 \text{ s/ft}^3 \quad (7.5)$$

or $b = 45 \times 10^{-2}$ s/ft^3

Now,

$P = 50$ lb/in^2 = 50×12^2 lb/ft^2

Filtrate viscosity = 2 cP = $2 \times 6.72 \times 10^{-4}$ lb/ft-s

Filter area $A = 20$ ft^2

$C = 5$ lb/100 lb liquid

or $C = 5$ lb/(100/62.5) ft^3

or $C = 5/1.6$ (lb/ft^3) = 3.12 (lb/ft^3)

We have,

$b = \mu\, \gamma\, C/2PA^2$

or $\gamma = 2PA^2b/\mu \times C$

$$\text{or } \gamma = \frac{2 \times 50 \times 12^2 \times 20^2 \times 45 \times 10^{-2}}{2 \times 3.12} \text{ s}^2/\text{lb}$$

$$\text{or } \gamma = \frac{100 \times 144 \times 400 \times 45 \times 10^{-2}}{6.24} \text{ s}^2/\text{lb}$$

$$\text{or } \gamma = \frac{144 \times 400 \times 45}{6.24} \text{ s}^2/\text{lb}$$

or $\gamma = 415384.62$ s^2/lb

Question 2 In Question 1, the observed specific resistance varied with pressure. Table 7.3 gives the observations of specific resistance with corresponding pressure.

Using Carman's equation (Equation 7.6), determine the coefficient of compressibility and the constant of variation.

$$r = r'P^S \quad (7.6)$$

Answer We have,

$r = r'P^S$

or $\log r = S \log P + \log r'$

Table 7.3 Observed pressure and specific resistance relationship

Pressure (psi)	Specific resistance (s^2/lb)
251.2	7.943×10^9
316.2	9.550×10^9
398.1	1.122×10^{10}
501.2	1.380×10^{10}
631.0	1.585×10^{10}
794.3	1.905×10^{10}

The plot of log r versus log P (Table 7.4) produces a straight line with slope S and intercept log r'.

In Figure 7.1, slope = coefficient of compressibility. That is

$$S = \frac{0.15}{0.2}$$

Thus,

$$S = 0.75$$

7.3 ULTRAFILTRATION

Question 3 Determine the flux rate expected in a tubular ultrafiltration system being used to concentrate milk. The following conditions apply: (1) density of milk = 1.03 g/cm^3, (2) viscosity = 0.8 cP, (3) diffusivity = 7×10^{-7} cm^2/s, (4) C_B = 3.1% weight per unit volume, (5) C = 22% weight per unit volume, (6) diameter of tube = 1.1 cm, (7) length = 220 cm, (8) number of tubes = 15, and (9) fluid velocity = 1.5 m/s.

Answer From the data given, we have

$$N_{Re} = \frac{\rho ud}{\mu} = \frac{1.03 \text{ g/cm} \times 150 \text{ cm/s} \times 1.1 \text{ cm}}{0.008 \text{ poise}} = 21243$$

$$N_{Sc} = \mu/\rho D$$

Table 7.4 Values of log r and log P

log r	log P
9.98	2.5
10.05	2.6
10.14	2.7
10.20	2.8
10.28	2.9

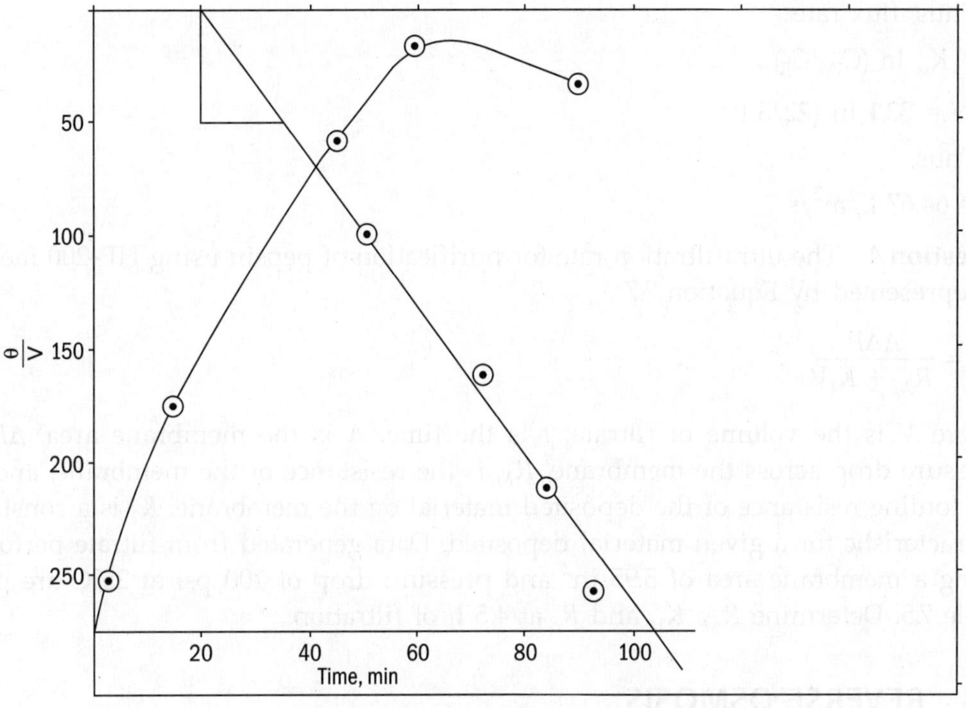

Figure 7.1 Plot of θ/V versus time

or $N_{Sc} = \dfrac{0.008 \text{ poise}}{1.03 \text{ g/cm}^3 \times 7 \times 10^{-7} \text{ m}^2/\text{s}}$

or $N_{Sc} = 1.11 \times 10^4$

$N_{Sh} = K_m d/D$

or $N_{Sh} = 0.023(N_{Re})^{0.8}\ (N_{Sc})^{0.33}$

or $N_{Sh} = 0.023\ (21\,243)^{0.8}\ (1.11 \times 10^4)^{0.33}$

or $N_{Sh} = 1440$

Thus, $K_m d/D = 1440$

or $\dfrac{K_m \times 1.1}{7 \times 10^{-7}} = 1440$

or $K_m = \dfrac{1440 \times 7 \times 10^{-7}}{1.1}$

or $K_m = 0.92 \times 10^{-3}$ cm/s

or $K_m = 0.92 \times 10^{-3}$ cm^3/m^2/s

or $K_m = 33.1$ L/m^2/s

Thus, flux rate

$N = K_m \ln (C_W/C_B)$

or $N = 33.1 \ln (22/3.1)$

Thus,

$N = 64.67 \text{ L/m}^2\text{/s}$

Question 4 The ultrafiltration rate for purification of pepsin using HF–200 membrane is represented by Equation 7.7.

$$\frac{dV}{dt} = \frac{A\Delta P}{R_M + K_1 V} \tag{7.7}$$

where V is the volume of filtrate, t is the time, A is the membrane area, ΔP is the pressure drop across the membrane, R_M is the resistance of the membrane, and K_1V is the fouling resistance of the deposited material on the membrane, K_1 is a constant and characteristic for a given material deposited. Data generated from filtrate performance using a membrane area of 5.95 in^2 and pressure drop of 100 psi at 25°C are given in Table 7.5. Determine R_M, K_1, and R_e at 4.5 h of filtration.

7.4 REVERSE OSMOSIS

Question 5 About 25 × 10^9 gal of a biological liquid is treated in a reverse osmosis system. The output rate of the liquid is 100 000 gal/day. Records show that the lowest liquid temperature expected for any prolonged period is 68°F (20°C). Over its two years of operation, the cellulose acetate membrane chosen for this system is expected to have an average flux of 15 gal/day/ft^2 operating at 600 psig. The average flux was determined at a base temperature of 77°F. Specify the membrane area for the reverse osmosis system and also the time required to treat the liquid.

Table 7.5. Data generated from filtrate performance

Filtration time (h)	Filtrate volume (ml)	Filtration rate (ml/min)
0.0	—	—
0.5	3.0	6.0
1.0	4.4	4.4
1.5	6.0	4.0
2.0	7.6	3.8
3.0	9.0	3.0
4.0	11.6	2.9
5.0	14.0	2.8
6.0	16.8	2.78
6.5	18.0	2.77

Answer The required membrane area is computed as follows.

$$\text{Membrane area} = \frac{\text{Output rate of liquid (gal/day)}}{\text{Flux of liquid (gal/day/ft}^2\text{)}}$$

or membrane area = 100000/15

or membrane area = 6667 ft^2

The correction factor, C_F, for feed at 20°C (68°F) is 1.15.

Thus,

area at 68°F = 6667 × 1.15 ft^2 = 7667 ft^2

Therefore,

$$\text{time required} = \frac{25 \times 10^9 \text{ gal}}{15 \times 7667 \text{ gal/day}}$$

$$\text{or time required} = \frac{25 \times 10^9}{15 \times 7667} \text{ days} \approx 25\,000 \text{ days}$$

7.5 FOAM FRACTIONATION

Question 6 A placental protein concentrate containing 0.001 kg/m^3 protease enzyme was separated at 10°C using foam fractionation in a column of diameter 0.04 m having a foam bed height of 0.42 m. Considering volume of feed 100 ml, surface tension of the fluid to be given by the ratio of mg protein/2L_0 and effluent protein reading to be 334.1 mg over a strip length (L_0) of 0.25 m and foam residence time 4 s, calculate the following.

(a) Surface tension of the fluid.
(b) Fold purification of the protease enzyme.
(c) Heat of desorption (λ).
(d) Mass transfer coefficient.
(e) Average Sauter mean bubbler diameter.
(f) Number of transfer units (NTU) required.
(g) Check per cent deviation in height of transfer units (HTU) calculation.

Use Figures 7.2 and 7.3 and Tables 7.6 and 7.7 and make suitable assumptions if required.

Answer

(a) Calculation of surface tension.

Placental homogenate of protein concentration 0.001 kg/m^3 gave a reading of 334.1 mg.

Length of the strip (L_0) = 0.025 m

From the experiment, we have

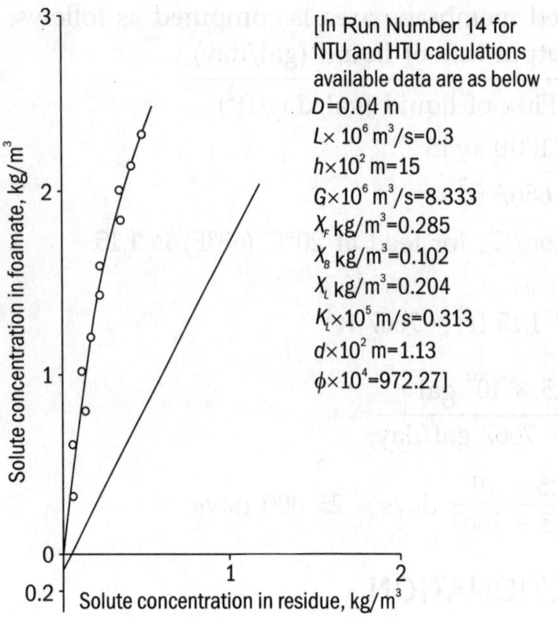

Figure 7.2 Equilibrium curve and operating line (for Run Number 14, for placental homogenate at 283 K)

Figure 7.3 Determination of the mass transfer coefficient $K\mu$

Table 7.6 Computation of integrand at various residence times

Residence time, S	X_F(kg/m^3)	X_B(kg/m^3)	$1/[\lambda - RT \ln(X_F/X_B)]$
0	0.310	0.310	2.143×10^{-4}
4	0.395	0.295	2.512×10^{-4}
6	0.490	0.270	3.063×10^{-4}
10	0.605	0.230	4.182×10^{-4}
16	0.728	0.198	6.237×10^{-4}
27	0.845	0.162	1.281×10^{-3}
35	0.900	0.150	2.217×10^{-3}

Table 7.7 Calculation of area by Simpson's rule

Residence time, S	X_{FO}–X_B (kg/m^3)	Area
4	0.310–0.395	0.198×10^{-4}
6	0.310–0.490	0.463×10^{-4}
10	0.310–0.605	0.880×10^{-4}
16	0.310–0.728	1.521×10^{-4}
27	0.310–0.845	2.635×10^{-4}
35	0.310–0.900	3.597×10^{-4}

Surface tension γ is given by $mg/2L_0$.

Hence,

$$\gamma = \frac{mg}{2L_0} = \frac{334.1 \times 10^{-6} \times 9.81}{2 \times 0.025} \text{ kg-m/m-s}^2$$

or $\gamma = 65.55 \times 10^{-3}$ N/m

(b) Calculation of fold purification of enzyme.

In the experiment, results obtained for $G = 5.00 \times 10^{-6}$ m^3/s in the foam separation column were as below.

Specific activity of enzyme in the feed = (0.49/0.204) U/mg = 2.402 U/mg

Specific activity of enzyme in the foamate = (5.27/0.372) U/mg = 14.167 U/mg

Therefore, fold purification of enzyme = 14.167/2.402 = 5.9

(c) Calculation of heat of desorption.

For calculating the heat of desorption, Equation 7.8 can be used.

$$\ln (V_L/V_B) = [1/(e^{\lambda}/RT - 1)] \ln (X_L/X_B) \quad (7.8)$$

where V_L is the initial volume and V_E is the bulk volume.

Putting the corresponding values, we get

$\ln [100/(100 - 12)] = 1/(e^{\lambda/RT} - 1)\} \ln (0.234/0.105)$

or $1/(e^{\lambda/RT} - 1) = 0.1595$

or $\lambda = 4667$ kJ/kmol

(d) Calculation of mass transfer coefficient.

From Table 7.7, for residence time = 4 s,

$1/[\lambda - RT \ln(X_L/X_B)] = 1/[4667 - (8.3143 \times 283) \ln(0.395/0.295)]$

or $1/[\lambda - RT \ln(X_L/X_B)] = 2.512 \times 10^{-4}$ kmol/kJ

Now, from Table 7.7, the integrand in Equation 7.8 for each time interval has been evaluated by Simpson's rule.

The area of the two faces of each lamella is 0.5 $\pi D^2 m^2$. About 6.05 ml of the foamate was collected in 60 s, and the number of lamellae flowing per second was found to be 5.

Hence,

$V = (6.05 \times 10^{-6})/(60 \times 5)\ m^3$

Therefore,

$A/V = (0.5\ \pi D^2)/V = 124\ 625\ m^2/m^3$

The slope of the straight line in Figure 7.3 is

$$\frac{(2.728 - 0.53) \times 10^{-4}}{27 - 7} = 1.099 \times 10^{-5}$$

Hence,

$K_\mu\ A/V = 1.099 \times 10^{-5}$ (from Equation 7.8)

or $K_\mu = 8.82 \times 10^{-11}$ kg-kmol/kJ-m^2-s

(e) Calculation of average foam bubble diameter d.

From Figure 7.3 (for Run Number 14) and according to Equation 7.9,

$$d = \frac{\Sigma n_1 d_1^3}{\Sigma n_1 d_1^2} \tag{7.9}$$

$$\text{or } d = \frac{2(1.1)^3 + 4(0.9)^3 + 4(1.0)^3 + 4(1.2)^3 + 2(1.3)^3 + 2(1.4)^3 + 3(0.6)^3 + 1(0.8)^3}{2(1.1)^2 + 4(0.9)^2 + 4(1.0)^2 + 4(1.2)^2 + 2(1.3)^3 + 2(1.4)^2 + 3(0.6)^2 + 1(0.8)^2} \text{ cm}$$

or $d = 1.13$ cm

or $d = 1.13 \times 10^{-2}$ m

(f) Calculation of NTU (for Run Number 14).

The slope and the intercept of the operating line are given by Equations 7.10 and 7.11.

Slope = L/F = 1.796 (7.10)

Intercept on y-axis = $-(B/F)\,X_B = -\,0.081$ kg/m^3 (7.11)

With this slope and intercept, the operating line has been superimposed on the equilibrium diagram as shown in Figure 7.2.

Now,

$$\text{NTU} = \int_{\overline{X}_B^*}^{X_F} \frac{d\overline{X}}{\overline{X}^* - \overline{X}} = \int_{X_F}^{X_B^*} \frac{d\overline{X}}{\overline{X}^* - \overline{X}}$$

$$\text{or NTU} = \int_{0.285}^{0.72} \frac{d\overline{X}}{\overline{X}^* - \overline{X}}$$

$\overline{X}_B^*$ is obtained with the help of the equilibrium curve corresponding to X_B = 0.102. Now, the integrand has been evaluated numerically by Simpson's rule. The corresponding values of $\overline{X}$ and $\overline{X}^*$ are obtained with the help of the operating line and equilibrium curve, respectively.

Following this procedure, the value of NTU obtained is 1.790.

(*g*) Per cent calculation of deviation (Run Number 14).

Per cent deviation = (actual HTU – predicted HTU) × 100/actual HTU

or per cent deviation = [(0.0838 – 0.08) × 100/0.0838] = 4.53

7.6 ADSORPTION

Question 7 Adsorption of gentamicin from water at pH 9.5 was carried out on non-ionic carbon and the data were recorded (Table 7.8). What adsorption isotherm fits the data?

Answer The data are not linear on Cartesian coordinates (Figure 7.4), but these are linear in a log y versus log q plot (Figure 7.5). The data are not linear in a q^{-1} versus y^{-1} plot (Figure 7.4). As a result, the data appears to fit the Freundlich isotherm but not

Table 7.8 Observed data on adsorption

Y (mg/L)	q (mg/cm^3)
0.01	5.3
0.02	7.1
0.03	8.3
0.06	11.1
0.10	13.7

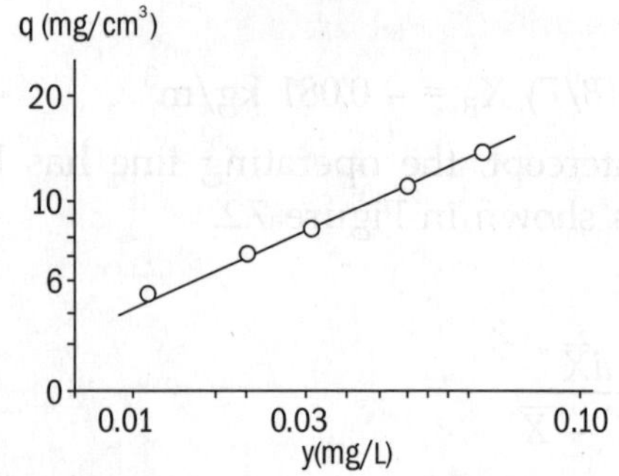

Figure 7.4 Plot of concentration versus volumetric concentration

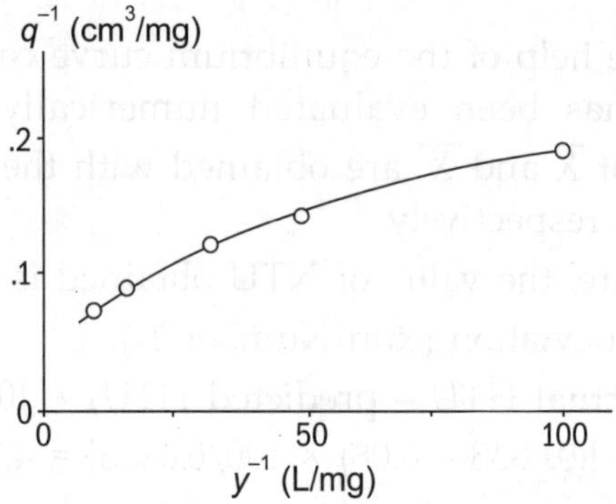

Figure 7.5 Double reciprocal relation of q and y

the Langmuir isotherm. It should be noted that in many cases, the data get scattered and there is no need to choose one isotherm over the other.

Question 8 About 80 cm^3 of modified dextran absorbed up to 7.8×10^{-6} mol of immunoglobulin G(*IgG*)/cm^3 of adsorbent. This adsorption followed a Langmuir isotherm, with a constant K of 1.9×10^{-5} mol/L. What concentration in 1.2 L of feed solution will exhaust 90% of the capacity of dextran?

Answer For 90% recovery, we expect that

$$y = 0.1\ y_F \tag{7.12}$$

From the adsorption data given, we have

$$q = \frac{(7.8 \times 10^{-6}\ \text{mol/cm}^3)\ 0.1\ y_F}{(1.9 \times 10^{-5}\ \text{mol/L} + 0.1\ y_F)} \tag{7.13}$$

From a mass balance,

$$1.2\ \text{L}\ y_F = 1.2\ \text{L}\ (0.1)\ y_F + 80\ \text{cm}^3\ q \tag{7.14}$$

Combining Equations 7.12, 7.13, and 7.14, we get

$y_F = 3.9 \times 10^{-4}$ mol/L

Question 9 The gentamicin adsorption on activated carbon as given in Question 7 followed Equation 7.15.

$$q = 0.351\ y^{0.41} \tag{7.15}$$

where q is in mg/cm^3 carbon and y is in mg/L.

If 10 cm^3 of previously unused carbon is added to 3 L of a fermentation been containing 46 mg/L antibiotic, what will be the per cent recovery?

Answer The operating line is obtained from Equation 7.16.

$$q = q_F + \frac{H}{W}(y_F - y) \tag{7.16}$$

Here,

$$q_F = 0$$

Thus,

$$q = \frac{H}{W}(y_F - y)$$

$$\text{or } q = \frac{3.0\text{L}}{10\ \text{cm}^2}(46\ \text{mg/L} - y)$$

That is,

$$q = 13.8 - 0.3\ y \tag{7.17}$$

At equilibrium,

$$q = ky^n \tag{7.18}$$

Plotting lines 7.16 and 7.17 gives the intersection (Figure 7.6).

Thus, per cent recovery is

$$\frac{46 - 0.105}{46} = 99.8\%$$

The per cent recovery is high because adsorption is strongly favourable.

7.7 EXTRACTION

Question 10 For a water-amyl acetate system, the values of K_i for penicillin K and penicillin F are 215 and 213, respectively, and that of pk_a are 2.77 and 3.51, respectively. If penicillin F is the desired product, will an extraction at pH 3 give a purer product than one at pH 4?

Answer From the values given, we have

$$K = \frac{K_i}{1 + \dfrac{K_a}{[H^+]_H}} \tag{7.19}$$

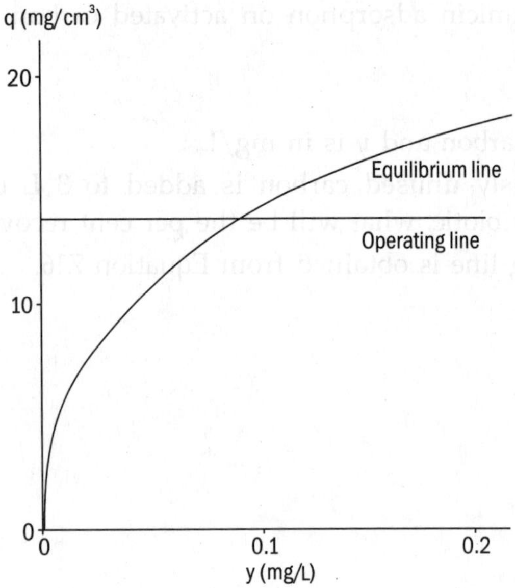

Figure 7.6 Equilibrium and operating time relation between q and y

$$\text{or } \log_{10}\left[\left(\frac{Ki}{K}\right) - 1\right] = \text{pH} - pK_a$$

where K_i is the intrinsic partition coefficient and β is the selectivity of separation given by

$$\beta = \left(\frac{K_i(A)}{K_i(B)}\right)\left(\frac{1 + [K_a(B)/H^+]}{1 + [K_a(A)/H^+]}\right)$$

The values of K_K and K_F are calculated accordingly (Table 7.9).

Thus, β = 1.3 at pH 3.0

and β = 2.7 at pH 4.0

Hence, the extraction at pH 4 may require certain stages but the isolated solution will be purer.

7.8 NOVEL BIOPRODUCT RECOVERY DESIGN PROBLEM

Question 11 A separation/purification unit was designed for β-hCG hormone recovery obtained from an upstream bioprocessing of a microcarrier-supported culture of

Table 7.9 Varitation of K_K and K_F in relation to pH

pH	K_K	K_F
pH 3	80	100
pH 4	12	32

Vero cell transfected by recombinant vaccinia virus by transfectional fermentation biotechnology (TFB). The production data and characteristics are as given.

Culture volume = 10 L

Extracellular β-hCG concentration = 4.62×10^{-3} g/L

Thus, total extracelluar β-hCG in 10 L = 4.62×10^{-2} g = 46.2 mg

Intracellular β-hCG concentration = 0.45 μg/ml equivalent cell

$$\text{Total volume of cells} = 10\ \text{L} \times \frac{0.393\ \text{cm}^3}{\text{ml}} \times \frac{10^3\ \text{ml}}{1\text{L}}$$

or total volume of cells = 3.93×10^3 cm^3

Thus, the total intracellular β-hCG is

$3.93 \times 10^3 \times 0.45$ μg = 1.8×10^3 μg = 1.8 mg

7.9 POSSIBLE CONTAMINANTS

Question 12 A culture medium contains, besides β-hCG, unused proteins, lipids, protease inhibitors, vitamins, lipopolysaccharides, and so on. The main source of these contaminants is the serum used in the culture medium. Besides unused glucose and glutamine, produced lactate and ammonia are other contaminants. These will not inhibit the purification of β-hCG since these are not proteins. In case of intracellular β-hCG, host DNA and virus particles are major contaminants present in the cell lysis medium. When somebody tries to recover the intracellular pool of β-hCG, major care has to be taken for these contaminants, because host cell DNA and virus particles can severely affect the immunogenic response of the body when injected with the vaccines. A tedious sepharose fast flow cation exchange chromatographic steps seem to be the only solution to get rid of these intracellular contaminants.

The potential contaminants are listed in Table 7.10.

The main protein impurity for DMEM media supplemented with 5% newborn calf serum (NBCS) is albumin.

The albumin concentration expected for 5% serum supplemented media is

40 × 0.60 × 0.05 mg/ml = 1.2 mg/ml = 0.12 wt%

Provide calculations in design steps for the separation and purification of β-hCG. The flowchart and schematic diagram of downstream processing are given in Figures 7.7 and 7.8, respectively.

Prior to step 1 (Figure 7.7), when both extracellular and intracellular β-hCG pools are mixed, the total amount of β-hCG present in the medium is 46.2 + (1.8 × 0.7). That is, the total amount of β-hCG is 47.46 mg.

Table 7.10 Potential contaminants

Source	Potential contaminant	Molecular kDa
NBCS	Total protein ~40 mg/ml	—
	Albumin (60%)	68
	Immunoglobulin (<0.1%)	150
	Protease inhibitor (910%)	—
	Transferrin (2.5%)	3
	Insulin (1%)	3
	Miscellaneous protein (20%–35%)	—
	Proteases	—
	Growth factors	—
	Lipids	—
	Vitamins	—
DMEM	Sugars	—
	Trace elements	—
Intracellular	Host cell DNA	—
	Virus particles	—

DMEM – Dulbecco's Minimal Eagle's Medium; kDa – kilo Dalton

We have,

medium volume = 10 L

Thus,

$$\beta\text{-hCG concentration} = \frac{47.46 \text{ mg}}{10 \text{ L}}$$

$$\text{or } \beta\text{-hCG concentration} = \frac{0.0004746 \text{ g}}{100 \text{ ml}}$$

β-hCG concentration = 0.0004746 wt%

Serum albumin concentration = 0.12 wt%

7.9.1 Ultrafiltration steps

Step 1: Ultrafiltration using 60 kDa membrane cut-off

Aim To exclude serum albumin from medium containing β-hCG.

In this case, (1) concentration of albumin = 0.12 wt%; (2) molecular weight = 69 000; (3) medium volume (V_0) = 10 L; and (4) diffusivity of albumin in water at 25°C (D_{H_2O}) = 5.94×10^{-7} cm^2/s.

The diffusivity of albumin in medium at 4°C (H_2O) is

$$(D_{H_2O}) = (5.94 \times 10^{-7}) \left(\frac{\mu_{water}}{\mu_{medium}}\right) \left(\frac{277}{298}\right) \text{ cm}^2/\text{s}$$

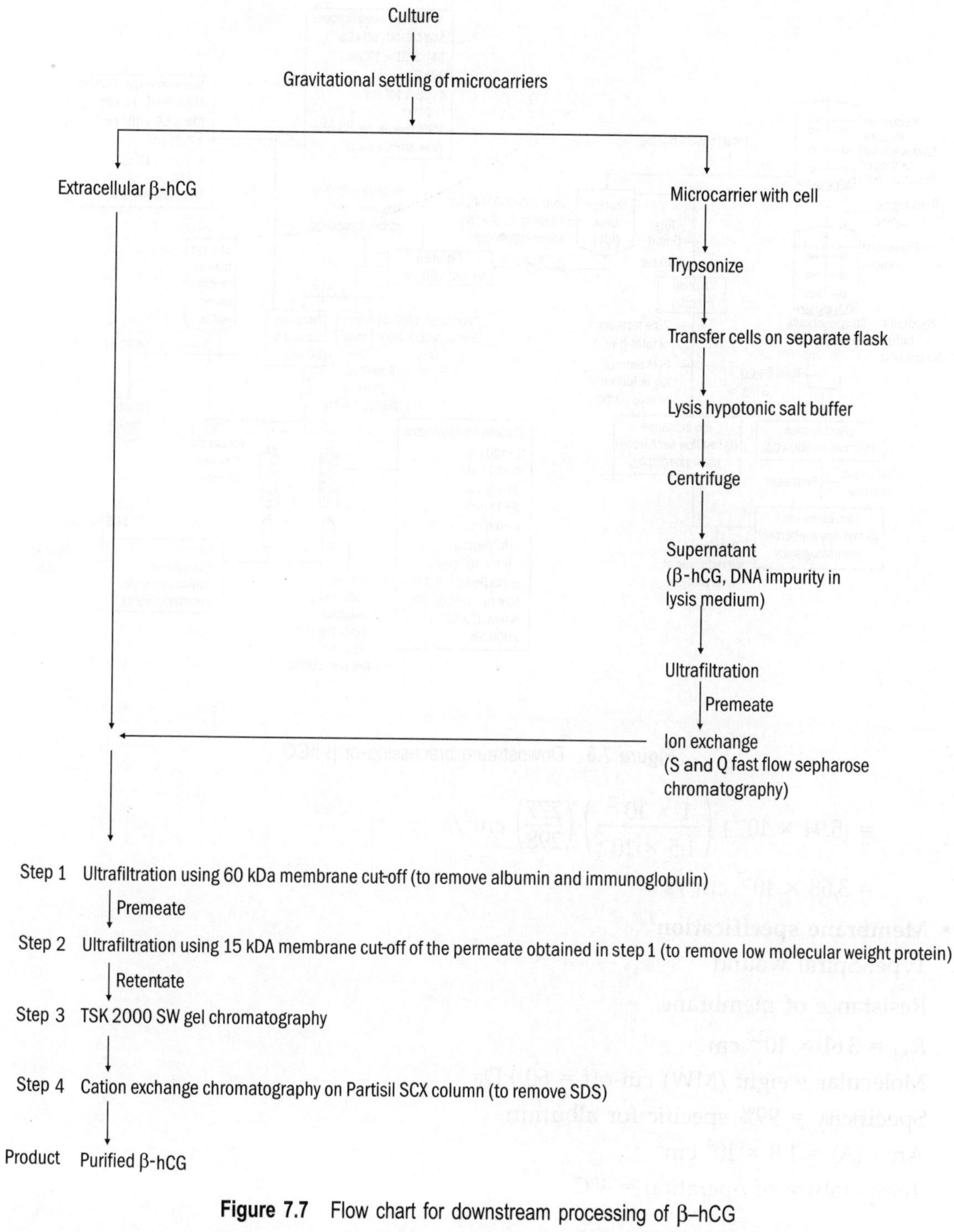

Figure 7.7 Flow chart for downstream processing of β–hCG

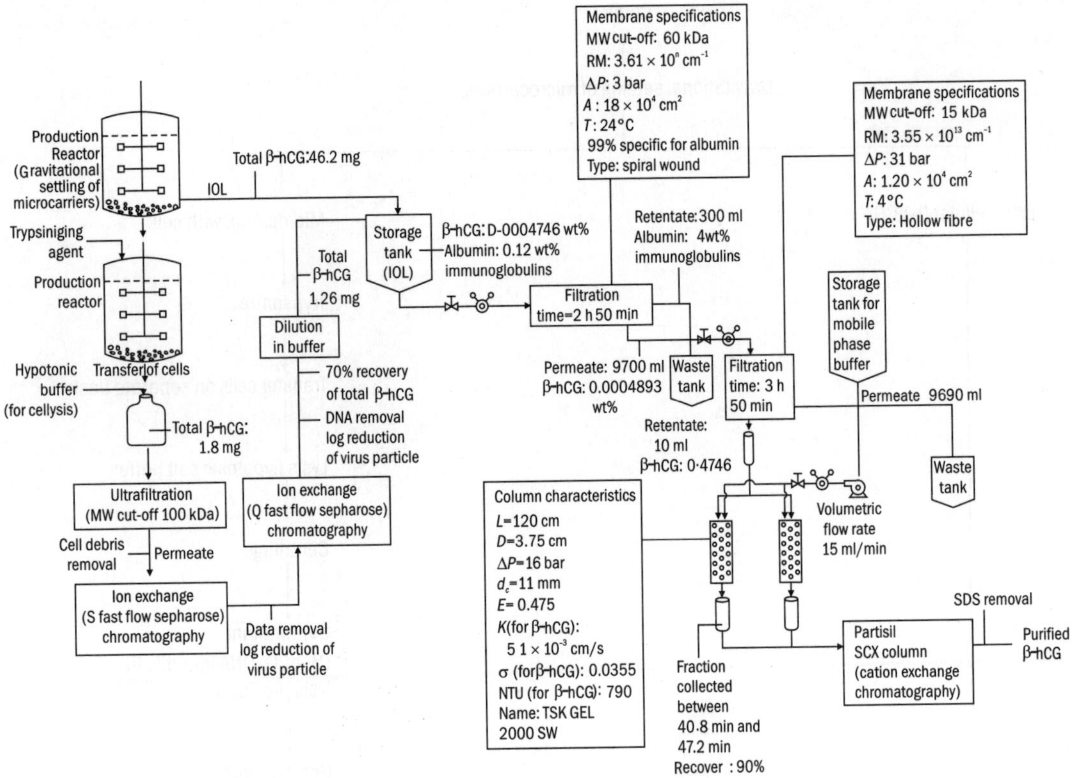

Figure 7.8 Downstream processing of β-hCG

$$= (5.94 \times 10^{-7}) \left(\frac{1 \times 10^{-2}}{1.5 \times 10^{-2}}\right) \left(\frac{277}{298}\right) \text{ cm}^2/\text{s}$$

$$= 3.68 \times 10^{-7} \text{ cm}^2/\text{s}$$

- **Membrane specification**[1]

  Type: Spiral wound

  Resistance of membrane,

  $R_M = 3.61 \times 10^{12}$ cm^{-1}

  Molecular weight (MW) cut-off = 60 kDa

  Specificity = 99% specific for albumin

  Area (A) = 1.8×10^4 cm^2

  Temperature of operation = 4°C

---

[1] In consultation with scientists at NCL, Pune.

Pressure = 3 atm

We have, $L_M = \dfrac{1}{R_M\mu}$

where L_M is the length of the membrane and $R_M\mu$ is the specific resistivity of the membrane

or $L_M = \dfrac{1}{3.61 \times 10^{12} \times 1.5 \times 10^{-2}}$

or $L_M = 1.846 \times 10^{-11}$ cm²/s/g

Now,

$j_v = L_P(\Delta P - \sigma\Delta\pi)$ (Ghosh and Mukhopadhyay 2000)

where j_v is the velocity of the solvent, L_P is the permeability, ΔP is the pressure difference, σ is the reflection coefficient, and $\Delta\pi$ is the osmotic pressure.

Assuming $\Delta\pi$ to be negligible,

$j_v = L_P\Delta P$

or $j_v = (1.846 \times 10^{-11})\ (3 \times 10.013 \times 10^6)$

or $j_v = 5.61 \times 10^{-5}$ cm/s

- **Final concentration of albumin in retentate (*V*) = 4 wt%**

$$V = \frac{10 \times 0.12}{4} = 0.3 \text{ L}$$

$$\frac{RTn_1}{\Delta P} = \frac{(0.082)\ (277)\ (10 \times 10^3 \times 0.12)}{3 \times 69\,000}$$

where R is the absolute gas constant and T is the absolute temperature.

That is, $\dfrac{RTn_1}{\Delta P} = 0.13$ L

$$t = \frac{1}{AL_P\Delta P}\left[(V_0 - V) + \frac{RTn_1}{\Delta P}\ln\left(\frac{V_0 - \dfrac{RTn_1}{\Delta P}}{V - \dfrac{RTn_1}{\Delta P}}\right)\right] \quad \text{(Ghosh and Mukhopadhyay 2000)}$$

$$\text{or } t = \frac{1}{(1.8 \times 10^4)\ (5.61 \times 10^{-4})}\left[(10\,000 - 300) + 130 \ln\left(\frac{1000 - 130}{300 - 130}\right)\right]$$

or $t = 10128.72$ s

Thus, ultrafiltration time, t = 2 h 49 min

- **Check for polarization**

$$\frac{D}{lj_v} = \frac{3.68 \times 10^{-7}}{(0.001)\ (5.61 \times 10^{-5})} = 6.56 \gg 1$$

Thus, the assumption that $\Delta\pi$ should be negligible holds good.

Step 2: Ultrafiltration using 15 kDa membrane cut-off

Aim To concentrate β-hCG in the retentate as well as exclude proteins of low molecular weight, growth factors, and hormones from the retentate.

In this case, (1) permeate volume obtained from Step 1 = 9700 ml; (2) per cent weight of β-hCG = $\frac{47.46 \text{ mg}}{9700 \text{ ml}}$ = 0.0004893 wt%; (3) molecular weight of β-hCG = 24 000; and (4) diffusivity of β-hCG in water = 9.4×10^{-7} cm^2/s at 25°C.

Thus, diffusivity of β-hCG in the medium at 4°C,

$$D = (9.5 \times 10^{-7}) \left(\frac{\mu_{\text{water}}}{\mu_{\text{medium}}}\right) \left(\frac{277}{298}\right)$$

$$\text{or } D = (9.5 \times 10^{-7}) \left(\frac{10^{-2}}{1.5 \times 10^{-2}}\right) \left(\frac{277}{298}\right)$$

$$\text{or } D = 5.88 \times 10^{-7} \text{ cm}^2/\text{s}$$

- **Membrane specification**

  In this case, (1) the membrane type is hollow fibre; (2) resistance of membrane, $R_M = 3.554 \times 10^{13}$ cm^{-1}; (3) cut-off = 15 kDa; (4) area = 1.20×10^4 cm^2; temperature of operation = 4°C; and (5) pressure drop = 31 atm.

  Now,

  $j_v = L_P(\Delta P - \sigma\Delta\pi)$ (Ghosh and Mukhopadhyay 2000)

  Assuming $\Delta\pi$ to be negligible,

  $j_v = L_P\Delta P$

  or $j_v = (1.87 \times 10^{-12})(31 \times 1.031 \times 10^6)$

  or $j_v = 5.87 \times 10^{-5}$ cm/s

  The final concentration of β-hCG in the retentate is 0.4746 wt%.

  $$\text{Thus, } V = \frac{9700 \times 0.0004849}{0.4746} = 10 \text{ ml}$$

  $$\frac{RTn_1}{\Delta P} = \frac{(0.082)\,(277)\,(9700 \times 0.0004893)}{31 \times 24000}$$

  where R is the absolute gas constant and T is the absolute temperature

  $$\text{or } \frac{RTn_1}{\Delta P} = 0.000145 \text{ L}$$

  $$\text{or } \frac{RTn_1}{\Delta P} = 0.145 \text{ ml} \ll V_0$$

Thus, $t = \frac{1}{AL_p\Delta P}(V_o - V)$

or $t = \frac{1}{(1.2 \times 10^4)(5.87 \times 10^{-5})}(9700 - 10)$

or $t = 13756.4$ s

Thus, ultrafiltration time, t = 3 h 50 min

- **Check for polarization**

$$\frac{D}{lj_v} = \frac{5.88 \times 10^{-7}}{(0.001)(5.87 \times 10^{-5})} 10 \simeq >> 1$$

Thus, the assumption of $\Delta\pi$ value used is valid.

A schematic of ultrafiltration and a proposed modification to avoid material loss at the second stage of ultrafiltration are shown in Figures 7.9 and 7.10.

Step 3: Chromatographic step: TSK-GEL 2000 SW gel chromatograph

Aim To purify β-hCG as a fraction depending on its molecular weight.

Laboratory scale gel specifications

Particle size (d) = 11 μm

Length (L) = 50 cm

Diameter (D) = 0.75 ml

Bed volume (V_B) = 26.5 ml

Void volume (V_0) = 12.6 ml

Void fraction (E) = 4.475

Total permeation

Volume (V_t) = 23 ml

Resolution power = 0.5 kDa – 60 kDa

Mobile phase = 0.05 M

Sodium phosphate buffer, pH 7.2, 0.3% w/v SDS

Flow rate = 0.3 ml/min

Sample load = 500 μL containing 0.002 wt% β-hCG

Distribution coefficient,

$$k_d = \frac{V_e - V_0}{V_t - V_0}$$

k_d of β-hCG = 0.07

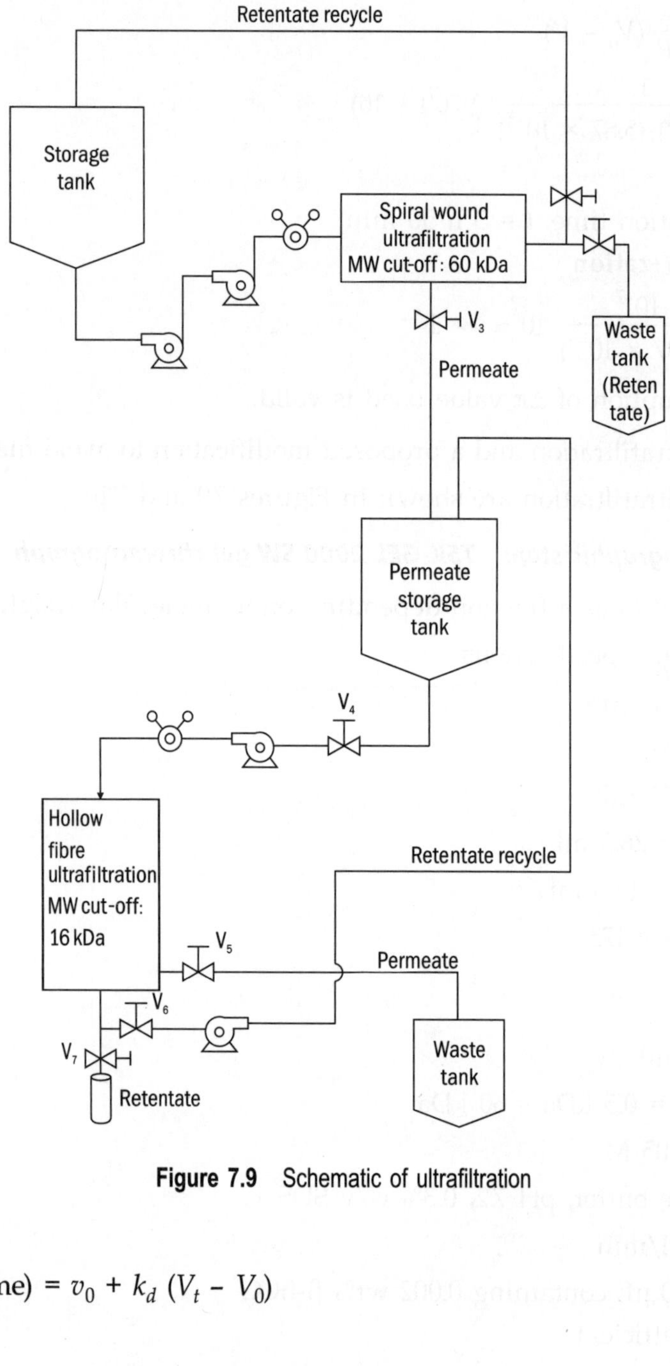

Figure 7.9 Schematic of ultrafiltration

Thus,

V_e (elution volume) = $v_0 + k_d (V_t - V_0)$

or V_e = 13.23 ml

Recovery = 91%

t_0 = 45.625 min

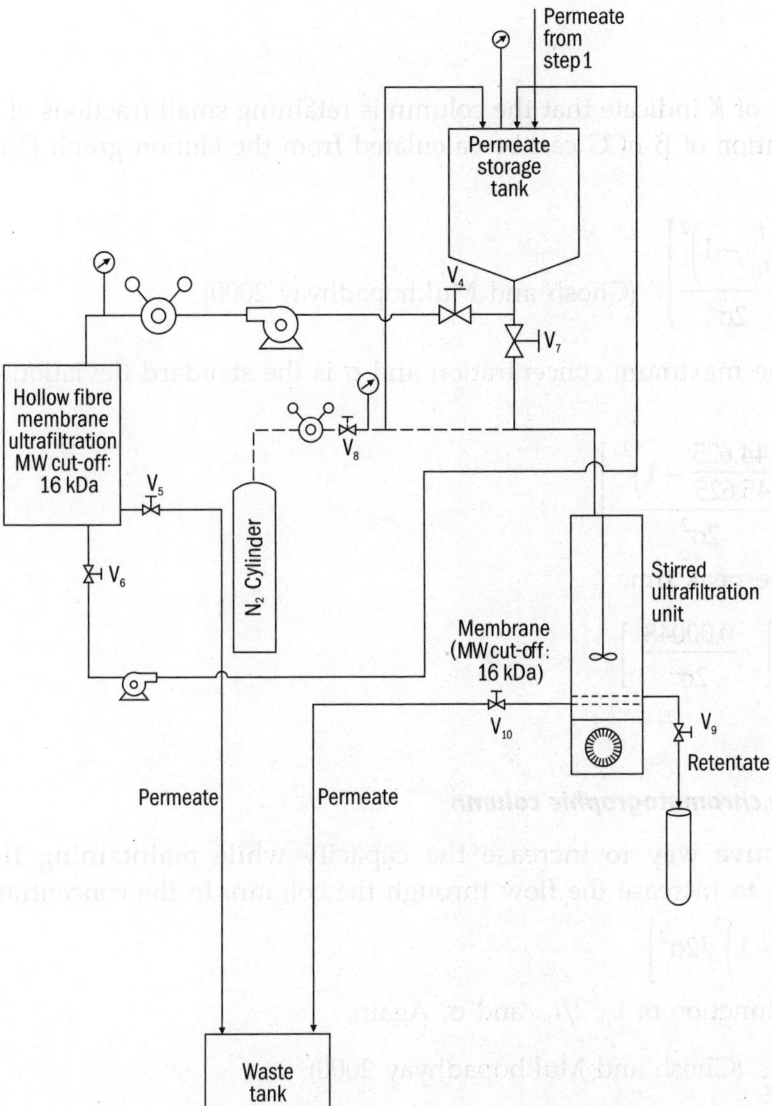

Figure 7.10 A proposed modification to avoid material loss at the second stage of ultrafiltration

Now,

$Ht_0 = [E + (1 - E)\ K]V_B$ (Ghosh and Mukhopadhyay 2000)

where Ht_0 is the total volume eluted at maximum peak and K is the linear isotherm constant.

Substituting the values for Ht_0, E, and V_B, we get

$13 = [0.475 + (1 - 0.475)\ K]\ 26.5$

Thus,

$K = 0.03.$

Low values of K indicate that the column is retaining small fractions of β-hCG loads. The concentration of β-hCG can be calculated from the elution graph (Table 7.11).

Now,

$$\frac{y}{y_0} = \exp\left[-\frac{\left(\frac{t}{t_0} - 1\right)^2}{2\sigma^2}\right] \quad \text{(Ghosh and Mukhopadhyay 2000)}$$

where y_0 is the maximum concentration and σ is the standard deviation of the peak. That is,

$$\frac{20}{22} = \exp\left[-\frac{\left(\frac{44.625}{45.625} - 1\right)^2}{2\sigma^2}\right]$$

where t_0 is the peak time

$$\text{or } \frac{20}{22} = \exp\left[-\frac{0.00048}{2\sigma^2}\right]$$

Thus,

$\sigma = 0.0355$

Scaling up the chromatographic column

A more effective way to increase the capacity while maintaining the yield and purification is to increase the flow through the column. In the concentration profile

$$y = y_0\left[-\left(\frac{t}{t_0} - 1\right)^2/2\sigma^2\right]$$

where y is a function of y_0, t/t_0, and σ. Again,

$$\sigma^2 \propto \frac{vd^2}{6\,(1 - E)L} \quad \text{(Ghosh and Mukhopadhyay 2000)}$$

where v is the velocity through the column and L is the column length.

If y_0, t/t_0, and σ remain constant, the concentration profile also remains constant.

Table 7.11 Varitation of concentration of β-hCG with time

Time (min) (after sample infection)	Concentration of β-hCG (in arbitrary units)
44.625 (t) 45.625 (peak) (t_0)	20 (Y) 22 (Y_0)

Scale up criteria

1. Increase column length twice.
2. Increase volumetric flow rate twice.
3. For 1 and 2, ΔP will increase four times.
4. Increase area 25 times. Thus, total throughput will increase 2 × 25 = 50 times, and column diameter will increase five times.
5. Keep d and E constant.
6. Use two such columns and load 5 ml of sample each from the retentate obtained in Step 2.

Assumption

While increasing ΔP by four times, E does not change.

Thus, in the scaled up column

- σ does not change as v/L and d are kept constant
- t_0 does not change as v/L remains constant
- $L = 2 \times 60 = 120$ cm
- $D = 5 \times 0.75 = 3.75$ cm
- Volumetric velocity = 3.5 × 25 × 2 = 15 ml/min
- $V = \dfrac{4 \times 15}{\pi \times (3.75)^2 \times 60} = 0.0227$ cm/s
- $\text{NTU} = \dfrac{1}{\sigma^2} = \dfrac{120}{790} = 0.152$
- $\text{HTU} = \dfrac{L}{\text{NTU}} = \dfrac{120}{790} = 0.152$
- Adsorption rate constant, $K = \dfrac{\text{NTU} \times \text{V}}{\text{aL}}$

  where

  $$a = \frac{6}{d}[1 - E]$$

  $$\text{or } a = \frac{6}{(11 \times 10^{-4})}[1 - 0.475]$$

  $$\text{or } a = 2863.63 \text{ cm}^2$$

  Thus,

  $$K = \frac{790 \times 1.36}{2863 \times 120 \times 60}$$

  $$\text{or } K = 5.1 \times 10^{-5}$$

- Time taken to get 91% yield

$$\text{Yield (91\%)} = \frac{1}{2}\left[1 + \text{erf}\ \frac{t - t_0}{\sqrt{2}\ t_0\ \sigma}\right]$$

$$\text{or } 0.91 = 0.5\left[1 + \text{erf}\ \frac{t - 45.625}{\sqrt{2}\ (45.625)\ (0.0355)}\right]$$

$$\text{or erf}\left(\frac{t - 45.625}{2.29}\right) = 0.81$$

Thus,

$t = 47.725$ min

- Time at which the sample has to be collected.

At 1% yield:

$$\text{Yield} = \frac{1}{2}\left[1 + \text{erf}\ \frac{t - t_0}{\sqrt{2}\ t_0\ \sigma}\right]$$

Thus,

$$0.01 = 0.5\left[1 + \text{erf}\ \frac{t - 45.625}{\sqrt{2}\ (45.625)\ (0.0355)}\right]$$

Thus,

$$\text{erf}\left(\frac{t - 45.625}{2.29}\right) = -\ 0.98$$

Thus,

$t = 40.816$ min

So for 90% yield, the sample fraction should be collected between 40.816 min and 47.725 min.

The volume eluted in this time is (47.725 – 40.816) × 15 = 104 ml

Purified β-hCG after Step 3

In total, 208 ml β-hCG content = 47.46 × 0.90 mg = 42.714 mg

Thus, weight per cent of β-hCG = 0.021 wt%

The purity cannot be calculated because the number of components in the starting media is not known.

Step 4: Eluted sample undergoes high performance cation exchange chromatography on partial SCX column

Aim To remove sodium dodecyl sulphate (SDS) from the sample.

Scope of Future Work The scaling up of column in Step 4 has to be done; that is, SDS has to be removed from the β-hCG sample collected after Step 3 using stepwise gradient elution with volatile buffer and Partisil SCX column.

Experiments should be done to check the validity of the separation process designed."

BIBLIOGRAPHY

Aiba S, Hamphrey A E, and Millis N F. 1973. ***Biochemical Engineering***, 2nd edn. Tokyo: University of Tokyo Press

Bailey J E and Ollis D F. 1986. ***Biochemical Engineering Fundamentals***, 2nd edn. New York: McGraw Hill Book Co. 918 pp.

Bungay H R and Belfort G (eds). 1987. ***Advanced Biochemical Engineering***. New York: John Wiley and Sons

Charm S E. 1971. ***Fundamentals of Food Engineering***. New York: AVI Publishers

Cheryan M. 1986. ***Ultrafiltration Handbook***. Lancaster, Pennsylvania: Technomic Publishing Co.

Dunnill P and Lilly M D. 1972. **Continuous enzyme isolation**. *Biotechnology and Bioengineering Symposium* **3**: 97

Ghose T K (ed). 1990. ***Bioprocess Computation in Biotechnology***, vol. 1. New York: Ellis Horwood

Ghosh A and Mukhopadhyay S N. 2000. ***Indian Chemical Engineering Congress Proceedings***, pp. 133–136, Calcutta: Indian Chemical Engineering

King C J. 1971. ***Separation Process***. New Delhi: Mc Graw Hill

Linko M. 1979. ***An Evaluation of Enzymatic Hydrolysis of Cellulosic Materials. Dechema Monographs***, pp. 209–217. Weinheim, Germany: Verlag Chemie

Mukhopadhyay S N and Malik R K. 1980. **Increased production of cellulase of *Trichoderma* sp. by pH cycling and temperature profiling**. *Biotechnology and Bioengneering* **22**(11): 2337–2250

Mukhopadhyay S N. 2007. ***Experimental Process Biotechnology Protocols***. New Delhi: Viva Books

Mukhopadhyay S N. 2010. ***Process Biotechnology Fundamental***, 3rd edn. New Delhi: Viva Books

Mukhopadhyay S N, Mukhopadhyay A, Talwar G P, Ghose T K. 1996. *Ind. Chem. Engr. Sec A* (4), 140–144

Prescott S C and Dunn C G. 1959. ***Industrial Microbiology***, 3rd edn. New York: McGraw Hill Book Co.

Rousseau R W (ed). 1987. ***Handbook of Separation Process Technology***. New York: John Wiley and Sons

Sen K. 1991. Separation of proteins from human placental extract by Foam Fractionation: its design aspects. Ph.D. Thesis, Jadavpur University, Calcutta

Volkenstein M V. 1983. ***Biophysics***. Moscow: MirPub

Vedajnananda S. 2000. Studies on the separation of proteolytic enzymes: trypsin and α-chymotrypsin using affinity ultrafiltration technique, Ph.D. Thesis, Jadavpur Univ., Calcutta

8

Unsolved Numerical Problems

8.1 MICROBIAL CELL SYSTEM

Question 1 Whole cells of *Saccharomyces cerevislae* containing invertase can be used to affect the sucrose inversion process. Recently, an immobilized whole cell process with this organism was developed. The activity of this preparation was only 600 IU/g catalyst. However, the half-life was found to be 7200 h. Compare the relative economics of the immobilized enzyme versus immobilized whole cell process.

Question 2 It is desired to model the growth of an individual bacterium. The cell transports S_1 into the cell enzymatically, and the permease is subject to product inhibition. S_1 is converted into precursors P, which are converted into the macromolecular portion of the cell M. In a chemical sense, the catalyst for all reactions is M.

$$S_1^* \xrightarrow[M]{} S_1 \text{ (per unit surface area)} \tag{8.1}$$

where S_1^* is the outside concentration of S_1.

$$S_1 \xrightarrow[M]{} P_M \tag{8.2}$$

$$\left.\begin{array}{l} \text{Energy} + P \xrightarrow[M]{} M \\ S_1 \xrightarrow[M]{} \text{Energy} \end{array}\right\} \text{Coupled reaction} \tag{8.3}$$

or

$$S_1 + P \xrightarrow[M]{} M \tag{8.4}$$

The dry weight of the cell X is given by Equation 8.5.

$$X = S_1 + P + M = PV \tag{8.5}$$

where P is the cell density and V is the cell volume.

Write the equations, and define all symbols necessary to describe the changes in S_1, P, M, and X within the cell. Note that the cell volume is always changing.

Question 3 During growth at temperatures above the optimum, cell death rate is not negligible. What effect does this have on cell yield in a chemostat culture? Make appropriate assumptions to derive necessary equations.

Hints

(a) Consider the maintenance requirements of both viable and dead cells.

(b) *A priori*, it is not necessary to assume that dead cells do not metabolize the limiting nutrient.

Question 4 An industry is considering the manufacture of single-cell protein (SCP) by growing the yeast *Torula utilis* on cane molasses. For a design estimate, one needs to know the approximate productivity of protein (g-protein/1000 L/h) expected from a typical fermenter unit, an estimate of the concentration (g-protein/L) of protein expected from the fermenters, and the rate (gal/h) of feeding cane molasses into the fermenter.

List all the assumptions to arrive at the estimate. Assume that the cane molasses run approximately 50% fermentable sugars on a weight basis.

Question 5 Many investigators have proposed that penicillin be produced in continuous culture. To address this issue, examine the effect of dilution rate on product concentration and yield based on the following data given.

q_p = 6 units penicillin g/mg cell-h

x (steady state) = 20 g/L

m = 0.022 g/g cell-h

$Y_{p/s}$ = 1.1 g penicillin g/g glucose

$Y_{x/s}$ = 0.5 g/g glucose

Question 6 *Pseudomonas* sp. bacteria have a culture mass doubling time of 2.4 h when grown on acetate. The saturation constant using this substrate is 1.3 g/L. The chemostat is operated on a feed containing 38 g/L acetate, and the cell yield on acetate is 0.46.

(a) Find the cell concentration when D is half the maximum value.

(b) Find the substrate concentration when D = 0.8 D_{max}.

(c) What is the value of maximum dilution rate?

(d) Find the productivity at 0.8 D_{max}.

Question 7 Two organisms, a yeast *Candida utilis* and a bacterium *Escherichia coli*, are grown in a chemostat together. Both organisms compete for the common limiting carbon source, glucose, which is fed into the reactor at 10 g/L concentration. The specific growth rates (μ) of yeast and bacterium can be expressed by Equations 8.6 and 8.7, respectively.

$$\mu_1 = 0.5\left(\frac{S}{0.03 + S}\right) \text{ per hour} \tag{8.6}$$

$$\mu_2 = 0.4\left(\frac{S}{0.01 + S}\right) \text{ per hour} \tag{8.7}$$

where S is the glucose concentration (g/L).

Construct a diagram of the concentration of each organism as a function of dilution rate (D/h) in the chemostat system. The yield coefficients are 0.4 g/g glucose and 0.5 g/g glucose for the yeast and the bacterium, respectively. What conclusions can you draw from your results regarding competitive interactions in ecosystems?

Question 8 Organism A grows on substrate S and produces product P, which is the only substrate that organism B can utilize. The batch kinetics is as follows.

$$\frac{dX_A}{dt} = \frac{\mu_A S X_A}{K_S + S}$$

$$\frac{dX_B}{dt} = \frac{\mu_B P X_B}{K_P + P}$$

$$\frac{dP}{dt} = \alpha\frac{\mu_A S X_A}{K_S + S} - \frac{\mu_B P X_B}{Y_B(K_P + P)}$$

$$\frac{dS}{dt} = -\frac{\mu_A S X_A}{Y_A(K_S + S)} - \frac{\alpha\mu_A S X_A}{Y(K_S + S)}$$

Assume the following parameter values.

μ_A = 0.18 per hour

K_S = 0.42 g/L

μ_B = 0.29 per hour

K_P = 0.30 g/L

Y_A = 0.3

Y_B = 0.5

Y = 1.0

α = 4.0

S_0 = 10 g/L

Determine the behaviour of these two organisms in a chemostat. Plot S, P, X_A, and X_B versus dilution rate. Discuss what happens to organism B as the dilution rate approaches the washout dilution rate for organism A.

Question 9 The dynamic behaviour of a given continuous fermenter can be represented by Equations 8.8 and 8.9.

$$\dot{X} = X\, \mu_m \left(\frac{S}{K_S + S}\right) - D \tag{8.8}$$

$$S = D\,(S_0 - S) - \frac{X}{Y}\, \mu_m \left(\frac{S}{K_S + S}\right) \tag{8.9}$$

Figure 8.1 shows an analog computer solution of these equations, in terms of a phase plane plot.

Illustrate the transient behaviour, that is, X/X_∞ versus time and S/S_∞ versus time for the following changes.

A drop of concentrated S is added to the fermenter when it is at equilibrium with no change (initially) in X.

The fermenter (initially at equilibrium) is instantaneously diluted by 50% with pure water, that is, $S_{\text{initially}} = (1/2)S_\infty$ and $X_{\text{initially}} = (1/2)X_\infty$.

Question 10 A continuous stirred tank reactor (CSTR) is used for an enzymatic biochemical conversion process, in which the reaction rate follows Equation 8.10.

$$R = \frac{kC}{(1 + KC)^2} \tag{8.10}$$

where R is the rate of depletion of the reactant. The reaction parameters are as follows.

C_0 = concentration of reactant in feed stream = 1.32 mol/ft^3

K = kinetic parameter = 10 ft^3/mol

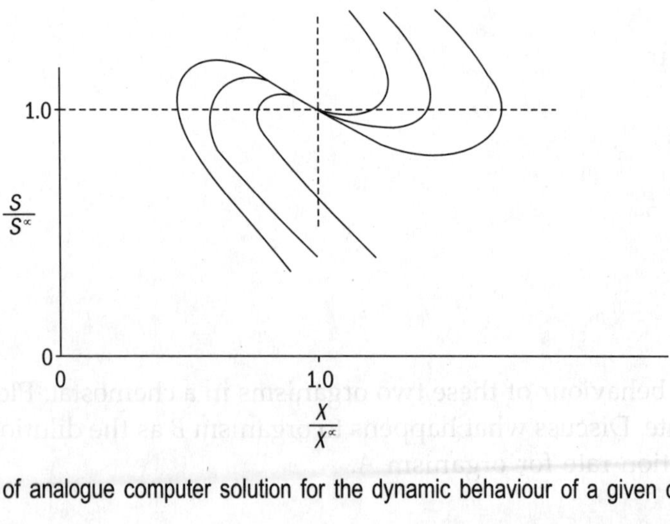

Figure 8.1 Graph of analogue computer solution for the dynamic behaviour of a given continuous fermenter

k = kinetic parameter = 1.75 per minute

V = fermenter volume = 90 ft^3

q = flow rate of feed to fermenter = 3 ft^3/min

Show that this CSTR will have multiple steady states, and determine the concentrations of reactant at which the steady states occur.

8.2 REACTOR KINETICS AND STABILITY

Question 11 A mixed culture of two microbial species is grown in a continuous culture. Equations 8.11 and 8.12 give the kinetic rate expressions for the production of each species. If the system is designed such that only a single steady state exists, establish the criteria for this in terms of the fermenter dilution rate q/v. Assume that the feed stream is sterile.

$$R_1 = 12x_1^2 - 2x_1^3 - 10x_1 - 8x_2 \tag{8.11}$$

$$R_2 = 4x_2 - x_1 \tag{8.12}$$

Question 12 When the system described in Question 11 is operated at a dilution rate $q/v = 2$, multiple steady states exist. One occurs at $x_1 = 2$ and $x_2 = 1$. Determine the character of this steady state (that is, stable node, focus or saddle point; unstable; indeterminate).

Question 13 In autohydrolytic conversion of rice husks xylan (x_n), under non-isothermal conditions, to xylodigomers (*O*), xylose (*X*), and furfaral (*F*), Vila *et al.* proposed a simplified model of degradation (Figure 8.2).

In the figure, X_{nF} indicates fast-reacting xylan and X_{nS} indicates slow-reacting xylan. Assume that the temperature dependence of kinetic parameters follow Arrhenius equation (Equation 8.13).

$$k_2 = k_{2\max} \exp\left(\frac{-E_{a2}}{RT}\right) [1 - \exp(-\beta t)] \tag{8.13}$$

Based on material balance analysis, formulate the xylan conversion in the form of a set of differential equations. Assume the following.

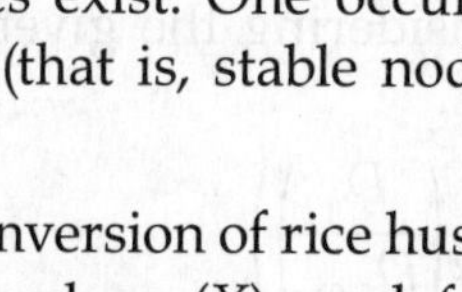

Figure 8.2 Simplified model of degradation

$k_{1S} = 0$

$k_2 = f(t) = k_{2max}\,[1 - \exp(-\beta t)]$

Consider X as the percentage of initial xylan converted to xylose; values of variable F (percentage of initial xylan converted into furfural) are calculated by a material balance to the xylan-derived compounds.

Question 14 Anaerobic digestion at 35°C of municipal solid waste (MSW) in a 28 L CSTR, having hydraulic retention time (HRT) 15 days, produced CH_4, CO_2, and H_2 in biogas. Stripping of biogas compounds was followed assuming that generated rising gas bubbles have similar composition. The mass of gaseous component I transferred from the liquid to gas phase (ρ_{Li}) is expressed by Equations 8.14 and 8.15.

$$\rho_{Li} = K_L a_i (C_L - C_{Li}) \quad (8.14)$$

$$\text{where } C_{Li} = \rho_i \left(\frac{MG_i}{H_i RT}\right) \quad (8.15)$$

In Equations 8.14 and 8.15, C_L is the bulk dissolved gas concentration. C_{Li} is the dissolved gas concentration at the boundary layer interface, ρ_i is the gas density, H_i is Henry's constant for 30–60°C, R is the gas constant (8.314×10^{-5} bar m^3 gas/mol/K) and T is the absolute temperature. Considering the given equations and data given below, determine the following.

$$\left(\text{Given } K_L a = \left(\frac{D}{d_\ell}\right) a,\ K_L a_i = K_L a_{CO_2}\left(\frac{D_i}{D_{CO_2}}\right)\right)$$

1. $K_L a_{CO_2}$, $K_L a_{CH_4}$, $K_L a_{H_2}$
2. Total volume specific dry biogas flow, q_{tot} (m^3 gas/m^3 $_{CSTR}$/d)
3. Computed values at thermophilic temperature (55°C)

Question 15 Many proteins are normally present in *E. coli* at concentrations of 2 mol/cell. What is the molar concentrations of such a protein? (The cellular data of *E. coli* are as follows: 10 L cultures of volume, maximum cell density 10^{10} cells/mt [a saturated culture], cylindrical cell 2 μm long and 1 μm in diameter.) Conversely, how many glucose molecules does an *E. coli* cell contain if it has an internal glucose concentration of 1.0 mM?

Question 16 The DNA of an *E. coli* chromosome measures 1.4 mm in length when extended and 2 Å in diameter. What fraction of an *E. coli* cell is occupied by its DNA?

A human cell has 700 times the DNA of an *E. coli* cell and is typically spherical with a diameter of 20 μm. What fraction of such a cell is occupied by its DNA?

Question 17[1] *Staphylococcus aureus* is a pathogen found in food and environment. It has B domain of protein A(BdpA), which is a factor responsible for its pathogenesis. Its native (*N*) and denatured (*D*) forms were examined. Under equilibrium conditions, proteins are distributed between its *N*(or folded) and *D*(or unfolded) forms.

$$\text{Unfolded } (D) \underset{k_u}{\overset{k_f}{\rightleftharpoons}} \text{Folded } (N)$$

For two-state folding proteins like BdpA from *S. aureus*, only *N* and *D* states are significantly populated under experimental conditions. For BdpA, the equilibrium constant for folding ($K_{eq.f}$) can be as high as 3000 under native conditions. Urea is a naturally occurring osmolyte and can denature proteins by stabilizing the *D* state relative to the *N* state. The process of denaturation occurs by disturbing the folding (k_f) and unfolding (k_u) rate constants. Protein denaturation in the presence of urea lowers the $K_{eq.f}$ to a value proportional to the concentration of the osmolyte under experimental conditions. Calculate the following for BdpA under native ($K_{eq.f}$ = 3000 in 0 M urea) and denaturing conditions ($K_{eq.f}$ = 26 in 2 M urea). (Given R = 1.987 cal/°K/mol)

(a) Concentration of *N* and *D* if starting protein concentration is 1 mM.
(b) Free energies of folding and unfolding (at 37°C).
(c) Unfolding rate constants if the folding rate constants in 0 M urea k_f = 160 000 per second and in 2 M urea k_f = 22 000 per second.
(d) Unfolding and folding rates.
(e) Time required to unfold and fold the protein.

Question 18 You have been asked to study the behaviour of mass transfer in laboratory and large-scale fermenters for scale up purposes. Outline a detailed approach you would take to obtain meaningful correlations of operating variables to the mass transfer coefficient. Take into consideration that the geometric similarity of the different scales of operation does not have to be maintained. Justify all reasoning.

Question 19 It has been proposed that a 14 L New Brunswick fermenter be operated continuously to produce single-cell protein. To maximize its productivity, the fermenter is operated at a dilution rate D = 1.0 per hour, with a medium volume of 10 L. To sterilize the medium continuously, a pilot plant fermenter is used as a continuous sterilizer. Steam in great excess is available to the fermenter jacket at 135°C. The overall heat transfer coefficient of the jacket is 113 kCal/m^2-h-°C. The maximum heat transfer area of the sterilizer is 1.4 m^2. The medium is charged to the pilot plant fermenter at 25°C. The density and specific heat of the medium are 1 g/ml and 1 cal/g-°C, respectively,

[1] Problem from Pooja Mukhopadhyay (Nee Arora),* Parag Mukhopadhyay,# and S N Mukhopadhyay.† (*formulated, #supported the concept, †developed the concept); Pooja Arora, PhD thesis, Duke University, North Carolina, 2006

and do not change significantly with temperature. Calculate the minimum volume of medium required in the sterilizer in order to achieve the following microbial destruction (state all reasonable assumptions).

Inlet contamination = 150 organisms per litre

Desired outlet contamination = 0.01 organism per litre

Figure 8.3 shows the thermal inactivation curve.

Question 20 A 10 000 gallon working capacity fermenter is used to carry out a fermentation process for four days. The volumetric air flow rate used is equivalent to 0.1 VVM. The incoming air contains, on an average, 150 organisms/ft^3 at an average bacteria diameter of 1 μ. Fibrous filter material having average fibre diameter of 19 μ and a void fraction of 0.95 is available for the construction of an air filter. An allowable risk of 0.001 has been designated in terms of bacterium penetration. Estimate the length of filter required.

Question 21 It is desired to increase the capacity of a rotary vacuum drum filter. Give, as accurately as possible, the per cent capacity increase that results from the following.

(a) Doubling the rate of rotation.

(b) Doubling the submergence, where submergence is the distance from the slurry pond surface to the bottom of the drum.

(c) Doubling the concentration of the feed.

(d) Doubling the pressure drop.

Question 22 The results obtained from constant filtration (at 50 psi) in a leaf filter with a filtering area of 20 ft^2 are given in Table 8.1.

Before filtration, the slurry contained 5 lb of solid/100 lb of liquid. The densities of the solid and the liquids were 78.5 lb/ft^3 and 62.4 lb/ft^3, respectively. The viscosity of the slurry was 2 cP, and the cake was found to be incompressible.

Determine the specific resistance, *r*, of the cake and the equivalent length of the filter cloth.

Question 23 A tubular centrifuge with a radius of 2.05 in and a height of 12 in is used to separate two liquids with densities 62.0 lb/ft^3 and 63.9 lb/ft^3, respectively. For satisfactory separation, it is necessary that the heavy fluid resides in the centrifuge for 5.5 s and the light fluid resides for 4 s. The volumetric ratio of the fluid is 1:1. The thickness of the heavy liquid is ¾ in.

(a) What is the capacity of the centrifuge of the total feed (in gal/min)?

(b) Determine the location of the outlet weirs for the two streams.

Question 24 What is the capacity (in gal/min) of a clarifying centrifuge operating under the following conditions?

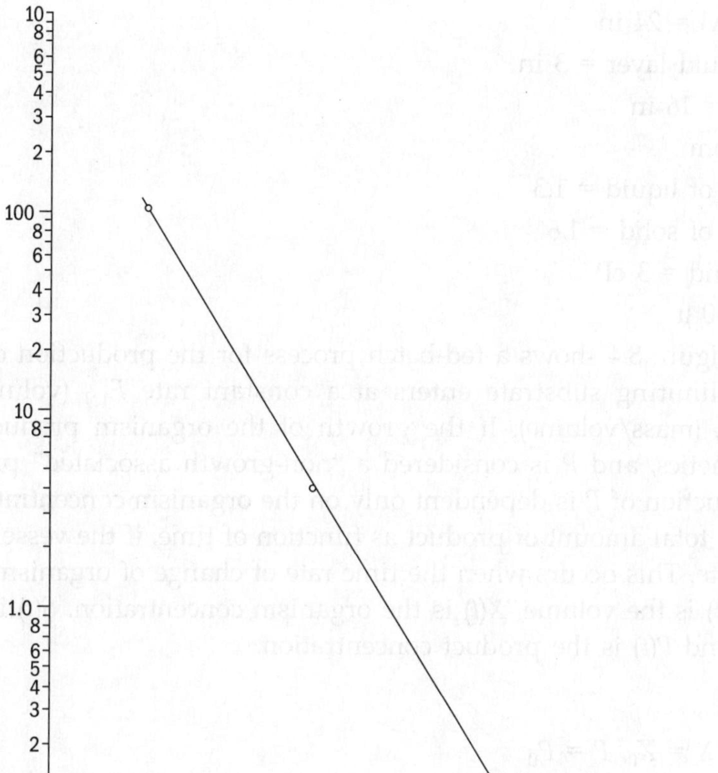

Figure 8.3 Thermal inactivation curve

Table 8.1 Results obtained during constant filter at 50 psi

Time (min)	Filtrate volume (ft^3)
15	100
30	163
45	213
60	258
90	235

Diameter of bowl = 24 in
Thickness of liquid layer = 3 in
Height of bowl = 16 in
Speed = 1000 rpm
Specific gravity of liquid = 1.3
Specific gravity of solid = 1.6
Viscosity of liquid = 3 cP
Particle size = 30 μ

Question 25 Figure 8.4 shows a fed-batch process for the production of a metabolite *P*. The growth-limiting substrate enters at a constant rate F_{1N} (volume/time), at a concentration S_f (mass/volume). If the growth of the organism producing *P* follows Monod-type kinetics, and *P* is considered a "non-growth associated" product (that is, the rate of production of *P* is dependent only on the organism concentration to the first power), find the total amount of product as function of time, if the vessel is operated at quasi-steady state. This occurs when the time rate of change of organism concentration is negligible. *V*(*t*) is the volume, *X*(*t*) is the organism concentration, *S*(*t*) is the substrate concentration, and *P*(*t*) is the product concentration.

At $t = 0$

$V = V_0,\ S = S_0,\ X = X_0,\ P = P_0$

8.3 PLANT CELL SYSTEM

Question 26 A horizontal rotary fermenter of 30 cm diameter and 1 m long rotates coaxially inside a fixed outer cylinder jacket of approximately same length and 30.5 cm in diameter. Hot water fills the space between the cylinders to control the temperature of the plant cell culture fermentation by *Cantharanthus* sp. A torque of 0.25 ft-lb is applied to the inner cylinder. Ignoring end effects, calculate the velocity gradients at the cylinder walls, the resulting rpm, and the power dissipated by fluid resistance after constant velocity is attained.

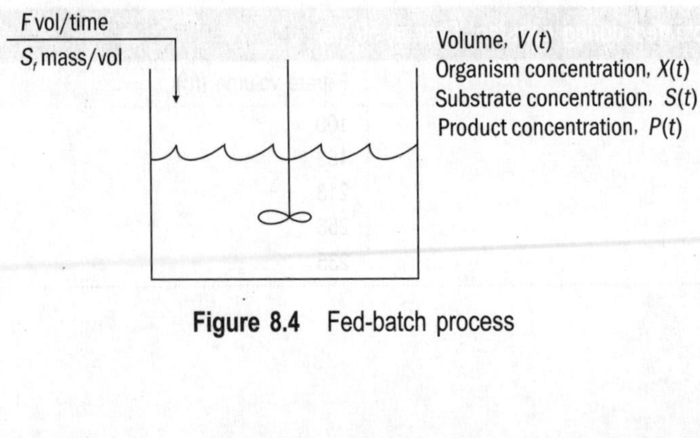

Figure 8.4 Fed-batch process

Question 27 *Cantharanthus roseus* cell cultivation mash (100 g/L wet weight) was exposed to a constant shear rate of 290 per second using a Contravas Rheomal 115 viscometer. Figure 8.5 shows the variation of apparent viscosity of the suspension. Examine the behaviour of the cell cultivation mash and write down your conclusion. Estimate the corresponding shear stresses (Figure 8.5).

8.4 ANIMAL CELL SYSTEM

Question 28 In chromatography, each solute has an equilibrium partition coefficient $K_i = s_i/c_i$, where s_i is the adsorbed species concentration and c_i is the bulk concentration. If a sample of width d at the injection point is to be separated into peaks of at least this peak-to-peak spacing, show the following.

(a) The mean residence time of any single component i is given by

$$t_i = \frac{L}{V_\varepsilon}(1 - \varepsilon)(1 + K_i)$$

where n is the fluid velocity and L is the column length.

(b) The required maximum column length L has the maximum value of

$$L = \frac{d(K_j - K_i)}{(1 + K_j)^2 (1 + K_i)}$$

where $K_j > K_i$.

(c) In liquid chromatography, the diffusion coefficient is typically 10^{-6} cm^2/s for larger molecules. If the chromatography column is packed with particles of 10^{-2} cm diameter, use the Einstein equation for the root-mean-square (rms) distance traversed

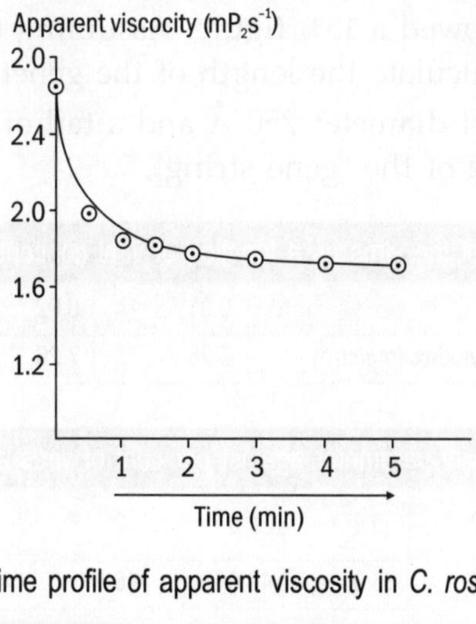

Figure 8.5 Time profile of apparent viscosity in *C. roseus* cell culture

by a molecule in time t, $<Z^2>^{1/2} = (Dt)^{1/2}$, to show that flow velocities larger than the order of 10^{-4} cm/s will require a longer column than indicated in part (*b*).

(d) Charm *et al.* used an affinity chromatography column to separate serum hepatitis antigen from pooled blood plasma. For such specific separation, $K_i \to \infty$ (i is the antigen), $K_j \to 0$ ($j \neq i$). If the column is packed with non-porous beads of radius R and the antigen concentration is n particles per litre, show that the maximum number of blood volumes that can be cleared continuously of antigen by a single unit of column volume is given by Equation 8.16.

$$\frac{3(1-\varepsilon)}{n\pi^2 R} \tag{8.16}$$

where ε is the column void volume and r is the antigen radius.

Question 29 Assuming the adsorption of vaccinia virus on the cytodex 1 microcarrier culture of Vero 76 cells as per Table 8.2 (at 37°C for 1.5 h), examine what adsorption isotherm fits these data.

Question 30 Adsorption of vaccinia virus on Vero 76 cells at different cellular activity parameter (P) provided results as given in Table 8.3.

What is the significance of P in this relation? Is there any linkage of P in terms of the data in Question 29? Comment on the profile (μ indicates specific growth rate).

8.5 MISCELLANEOUS BIOSYSTEMS

Question 31 Horshey and Rotman were able to observe 0.15% separation and recombination of two genes in a bacteriophage.

(a) Assuming that 10^{-5} Morgan is 30 Å, calculate the separation between the genes.

(b) Another observation showed a 15% figure. Assuming these to lie at opposite ends of the linkage group, calculate the length of the genetic time required.

(c) The phage has a head of diameter 750 Å and a tail of the same length. Comment on the molecular nature of the "gene string".

Table 8.2 Variation of adsorption of the virus as a function of viral concentration

Virus concentration (mg/L)	0.01	0.02	0.03	0.06	0.10
Adsorption on vero cells anchored on cytodex (mg/cm^3)	5.28	7.26	8.49	11.08	13.68

Table 8.3 Variation of cellular activity parameter against per cent of virus adsorbed

$P = \left(\frac{\mu_t}{\mu_{max}} \times 10\right)$	0.1	0.2	0.5	1.0	2	4	6	8	10
Virus adsorbed (%)	20.0	30.0	38.0	40.0	46	46	44	45	44

Question 32 Novick and Szilard found that at 37°C a culture of *E. coli* gave 0.67 mutations per 10^8 bacteria per hour. Assuming that there is only one gene to mutate per bacterium, calculate the half-life of the gene in years.

Question 33 Dunnill and Lilly observed that protein recovery from a passage of baker's yeast cell suspensions through a large pressure drop confined to a small volume (homogenizer) followed Equation 8.17.

$$\ln \frac{R_m}{R_m - R} = KN \tag{8.17}$$

where K is the first-order rate constant, N is the number of passes through the homogenizer, R_m is the maximal achievable protein release, and R is the protein release after N passes. Here

$K = kp^a$

where k and a depend on microbe and p is the operating pressure.

(a) Show that a first-order rate law gives the expression observed, treating N as a continuous variable.

(b) If the power input per pass, W, varies linearly with operating pressure p, show that a maximum exists in the curve of Q (per cent protein released per kilowatt power input) versus pressure.

(c) If a = 2.9 for bakers' yeast and the maximum in Q is 7.0 at 570 kg_f/cm^2 for N = 1, evaluate k and W/P and calculate the profiles of Q versus p for N = 1.2.

Question 34 Around 1000 lbs of egg white, containing 2.7% glucose, is to be desugared by adding 34 000 units of glucose oxidate, which also contains some catalase. If 6 lbs of 30% H_2O_2 is to be added slowly to the mixture (pH 7, 24°C) to provide the substrate for catalase action, how long will it take to reduce the glucose level to 0.1%?

Question 35 An aerobic fermentation using fungal culture formed pellets. The average mycelia density is 0.02 mg of cell/mm^3 of pellet volume. The average dissolved oxygen concentration in the bulk broth is 5 µg/ml. If the specific oxygen uptake rate by the mould is 1.2 mmol O_2/h/g of cell and if the O_2 diffusion coefficient is 1.8×10^{-5} cm^2/s in broth, estimate the minimum size of the pellet that will cause O_2 deficiency in the pellet.

Question 36 In the bioconversion of glucose containing cellulose hydrolysate into gluconic acid by *Pseudomonas ovalis*, the change in the total substrate followed Equation 8.18.

$$V_e \frac{dS_i}{dt} = J_{in}\left(\frac{S}{K_1 + S}\right) V_e - J_{out}\left(\frac{S}{K_2 + S_i}\right) V_e - J\left(\frac{S_i}{K_3 + S_i}\right) V_e \tag{8.18}$$

where V_e is the unit cell volume, J_{in} is the flux of substrate uptake, J_{out} is the flux of substrate discharge, S is the substrate concentration in the system, S_i is the substrate concentration in the cells, J is the maximum reaction rate of substrate metabolism in cells, and K_1 and K_3 are constants ($K_1 \approx K_3$).

Assuming $K_2 \approx K_3$, $S \ll K_1$, $S_i \ll K_2$, and $\mu = kJ\left(\frac{S_i}{K_3 + S_i}\right)$ (where k = constant), derive the relation between μ and S as a first-order lag.

Mammalian versus microbial cell function

Question 37 Assuming metabolism to be proportional to the surface area, compare the rate of metabolism in man (surface area ~260 cm^2/kg of body weight) with that of a cylindrical bacteria (size = 1 × 2 μ^2). Assume that the specific gravity of the bacteria is 1. Comment on the result.

Question 38 In a *Trichoderma reesei* cell cultivation experiment, soft pellets could be noticed during cellulose production. Assuming zero-order oxygen uptake kinetics with suitable boundary conditions, show that the radius (r_p) of the pellet may be related to the operating parameters by Equation 8.19.

$$r_p = \sqrt{\frac{6DC_L}{\rho_p K}} \tag{8.19}$$

Question 39 In an in vitro *lac* (molecular weight 150 000) system of *E. coli*, the interactions between operator (O) and repressor (R) followed Equation 8.20.

$$R + O \underset{K_b}{\overset{K_f}{\rightleftharpoons}} RO \tag{8.20}$$

Membrane filter assay technique showed that the rate of labelled carbon transmittance sensitivity (CTS) bound to the filter increased linearly with repressor concentration (Figure 8.6a). The results of the rate of formation and rate of dissociation of RO complex with 20-fold excess unlabelled λψ 80 d lac DNA are shown in Figures 8.6b and 8.6c, respectively. As indicated in Figure 8.6c, the control experiment confirmed that addition of excess cold λψ 80 DNA did not promote the dissociation of labelled RO complex. From the initial rate of decay of RO complex, calculate the following.

(a) Reaction order and association rate constant (K_f) with respect to operator concentration.
(b) Half-life of the RO complex.
(c) Rate constant for the backward reaction (K_b).
(d) Value of the overall equilibrium constant (K) from the results obtained from (a) and (b). Comment on the result.

Question 40 The results obtained from an aerobic cultivation of *Aerobacter acrogenes* in minimal medium with different substrates are given in Table 8.4. Comment on the results. For each substrate to undergo cellular biochemical reactions, estimate the following from the given data.

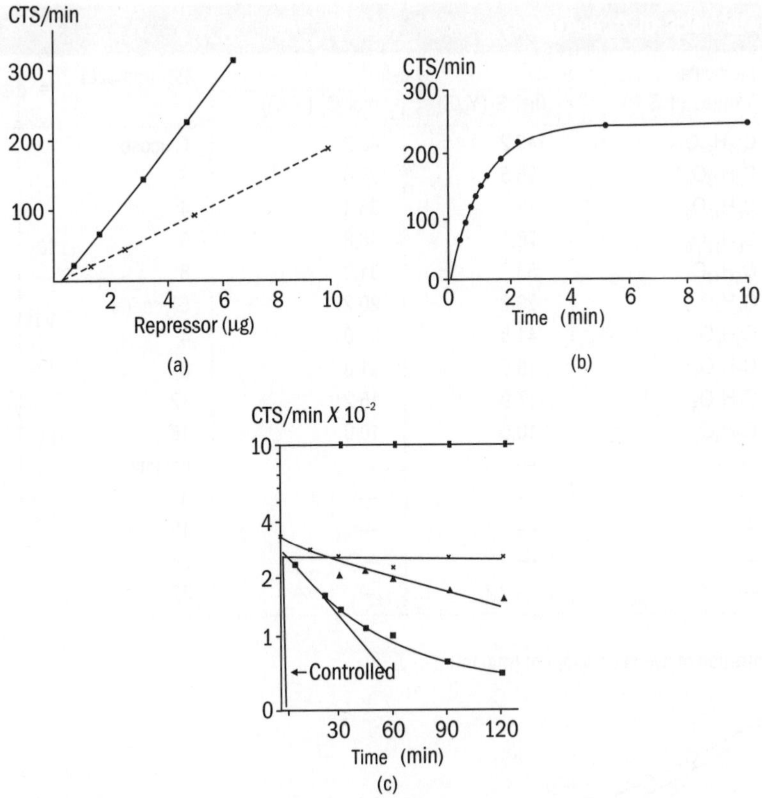

Figure 8.6 Relation between repressor concentration and carbon transmittance sensitivity

(a) Available electrons.

(b) Enthalpy changes.

Question 41 Fermentations for gluconic acid production by *P. ovalis* B1486 were carried out in media containing 51 mg/ml and 90 mg/ml glucose. The specific glucose uptake rate in the first medium was 5.5 mg/UOD/h and in the second, 3.0 mg/UOD/h. The cells were filtered and killed by dettol vapour disinfection and dried at 60°C for 3 h. The numbers of cells per mg in the first and second media, counted by haemocytometer, were 2.5×10^{13} and 3.2×10^{13}, respectively. The surface area of the dried cells was measured after keeping them for 2 h in aqueous methylene blue dye solution, having an initial concentration 1.03 mg/L with optical density, $OD_{480} = 0.32$. The dye adsorbed on the cells reached an equilibrium with respect to each cell concentration. The OD of each dye solution after centrifugation was measured and the adsorption pattern is shown in Figure 8.7. The standard relation of dye concentration with corresponding OD_{480} is given in Table 8.5.

Table 8.4 Growth yields of *A. aerogenes*

Substrate (S)	Molecular formula of S	g cell / mol S (Y_x/S)	g cell / mol O_2 (Y_x/O)	ΔS (mmol/L)	ΔX (mg-cell/ L)
Maltose	$C_{12}H_{24}O_{12}$	149.2	48.2	Glucose	—
D-Manitol	$C_6H_{14}O_6$	95.5	38.0	2	97
D-Glucose	$C_6H_{12}O_6$	72.7	35.4	4	150
D-Fructose	$C_6H_{12}O_6$	76.1	46.8	6	210
D-Ribose	$C_5H_{10}O_4$	53.2	31.2	8	280
Succinate	$C_4H_6O_4$	29.2	20.2	Glycerine	—
Glycerine	$C_3H_6O_3$	41.8	31.0	4	100
DL-Lactate	$C_3H_6O_3$	16.6	11.8	8	165
Pyruvate	$C_3H_4O_3$	17.9	15.2	12	250
Acetate	$C_2H_4O_2$	10.5	10.0	16	310
—	—	—	—	Lactate	—
—	—	—	—	8	70
—	—	—	—	16	110
—	—	—	—	24	200
—	—	—	—	32	225

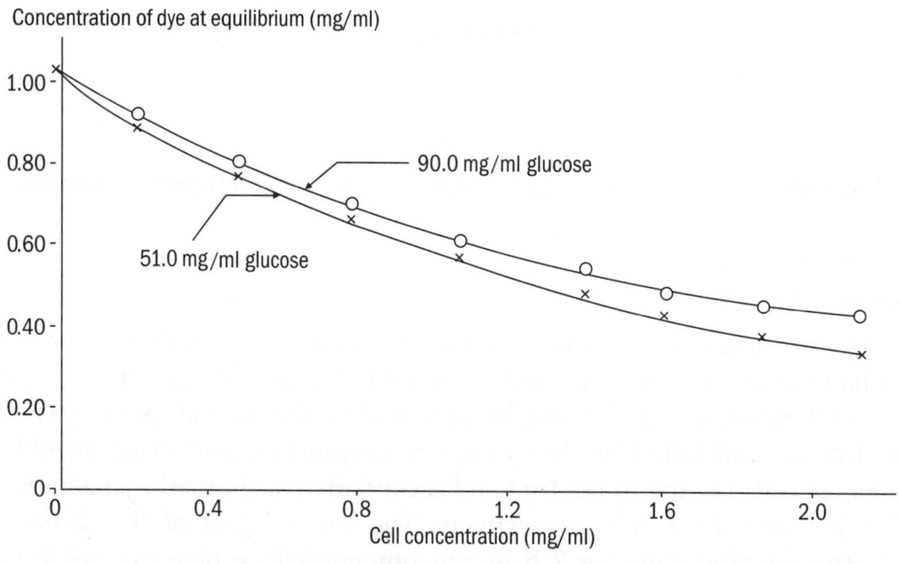

Figure 8.7 Dye adsorption as a function of cell concentration

Assuming Langmuir adsorption isotherm relation, the area covered by unit mass of dye is 10^4 cm^2/mg. Calculate the average surface area of *P. ovalis* B1486 cells from the two fermentation media and explain the observation on glucose uptake rate.

Table 8.5 Standard relation of dye concentration versus OD_{480}

Dye concentration (mg/ml)	0.02	0.05	0.08	0.10
OD_{480}	0.055	0.158	0.26	0.34

OD – optical density

BIBLIOGRAPHY

Alberts B, Johnson A, Lewis J, Raff M, Roberts K, Walter P. 2008. ***Molecular Biology of the Cell***, 5th edn. New York: W H Freeman and Co.

Atkinson B and Mavituna F. 1991. ***Biochemical Engineering and Biotechnology Handbook***, 2nd edn. Hampshire, England: Stockton Press

Bailey J E and Ollis D F. 1986. ***Biochemical Engineering Fundamentals***, 2nd edn. New York: McGraw Hill Book Co. 918 pp.

Chion S, Chem P C Y, and Fung Y C (eds). 2008. ***An Introductory Text to Bioengineering***. UK: World Scientific Imperial College Press

Clark D P and Pardernik N J. 2009. ***Biotechnology Applying the Genetic Revolution***. Amsterdam: Acad Press

Day E D (ed). 1990. ***Advanced Immunochemistry***, 2nd edn. New York: Wiley–Liss

Ghose T K (ed). 1990. ***Bioprocess Computation in Biotechnology***, vol.1. New York: Ellis Horwood

Ghosh A and Mukhopadhyay S N. 2000. ***Indian Chemical Engineering Congress Proceedings***, pp. 133–136. Kolkata: Indian Chemical Engineering

Ho C S and Wang D I C (eds). 1991. ***Animal Cell Bioreactor***. Boston: Butterworth Heinemann

Mukhopadhyay A, Mukhopadhyay S N, and Talwar G P. 1995. **Studies on the synthesis of β-hCG hormone in vero cells by recombinant vaccinia virus**. *Biotechnology and Bioengineering* **48**: 158–168

Mukhopadhyay S N. 2007. ***Experimental Process Biotechnology Protocols***. New Delhi: Viva Books

Mukhopadhyay S N. 2010. ***Process Biotechnology Fundamentals***, 3rd edn. New Delhi: Viva Books

Index

A

abiospheric world 50
abiotic 4
absolute pressure
activated sludge processing (ASP) 54
adenosine triphosphate (ATP) 55
adsorption 369, 371, 377, 557
  activated 370
  chemical 371
  isotherms 373
  monomolecular layer theory 371
  multimolecular layer theory 372
  physical 369
  theory of 369
  types of 369
Aerobacter aerogenes 277
aerobic environment 15
aerobic–anaerobic growth transitions 150
aerobiosis 163
Agrobacterium tumefaciens 20, 134
agrochemicals 20
alcoholic fermentation 57
Almighty Supra Nature (ASN) 3
anaerobic
  digestion 54
  life 11
  pathways 55
angular quantum number 32
animal cell system 585
animals 4
anoxic
  bioprocessing 248
  environment 4
  pathways 55
antibodies, valence and affinity of 73
aquatic ecosystems 11
archaebacteria 116, 127
atomic number 33
atomic structures 31
Azimuthal quantum number 32
Azotobacter vinelandii 160

B

bacteria 95, 99, 101, 104, 107, 108
  Gram-negative 102, 103,
  Gram-positive 102, 103
  methane-utilizing 154,
  parameters for existence 105
Bernoulli equation
  correction of 306
  in pump work 306
Big Bang 4, 7
bioassay of
  amino acids 168
  vitamins 168
biocatalysts 547
biocell
  definition of 95
  types 95
  microbes 98
  growth principles 137
  products 547
biochemical
  engineering principles 229
  engineering process 173
biochemicals
  productivity 165
bioconversion 71
biodegradation 444

bioenergetics 507
biofertilizer 11, 451
biogas 192, 449
bioleaching 391
  factors governing 392
biological nitrogen fixation (BNF) 16, 17
biological products
  drying of 413, 415–420
  freeze drying 420, 422
biological reaction systems, modelling 249
biological sludge 187
biomanure 192
biometric condenser, design of 340
biomolecular design engineering tools 173
bioorganic manure 186
biopolymeric engineering material 432
  functions of 432
  properties of 432
biopolymers 442
bioprocessing,
  aerobic 196,
  anaerobic 196
  of cell cultivation 199,
bioreactors 178, 201, 235
  anoxic thermophilic digestion (ATD) 186, 187, 190, 191, 192
  conventional 173
  cross-flow hollow fibre 194
  fluidized bed 193
  hollow fibre 194
  horizontal rotary thin film, 178
  membrane 178
  novel 177
  oxic thermophilic digestion (OTD) 184, 185, 188, 189
  perfusion 181, 182
biospheric world 50
biosystem 586
  applications in 455
  immobilized 519
  use of computer 453
biotechnology
  cellular materials 136
  metabolism in 136
biothermodynamics 507
biotic 4
biotic chemistry 31

C

cacterial biotechnology, engineering biology 479
Canida utilis 161
carbohydrate 64
carbohydrate metabolism 112
carbon 4, 9, 11
  properties of 35
carboxylase 108
cathode ray tube (CRT) 454
cell cultivations 174
  fed-batch culture 208, 209
  hardware 174
  modes 208
cell growth 254
cell sedimentation 237
cellular activity, mutation 164
chemical bonds 31
chemical engineering principles 229
chemical nucleases 172
chemotaxonomy 108
cholera red test 113
chromium, recyclable 451
Clostridium thermocellum 68, 71
Clostridium thermosaccharalyticum 68
coal 307
collagen fibrils 54
condenser 334
conductors
  electrical 319
  thermal 319,
continuous culture
  devices 256
  dynamics of growth 256
  in selection of mutants 258
  selection 263,

continuous gravity decanter 282
control system 426
conventional continuous stirred tank bioreactor (CSTBR) 175
cosmos, heat death in 5
covalent bonds 31, 37, 38, non-polar 41
crude oil 307
Cyanobacteria 8
cytochromes 159

D

data acquisition system (DAS) 204
data analysis computer software (DAN) 21
declining/death phase 148
Delhi Jal Board 187
deoxyribonucleic acid (DNA) 5, 107, 119, 173
diffusion laws 308, 310, 311,
digital computer simulation 509
dimensionless groups 233
dimension
  analysis 232, 234
  components 232
  concept of 229
  principles of 233,
distillation 402 407
dynamic in motion 304

E

earth 4
eco-biotechnology
  alien species stresses 18
  asymbiotic non-rhizobial MPI 19
  environmental pollution 18
  industrial production of agrochemicals 20
ecological
  balance 54
  domain 11, 13
  engineering principle (EEP) 11
  habitats 13
ecology and environment 13
Eijkman test 113
ejector 334
  steam jet 343
electrolysis 51
electron acceptor 157
electron cloud diagrams 32
electronegative atom 39
electro negativity 38, 39
electrons 31
emissivity 325
endonucleases 172
energy balance 363
  basics of 247
engineering polymers
  classification of 453
  as plastics 434
enzyme 450
  kinetics 629
  production strategies 542
  reactors 539
ethanol 266
  production 251, 268
eubacteria 116
eukaryotes 116
evaporation 330
  material balance 333
  heat balance 334
extraction 378, 559
  equilibrium conditions 380,
  liquid–liquid 378

F

fat 64
fermentation 62, anoxic 62
fermenter 321
film theory 313
filtration 548
flagellum 106
flash vaporization 405
fluid flows 302
  in boundary layers 298, 301
  laminar 295, 297, 298, 299

Reynolds number 297
turbulent 295, 297, 298, 299
fluid statics 279
applications 281
foam fractionation 553
force and weight 230
Fourier's Law 318
fungi 95

G
gas absorption 307
gas pressure 279
genetic engineering, use of 127
genetically modified organisms (GMOs) 16
gluconic acid 200
glycolysis 62

H
heat exchangers 361
design of 361
double pipe 361
shell and tube 361
swept surface 362,
heat transfer 422
by conduction 318
by convection 322
by radiation 324
coefficient 323, 364
dimensional analysis 323
influential factors 324
three-dimensional 319
Holarrheria antidysenterica 131
Homo sapien 3, 23
homogenic cell–cell interaction 486
Horizontal rotary thin-film bioreactor (HRTFB) 133
hexose monophosphate (HMP) 66
human body design 26
human body design evolution 23
belief and biblical records 23
biocosm theory 24
intelligent descent theory 24
NADE/RADE theory 25
scientific Darwinian theory 23
human chorionic gonadotropin (hCG) 79
hybridization 34
hydrocarbon fermentation 154
hydrogen 4, 9, 11
hydrogen bond 31, 39
hydrogen donor 39
hydrogen sulphide, production 110
hydrolysed protein (HP) 448, 449
hydrostatic equilibrium 280
in a centrifuged field 280,

I
immunity, duration of 74
immunization
active 74, 75
passive 74
immunobiotechnology 511
immunochemistry, fundamentals of 71
ionic bond 37
isoschizomers 106

K
Krebs Cycle 64

L
lactic acid bacteria 57
lag phase 139
laminar flow 284
leaching 385
methods of 386, 387
leather industry, process biotechnology in 442
life forms, biochemical pathways 55
lipid peroxidation 162
living body reactor engineering (LBRE) 21
living cells
basics of 97
classification 97
long phase 141
lyophilization 265, 266
lyophilizer 423

M
manometer 281, 303
mass transfer 422
  coefficients 313, 315, 317
  principles 312
material balance
  fundamentals of 241, 244
  in microbial reaction 245,
metabolism regulator 159
Methanobacterium cells 125
Methanobacterium formicicum 124
Methanobacterium ruminantium 125
*Methanobacterium thermoautotrophicum*124
Methanosarcina barketi 119
Methanospirillum 125
Methanothermus 126
Methonococcus 126
microbe–plant interaction (MPI) 11, 13, 15, 18
  advances 16
  agrochemical production 20
  asymbiotic non-rhizobial 19
  non-rhizobial factors 21
  rhizobial factors 21
  Rhizobium-participated 17
  symbiotic 17
microbes 4, 108, 137
  oxygen on 149,
microbial biotechnology 475
  mixed culture 514
microbial cell 171
  growth principles 137
  cell system 575
microbial leaching 395
  mechanism 396
  merits of 399
microbial sword 172
microbiology 137
microbioreactor 171
Micrococcus radiodurans 124
microcomputers 453
microorganisms 96
  metabolic pathway 68
mixed liquor suspended solids (MLSS) 187
molecular biology
  advances in 71
molecular biology, use of 127
molecular structures 31
molecules 4
most probable number test (MPN) 113
mushrooms 95
mutagenesis 263
Mycobacterium bovis 76

N
natural environment 4
natural gas 307
naturally acquired design engineering (NADE)
  3, 4, 15, 77, 95
neutrons 31
nitrate reduction 111
nitrogen 4, 9, 11, 13
nitrogen supplement 21
numerical taxonomy 108

O
oligomers 46
ontogenetic development 11
origin of life 3
  on earth 3
oxic environment 4
oxidation 475
oxidation–reduction balance 246
oxygen 9, 33
oxygen transfer rate (OTR) 162
oxygenation in plants 17
oxygen–microorganism interactions 152

P
Pachysolen tannophilus 259, 261, 263, 267
Pasteur effect 57
pH control 207
photic test 114

phylogenetic evolution 11
Pi theorem, illustration of 234
pigmentation or chromogenic test 114
planets 4
plant breeding 134
plant genetic engineering 134
plant tissue culture 131
plants 4
plastics
  general purpose 434
  engineering 436, 437
polar bond 38
polio 71
polycarbonate 438
polyethylene 434
polymer, biodegradable 448, 449
polymeric engineering material 432
  properties of 432
  functions of 432
polyphenylene sulphide 440
polyphenylene sulphone 439
polyphenylene oxide 440
polypropylene 435
polyesters 438
polystyrene 435, high impact 435
polysulphones 439
polyurethane 441
prebiotic natural chemistry stage 4
prebiotic universe 9
process biotechnology 494
  modern instrumental techniques 463
  catabolic repression 484
  batch culture 494
  continuous culture 496,
process control, principles 424
product formation
  growth-associated model 148
  non-growth-associated model 148
  combined model 148
Prokaryotes 17
prokaryotic 116
protein 64
  hydrodynamic effect 85, 86
  kinesis engineering (PKE) 78, 80, 82, 85
  oxygenation 8
  physiological status of cell 85, 129,
  role of oxygen 149
  substrate-transient response 86
proteins, purification of 461
protons 31
Pseudomonas saccharophiles 160
pure microbial culture 115

R

Raoult's Law 404
rDNA technology 71
reactor kinetics 579
recombinant cells, genetic engineering 458,
  production of 458
recominatorially acquired design engineering
  (RADE) 3, 4, 8, 19, 95
recycle 242
redox potential controller 162
refrigeration 358
  basic components 359,
respiratory quotient (RP) 64
Rhizobium 11, use of 21
ribosome 107
rhizobia 16

S

Saccharomyces carsbergensis 161
Saccharomyces cerevisiae 259, 267
Salmonella typhimurium 218
Sarcina ventriculi 119
Schizosacchatomyce pombe 217
sewage treatment plant (STP) 54
shear–stress field 285
single-cell isolates, preparation of 264
small pox 71
specific heat 14
spectrophotometer 265
stars 4
steak plate method 115

stem cell, system 135, criteria 135,
stirred tank bioreactor (STBR) 82, 85
stoichiometry 507
styrene acrylonitrile copolymer 436
sulfolbus, isolation and characterization 118
surface tension 14
survival of the fittest 8

T

tannery waste 444
taxonomy 108
teflon 439
terrestrial evolution 4, 5
theories of controversies 26
thermal equilibrium 325
thermophiles 68
thermophilic biokinetics 190
thermoplasma, isolation and characterization 118
Ti plasmid 20, 134
tobacco mosaic virus (TMV) 77
total dissolved solids (TDS) 187
transfectional fermentation 80
transfectional fermentation biotechnology (TFB) 78
transfectional process biotechnology (TFPB) 3
transient and stationary phase 146
tricarboxylic acid (TCA) 57, 58
trickling filtration (TF) 54

U

ultrafiltration 550, 542, 546
units, system of
  base 229
  CGS 229
  concept of 229
  FPS 229
  multiple 230, derived 230, Newton 231,
urease test 113

V

vaccines 72, 78, 80, types 71, varieties 76
vapour pressure 404
vapour–liquid equilibrium 405
viability test 115
virus 77, 78
  disadvantages 78
  industrial and commercial importance 78
  vaccinia 81
viscosity 284, 286
  Newton's Law 286
  of gases 287
  of liquids 287
vivo biosafety 15
volatile dissolved solids (VDS) 187

W

waste water 514, treatment 514
waste water treatment 253
waste, value-added products from 447
water on earth 11
water
  aquatic transport 49
  biogenic properties 44, 54
  biological functions 47
  chemo-physical properties 43, 54
  chlorination of 53
  dimeric 43
  ecological acceptability 52
  ecology 50
  in cells 44
  management 50
  monomeric 43
  pollution 50
  pollution abatement 50
  properties 14, 42, 47
  purification 51
  structure 40
  tetrameric 47
  toxic chemicals 52
  treatment 50
  trimeric 43
  values 14

weirs 294
  rectangular 294
  v-notch 294
  notch 295
wild-type isolates 266

Y

yeasts 128, 154

Z

Zymomonas 63

About the Author

Dr S N Mukhopadhyay is currently Professor at the School of Biotechnology, Gautam Buddha University, Greater Noida (Uttar Pradesh) and Adjunct Professor at the Department of Biological Sciences, Birla Institute of Technology and Science Pilani. He was formerly Professor, Department of Biochemical Engineering and Biotechnology (DBEB), at the Indian Institute of Technology (IIT) Delhi, India. He is one of the founder members of Biochemical Engineering Research Centre (BERC) and DBEB and was Head of BERC during 1985–90.

He has been a Visiting Expert Scientist at LANFI, Mexico, and SFIT, Zurich, Switzerland; and Visiting Professor at CIEA del. IPN, Mexico. He was awarded JSPS Fellowship at Osaka University, Japan; and Fulbright Fellow, at UCHC, USA.

Dr S N Mukhopadhyay was an Overseas Associate (Government of India) at Caltech, Chemical Engineering Department, USA; and FNAE, India. He was also an invited Visiting Scientist, Tomas Bata University, Zlin, Czech Republic. He is an Honorary Member to International Division of Research Board of Advisors, ABI, USA, and was elected Life Fellow of Indian Institute of Chemical Engineers. He was also selected and nominated for International Engineer of the year 2008 by IBC, Cambridge, England. He has authored four books, published by CRC Press and Viva Books. He holds more than 131 publications to his name.

Answer Keys

Answers to first 200 multiple choice questions given in *Process Biotechnology Fundamentals* 1st, 2nd, and 3rd editions (2000, 2004, and 2010)

1. (d)	2. (c)	3. (d)	4. (b)	5. (b)	6. (c)	7. (d)	8. (c)	9. (c)	10. (c)
11. (c)	12. (c)	13. (b)	14. (b)	15. (b)	16. (d)	17. (c)	18. (b)	19. (c)	20. (b)
21. (c)	22. (c)	23. (c)	24. (c)	25. (e)	26. (c)	27. (b)	28. (c)	29. (d)	30. (c)
31. (b)	32. (c)	33. (b)	34. (b)	35. (a)	36. (d)	37. (d)	38. (b)	39. (c)	40. (a)
41. (a)	42. (a)	43. (a)	44. (b)	45. (c)	46. (a)	47. (d)	48. (c)	49. (b)	50. (d)
51. (c)	52. (d)	53. (c)	54. (b)	55. (c)	56. (a)	57. (d)	58. (c)	59. (c)	60. (b)
61. (c)	62. (c)	63. (a)	64. (c)	65. (d)	66. (b)	67. (d)	68. (d)	69. (c)	70. (a)
71. (b)	72. (a)	73. (b)	74. (b)	75. (b)	76. (b)	77. (c)	78. (c)	79. (b)	80. (a)
81. (b)	82. (c)	83. (d)	84. (a)	85. (c)	86. (c)	87. (d)	88. (c)	89. (c)	90. (d)
91. (a)	92. (c)	93. (c)	94. (c)	95. (c)	96. (c)	97. (c)	98. (b)	99. (b)	100. (b)
101. (b)	102. (c)	103. (a)	104. (d)	105. (c)	106. (b)	107. (a)	108. (c)	109. (c)	110. (a)
111. (b)	112. (d)	113. (b)	114. (c)	115. (a)	116. (b)	117. (c)	118. (c)	119. (d)	120. (b)
121. (b)	122. (b)	123. (b)	124. (c)	125. (c)	126. (d)	127. (a)	128. (c)	129. (c)	130. (b)
131. (d)	132. (a)	133. (b)	134. (c)	135. (c)	136. (a)	137. (c)	138. (b)	139. (a)	140. (d)
141. (a)	142. (d)	143. (c)	144. (d)	145. (d)	146. (b)	147. (c)	148. (d)	149. (d)	150. (b)
151. (c)	152. (c)	153. (c)	154. (c)	155. (b)	156. (b)	157. (b)	158. (a)	159. (a)	160. (c)
161. (d)	162. (c)	163. (a)	164. (b)	165. (b)	166. (b)	167. (c)	168. (d)	169. (a)	170. (c)
171. (b)	172. (a)	173. (c)	174. (b)	175. (a)	176. (b)	177. (d)	178. (c)	179. (d)	180. (c)
181. (c)	182. (d)	183. (b)	184. (c)	185. (b)	186. (d)	187. (d)	188. (b)	189. (d)	190. (b)
191. (a)	192. (d)	193. (c)	194. (d)	195. (a)	196. (a)	197. (c)	198. (a)	199. (d)	200. (a)